BRITISH GEOLOGICAL SURVEY

MEMOIRS OF THE GEOLOGICAL SURVEY OF GREAT ⌐

Geology of th
between Durham ⌐   West
Hartlepool

Explanation of one-inch geological sheet 27, New Series

D B Smith, BSc, and E A Francis, BSc

with contributions by

M A Calver, MA, A H Edwards, BSc, G D Gaunt, BSc,
R K Harrison, MSc, and J Pattison, BSc

LONDON   HER MAJESTY'S STATIONERY OFFICE   1954
1989 reprint

HER MAJESTY'S STATIONERY OFFICE

HMSO publications are available from:

HMSO Publications Centre
(Mail and telephone orders)
PO Box 276, London SW8 5DT
Telephone orders (01) 873 9090
General enquiries (01) 873 0011
Queueing system in operation for both numbers

HMSO Bookshops
49 High Holborn, London WC1V 6HB
    (01) 873 0011 (Counter service only)
258 Broad Street, Birmingham B1 2HE
    (021) 643 3740
Southey House, 33 Wine Street, Bristol
    BS1 2BQ   (0272) 264306
9 Princess Street, Manchester M60 8AS
    (061) 834 7201
80 Chichester Street, Belfast BT1 4JY
    (0232) 238451
71 Lothian Road, Edinburgh
    EH3 9AZ   (031) 228 4181

HMSO's Accredited Agents
(see Yellow Pages)

And through good booksellers

BRITISH GEOLOGICAL SURVEY

Keyworth, Nottingham NG12 5GG
    Plumtree (060 77) 6111

Murchison House, West Mains Road,
Edinburgh EH9 3LA   (031) 667 1000

BGS London Information Office at the
Geological Museum, Exhibition Road,
London SW7 2DE   (01) 589 4090

The full range of Survey publications is available through the Sales Desks at Keyworth and Murchison House. Selected items can be bought at the BGS London Information Office in the Geological Museum, and orders are accepted here for all publications. The adjacent Geological Museum Bookshop stocks the more popular books for sale over the counter. Most BGS books and reports are listed in HMSO's Sectional List 45 and can be bought from HMSO and through HMSO agents and retailers. Maps are listed in the BGS Map Catalogue and the Ordnance Survey's Trade Catalogue, and can be bought from Ordnance Survey Agents as well as from BGS.

The British Geological Survey carries out the geological survey of Great Britain and Northern Ireland (the latter as an agency service for the government of Northern Ireland), and of the surrounding continental shelf, as well as its basic research projects. It also undertakes programmes of British technical aid in geology in developing countries as arranged by the Overseas Development Administration.

The British Geological Survey is a component body of the Natural Environment Research Council.

First published 1967
Second impression 1989

Printed in the United Kingdom for Her Majesty's Stationery Office

C5   2/89   Dd 240445

ISBN 0 11 884242 0

Bibliographical reference
Smith, D B, and Francis, E A. 1967.
Geology of the country between Durham and West Hartlepool. *Memoir of the Geological Survey of Great Britain*, Sheet 27 (England & Wales).

PREFACE

The country between Durham and West Hartlepool was originally surveyed on the six-inch scale by H. H. Howell, who was District Surveyor, and A. G. Cameron. Two of the six-inch whole sheets were published—namely Durham 20 and 35, which appeared in 1871 and 1876 respectively. They were followed in 1881 by publication of both Solid and Drift editions of the Old Series One-inch Map Sheet 103 N.E. Revised editions of the Drift map were printed in 1889 and 1892, but no explanatory memoir was published.

The resurvey of the district was begun in 1954 by Mr. D. B. Smith working on the Permian rocks in the east. He was joined in 1955 by Messrs. E. A. Francis and G. D. Gaunt, who mapped the ground covered by the western and southern parts of the sheet respectively. The field work was supervised by Mr. W. Anderson and was completed in 1958. The six-inch standards were drawn on the National Grid and published during 1963 and 1964. They are listed on p. x. Solid and Drift Editions of the new One-inch Map, Sheet 27, were printed in 1965.

The task of writing this memoir has fallen almost entirely to Messrs. Smith and Francis, who have incorporated contributions from Mr. Gaunt into the chapters on the Permian and Pleistocene and Recent. The fossils obtained from the Coal Measures have been identified by Mr. M. A. Calver, and from the Permian by Mr. J. Pattison, Dr. F. W. Anderson and Sir James Stubblefield. Thanks are due to Dr. A. Logan for permission to use his unpublished revision of the Permian brachiopods and lamellibranchs. Comments on the fauna of the Easington Raised Beach have been made by Mr. D. F. W. Baden-Powell of Oxford University, and isotopic age determinations on shells from this beach have been provided by Drs. Meyer Rubin and Irving Friedman of the United States Geological Survey. Most of the sections on petrography have been written by Mr. R. K. Harrison, but also included in the text are brief descriptions of rocks by Dr. R. Dearnley, myself (when Chief Petrographer) and Dr. J. Phemister. The chapter on mineral products and water supply contains a contribution on rank and quality of coal written by Mr. A. H. Edwards, Coal Survey Officer, National Coal Board, Northumberland and Durham Division. The memoir has been edited by Mr. E. H. Francis.

Assistance obtained during the resurvey and writing of the memoir from officials of the National Coal Board, water works engineers and local government officers is gratefully acknowledged.

This Memoir received final approval from Sir James Stubblefield, F.R.S., as Director before his retirement in December 1966.

K. C. DUNHAM,
Director

Institute of Geological Sciences
Geological Survey Office
Exhibition Road
South Kensington
London S.W.7
14th March 1967

CONTENTS

(References are listed at the end of each chapter)

ILLUSTRATIONS

TEXT-FIGURES

vi

EXPLANATION OF PLATES

1. *Lingula* cf. *elongata* Demanet; Kirkby's Marine Band, Hawthorn Colliery Shaft, South Hetton, depth 436 ft. (YGF 1986) × 10. See p. 29.

2. *Lingula mytilloides* J. Sowerby [broad form]; High Main Marine Band, Easington Colliery E.11 Up-bore, 350 ft 6 in above floor of Durham Low Main Coal. (YGF 3213) × 9. See p. 28.

3. *Anthraconaia* aff. *librata* (Wright); Lower *similis-pulchra* Zone, 26 ft above High Main Marine Band, locality as for Fig. 2, 377 ft above floor of Durham Low Main Coal. (YGF 3219) × 2. See p. 28.

4. *Lioestheria vinti* (Kirkby); Lower *similis-pulchra* Zone, 2 ft 3 in below High Main Marine Band, Horden Colliery H.19 Up-bore, 365 ft 4 in above floor of Durham Low Main Coal. (YGF 2907) × 6. See p. 28.

5. *Anthracosphaerium propinquum* (Melville); Lower *similis-pulchra* Zone, 8 ft below High Main Coal, Shotton Colliery No. 6 Up-bore, 275 ft 4 in above floor of Durham Low Main Coal. (YGF 1856) × 2. See p. 28.

6. *Anthraconaia sp. nov.* cf. *wardi* (Hind *pars non* Salter); horizon and locality as for Fig. 5, 275 ft 7 in above floor of Durham Low Main Coal. (YGF 1857) × 2. See p. 28.

7. *Anthracosphaerium* aff. *turgidum* (Brown); Upper *A. modiolaris* Zone, 11 ft above horizon of Top Brass Thill Seam, Shotton Colliery No. 6 Down-bore, 17 ft 9 in below Durham Low Main Coal. (YGF 1906) × 2. See p. 27.

vii

8. *Anthracosia* aff. *disjuncta* Trueman and Weir; Upper *A. modiolaris* Zone, 24 ft above horizon of Top Brass Thill Coal, South Hetton No. 7 Downbore, depth 173 ft below Main Coal. (YAF 121) × 2. See p. 27.

9. *Anthracosia sp. nov.* aff. *phrygiana* (Wright); Upper *A. modiolaris* Zone, 27 ft above Top Brass Thill Coal, Blackhall Colliery No. B.33 Underground Bore, depth 38 ft below Durham Low Main. (YGF 527) × 2. See p. 27.

10. *Anthracosia ovum* Trueman and Weir; Upper *A. modiolaris* Zone, 27 ft 10 in above Harvey Marine Band, locality as for Fig. 7, 183 ft 6 in below Durham Low Main Coal. (YGF 1932) × 2. See p. 27.

11. *Anthracosphaerium* cf. *dawsoni* (Brown); base of *A. modiolaris* Zone, 8 ft above Top Busty Coal, Fishburn Colliery No. 3 Bore, depth 612 ft. (NEW 2102) × 2. See p. 26.

12. *Carbonicola declivis* Trueman and Weir; Holotype (figured Trueman and Weir 1946, pl. ii, figs. 15, 17, 18), *C. communis* Zone, Brockwell Ostracod Band, Fishburn No. 2 Bore, depth 418 ft. (NEW 1814b). See p. 24.

13. *Curvirimula candela* (Dewar); *C. communis* Zone, 12 ft below Three-Quarter Coal, Eppleton No. 2 Underground Bore, depth 68 ft 5 in below Busty Coal. (YGF 2071) × 2. See p. 25.

14, 15. *Carbonicola* cf. *declivis* Trueman and Weir; *C. communis* Zone, Brockwell Ostracod Band. 14, Fishburn No. 3 Bore, depth 729 ft. (NEW 2149) × 2. 15, Bowburn Underground Bore, depth 71 ft 11 in. (DB 299) × 2. (Compare also with the holotype of *C. browni* Trueman and Weir as refigured by Lumsden and Calver 1958, pl. iv, fig. 12). See p. 24.

All the specimens figured are in the Geological Survey Collection, Leeds; registered numbers are given in round brackets.

1. *Liebea squamosa* (J. de C. Sowerby); internal mould of left valve; Upper Magnesian Limestone, ?Hartlepool and Roker Dolomites, base of cliffs on northern side of Nesbitt Dene [460 370]. (YGF 1641) × 4. See p. 153.

2. *Tubulites permianus* (King); Concretionary Limestone, base of coastal cliffs [4824 3745], 1 mile S.E. of Black Halls Rocks. (YGF 1816) × 2. See p. 153.

3. *Schizodus schlotheimi* (Geinitz); left valve; Concretionary Limestone, locality as for Fig. 2. (YGF 1728) × 1. See p. 153.

4–5. *Dielasma elongatum sufflatum* (Schlotheim); Middle Magnesian Limestone, reef facies, Hawthorn Quarry [437 463], Hawthorn. 4. Ventral valve. 5. Side view of both valves. (F 3203) × 4. See p. 134.

6. *Horridonia horrida* (J. Sowerby); internal mould of dorsal valve; Middle Magnesian Limestone, reef talus, horizontal tunnel [4367 4412], 70 yd W.S.W. of West Pit, Easington Colliery. (YGF 2534) × 1. See p. 169.

7. *Parallelodon striatus* (Schlotheim); Middle Magnesian Limestone, reef facies, Townfield Quarry [4343 4380], Easington Colliery. (YPF 323) × 2. See p. 139.

8. *Naticopsis minima* (Brown); Middle Magnesian Limestone, reef facies, Hesleden Dene [4444 3773], 330 yd S. 40° E. from Hesleden station. (YPF 379) × 10. See p. 141.

9. *Cardiomorpha modioliformis* (King); Middle Magnesian Limestone, lagoonal facies, Nesbitt Dene [4416 3776], 170 yd S. of Hesleden station. (YGF 1676) × 3. See p. 128.

10. Polyzoan ?*Gen. et sp. nov.*; obverse and reverse sides of main stem. Also shown in the top left corner is a fragment of *Fenestrellina*. Lower Magnesian Limestone, Raisby Hill Quarry [347 352]. (WEG 236) × 10. See p. 111.

11. *Janassa bituminosa* (Schlotheim); Middle Magnesian Limestone, basinal facies, Hesleden Dene No. 3 Bore [4660 3703], depth 211 ft 6 in. (NEW 4004) × 2. See p. 173.

12. *Astartella vallisneriana* (King); left valve; Middle Magnesian Limestone, lagoonal facies, Castle Eden Dene [4239 3878], 360 yd W. of the Castle. (YGF 1704) × 4. See p. 126.

13. *Acanthocladia anceps* (Schlotheim); Lower Magnesian Limestone, Raisby Hill Quarry [347 352]. (WEG 237) × 2. See p. 111.

2. Similar laminae of clay (dark) and silty clay show marked depressions caused by a subrounded pebble (2·5 mm across) of argillaceous quartz-sandstone. The dark laminae consist of kaolinite, illite, chlorite, micrinite and ferric oxide dust, while the lighter laminae are composed of subangular quartz grains (averaging 0·05 mm), illite, kaolinite, chlorite and coal particles (0·01–0·02 mm). Microscopic shears have been filled with illite. Sand-grade particles include subangular quartz (0·2 mm), coal debris, and labile rock grains with muscovite flakes. E 32395 (MLD 1941) × 7·5

3. Subrounded grains (2 mm average diameter) of sandy argillaceous siltstone, with finer particles of subangular quartz, orthoclase, quartzite and ankerite, are strewn through a clay of illite, kaolinite, ankerite, quartz and micrinite dust. Laminae are weakly developed and arise through varying concentrations of clay-minerals. E 32396 (MLD 1942) × 7·5

4. Pebbles up to 3 mm in length of derived till, consist of fractured quartz, feldspar, carbonate and coal particles in a clayey matrix, and are scattered through highly contorted laminated clay charged with fine ankerite, quartz and coal dust. This band is followed by relatively undisturbed clay composed of mainly illite, kaolinite, quartz and chlorite. E 32397 (MLD 1943) × 7·5

Some differences between the lower and upper parts of the succession thus emerge, such as the marked laminae and the virtual absence of ankerite in the upper part, perhaps indicating a change in provenance. The laminae may reflect relatively rapidly changing sedimentation of clay under tranquil conditions followed by silt and coarser detritus derived, perhaps, through surges of melt-water.

R.K.H.

LIST OF SIX-INCH MAPS

Published geological six-inch maps included in one-inch Sheet 27 are listed below with initials of the surveyors and dates of survey. The surveyors were: E. A. Francis, G. D. Gaunt and D. B. Smith. Strips of ground inside the western and southern margins of the one-inch map are covered by parts of the following incomplete and unpublished six-inch sheets: NZ 22 N.E., 23 N.E., 23 S.E., 24 N.E., 24 S.E., 32 N.W., 32 N.E., 42 N.W., 42 N.E., 52 N.W.

NZ 33 N.W.	Bowburn, Coxhoe, Quarrington and Kelloe	E.A.F.... ...	1955–58
33 N.E.	Thornley, Trimdon Grange and Wingate (west)	D.B.S.... ...	1955–59
33 S.W.	Cornforth, Mainsforth and Bishop Middleham	E.A.F., G.D.G.	1957–60
33 S.E.	Fishburn, Trimdon and Butterwick ...	G.D.G. E.A.F.	1956–60
NZ 34 N.W.	Rainton and Moorsley	E.A.F., D.B.S.	1953–56
34 N.E.	Hetton and East Murton	D.B.S., E.A.F.	1953–56
34 S.W.	Sherburn, Pittington and Shadforth ...	E.A.F.... ...	1955–58
34 S.E.	Ludworth, Haswell and Shotton Colliery	D.B.S., E.A.F.	1954–56
NZ 43 N.W.	Wingate (east) and Hesleden	D.B.S.... ...	1956–61
43 N.E.*	Blackhall Colliery and Hart (north) ...	D.B.S.... ...	1953–59
43 S.W.	Embleton and Sheraton	D.B.S., G.D.G.	1956–61
43 S.E.	Dalton Piercy and Hart (south)	G.D.G., D.B.S.	1956–61
NZ 44 N.W.	Seaham and Hawthorn	D.B.S.... ...	1953–56
44 S.W.	Easington and Horden	D.B.S.... ...	1954–59
NZ 53 S.W.†	Hartlepool and West Hartlepool ...	G.D.G. ...	1955–56

* Includes part of NZ 44 S.E.

† Includes part of NZ 53 N.W.

Chapter I

INTRODUCTION

THIS MEMOIR describes the geology of the district covered by the Durham (27) Sheet of the One-inch Geological New Series of maps of England and Wales[1]. This district (Fig. 1) lies entirely within the County of Durham and includes the south-eastern part of the Durham Coalfield. Coal Measures crop out in the west, but elsewhere they are overlain by Permian rocks and these are succeeded in turn by Triassic strata over a small area in the south-east. In much of the district, the solid rocks are obscured by extensive spreads of superficial deposits.

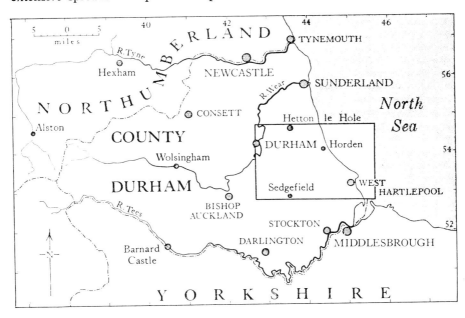

FIG. 1. *Sketch-map showing location of the Durham–West Hartlepool district (inner rectangle)*

The district can be divided on the basis of topography and geology into four areas, namely western, central, eastern and southern. The western area coincides approximately with the outcrop of the Coal Measures (Fig. 2). It lies mainly between 150 and 400 ft O.D. though it falls to as little as 25 ft in the northern part of the entrenched valley of the River Wear. This river, and its tributaries, drain the area and locally flow through gorges which are cut through thick deposits of glacial sands and clays and

---

[1] Referred to as ' the district ' throughout this memoir.

1

which form important topographical barriers. The horseshoe-shaped gorge at Durham (Pl. I) forms a fine defensive site for the Castle and Cathedral ; and between the city and the northern margin of the district, the river is crossed only by one railway bridge, one footbridge and one road bridge. The area is bounded to east and south by a prominent feature frequently referred to as the ' Permian Escarpment ', though in general only the capping rocks are of Permian age. The escarpment is dissected by re-entrant valleys, one of which, at Shadforth, is nearly two miles long and is floored, east of the village, by at least 100 ft of drift. Most of the other re-entrants have also been proved to contain buried valleys and these fall to the main buried valley of the River Wear which lies below present sea-level downstream from near Shincliffe. The escarpment is further breached in the south by the Ferryhill Gap and this too contains a buried valley. The gap links the Wear and Tees drainage basins and provides an easy route for the Edinburgh–London railway.

The central area, lying immediately to the east and south of the escarpment, is formed by a rolling plateau, generally over 400 ft high, and including the highest ground in the district, reaching over 600 ft around Cassop Vale, Raisby Hill, Trimdon, and west of Ferryhill. Drift is mostly thin or absent, though it becomes thicker towards the east.

The eastern area, which lies roughly east of a line through Easington and Embleton, is generally below 400 ft and falls gently eastwards to coastal cliffs north of Black Halls Rocks and to low ground around West Hartlepool. The streams in the northern part drain directly to the North Sea through steep-sided gorges (' denes ') which locally cut down through the drift into the Permian rocks. Farther south, the streams flow to the estuary or tributaries of the River Tees. Drift is more than 60 ft thick over much of the area and buried valleys have been located, though their courses are not known in detail.

The southern area lies south of a line through Ferryhill Station and Fishburn, and is largely covered by thick drift. It merges with the eastern area between Butterwick and Embleton. The streams drain mainly into the River Skerne, a tributary of the Tees, and the ground is generally below 350 ft. A feature of the area is the linked series of flats (' carrs ') which mark the sites of former lakes. Peat and other organic deposits have been proved in some of these former lake-basins.   E.A.F., D.B.S.

HUMAN SETTLEMENT

The two main centres of population within the district are Durham, in the west, and West Hartlepool, in the east. The distribution of the more important smaller towns and villages is shown in Fig. 2.

Old Durham, which lies nearly one mile east-south-east of the Cathedral, was sited on a well-drained terrace and on glacial sands and gravels above the level of the river alluvium. It was settled at least as early as Roman times, when a villa, said to be the northernmost farming estate of the Roman world, was occupied (Richmond 1949, p. 64). Durham itself flourished as a monastic centre during the Norman period, when the Cathedral and Castle were built. It later became a market town and in

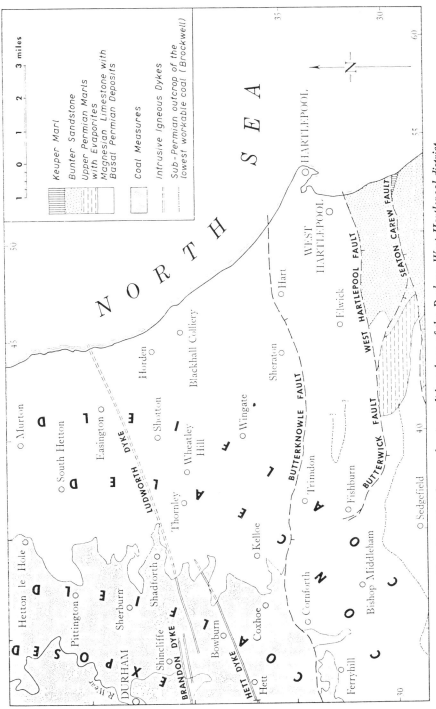

FIG. 2. *Sketch-map showing solid geology of the Durham–West Hartlepool district*

97544

B

the 19th Century was a natural site for a university. It remains the administrative capital of the county, and supports several light industries, which are becoming increasingly concentrated in the Dragonville area. West Hartlepool, though larger, is younger, having expanded mainly in Victorian times as an industrial centre. Its coal-exporting and ship-building activities have declined, but timber-importing and steel-making are still important. In recent years the field of employment has been widened by the introduction of light industries, including a plant, established in 1937, for the extraction from sea-water of magnesia which is used in the manufacture of refractories.

Some of the other settlements, including Shadforth, Easington, Fishburn, Sedgefield, Hett, Cornforth, Hutton Henry and Trimdon, are set around village greens and were originally agricultural. In later times, mining villages such as Sherburn Hill were newly formed from rows of miners' houses erected around local pits, and existing villages like Pittington were rapidly enlarged in the same way.

In the inter-war and post-Second World War years, new housing estates have been built around Durham and West Hartlepool and also at some of the mining villages, such as Thornley and Wheatley Hill. Along the coast, where mining prospects are good, the villages of Easington Colliery, Horden and Blackhall now form an almost continuous settlement about four miles long, and the concentration of population is made greater by the junction with Easington at the northern end, and by the establishment of the new town of Peterlee near Horden. Inland, where the future of mining is less secure, some of the villages have taken on a secondary function as dormitories for Durham and Darlington. Further changes in the distribution of population and light industry are likely to follow from the construction of the Durham Motorway through the western part of the district.

Farming has been influenced to a considerable extent by the development of mining and other industries in the district. There is a large arable acreage which is partly devoted to the production of potatoes and other vegetables. Milk production is also important.   E.A.F.

HISTORY OF MINING

The Northumberland and Durham Coalfield was the site of some of the earliest mining for coal and was certainly the first British coalfield to be developed commercially. Although working by the Romans seems to have been confined to the Tyne area, much more widespread exploitation took place during the monastic period from the twelfth to the sixteenth centuries. The earliest mention of coal-mining by the monks of Durham is in 1188, but workings only seem to have become important during the fourteenth century, when a considerable revenue began to be derived. The most important places in the Durham district and the dates mentioned are : Hett (1293), Coxhoe (1327), Ferryhill (1327), Rainton (1347), Moorhousefield (1360), West Rainton (1430), Trillesden, i.e. Tursdale (1447), Cassop (1456). There seems to have been extensive mining from 1402 to 1528 at Moorhousefield (near Moor House), south of West Rainton, associated with

Finchale Priory, and at Rainton by the Durham monks from 1351 to 1431. Because there were not more than 60 monks at Durham and Finchale, some of the workings were leased ; during 1447 land and a coal pit at Tursdale were let to three or four people by the Prior of Durham in one of the earliest colliery leases on record (Simpson 1910, pp. 589, 595).

Although coal was shipped from the Tyne as early as the thirteenth century, it was mostly burned near outcrops until Elizabethan times. Then, with the depletion of timber resources, and consequent rise in prices, it became economic to transport the easily worked household coal of Northumberland and Durham by sea to London. Thus the coalfield gained a dominant position which was unrivalled until the Industrial Revolution. The collieries which produced the so-called sea-coal were at first mainly near the Tyne and based on the High Main Seam. But by the beginning of the seventeenth century, this seam was apparently nearly exhausted near Newcastle, due to difficulties with water, and though its life was prolonged by the introduction of new methods, including steam-driven pumping, the difficulties became progressively greater, until by 1763 the central part of the Tyne basin was flooded out.

The flooding helped to promote the coal export trade in Durham. It was first based on staithes on the River Wear, near Penshaw and Fatfield, but by 1670 short wagonways were in general use linking these with the nearby pits. Elsewhere in the coalfield there were only a number of landsale pits which worked various seams for local use.

One of the earliest mine-plans of the district refers to Flintoff's Engine Pit, which worked the Hutton Seam 550 yd north of Oswald House, Durham, possibly in the early part of the eighteenth century. The earliest records of bores in the district are those put down on Cornforth Moor in 1748 and in the Coxhoe and Quarrington area between 1750 and 1785. A bore was also put down near Moor House, Rainton, in 1751. By 1800, the system of wagonways leading to the staithes near Penshaw had reached back into the Rainton area on the northern margin of the district, but the record of the earliest sinking within the district, Rainton Adventure Pit, is dated 1816. At the beginning of the nineteenth century, much of the production from the Wear collieries was derived from the Main and Five-Quarter seams, which yielded only a moderate household coal, but a first-class household coal was gained by sinking to the Hutton Seam when the Tyne High Main was finally nearing exhaustion (Smailes 1935, p. 205). The Hutton had, however, been exploited near Lambton somewhat earlier. The Rainton pits were developed from 1816 to 1823, and at Seaham, just beyond the northern margin of the district, the Marquis of Londonderry built an artificial harbour which began to ship coal from these pits in 1831, and from Murton a few years later.

Before 1820 it was generally thought that no valuable coal existed under the Permian rocks, but in that year a new era opened when Hetton Colliery shafts were sunk through the Magnesian Limestone to the Hutton Seam, which was of comparable thickness and quality to the same seam worked in the Rainton pits. The chief problem in sinking through the Permian rocks lay in the water-bearing Yellow Sands ; though only 27 ft thick in the Murton sinking of 1837, these sands introduced 9,000 gallons of water per minute into the shaft (Poole and Raistrick 1949, p. 88).

The development of the railway system was also an important factor in the exploitation of the coal resources of the district (Smailes 1935, p. 206). In 1830, the Clarence Railway and the Middlesbrough extension of the Stockton and Darlington Railway connected the southern part of the coalfield and the Tees estuary. A branch line provided access through the Ferryhill Gap to the Coxhoe area, and the small Bell's Pit sunk in 1827 was superseded by Coxhoe Colliery in 1835. Soon after, the Hartlepool Docks and Railway Company developed Hartlepool as a coal-shipping port and opened a more direct route from this part of the coalfield to the coast. The company also opened a main line to Haswell with a number of branch lines into the surrounding area, so that by 1850 this part of the concealed coalfield was further developed than the exposed coalfield west of Durham City. The Durham and Sunderland Railway (1836) also provided access from the Rainton–Pittington area to the docks which were built at the mouth of the Wear in 1838 and 1840. Many more pits were opened during this period, as for instance, Sherburn Hill (1835), East Hetton (1836), Shincliffe (1837), Houghall (1841), Kepier Grange (1844) and Bishop Middleham (1846).

After the middle of the nineteenth century, London gained access to other coalfields by rail and the household coal trade in this coalfield declined. However, the coking coal trade grew in its stead and in central and east Durham the decline was offset by an increase in demand for gas coal. In this period few pits were sunk in east Durham because of the financial risk associated with the water in the Yellow Sands. These sands are over 100 ft thick at Haswell, and the first attempt to sink a shaft there was abandoned after a loss of £60,000 (Galloway 1898, p. 452). Soon after the beginning of the twentieth century, however, the technique of freezing water-bearing strata during sinking was introduced, and this method was used at Dawdon (1903) and Horden (1904) and also at Easington and Blackhall. In 1905 these coastal collieries were joined by a railway which provided them with access to Seaham and Hartlepool. The last major colliery to be opened in this district was Fishburn in 1912. After the First World War the industry contracted and in the succeeding years, the only expansion of importance took place in the coastal area.

Since nationalization in 1947 there has been reorganization of the collieries that are economically viable and closure of those pits that have either exhausted their resources or are too expensive to maintain. One result of this programme has been the joint running of two or more collieries as a unit, illustrated in this district by the amalgamation of Dean and Chapter and Chilton, and of Elemore, Murton and Eppleton. For the last three, the new coal-drawing Hawthorn Shaft has been sunk, and now produces the largest total in the district, though the takes of Eppleton and Murton lie partly beyond the northern margin of this district.

Including Dawdon, which lies a few yards beyond the northern margin of the district, but draws coal partly from within it, the total production for 1962 was nearly 9 million tons, of which more than half was raised in the north-eastern and coastal collieries. The latter are the collieries with the most assured futures because they have large undersea reserves. Most of the inland pits, by contrast, have only limited reserves of relatively thin seams, and in consequence there has been a movement of labour either to the coastal collieries or to other coalfields.   E.A.F.

GEOLOGICAL SEQUENCE

The divisions represented on the One-inch Geological map and sections of the Durham district are summarized below:

SUPERFICIAL FORMATIONS (DRIFT)

RECENT AND PLEISTOCENE

Landslip

Blown Sand

Peat

Alluvium, lacustrine and flood-plain

River Terrace Deposits

Marine Beach Deposits

Marine or Estuarine Alluvium

Submerged Forest

Terrace of Warp

Glacial Sand and Gravel

Boulder Clay and Glacial Drift, undifferentiated

Glacial Lake Deposits

Morainic Drift

SOLID FORMATIONS

Generalized thickness in feet

PERMO-TRIASSIC

Keuper Marl: Red-brown mudstone	to about	50
Bunter Sandstone: Brown-red sandstone		700
Upper Permian Marl: Brown-red mudstones and siltstones with thin gypsum beds in lower part: Billingham Main Anhydrite at base		430
Magnesian Limestone with interbedded anhydrite and with Marl Slate at base		1000
Basal Permian Sands (Yellow Sands) and Breccias	up to	160

unconformity

CARBONIFEROUS

Middle Coal Measures

Mudstones and shales, sandstones, numerous coal seams and seatearths; Harvey Marine Band at base	1100

Lower Coal Measures

Mudstones and shales, numerous coal seams in upper 500 ft, sandstones especially in lower part	600

Millstone Grit Series (not shown on One-inch map)

Mudstones, shales, sandstones and limestones, a few thin coals. Recorded in a few deep bores, but not known in detail. .	400

INTRUSIVE IGNEOUS

Quartz-dolerite dykes and sills of late Carboniferous or early Permian age.

GEOLOGICAL HISTORY

Nothing is known of the pre-Carboniferous history of the district and the Carboniferous strata are thought to lie unconformably on older rocks, such as the Skiddaw Slates or Caledonian granite proved in neighbouring areas (Woolacott 1923, p. 60 ; Dunham and others 1961 ; 1965). In the adjacent parts of northern England, the pattern of Lower Carboniferous sedimentation is thought to be linked closely with the structure : the Lower Palaeozoic rocks form a block outlined on the north, west and south by faults, and flanked to north and south by troughs in which a much greater thickness of Carboniferous strata accumulated than on the block. According to Bott and Masson-Smith (1957), the Butterknowle Fault (see Fig. 2), crossing the southern part of the district, marks the line of a hinge to the south of which thick Carboniferous sedimentation took place.

Such differential subsidence seems to have ceased by Coal Measures times, since these beds show no significant difference in thickness across the fault. The Coal Measures seem to represent a complex history of limited inter-mittent subsidence over a long period interrupted by phases of relative stability, sometimes only local, which permitted the spread of coal-forming vegetation across the depositional surface. It is probable that Upper Coal Measures, as well as the Lower and Middle divisions, were deposited in the district, for they have been preserved from erosion in the district to the north.

During the Armorican orogeny, the Carboniferous rocks were subjected to folding and faulting, and quartz-dolerite dykes and sills were intruded. Then followed a period of erosion during which Upper Coal Measures, and in places older rocks, were removed.

By the end of Lower Permian times, the surface of erosion probably took the form of a low desert plain and on it sands and breccias were deposited. The sands were shifted by wind, and the breccias, in the southern part of the district, were probably redeposited in low fans by occasional floods. Both sand and breccias were probably partly redistributed by the ensuing trans-gressive Upper Permian or Zechstein Sea in which the Marl Slate was the first undoubted marine deposit followed by carbonates and by sulphates and chlorides to the east.

Permian deposition ended with shallow-water clastic, fine-grained red beds with thin evaporites. The division between them and the later continental sandstones and mudstones of the Trias is entirely lithological and may not be contemporaneous with that drawn in other areas in north-eastern England.

No other deposits are known in the district until the drifts of the Pleistocene. It is certain that the Permo-Triassic rocks of the district were disturbed by later earth-movements, particularly along the Butterknowle Fault and the adjacent anticline. Similar movements, generally held to be Tertiary in age, affected Jurassic and Cretaceous rocks in East Yorkshire, and the present cycle of erosion began when these movements led finally to uplift and tilting of the whole region. Remnants of a surface at about 600 ft above O.D. probably belong to the stage reached during the later Pliocene. The lines of the present major buried river valleys were established, and sub-sequently these were partly obliterated by the invasion of the great Pleistocene ice-sheets. Upon the melting of the ice great quantities of material were left, mainly in the valleys, and these are at present in process of removal. Erosion was rapid at first, but most of the streams are now fairly well graded and erosion is only significant during abnormal floods, though in places landslides are still active. E.A.F., D.B.S.

REFERENCES

BOTT, M. H. P. and MASSON-SMITH, D. 1957. The geological interpretation of a gravity survey of the Alston Block and the Durham Coalfield. *Quart. J. Geol. Soc.*, **113**, 93–117.

DUNHAM, K. C., BOTT, M. H. P., JOHNSON, G. A. L. and HODGE, B. L. 1961. Granite beneath the northern Pennines. *Nature,* **190**, 899–900.

——, DUNHAM, A. C., HODGE, B. L. and JOHNSON, G. A. L. 1965. Granite beneath Viséan sediments with mineralization at Rookhope, northern Pennines. *Quart. J. Geol. Soc.*, **121**, 383–414.

GALLOWAY, R. L. 1898. *Annals of Coal Mining and the Coal Trade.* London.

POOLE, G. and RAISTRICK, A. 1949. Extractive industries. *Scientific Survey of North-Eastern England,* 87–98, Brit. Assoc., Newcastle upon Tyne.

RICHMOND, I. A. 1949. Roman settlement. *Scientific Survey of North-Eastern England,* 61–8, Brit. Assoc., Newcastle upon Tyne.

SIMPSON, J. B. 1910. Presidential address. Coal-Mining by the monks. *Trans. Inst. Min. Engrs.,* **39**, 572–600.

SMAILES, A. E. 1935. The development of the Northumberland and Durham coalfield. *Scott. Geogr. Mag.,* **51**, 201–14.

WOOLACOTT, D. 1923. On a boring at Roddymoor Colliery, near Crook, Co. Durham. *Geol. Mag.,* **60**, 50–62.

Chapter II

UPPER CARBONIFEROUS

GENERAL

IN THE Durham–West Hartlepool district the proved thickness of Carboniferous rocks amounts to over 2000 ft including 400 ft of Millstone Grit Series, 600 ft of Lower Coal Measures and about 1100 ft of Middle Coal Measures. From comparison with neighbouring areas it is likely that a further thickness of about 450 ft of Millstone Grit Series lies at depth and that the full thickness of Middle Coal Measures before erosion was about 1600 ft.

Carboniferous strata are unconformably overlain by Permian over all but the north-western part of the district where Middle Coal Measures crop out in a strip about 9 miles long by about 3 to 5 miles wide. Southward, the Permian oversteps Coal Measures on to Millstone Grit Series, but eastward the Coal Measures continue beneath the younger rocks to form the south-eastern part of the Durham Coalfield. At the time of writing, mining operations in these measures extend for $3\frac{1}{2}$ miles beyond the coast-line and the coalfield has been proved by deep offshore borings to extend seaward for at least one further mile.

The beds underlying the worked Coal Measures in the Durham district have not been described previously though a general account of the equivalent measures farther west has been given by Dunham (1948). The Coal Measures of Durham and Northumberland have been described in some detail by Hopkins (1954) and those of this district have also been mentioned by Hickling (1949) and Armstrong and Price (1954) in papers dealing with the coalfield farther north in the Sunderland (21) district. The Coal Measures of the south-eastern part of the district are described in a recent paper by Magraw and others (1963).

CLASSIFICATION

The traditional three-fold division of the British Carboniferous into Carboniferous Limestone Series, Millstone Grit Series and Coal Measures has been tacit in Durham for nearly 150 years. The boundaries of the divisions have been defined at different horizons by different authors, though from Conybeare and Phillips (1822) onwards all have agreed on the inclusion in the Millstone Grit Series of the predominantly arenaceous sequence underlying the lowest worked seam of the Coal Measures. This arenaceous sequence was subdivided by the Primary Geological Survey into First, Second and Third grits, in ascending order.

These classifications have been based entirely on lithology and it has been apparent for some time that they do not accord with the standard

10

classification of central and southern England and Wales. In order to reconcile this discrepancy it is proposed in this memoir to use the palaeontological boundaries which are now generally accepted throughout north-western Europe. Thus the base of the Millstone Grit Series is now placed at the base of the Great Limestone since it is the most convenient mappable horizon approximating to the first appearance of *Cravenoceras leion* Bisat, at the base of the Namurian (Johnson and others 1962). This limestone lies well below the lowest proved horizon in the Durham–West Hartlepool district. The boundary between the Millstone Grit Series and the Coal Measures, that is the Namurian–Westphalian boundary, is internationally accepted to be the marine horizon with *Gastrioceras subcrenatum* (Frech). This goniatite has yet to be found in north-eastern England, but there is now indirect evidence (Mills and Hull, *in press*) to suggest that the equivalent horizon is the lowest of a group of marine bands between the Second and Third grits of the earlier surveyors. The boundary between the Millstone Grit and Coal Measures is consequently drawn at the base of this lowest band.

In other parts of the country the division between the Lower and Middle Coal Measures is now taken (Stubblefield and Trotter 1957) at the base of the marine band characterized by the goniatite *Anthracoceras vanderbeckei* (Ludwig) and lying near the middle of the non-marine lamellibranch zone of *Anthraconaia modiolaris*. The Durham equivalent of this bed is the Harvey Marine Band. The top of the marine band characterized by *Anthracoceras cambriense* Bisat is taken as defining the upper limit of the Middle Coal Measures. In the Sunderland district, to the north, this band lies about 550 ft above the Usworth Coal, but neither it nor the overlying Upper Coal Measures are known in this district.

LITHOLOGY

The Carboniferous succession is generally regarded as a sequence of cycles or cyclothems, each of which might be defined ideally as starting at the base with limestone and continuing upwards through marine mudstone, non-marine mudstone, sandstone and seatearth to coal. For reviews of the extensive literature dealing with this aspect of Carboniferous sedimentation the reader is referred to Wells (1960) and Duff and Walton (1962). Limestones are restricted to the cycles below the Coal Measures. In the lower part of the Lower Coal Measures the sandstone phase predominates, but the remainder of the Coal Measures is characterized by thick beds of coal.

The coals of the Durham Coalfield are almost entirely bituminous, though cannels are found locally, particularly at the tops of certain seams. The properties of the principal coals of Durham are described by Mr. A. H. Edwards of the National Coal Board on p. 257. The shales and mudstones, mainly dark grey in colour, pass upwards through paler silty mudstone to siltstone. In the Millstone Grit Series many of these argillaceous bands contain marine faunas, but in the Coal Measures non-marine lamellibranchs or 'mussels' are more common and are best preserved in clay-ironstone nodules. The few marine bands in the Coal Measures are for the most part thin and are found at or near the base of mudstone and siltstone units. The ecological distribution of the Coal Measures fossils is more fully described on pp. 19–22. Here it will suffice to say that ostracods are particularly abundant at specific horizons such as the Hopkins Band in the roof of the Harvey Coal (p. 20 and Fig. 3), and that '*Estheria*' is of similar stratigraphical significance in being more persistent at certain horizons than at others. Fossils of less use as indices include annelid burrows and trails, fish remains and plant debris.

The term siltstone has no universally accepted definition and does not appear in older records. As used in modern borehole logs and sections quoted in this memoir it refers to rocks of grain size between 0·002 and 0·06 mm (cf. British Standard, CP 2001 of 1957). Rocks of this grade occur interlaminated with fine sandstone and also as hard grey massively bedded rocks with conchoidal fracture.

Most of the Carboniferous sandstones in Durham are fine- to medium-grained rocks composed mainly of quartz with subordinate feldspars such as orthoclase, microcline and oligoclase. Among the heavy minerals, zircon, garnet, rutile, tourmaline, apatite, magnetite, chlorite, epidote, ilmenite, sphene and sapphire have been recorded by Kellett (1927) who has suggested that the Durham Coal Measures were derived principally from Cheviot igneous rocks and Silurian, Old Red Sandstone and Carboniferous Limestone sedimentary rocks of the Border district. In some sandstones, muscovite, lying parallel to bedding planes, is so abundant as to impart fissility. Most sandstones, however, are massive and cross-bedded and provide good freestones which have been used extensively in building. Coal Measures sandstones have been used in the construction of Durham Cathedral and Castle, and for walls in the old city. The harder, more compact sandstones are cemented by silica or ankerite, but a more common cement is kaolinite which gives rise to softer, more porous rocks. Where fresh the sandstones are off-white to light grey, but where weathered they are yellow or rusty brown with the iron hydroxide concentrated to form ' boxstones '. In places, particularly in the bottom layers of thick sandstones, conglomerates are seen. The pebbles in these are mainly subangular fragments of siltstone or mudstone, or clay-ironstone nodules dissociated from their argillaceous matrix. Evidence of slumping, sliding and ' balling-up ' is most common in interlaminated and interbedded fine sandstones and siltstones, but slumping or overturning of cross-bedding is also seen in relatively homogeneous sandstones. Generally the slumped beds are not appreciably thicker than individual beds in a cross-bedded sandstone, and like these range from about 1 in to 3 ft in thickness, averaging about 12 to 18 in. Their extent is rather limited, although slumping seems to be concentrated at particular horizons.

Seatearths range from soft apparently unbedded brownish or light grey fireclays to hard siliceous ganisters and are characterized by rootlets either penetrating the rock from top to bottom or radiating from Stigmarian rhizomes. Ferruginous concretions are common. Some take the form of sphaerosiderite, which are small spherical grains of iron carbonate, about the size of a large pin-head, usually scattered through the bed. Others are clay-ironstone nodules which are noticeably more irregular than the bedded or sub-spherical varieties found in the mudstones and are commonly concentrated in the lower part of the seatearths. Most of the seatearths contain too much iron to be of use as refractories, but farther west some of the Lower Coal Measures ganisters have been extensively worked and the lateral equivalents of some of these are recognizable in the Durham–West Hartlepool district.

STRATIGRAPHY

Millstone Grit Series. It is inferred, from evidence in other parts of Durham, that the Millstone Grit Series in this district is probably between 900 and 1100 ft thick, though only the uppermost 375 to 400 ft have been proved (Fig. 5). The fullest section is given by the Ryal Bore, near Sedgefield, where an 11-ft limestone near the base is correlated with the Upper Felltop Limestone of West Durham. Above the limestone there is no obvious equivalent of the Grindstone Sill, but thin limestones a little higher in the sequence are thought to correspond with marine beds overlying that bed in West Durham. About 115 ft above the supposed Upper Felltop Limestone lies an $11\frac{1}{2}$-ft limestone which is correlated with a $12\frac{1}{4}$-ft limestone near the bottom of Littletown Lady Alice Pit Bore. At Bishop Middleham the position of this limestone appears to be marked by a pebbly sandstone.

The remainder of the succession up to the top of the Millstone Grit Series is 250 to 275 ft thick and consists mainly of sandstone with subordinate, probably marine, shales and a few thin limestones. The sandstones in the lower part of this sequence cannot be followed individually from one borehole to another in this district and it is not therefore possible to define the local equivalent of the ' First Grit '. In the upper part of the sequence, however, the ' Second Grit ' can be recognized as a persistent sandstone 45 to 65 ft thick. Argillaceous strata underlying the ' Second Grit ' include a marine band proved in a boring at Middle Stotfold, just inside the southern margin of the district.

Lower Coal Measures (Fig. 3 and Pl. III). A marine band interpreted as overlying the ' Second Grit ', and therefore marking the base of the Coal Measures as here defined (p. 11), has been proved only in the Middle Stotfold Bore, where it contains fragments of costate Productids. Between this horizon and the Brockwell Coal there are 325 ft of strata including several thin coals. Some of the latter are correlatives of the Gubeon, Ganister Clay, Marshall Green and Victoria coals of other districts, but none has been worked in this part of Durham. A second marine band, the Gubeon Marine Band, lies in the roof of the Gubeon Coal. It has been proved at two localities, and contains *Lingula*, foraminifera and sponge spicules. Non-marine lamellibranchs are not common below the Victoria Coal, but they have been found about 16 ft below and about 30 ft above the Ganister Clay Coal, and between about 6 and 22 ft below the Victoria Coal. The junction between the *Anthraconaia lenisulcata* and *Carbonicola communis* zones is taken, tentatively, at the Ganister Clay Coal (p. 23). Persistent ' mussel '-bands forming the lowest well-defined marker bands of the Durham Coal Measures lie in the roof of the Victoria Coal and between about 15 and 30 ft above it. About 40 ft beneath the Brockwell Coal, or at intermediate positions up to that seam, a seatearth commonly underlies one or more thin coals. In places, these thin coals thicken and unite to form a single banded seam, which is as much as 47 in thick in Offshore No. 1 Bore. This seam is known as the ' Bottom

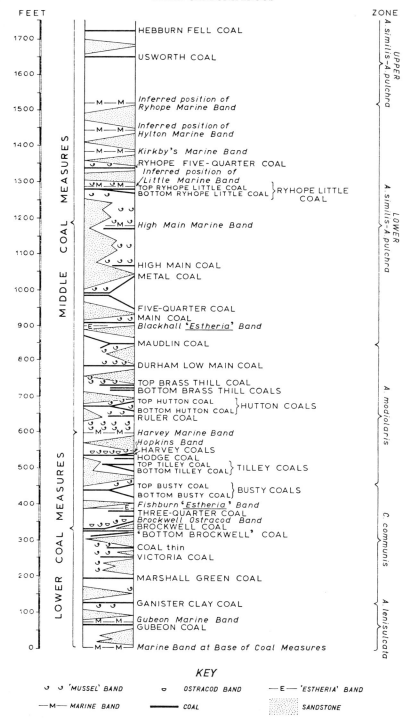

FIG. 3. *Generalized vertical section of the Coal Measures, showing the principal coal seams, sandstones and fossil bands*

Brockwell ', but it is regarded here as distinct from the true Brockwell Seam.

The Brockwell Seam has been worked in the south and south-west where it is 4 to 6 ft thick, but in much of the central part of the district it is thin and poor in quality. The roof measures, up to 65 ft thick, include thin coals and the uppermost of these underlies the distinctive Brockwell Ostracod Band, which, in places, especially in the southern part of the district, contains abundant ostracods and non-marine lamellibranchs. The roof of the Three-Quarter Coal—a seam of no economic importance in this district—contains the Fishburn ' *Estheria* ' Band, which is the equivalent of the Low ' *Estheria* ' Band of the upper part of the *C. communis* Zone in the East Pennine Coalfield. Up to about 60 ft of sandstone overlie the Three-Quarter Coal in some south-western parts of the district, but this is not persistent. Over most of the district, the Bottom and Top Busty coals are distinct seams, but in the west central area they are close together and in places form a single seam 4 to 6 ft thick. A general deterioration in quality sets in eastwards of a north-north-westerly line across the centre of the land-area, and there is a further area of impoverishment trending north-north-east through the north-western portion of the district. The Top Busty Coal is provisionally taken as the top of the *C. communis* Zone (p. 26).

The strata between the Busty and Harvey seams comprise a complex sequence 70 to 175 ft thick including three or more coal seams which are persistent in the western and south-western parts of the district but are generally absent in central and eastern areas. The nomenclature of these seams is confused, since it has been the custom to name the locally thickest coal the Tilley, after a worked seam of this name in West Durham. Correlation of this part of the sequence with West Durham, however, is uncertain. Thus in some western parts of the district there are three impersistent coals, and the lowest two have been named Bottom Tilley and Top Tilley seams ; it is assumed that the highest is equivalent to the Hodge Seam of north-west Durham. The use of the terms Bottom and Top Tilley should not be taken to imply that these seams are splits from the Tilley Seam of the type area, for the possible alternative is that the local Bottom Tilley may correspond with the Hand Seam of West Durham. Both Top and Bottom Tilley seams, as here defined, split into two or more thin coals. In parts of the district where a sandstone up to 106 ft thick overlies the Top Busty Coal, the Bottom Tilley Seam is absent. With the exception of the west-central area the Top Tilley Coal is also impoverished. The measures between the ' mussel '-band over the Top Busty Coal and the Harvey Coal are relatively unfossiliferous.

The Harvey Coal has been extensively worked. In the north-western area it is less than 2 ft thick, and there is also an impoverished belt in the coastal area where thick sandstones appear in the overlying measures. Farther east, in the offshore area, the seam splits into Top and Bottom Harvey coals and the line of split can be traced south-westwards through the Fishburn area at least as far as Windlestone. Where split there is scattered information to suggest that the Bottom Harvey is generally over 2 ft thick and is therefore an important reserve for the East Durham collieries. The Top Harvey is overlain by the Hopkins Band, which is the most persistent ostracod-bearing marker band in the Durham Coal Measures. Non-marine lamellibranchs in this band are representative of the *Anthracosia regularis* fauna which is typical of the measures somewhat below the marine band in the middle of the *Anthraconaia modiolaris* Zone. The uppermost widespread member of the Lower Coal Measures is a seat-earth or thin coal horizon.

Middle Coal Measures (Fig. 3 and Pl. IV). The base of the Middle Coal Measures is marked by the Harvey Marine Band, lying near the middle of the *A. modiolaris* Zone. It is generally thin, and in places absent : well-preserved *Lingula* are common and *Orbiculoidea* and *Hollinella?* have also been recorded. Above the marine band lies an important ' mussel '-band which is over 20 ft thick and forms one of the most persistent marker bands of the Durham Coal Measures.

At Ferryhill, the measures between the marine band and the Hutton coals are about 150 ft thick and consist largely of sandstone. Elsewhere they average about 100 ft and include several thin coals, the most persistent of which is here correlated with the Ruler Seam. Scattered non-marine lamellibranchs in the roof of the Ruler Coal include *Anthracosia aquilina* (J. de C. Sowerby) associated with *A. ovum* Trueman and Weir and *A. phrygiana* (Wright). Shells are not abundant and there is no bed comparable with the Plessey Shell-bed (Hopkins 1929, p. 132 ; Fowler 1936, p. 67) of the Northumberland Coalfield.

In the northern half of the district, the Hutton is a single 4- to 6-ft seam of good quality which has been widely worked. In the southern half of the district, it is split into Top and Bottom Hutton coals, of which the Bottom is the more valuable, while in the east further splitting results in a number of thin seams of no economic value. The strata between the Top and Bottom Hutton seams are generally 20 to 40 ft thick, and locally contain at the base poorly preserved ' mussels ', ostracods and fish remains. The ' mussels ' in the roof of the Top Hutton are also poorly preserved. The faunas found at low horizons in the Middle Coal Measures are rather similar, but good collections over a wide area suggest that the assemblages have different dominant forms (p. 27).

The Bottom Brass Thill Coal consists of two thin and impersistent leaves which are, in most places, separated by up to 10 ft of strata from the thicker, more persistent, Top Brass Thill Coal. Near Finchale, however, all three leaves join to form a single seam about 3 ft thick. The overlying Brass Thill Shell-bed is a significant marker band containing an abundant *A. beaniana* King-*A. phrygiana* assemblage.]

In the central part of the district, the Durham Low Main Coal is an important worked seam 2 to 4 ft thick, but there are belts in which the coal is thin, for it is more affected by ' washouts ' than any other seam in the district. It is separated from the Maudlin Coal by measures which include a sandstone locally over 100 ft thick—the ' Low Main Post '. Towards the northern margin of the district, however, the Durham Low Main and Maudlin coals are united to form a seam which is locally more than 8 ft thick. As noted by Hopkins (1933, p. 216) there is a general relationship between the thickness of the Low Main Post and that of the Maudlin Coal: where the post is thick the coal is thin, and *vice versa*. There are exceptions, however, as at Trimdon Grange Colliery, where sandstone and coal are both thin, and in the offshore area where there is no obvious relationship between the two. The top of the *A. modiolaris* Zone is drawn at the Maudlin Coal position (p. 27). In western parts of the district a fairly persistent sandstone overlies the Maudlin Coal; in eastern areas the measures somewhat higher in the sequence, but below the Main Coal, include the Blackhall '*Estheria*' Band (Magraw and others 1963, p. 164).

The Main Coal which, as its name suggests, has been one of the richest seams in the district, has been removed by erosion from large tracts in the west. It is 6 to 9 ft thick in the north, but is only 2 ft or less in other parts of the district. Immediately west of Blackhall Colliery it lies less than 2 ft below the Five-Quarter Coal, but the interval increases in other parts of the district to over 100 ft including 70 ft of sandstone at the base. In the northern part of the district, this sandstone unites with another overlying and cutting out the Five-Quarter Coal to make up a combined thickness of over 140 ft. Generally, however, the Five-Quarter Coal is 2 to 7 ft thick and has been widely worked. Over large areas in the central part of the district it is closely overlain by the Metal Coal which contains several bands of dirt. The coal to which the name 'Metal' is now applied by the Geological Survey is locally known as the Three-Quarter, and by the National Coal Board as the Bottom High Main. In order to avoid confusion with the Lower Coal Measures Three-Quarter Coal on the one hand, and to avoid accepting an unproved split in this upper coal from the High Main on the other, the name 'Metal' has been adopted from the Newcastle upon Tyne area. In part of the coastal area the Metal and High Main seams are united to form a thick banded seam which includes about 13 ft of coal, but farther east, in offshore areas, they are again separate.

In central parts of the district a sandstone overlies the Metal Coal and this unites in places with another sandstone overlying the High Main Coal to form a thick bed occupying much of the interval up to the High Main Marine Band. In such areas, therefore, the High Main Coal is absent. Elsewhere it is generally less than 2 ft thick, but it has been worked extensively in the north-eastern part of the district, where it is 5 to $7\frac{1}{2}$ ft thick and is commonly banded.

A maximum of about 700 ft of Middle Coal Measures above the High Main Coal has been preserved beneath the Permian. In the Durham Coalfield as a whole these measures contain five marine bands, named in upward succession the High Main, Little, Kirkby's, Hylton and Ryhope (Fig. 3). Of these only the High Main and Kirkby's marine bands have so far been proved within the district. The High Main Marine Band consists of a few inches of dark shale with *Lingula*. It commonly overlies a thin coal separated from the High Main Coal either by a sandstone—the High Main Post—which is up to 80 ft thick in the north-central part of the district, or by siltstones and mudstones containing plant debris and sparse non-marine lamellibranchs. The strata between the High Main Marine Band and the Ryhope Little Coal include three 'mussel'-bands. The Ryhope Little Coal is locally split to form the Top and Bottom Ryhope Little coals. All are thin and therefore unworked. Above the coal a thin bed of dark shale marks the position of the Little Marine Band which has yielded *Lingula* in adjacent districts.

The next seam above is the Ryhope Five-Quarter Coal which is $2\frac{1}{2}$ to 3 ft thick and is worked only in a small area near the northern margin of the district. It is immediately overlain by a ' mussel '-band and about 50 ft above it is Kirkby's Marine Band which contains scattered *Lingula* and abundant foraminifera. The succeeding 100 ft of strata probably include the Hylton and Ryhope marine bands, though for want of modern sections neither band has yet been proved in the district. Near the top of the known sequence of Middle Coal Measures, the Usworth and Hebburn Fell coals are preserved in a restricted area in the north-east. Though thick, both coals are thought to lie too close to the unconformity to be workable. E.A.F., D.B.S.

REDDENED CARBONIFEROUS STRATA

Reddened sandstones crop out in the lower part of the escarpment capped by Permian strata, and have been sunk and bored through in the search for coal. Sedgwick's (1829) belief that they were New Red Sandstone in age remained unchallenged until Howse (1857), citing floral evidence, assigned them to the Coal Measures—a view which was supported on other grounds by Daglish and Forster (1864).

More recently Anderson and Dunham (1953) recorded that beneath the Permian of Durham there is a reddened zone. This zone, which is transgressive and in no way attributable to original red minerals in Carboniferous rocks, does not lie directly beneath the Permian, but is separated from it by grey rock. The latter is generally 3 to 10 ft thick, though less commonly it amounts to between 10 and 30 ft, and in extreme cases is absent or, as in a bore near Black Hurworth, as much as 53 ft thick.

This grey zone, especially in its upper part, is in most places pyritic, and it seems reasonable to suppose that it owes its colour, in part at least, to the reduction of secondarily oxidized and reddened Coal Measures by waters percolating down from the Marl Slate sea (p. 96). This explanation is satisfactory where the grey zone is shallow, but where it is deep original non-reddening, rather than reduction, may be responsible. At the top of the grey zone there is commonly what appears to be a leached layer, generally only a few inches thick, but in places extending down to a foot or two. Where the rock at the sub-Permian surface was formerly mudstone it has been changed in places to a very pale grey or whitish dust. The superficial pale grey zone, though apparently leached, is now commonly pyritic. Where mudstone in this zone contains moisture, it resembles certain types of seatearth-mudstone and it has consequently been termed ' seggar ' in some old records. In some respects it is indeed comparable, for it seems to be deficient in iron and it may have a rather greasy texture, but it does not contain the characteristic rootlets of normal seatearth, and this enables the distinction to be made.

The reddened zone below the grey zone can be loosely divided into two parts. In the upper of these all the rocks are coloured except where they are near to coals or other beds with a high organic content. Mudstones and siltstones are generally purplish red, clay-ironstones exhibit colour-banding with brownish yellow cores, and coals, in some cases, have been totally oxidized. This subzone extends in places down to about 70 or 80 ft below the unconformity, but it may be as little as 10 ft or even absent. It is underlain by a subzone in which reddening is confined to sandstones. Micaceous partings in these are commonly picked out in bright shades of red, while the body of the rock is only pale pink. In many bores this restricted reddening extends to more than 100 ft below the unconformity, and the maximum depth so far proved is 292 ft in the Dalton Nook Bore, closely approached by 285 ft in Fishburn 'D' Bore.

The origin of the reddening in Durham is thought by Anderson and Dunham (1953, p. 31) to be due to oxidation of pyrite and chalybite *in situ* and to introduction of red iron oxide along joints and into pore-spaces of sandstones. The pre-Zechstein age of at least some of the reddening can now be demonstrated by the incorporation of fragments of reddened Carboniferous rocks in the Basal Permian Breccias. Since some of the breccias have a general pink tinge, however, the possibility arises that reddening continued while they were being formed.   E.A.F.

PALAEONTOLOGY OF THE COAL MEASURES

In the predominantly non-marine environment under which the greater part of the Coal Measures was deposited, the most common inhabitants of the waters of the swamps were non-marine lamellibranchs, crustaceans such as ostracods, and worms and fish. The periodic but short-lived incursions of the sea profoundly affected the endemic swamp fauna, causing it to withdraw from the area during the duration of marine conditions, and in turn introducing a marine fauna of foraminifera, sponges and brachiopods. On the withdrawal of the sea the swamp fauna repopulated the area until driven out by the emergence of the coal-forming forests or further marine invasions. 'Estheria' is found interbedded at intervals in the non-marine facies and also in strata associated with the marine bands ; the appearance of these fossils in the sequence is thought to be dependent on the extension of special and somewhat brackish conditions into the swamps (see below).

NON-MARINE FOSSILS

Lamellibranchs. The non-marine lamellibranchs or 'mussels' are the best known of the Coal Measures fossils ; they occur abundantly at certain levels forming characteristic 'mussel'-bands, which are valuable as marker-horizons. The 'mussels' are most commonly found in the roof measures of coals, but they also occur in mudstones in other parts of the sequence. They inhabited the relatively shallow tracts of water which spread over extensive areas with the drowning of the Coal Measures forests. The mutual exclusion of the 'mussels' and typical marine fossils is accepted as evidence that the habitat of the 'mussels' was not marine (Trueman 1946, p. lxii), but whether they favoured fresh-water or brackish conditions is not definitely established. It is evident that the ecological ranges of the different genera varied to some degree, particularly their tolerance of salinity (Weir 1945). In this respect the habitat of *Curvirimula,* and to a lesser extent *Naiadites,* approached most closely to near-marine conditions (see p. 28). *Carbonicola,* from its faunal associations in the Lower Coal

Measures, appears to have favoured slightly more brackish conditions than the typical Middle Coal Measures genera *Anthracosia, Anthracosphaerium* and *Anthraconaia*. The habitat of *Anthraconauta*, though not known for certain, was apparently different from that of the other genera and related to the special conditions arising from the withdrawal of marine influence from Western Europe in Upper Coal Measures times (Weir 1960, p. 299).

In the Durham–West Hartlepool district, of the seven genera of Coal Measures non-marine lamellibranchs, only *Anthraconauta* is not recorded and this can be explained by the absence of the Upper Coal Measures to which it is virtually confined. The distribution of the other six genera is as follows : *Carbonicola* is characteristic of, and confined to, the Lower Coal Measures. During the resurvey the highest position at which it has been found is in the roof measures of the Top Busty Coal, but in adjacent districts *Carbonicola* is found up to the band above the Harvey Coal. *Anthracosia* replaces *Carbonicola* as the dominant genus, appearing in the highest part of the Lower Coal Measures and continuing up to the beds associated with Kirkby's Marine Band, which is the highest faunal horizon proved in the district. Elsewhere in the Durham coalfield, *Anthracosia* ranges up to just below the Ryhope Marine Band. *Anthracosphaerium* is known rarely in the Lower Coal Measures, but is more common in the lower part of the Middle Coal Measures ; the highest record obtained during the resurvey is from just below the position of the High Main Marine Band ; in other districts the upper limit is similar to that of *Anthracosia*. *Anthraconaia* is generally rare in the district but ranges from the beds above the Brockwell Coal to the ' mussel '-band above the High Main Marine Band. *Naiadites* is known from most of the ' mussel '-bands and also in isolation at numerous intervening horizons. Finally, *Curvirimula* is typical of the Lower Coal Measures faunal horizons up to the ' mussel '- band 15 ft above the Three-Quarter Coal ; it is then absent from the sequence until it reappears in the ' mussel '-band overlying the High Main Marine Band and in the upper part of Kirkby's Marine Band. The inference from this close association with the regressive phase of the marine incursion is that *Curvirimula* favoured a more saline environment than the other ' mussel ' genera (see also Weir 1960, pp. 298–9).

Crustaceans. The commonest crustaceans in the Coal Measures are the ostracods which are mainly found either closely associated with the lamellibranchs or occupying distinct layers within the ' mussel '-bands. The two principal genera are *Geisina* and *Carbonita ;* the former is characteristic of the Lower Coal Measures, but is absent from the Middle Coal Measures except for rare examples in the regression stage of the High Main Marine Band (p. 28). This distribution of *Geisina* accords with that found in the other Pennine coalfields, and also with the view that the habitat of *Geisina* was more brackish than that of *Carbonita*. The latter genus ranges throughout the British Coal Measures particularly in the higher beds where the marine influence has declined. The two characteristic bands in this district are the Brockwell Ostracod Band (p. 24) containing *Carbonita humilis* (Jones and Kirkby), *C.* cf. *pungens* (Jones and Kirkby) and numerous immature forms, and the Hopkins Band (p. 26) in the roof measures of the Harvey Coal, where *Geisina arcuata* (Bean) is abundant. *C. humilis* and *C. pungens* also occur in the Middle Coal Measures, and there are records of elongate forms comparable with *C. scalpellus* (Jones and Kirkby) from above the Low Main Post and close below the High Main Marine Band. An elongate ostracod with a finely striated carapace, here referred to as *Hilboldtina sp.,* is known from the ' mussel '-band just below the Hutton Coal and also above the Bottom Brass Thill Coal.

' *Estheria* ' is a small bivalved crustacean related to the water fleas. The isolated valves of the carapace are distinctive fossils and commonly show an iridescent preservation. In such circumstances the precise generic identity cannot always be established and the general term ' *Estheria* ' is employed. The tendency for ' *Estheria* ' to be restricted to discrete horizons of wide lateral extent is comparable to the behaviour of the marine bands with which ' *Estheria* ' is commonly associated. The implication of this distribution is that ' *Estheria* ' was not a member of the swamp fauna, but favoured special conditions found in the brackish regions fringing the marine environment (see also Defrise-Gussenhoven and Pastiels 1957, p. 65). Periodically these brackish conditions extended into the swamps, in some cases as a prelude or aftermath of a marine incursion, and brought in the floods of ' *Estheria* ' which characteristically occur in the initial or regressive phases of a marine incursion. In general, ' *Estheria* ' is not found with other fossils, although at some horizons abundant fish remains are associated with it. In the Durham–West Hartlepool district the main occurrences of ' *Estheria* ' are in the Fishburn ' *Estheria* ' Band in the roof measures of the Three-Quarter Coal, in the Blackhall ' *Estheria* ' Band a short distance below the Main Coal in the offshore bores, immediately below the High Main Marine Band, and associated with Kirkby's Marine Band.

Worms. Although worms are seldom preserved as fossils, their former presence in a sediment is indicated by tubes, trails or burrows. The coiled tubes of the serpulid *Spirorbis* occur either isolated or attached to ' mussels ', notably *Naiadites,* or to plant remains. The trails and burrows, which are collectively known as ' trace-fossils ', are found mainly in the argillaceous rocks ; the sinuous trails of *Cochlichnus [Belorhaphe] kochi* (Ludwig) occur in the non-marine facies, notably in the beds above the Fishburn ' *Estheria* ' Band. The burrows of *Planolites ophthalmoides* Jessen are characteristically found in proximity to marine incursions (see p. 22), but are also found, though less commonly, associated with, or between, layers of ' mussels ' such as *Curvirimula* and *Cárbonicola.* The inference is that the layers containing *Planolites* represent a slightly more brackish environment than that of the ' mussel '-bands lying higher in the sequence. Such rare occurrences of *P. ophthalmoides* are known above the Top Marshall Green Coal, above the Victoria Coal and in the ' mussel '-band overlying the Brockwell Ostracod Band.

Fish. Isolated scales, teeth and bones of fish are common in both the ' mussel '-bands and marine horizons. The remains tend to be concentrated at the base of the faunal cycles immediately above a coal or seatearth, indicating that the fish were amongst the first animals to invade the area at the periodic flooding of the coal forests. Complete fish are rare, but their debris is notably abundant at certain horizons such as the Harvey Marine Band and in the beds associated with the High Main Marine Band. The common genera are *Elonichthys, Rhabdoderma, Rhadinichthys, Rhizodopsis* and Platysomids ; all these forms occur in both the non-marine and marine facies, particularly the *Lingula* phase (see below). It seems possible that these fish were able to live in the shallow coastal waters and estuaries, and also in the swamp facies, but their wide distribution may also reflect transport of disintegrated remains far from their place of death.

MARINE FOSSILS

Only five marine bands have been proved in the Durham–West Hartlepool district, namely the band in the Middle Stotfold Bore taken as the base of the Coal Measures (p. 35), the Gubeon a short distance above the base, the Harvey at the base of the Middle Coal Measures, and the High Main and Kirkby's marine bands near the middle of the Middle Coal Measures. Costate Productids are recorded from the basal Coal Measures band, and the remainder contain *Lingula*. In addition, the fauna of the Harvey Marine Band at one locality (p. 58) includes fragmentary *Orbiculoidea* and *Hollinella?* ; and the Gubeon and Kirkby's marine bands also contain foraminifera, sponge spicules and *Planolites ophthalmoides*. The bands are thin and the faunas impoverished and they do not display clearly the sequence of faunal phases found in fully developed marine bands such as those of the central and south Pennines (Calver *in* Smith and others 1967, pp. 91–4). However, some of the phases can be recognized in the district, particularly when augmented by evidence gained from adjacent parts of the coalfield. An idealized sequence of stages in a marine incursion may be envisaged as a progress from a basal fish phase through layers dominated respectively by *P. ophthalmoides* (p. 21), foraminifera, and *Lingula* ; the acme of the incursion is represented by the presence of the cephalopod/lamellibranch fauna such as is found in the Ryhope Marine Band farther north in the Sunderland (21) district. The regression stage of the incursion mirrors the advance, except that the equivalent of the basal fish phase is not often present, its place being taken by non-marine lamellibranchs such as *Naiadites*. In many cases the advance of the marine invasion appears to have been more rapid than the retreat, since the initial phases are condensed or may even be absent, whereas in the regression stage the different phases are thicker and more clearly distinguishable.

The typical lithology for both the *Planolites* and foraminifera phases is a pale grey to dark grey, smooth or slightly silty mudstone. Increase in carbonaceous matter or arenaceous material inhibited the existence of *Planolites* and foraminifera and they are seldom recorded from beds of this lithology. These factors probably explain their absence from the Harvey and High Main marine bands as developed in the district. The common foraminifera genera are *Ammodiscus, Glomospira* and *Tolypammina*.

FAUNAL SUCCESSION

The general sequence of faunas in the Coal Measures of the Durham and Northumberland coalfields has been described in a series of papers by Hopkins (1929, 1930 ; see also 1954) but no systematic investigation of their detailed distribution in the Durham district had been undertaken prior to the resurvey. A brief reference to fossils from the Fishburn boreholes was made by Trueman and Weir (1946, p. 14) in their discussion of *Carbonicola declivis* Trueman and Weir, while Magraw and others (1963) have referred to collections made from bores in the Elwick–Hartlepools area.

The application of the non-marine lamellibranch zonal scheme to the local sequence is shown in Fig. 3, which also gives the position of the more important faunal horizons. In the following account the main characteristics of the faunal sequence are described, and interpreted in terms of zonal position. The detailed fauna of individual bands is included in the detailed stratigraphical account (pp. 35–86) ; representative fossils from the district are figured on Pl. II.

Lower Coal Measures. In the absence of the characteristic *Gastrioceras subcrenatum* fauna, the base of the Coal Measures has been taken arbitrarily at the marine horizon containing costate Productid fragments at 942 ft in the Middle Stotfold Borehole. Os noted on pp. 11, 35 this decision is based on indirect faunal evidence, particularly on information gained from adjacent areas, but it accords with the known facies changes affecting the *G. subcrenatum* horizon in the north Pennines. Apart from this Middle Stotfold band the lowest fossiliferous horizon recorded from definite Coal Measures of the district is the Gubeon Marine Band of the East Hetton Colliery Busty Borehole and Offshore No. 1 Borehole which contains rare foraminifera, pyritized sponge spicules, *Planolites ophthalmoides, Lingula mytilloides* J. Sowerby, *Rhizodopsis sp.* and undetermined Palaeoniscid scales. There is insufficient faunal evidence to determine the precise equivalent of this marine band in terms of the standard south Pennine succession, but the presence of foraminifera suggests that this band is unlikely to be lower than the *Gastrioceras listeri* horizon. Although in the absence of non-marine lamellibranchs direct evidence is lacking, by inference the *Anthraconaia lenisulcata* Zone is represented by these lower measures ; the junction with the overlying *C. communis* Zone cannot be determined with confidence, but it is taken at the Ganister Clay horizon pending further evidence. The lowest fauna referred to the *C. communis* Zone is that containing *Carbonicola* cf. *communis* Davies and Trueman and *Curvirimula sp.* from 30 ft above the Ganister Clay Coal (see p. 37). There is some resemblance between the *Carbonicola* and the shells known from the 'mussel'-band which lies between the Albrighton and Lower Three-Quarter coals of the Cumberland Coalfield (Taylor 1961, pp. 7, 20 ; cf. Trueman and Weir 1947, pl. v, fig. 19) and which is placed in the lower part of the *C. communis* Zone and correlated with the *Carbonicola torus* Eagar fauna of the Lancashire Coalfield (Eagar 1956, p. 356).

Between the Top Marshall Green and Victoria coals poorly preserved lamellibranchs have been found at three horizons. The fauna includes *?Anthracosphaerium dawsoni* (Brown), a small form of *Carbonicola pseudorobusta* Trueman, *Curvirimula sp.* and *Naiadites flexuosus* Dix and Trueman. This assemblage is not diagnostic but gives the first indication of the incoming of *C. pseudorobusta* and is therefore likely to be near the base of the middle *C. communis* Zone, and approximately equivalent to the beds overlying the horizon of the Arley/Kilburn coals of the central Pennine coalfields. The *C. pseudorobusta* group appears in force above the Victoria Coal. The shells are smaller than typical forms and usually confined to the basal foot or so of the band in which *Curvirimula* and fish remains are commonly found. At some localities *Carbonita humilis* and *Geisina arcuata* also occur but are far less common than in the higher Brockwell Ostracod Band. In the Fishburn series of boreholes the basal layer is overlain by up to 5 ft of pale mudstones in which the shells are small and less common and are interbedded with *Planolites ophthalmoides*. The *Carbonicola* shells include *C.* cf. *rhindi* (Brown) and *C. declivis* Trueman and Weir *sensu lato* (cf. Trueman and Weir 1946, pl. ii, fig. 28).

A higher band overlying the seatearth about midway between the Victoria and Brockwell coals is typically represented by small poorly preserved *Carbonicola* (cf. Trueman and Weir 1946, pl. ii, figs. 24, 27). Those showing straight ventral margin, posterior truncation and carina are termed *C.* aff. *declivis* ; others, which form the majority, have curved ventral margin, higher Height/Length ratio, shorter anterior end and lack the carina and truncation, and are therefore compared to stunted forms of *Carbonicola* cf. *acuta* (J. Sowerby) and *C.* cf. *communis*. This horizon has been referred to as the *C. declivis* Band (Magraw and others 1963, pl. xix) but it is inadvisable to continue to use this name for this horizon as the acme of typical *C. declivis* is in the band above the Brockwell. *Curvirimula* is represented by *C. subovata* (Dewar) and *Naiadites flexuosus* is also present. The absence of *Carbonita* is a further distinction between this horizon and that above the Brockwell Ostracod Band.

The detailed faunal sequence in the Brockwell Ostracod Band provides a readily recognized marker horizon. As its name implies, the characteristic feature of this band over much of the district is the abundance of ostracods in the ferruginous basal layers, where they are associated with *Curvirimula subovata* and fish remains. *Carbonita humilis* is the most common ostracod, but *C. pungens* and undescribed species of *Carbonita* are also present ; the ostracods are in general of small size and immature forms are common ; *Geisina arcuata* is recorded at some localities, but is rare. In the upper part of the band lamellibranchs are dominant, but the fauna varies in different parts of the district. In the Fishburn area, where the band is approximately 6 ft thick, the majority of the *Carbonicola* belong to an assemblage of *C. declivis,* the type of which comes from this band in the Fishburn No. 2 Bore (Trueman and Weir 1946, p. 14, pl. ii, figs. 15, 17, 18 ; refigured in this memoir, Pl. II, fig. 12). Trueman and Weir also figured three specimens from the equivalent horizon in the Fishburn No. 1 Bore as *Carbonicola declivis* (pl. ii, figs. 21–22) and *C.* aff. *declivis* (pl. ii, figs. 25, 26). The type of *C. declivis* measures 14·5 mm but Trueman and Weir stated (p. 13) in their diagnosis that this species may attain 30 mm, which is about the size of the largest shells occurring in this Brockwell fauna ; the implication is that the type of *C. declivis* is either a juvenile or is stunted. However, certain of the larger shells compare with *Carbonicola browni* Trueman and Weir which has a similar form to *C. declivis,* but differs mainly in its greater size, the holotype measuring 54 mm (Trueman and Weir 1946, pl. ii, fig. 8 ; refigured Lumsden and Calver 1958, pl. iv, fig. 12). Since there is no evidence to link these larger shells (Pl. II, figs. 14, 15) in a continuous series with the type of *C. browni,* it seems preferable to include them with *C. declivis,* although with the emphasis on certain morphological features they could also be referred to

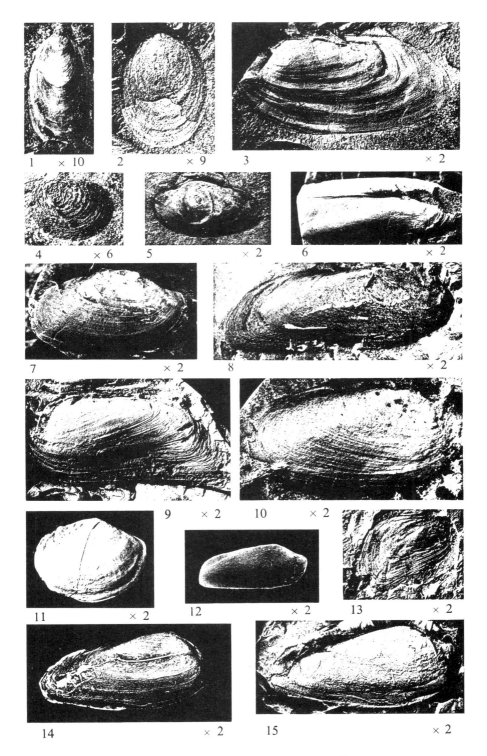

LOWER AND MIDDLE COAL MEASURES FOSSILS

(For explanation see pp. vii–viii)

as *C.* cf. *browni*. Associated fossils include cf. *Carbonicola circinata* Pastiels, *Curvirimula subovata* and *Anthraconaia fugax* Eagar. The latter fossil provides a link with the Yorkshire Coalfield where *A. fugax* occurs a short distance below the Low ' *Estheria* ' Band, and also with the Burnley Coalfield where the same species is known from above the Blindstone Rider Mine (Calver *in* Earp and others 1961, p. 200 ; Eagar 1962, p. 331). In Fishburn No. 1 and ' B ' bores a slightly higher band lies 10 ft above the *C. declivis* fauna and contains a different assemblage. The fossils identified include *Spirorbis sp.*, *Carbonicola pseudorobusta*, *Curvirimula candela* (Dewar), *C. subovata*, *Naiadites sp.*, and fish remains including *Elonichthys sp.* and *Rhabdoderma sp.*

Farther east, in the Elwick series of boreholes, e.g. at Low Throston, the fossils in the Brockwell Ostracod Band range over some 13 ft of strata, but *C. declivis* is absent. A basal layer with fish remains is overlain by a few inches of mudstone with abundant *Carbonita humilis* and rare *Geisina arcuata*. The succeeding measures contain a fauna of large shells including *Carbonicola pseudorobusta?*, *C.* cf. *rhomboidalis* Hind, *C.* cf. *robusta* (J. de C. Sowerby), *Curvirimula candela*, *C. subovata*, *Naiadites sp.*, in addition to *Geisina arcuata* and fish remains including *Elonichthys sp.* and *Rhadinichthys sp.* About the middle of the band are several examples of *Planolites ophthalmoides* associated with *Curvirimula* and *Carbonicola*. This fauna from the Elwick area is therefore more like that of the higher of the two bands recorded in Fishburn No. 1 and ' B ' bores (see above).

The generic assemblage from the Brockwell Ostracod Band is similar to that from immediately above the Victoria, although at the latter horizon ostracods are far less common, and the *Curvirimula* are smaller ; on the other hand, *Naiadites* is rare in the Brockwell fauna, but is recorded at several localities in the measures overlying the Victoria Coal.

Bores in the northern part of the district have shown that two faunal bands occur between the Brockwell and Three-Quarter coals. The lower, which is probably the equivalent of the Ostracod Band of the south, contains small *Carbonicola sp.*, *Carbonita spp.*, *Geisina arcuata* and fish remains. In the higher band, as proved in Eppleton No. 2 Bore, the fauna comprises *Curvirimula candela* (Pl. II, fig. 13), *C. subovata*, *Carbonita humilis*, *C.* cf. *pungens* and fish remains. In Murton No. 1 Bore, this higher band contains *Anthraconaia sp. nov.* of *A. modiolaris* (J. de C. Sowerby) group, *Curvirimula candela* and *C. subovata*. The presence of *Anthraconaia* is of interest for the genus is rarely found in the *C. communis* Zone, but a comparable record is noted by Hind (1895, p. 163) and Hopkins (1930, p. 103) in a reference to ' *Anthracomya adamsi* (Salter) from the Brockwell horizon near Wylam ' in the Newcastle upon Tyne (20) district.

The records of ' *Estheria* ' in the roof measures of the Three-Quarter Coal at several localities in the southern part of the district justify the term Fishburn ' *Estheria* ' Band (p. 45) for this occurrence. The specimens measure up to 4 mm in length and have the widely-spaced growth lines seen also in examples from the Low ' *Estheria* ' Band of the East Pennine Coalfield. The stratigraphical position of these respective bands in the two areas and their relationship to the non-marine faunas strongly suggest that they are the same horizon, which is a convenient boundary between the middle and upper parts of the *C. communis* Zone. Preservation of these fossils is generally poor, and pending generic revision, the non-committal designation 'Estheria' is employed for this form. However, the wide spacing of the growth lines can usually be discerned, a feature which distinguishes this form from the *Lioestheria vinti* (Kirkby) of the Middle Coal Measures. At some localities the pale-coloured mudstones overlying the 'Estheria' Band contain *Cochlichnus* [*Belorhaphe*] *kochi* and cf. *P. ophthalmoides*.

A 'mussel'-band about 15 ft above the Three-Quarter Coal is known in the Fishburn–East Hetton area and contains *Carbonicola pseudorobusta*, *Curvirimula candela, C. subovata, Carbonita* cf. *humilis, Geisina arcuata* and fish scales including *Rhabdoderma*. This is the highest band in the district from which *Carbonicola pseudorobusta* and *Curvirimula spp.* are recorded. The occurrence of large *C. candela* recalls the fauna from above the Cannel Mine of West Lancashire and above the Lower Yard Rider Mine of Burnley (Calver *in* Earp and others 1961, p. 201); on this comparison the zonal position is placed near the top of the *C. communis* Zone.

Several examples of *Anthracosphaerium* cf. *dawsoni* (Brown) (Pl. II, fig. 11), preserved as internal moulds in ironstone, were found associated with small *Carbonicola sp.* and *Naiadites flexuosus* in the roof measures of the Top Busty Coal of the Fishburn No. 3 Borehole. Elsewhere in the district fossils have rarely been found at this horizon ; exceptions are Fishburn 'B' Bore and a bore in the Butterwick area where *Carbonicola rhomboidalis?* is recorded. The *A.* cf. *dawsoni* recall the similar forms known as a constituent of the *Carbonicola cristagalli* Wright fauna at the base of the *Anthraconaia modiolaris* Zone in several southern Pennine coalfields, including the suite figured by Trueman and Weir (1955, fig. 27) from the roof of the Middle Lount Coal of the Leicestershire Coalfield. Although in the Durham–West Hartlepool district the *C. cristagalli* fauna is not recognized as such, an equivalent horizon is thought to be represented by this band containing *A.* cf *dawsoni*. The Top Busty Coal is therefore taken as the provisional junction of the *C. communis* and *A. modiolaris* zones.

The measures between the 'mussel'-band above the Top Busty Coal and the Harvey Coal are relatively unfossiliferous ; consequently there is little evidence of the fauna found in other coalfields in measures representing the lower part of the lower *A. modiolaris* Zone. The next higher fauna which can be assigned to a zonal position is that overlying the Harvey Coal where the *Anthracosia regularis* (Trueman) fauna becomes dominant, although the genus first appears a short distance below this coal (p. 53). As well as *A. regularis* the lamellibranchs include rare *Anthraconaia modiolaris, Naiadites sp.* intermediate between *productus* (Brown) and *quadratus* (J. de C. Sowerby), associated with *Spirorbis sp., Carbonita sp., Geisina arcuata* and Palaeoniscid scales. This assemblage is characteristic of the measures in the upper part of the lower *Anthraconaia modiolaris* Zone in the majority of the British coalfields. The distinctive faunal profile (see p. 56) described by Hopkins (1929, p. 8) as a feature of this band in the Northumberland and Durham coalfields cannot always be recognized, although the abundance of *G. arcuata* in the basal layers is characteristic. This abundance is also found in the *A. regularis* faunal belt of the whole Pennine area including Cumberland (Calver *in* Taylor 1961, p. 22), the East Pennine Coalfield (Eagar 1960, pp. 146–7) and also in the nearby Midgeholme Coalfield.

Middle Coal Measures. The Harvey Marine Band at the base of the Middle Coal Measures is the equivalent of the *Anthracoceras vanderbeckei* marine horizon of the British coalfields, but displays the impoverished development characteristic of the coalfields of the north and west of the Pennine Province. The fauna is usually restricted to *Lingula mytilloides* (up to 6 mm in length) and fish remains such as scales and teeth of *Elonichthys, Helodus, Rhadinichthys* and *Rhizodopsis*. In the Elwick No. 3 Bore sponge spicules were found in addition, and in the Hartlepools Lighthouse Point Bore the fauna consisted of fragmentary *Orbiculoidea* and *Hollinella?* The fossils occur characteristically in a finely-micaceous, silty mudstone, the maximum thickness of the band not exceeding 6 in. At some localities the *Lingula* are found with both valves in contact and undamaged suggesting deposition under quiet conditions ; elsewhere, e.g. Hartlepool Lighthouse Bore and Offshore No. 2 Bore, the brachiopods are fragmentary and largely confined to the more sandy patches of the core samples, implying erosion and re-deposition. The absence of foraminifera and *Planolites ophthalmoides* from this marine band can be explained by the prevalence of the silty lithology.

The measures between the Harvey Marine Band and the Maudlin Coal are placed in the upper *A. modiolaris* Zone, which is conveniently subdivided at the Bottom Brass Thill Coal into upper and lower parts. In the lower part, which includes the ' mussel '-bands above the Harvey Marine Band and those above the Ruler and Hutton coals, the *Anthracosia spp.* are mainly dominated by *A. ovum* Trueman and Weir (Pl. II, fig. 10) and *A. aquilina* (J. de C. Sowerby) with subordinate *A. phrygiana* (Wright) and *A. subrecta* Trueman and Weir. *Anthraconaia* is represented by *A. modiolaris* and *A.* cf. *curtata* (Brown) in the beds below the Hutton, but above this coal only small or fragmentary *Anthraconaia* has been obtained. The characteristic *Naiadites* are *N. triangularis* (J. de C. Sowerby) and *N. quadratus* ; the *Anthracosphaerium* species recorded include *A.* cf. *affine* (Davies and Trueman), *A. exiguum* (Davies and Trueman) and *A. turgidum* (Brown). There is little to distinguish the assemblages from above both the Bottom and Top Hutton coals, and in the absence of diagnostic *Anthraconaia* the precise position of the Hutton faunas in terms of the upper *A. modiolaris* sequence of faunas has not been established.

In the highest part of the *A. modiolaris* Zone the faunal records are largely confined to the ' mussel '-bands above the Bottom and Top Brass Thill coals. These two horizons possess similar assemblages, in which the *Anthracosia phrygiana* group (Pl. II, fig. 9), including *A. beaniana* and *A. disjuncta* Trueman and Weir (Pl. II, fig. 8), is more in evidence than in the lower beds. The *Anthracosphaerium* species are similar to those below except that *A. turgidum* (Pl. II, fig. 7) is more common. *Naiadites quadratus* is the typical species, but *Anthraconaia* is rare.

The Maudlin Coal is taken as the base of the Lower *similis-pulchra* Zone largely on the evidence obtained in adjacent areas to the north where the *Anthraconaia pulchella* Broadhurst fauna is well developed above the coal. An important faunal marker band near the base of the zone is the Blackhall ' *Estheria* ' Band (Magraw and others 1963, p. 164) occurring just below the Main Coal in the Offshore Nos. 1 and 2 boreholes. The ' *Estheria* ' are preserved as iridescent films and are invariably associated with fish debris. A similar ' *Estheria* ' band associated with the *Anthraconaia pulchella* fauna is known from the base of the Lower *similis-pulchra* Zone in Lancashire (Magraw 1961, p. 442), in the east Pennines (Calver *in* Smith and others 1967, p. 173) and in Ayrshire (Mykura 1967, p. 64); the horizon is considered to be the same in all four areas.

In the measures between the Maudlin and the High Main coals, corresponding to the lower part of the Lower *similis-pulchra* Zone, lamellibranchs are not common. Several horizons contain *Naiadites productus* and a winged variant, as well as *N*. aff. *alatus* Trueman and Weir. The *Anthracosia* species are largely confined to the faunal bands between the Metal and High Main coals from which *A*. cf. *aquilinoides* (Chernyshev) and *A*. *sp*. cf. *caledonica* Trueman and Weir were collected, together with *Anthraconaia sp. nov.* cf. *wardi* (Hind) (Pl. II, fig. 6) and *Anthracosphaerium propinquum* (Melville) (Pl. II, fig. 5).

The important *Anthracosia atra* (Trueman) group makes its appearance in the 'mussel'-band between the High Main Coal and the overlying marine band where it is associated with *Anthracosphaerium radiatum* (Wright) and other representatives of the *Anthracosia acutella* (Wright)/ *concinna* (Wright) fauna. This assemblage is similar in many respects to that from a slightly higher horizon overlying the High Main Marine Band in adjacent areas to the north (Hopkins 1933).

A distinctive faunal profile is recognizable in the measures associated with the High Main Marine Band as the following composite sequence illustrates. The seatearth or coal at the base is overlain by abundant fish remains; at some localities *Anthracosia* cf. *atra*, *Naiadites sp.*, *Carbonita humilis* and *C*. cf. *scalpellus* are found in addition. This basal layer is succeeded by 1 to 2 ft of mudstone with abundant *Lioestheria vinti* (Pl. II, fig. 4) and fish remains, and is in turn overlain by dark silty mudstone containing *Lingula sp.* The marine band is generally only a few inches thick, but in the Easington E. 11 Underground Bore the *Lingula* range over 1 ft 7 in. They are up to 4 mm in size and include a broad form of *L. mytilloides* (Pl. II, fig. 2); no other marine fossils have been found at this horizon in the district. A thin band with abundant fish debris overlies the marine band and is in turn succeeded by some 10 to 15 ft of mudstone with a characteristic fauna at the top. The fauna includes *Anthraconaia sp. nov.* cf. *wardi*, *Anthracosia* cf. *atra*, *A*. cf. *simulans* Trueman and Weir, *Curvirimula sp.*, *Naiadites* cf. *obliquus* Dix and Trueman, *N*. cf. *productus*, *Carbonita humilis* and *Geisina sp.* *Spirorbis sp.* is locally abundant at the base. Many of the lamellibranchs are small and possibly stunted, and, coupled with the presence of *Curvirimula* and *Geisina,* support the interpretation that the environment at this period was more brackish than that prevailing during the deposition of a typical 'mussel'-band. As noted above (p. 20), both *Curvirimula* and *Geisina* are found in the regression stages of the marine bands of the Lower and Upper *similis-pulchra* zones, although absent from intervening strata (see Calver *in* Smith and others 1967, pp. 88–9). The typical well-developed 'High Main shellbed' known elsewhere in Durham and Northumberland (Hopkins 1933) has not been noted from the district but it may be represented by the sparse *Anthraconaia* aff. *librata* (Wright) (Pl. II, fig. 3) found in Easington E.11 Bore 25 ft above the marine band.

Kirkby's Marine Band (the highest faunal band so far proved in the district) is distinguished by the close relationship between the marine fossils and non-marine lamellibranchs, and the association with *Lioestheria*. The fauna obtained from this horizon includes foraminifera, sponge spicules, *Planolites ophthalmoides, Lingula spp., Anthracosia* cf. *atra, Curvirimula sp.* and fish debris. From the evidence of the collections from Hawthorn Colliery Shaft and also from bores just to the north of the district, the non-marine lamellibranchs tend to be concentrated in a band between two layers of marine fossils, although both foraminifera and *P. ophthalmoides* occur on the same bedding plane as the *Anthracosia*. The basal marine phase occurs in a dark grey, slightly silty mudstone, and contains in addition to *Lingula,* rare foraminifera and pyritized sponge spicules. The succeeding grey mudstones have *Anthracosia* cf. *atra* and foraminifera at the base, but in their upper part the fauna is largely *P. ophthalmoides,* abundant foraminifera including *Glomospira* and *Tolypammina,* and *Lingula.* In these upper beds, the pyritized strap-like markings and pyrite-filled burrows are possibly atypical preservations of *P. ophthalmoides.* Some of the foraminifera occur massed in elongated patches in the plane of the bedding ; these accumulations may represent infilled burrows or it is possible that the tests have been used for the formation of the walls of worm-tubes similar to the present day Terebelloids. An elongate grouping is also shown by specimens of *Lingula* forming a chain-like arrangement. Although the majority of the *Lingula* are referred to *L. mytilloides* there are several examples which are proportionately narrower and are here compared with *L. elongata* Demanet (Pl. II, fig. 1), which is also recorded from the Haughton Marine Band of the East Pennine Coalfield (Calver *in* Smith and others 1967, p. 184). The presence of *Curvirimula* associated with foraminifera in the upper part of the band recalls the comments made in the description of the High Main Marine Band (see above). It is also significant that *Curvirimula* is characteristically found in strata associated with the presumed equivalent Bradford/ Haughton Marine Band in other coalfields of the Pennine province (Calver *in* Magraw 1960, p. 482 ; Calver *in* Smith and others 1967, p. 183.

Lioestheria is present in isolation in grey mudstones ; the precise position of this phase in the sequence of Kirkby's Marine Band based on the collections from Hawthorn Colliery Shaft has not been ascertained, but evidence from other areas suggests that the *Lioestheria* occur in the upper part of the marine band, or immediately overlying it.   M.A.C.

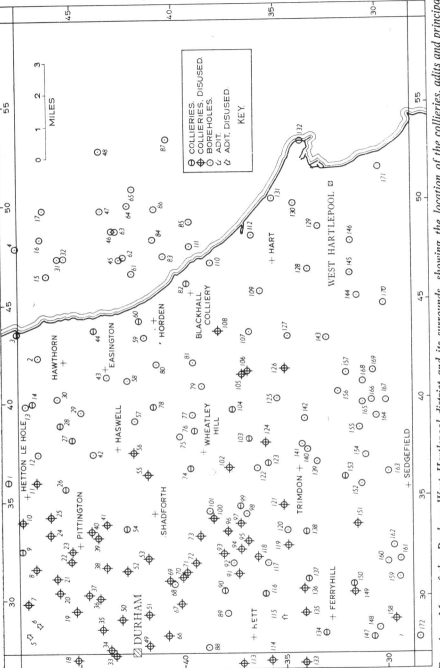

Fig. 4. Map of the Durham–West Hartlepool district and its surrounds, showing the location of the collieries, adits and principal bores. For key, see facing page

LIST OF COLLIERIES, BORES AND ADITS SITED ON FIG. 4

*Numbers marked with an asterisk correspond with sections under the same numbers in Plates III and IV.

1, Eppleton Colliery. *2, Murton No. 2 Bore. *3, Dawdon Colliery. *4, Dawdon No. 3 Bore. 5, Harbour House Drifts. 6, Cocken Drifts. 7, Cocken Colliery. *8, Rainton Colliery, Adventure Pit. 9, Rainton Colliery, Meadows Pit. 10, Rainton Colliery, Hazard Pit. 11, Hetton Colliery. *12, Eppleton No. 2 Bore. *13, Murton No. 1 Bore. *14, Murton Colliery. *15, Easington Colliery No. 11 Bore. *16, Easington Colliery No. 3 Bore. *17, Easington Colliery No. 9 Bore. 18, Framwellgate Colliery. *19, Frankland Colliery, Furnace Pit. 20, Brasside Colliery. 21, Woodside Colliery. *22, Belmont Colliery, Furnace Pit. *23, Rainton Colliery, Lady Seaham Pit. *24, Rainton Colliery, Alexandrina Pit. *25, North Hetton Colliery, Moorsley Winning. *26, Elemore Colliery. *27, South Hetton Colliery. 28, Hawthorn Colliery. *29, South Hetton No. 7 Bore. *30, Murton No. 3 Bore. *31, Easington Colliery No. 5 Bore. *32, Easington Colliery No. 12 Bore. 33, Aykleyheads Pit. *34, Durham Main Colliery. *35, Kepier Colliery, Florence Pit. 36, Kepier Grange Colliery. *37, Grange Colliery, New Winning. 38, Broomside Colliery. 39, North Pittington Colliery, Buddle Pit. *40, North Pittington Colliery, Londonderry Pit, and Littletown Colliery, Engine Pit. *41, Littletown Colliery, Lady Alice Pit. *42, South Hetton No. 6 Bore. *43, Murton No. 5 Bore. *44, Easington Colliery. *45, Horden No. 5 Bore. *46, Horden No. 20 Bore. *47, Horden No. 15 Bore. *48, Offshore No. 2 Bore. 49, Elvet Colliery. 50, Kepier Colliery. *51, Old Durham Colliery. *52, Sherburn Colliery, Lady Durham Pit. *53, Sherburn House Colliery, North Pit and Pit Bore. *54, Sherburn Hill Colliery, East Pit. *55, Ludworth Colliery. *56, Haswell Colliery. *57, South Hetton No. 10 Bore. 58, Mill Hill Reservoir Bore. *59, Horden No. 4 Bore. *60, Horden Colliery. *61, Horden No. 1 Bore. *62, Horden No. 19 Bore. *63, Horden No. 12 Bore. *64, Blackhall No. 33 Bore. *65, Blackhall No. 43 Bore. *66, Houghall Colliery, and No. 1 Bore. *67, Shincliffe Colliery. *68, Whitwell House Bore. 69, Whitwell Colliery, 'A' Pit. 70, Whitwell Colliery, 'C' Pit. 71, Whitwell Colliery, 'B' Pit. 72, Cassop Moor Colliery. 73, Cassop Vale Colliery. *74, Thornley Colliery. *75, Wheatley Hill No. 5 Bore. *76, Wheatley Hill Colliery. *77, Wheatley Hill No. 2 Bore. *78, Shotton Colliery. *79, Shotton Colliery No. 7 Bore. *80, Shotton Colliery No. 1 Bore. *81, Shotton Colliery No. 6 Bore. *82, Blackhall Colliery. *83, Blackhall No. 9 Bore. *84, Blackhall No. 16 Bore. *85, Blackhall No. 28 Bore. *86, Blackhall No. 36 Bore. *87, Offshore No. 1 Bore. *88, Low Butterby No. 2 Bore. *89, Butterby No. 4 Bore. *90, Bowburn Colliery, New Pit. *91, Coxhoe, Bell's Pit. *92, Heugh Hall No. 1 Bore. 93, Heugh Hall Colliery. 94, West Hetton Colliery. *95, Coxhoe Colliery, Engine Pit. 96, Crowtrees Colliery. 97, South Kelloe Colliery. *98, East Hetton Colliery, Busty Bore. *99, East Hetton Colliery, North Pit. *100, Cassop Colliery, 'A' Pit. *101, East Hetton Colliery, No. 32 Bore. 102, Kelloe New Winning. 103, Deaf Hill Colliery. 104, Wingate Grange Colliery. 105, Hutton Henry Marley Pit. *106, Hutton Henry Perseverance Pit. *107, Castle Eden No. 1 Bore. *108, Castle Eden Colliery. 109, Thorpe Bulmer Bore. *110, Black Halls Bore. *111, Blackhall Nos. 18 and 19 bores. 112, Hart Bore. 113, Croxdale Colliery, Thornton's Pit. *114, Tudhoe Colliery, West Pit. 115, Metal Bridge Drifts. *116, Tursdale Colliery, South Pit. *117, Coxhoe Brickyard Bore. 118, Clarence Hetton Pit. 119, Cornforth Colliery, George Pit. *120, Garmondsway 'B' Bore. 121, Garmondsway Moor Colliery. *122, Trimdon Grange Bore. *123, Trimdon Grange Colliery. 124, Trimdon Colliery. *125, Deaf Hill No. 4 Bore. *126, South Wingate Colliery. 127, Fifty Rigs Plantation Bore. 128, Naisberry Waterworks Nos. 1 and 2 bores. 129, Low Throston Bore. 130, Howbeck Hospital Bore. 131, North Sands Bore. 132, Hartlepool Lighthouse Bore. 133, Tudhoe Grange Colliery. *134, Dean and Chapter Colliery, No. 3 Pit. *135, East Howle Colliery, Catherine Pit. *136, West Cornforth Colliery, Thrislington Coke Oven Pit. *137, Thrislington Colliery, Jane Pit. *138, Garmondsway Pit. 139, Fishburn No. 1 Bore. *140, Fishburn No. 2 Bore. 141, Fishburn 'B' Bore. 142, Dropswell Bore. 143, Cotefold Close Bore. *144, Elwick No. 1 Bore. 145, Dalton Piercy Waterworks bores. *146, Dalton Nook Bore. *147, Chilton Colliery. *148, Chilton Colliery, 'E' Bore. 149, Little Chilton Colliery,

Cragg Pit. *150, Mainsforth Colliery, East and West Pits. *151, Bishop Middleham Colliery, East Pit. 152, Fishburn 'D' Bore. *153, Fishburn Colliery, North Pit. *154, Fishburn No. 3 Bore. *155, Fishburn No. 4 Bore. *156, Murton Blue House Lane Bore. *157, Murton Hall Farm Bore. *158, Windlestone Colliery. 159, South Mainsforth 'J' Bore. 160, Nunstainton No. 2 Bore. 161, Nunstainton No. 1 Bore. 162, Stony Hall Bore. 163, Ryal Bore. *164, Ten-o'Clock Barn No. 2 Bore. *165, Whin Houses Bore. *166, Whin Houses Farm Bore. *167, Ten-o'Clock Barn No. 1 Bore. *168, Embleton Old Hall Bore. 169, Tinkler's Gill Bore. 170, Middle Stotfold Bore. 171, Seaton Carew Bore. 172, Rushyford Bore.

DETAILS OF STRATIGRAPHY

During the long history of mining and exploration in this district about one thousand shafts and bores have been sunk. Representative sections are shown graphically in Fig. 5 (Millstone Grit Series), Plates III and IV (Lower and Middle Coal Measures) and Fig. 16 (uppermost Middle Coal Measures). The sites of these are indicated on Fig. 4 which also shows the localities of other collieries and bores. Details of the more important coal seams are summarized in Figs. 6 to 15, which show the principal lines of split and include isopachs of some of the thicker sandstones.

In many of the older records local terms are used to describe rock types and these are here quoted verbatim. Among these, ' post ' generally refers to sandstone, ' metal stone ' or ' fakes ' to strata consisting of alterations of mudstone, siltstone and sandstone, ' post girdles ' to thin bands of sandstone commonly present in ' metal stone ', and ' metal ' or ' blaes ' to mudstone or shale. ' Whin ' has been applied to any hard stone including certain sandstones and ironstones, though it is nowadays applied mainly to dolerite. The term ' thill ' embraces all types of seatearth, though seatclays approaching fireclay in lithology are more commonly known as ' seggar ' or the variants ' sagger ' and ' sagre '. Coal of poor quality is called ' splint ', and if dirty it is referred to as ' dant ''. Cannel or cannelly shale which gives out a ' chattering ' sound when burnt has been called ' parrot '' and pyrite is usually known as ' brass '. In sections of coal the term ' band ' is used for any interbedded layer other than coal and corresponds to the term ' dirt ' used in other coalfields ; ' banded coal ' thus signifies coal with layers of dirt.

Description of outcrop, shaft, bore and mine-plan information is, as far as possible, related to localities shown on the One-inch Geological Map, but the exact positions are indicated either in the text or in Appendix I by reference to the National Grid. Records of selected bores drilled since the publication by the Institution of Mining Engineers of the volumes of Borings and Sinkings in Northumberland and Durham have been summarized in Appendix I, p. 266, which also includes summaries of the full records of older bores and shafts published in those volumes.

MILLSTONE GRIT SERIES

Records of sinkings through the Millstone Grit Series of the district are restricted to a few borers' logs compiled during the nineteenth century. Some are shown graphically in Fig. 5, which indicates tentative correlation with the west Durham sequence.

The site of Ryal Bore (1874), which reaches the lowest stratigraphical horizon in the district, is not accurately known, but its position is thought to be in the vicinity of the swallow-hole [3633 2976], 300 yd S.W. of Ryal Farm, near Sedgefield. At a depth of

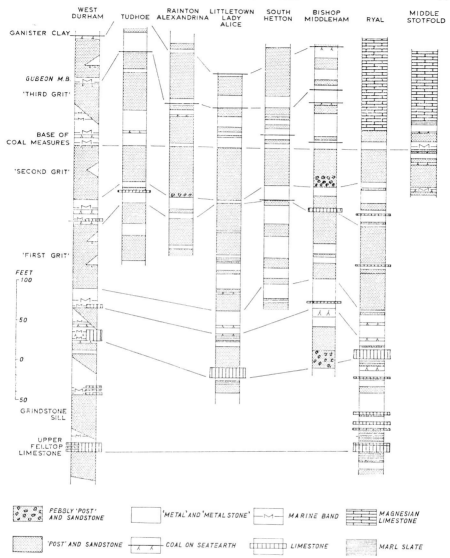

FIG. 5. *Comparative sections of the Millstone Grit Series and lowest Coal Measures, showing the suggested correlation with West Durham*

688 ft in this bore there is a limestone, 10 ft 9 in thick, with a 4-in shale band 1 ft 7 in from the top. The limestone is here correlated with the **Upper Felltop Limestone** of south-west Northumberland and west Durham (Fig. 5). Its lower part is reported to contain 'cockle shells', a term which by analogy with the Cockle Shell Limestone (Dunham 1948, p. 19) may imply Productids.

The Upper Felltop Limestone is overlain in the Ryal Bore by 1 ft 5 in of shale with 'blue beds' (presumably thin layers of calcareous shale or limestone), above which are 14 ft 8 in of interbedded 'post' and 'shale'. The equivalent beds in west Durham comprise up to 50 ft of shale overlain by the Grindstone Sill, a flaggy sandstone 30 to 50 ft thick (Dunham 1948). In the 21 ft of strata above the 'post' and 'shale' are three unnamed limestones, measuring, in upward succession, 5 in, 2 ft 4 in, and 2 ft 2 in. They are correlated here with a group of six thin limestones which lie above the Grindstone Sill in the Woodland Bore in the Barnard Castle (32) district (Mills and Hull, *in press*). Two further unnamed limestones, 9 in and 11 ft 7 in thick, are recorded 43 and 66 ft higher respectively in the Ryal Bore. The lower limestone may be the equivalent of 10 ft 11 in of 'grey shales with fossils' lying at the bottom of a deep bore at Littletown Lady Alice Pit; the upper thicker bed is here correlated with $12\frac{1}{4}$ ft of 'blue limestone with shells' lying at a depth of 1538 ft 9 in in the Littletown Bore. This $12\frac{1}{4}$-ft limestone is immediately underlain by beds of 'dark grey shale with fossils', $3\frac{1}{2}$ and 2 ft thick and separated by 2 ft of 'grey post'. In the Bishop Middleham Colliery (? East Pit) Bore, the limestone is absent and its position is assumed to lie at or near the base of 52 ft 5 in of white sandstone, the lower $22\frac{1}{2}$ ft of which is 'full of pebbles'.

The strata succeeding the thick limestones at Ryal and Littletown and the sandstone at Bishop Middleham, consist of 20 to 40 ft of 'shale' and 'post' overlain by about 10 to 20 ft of argillaceous beds which include seatclays. At Bishop Middleham (Fig. 5), seatclays are succeeded first by about 5 ft of 'dark grey shale' followed by 5 in of 'cherty or flinty limestone'. This bed is not recorded at Ryal and Littletown but appears to be represented by 10 in of hard dark grey argillaceous limestone in the Woodland Bore.

In West Durham, this thin limestone is overlain by up to 20 ft of generally argillaceous beds, separating it from the 'First Grit', which reaches over 100 ft in thickness. In this district, however, the 'First Grit' is less clearly defined, and in the absence of geologically examined sections, correlation is conjectural In the Ryal Bore, the 'First Grit' may be represented by two sandstones, a lower 48-ft and an upper 47-ft bed being separated by about 13 ft of strata including a 14-in limestone. In the Littletown Lady Alice Pit Bore, 61 ft 10 in of 'very coarse grey post' is recorded with

a 7-in 'grey shale' parting 27 ft above the base, but elsewhere most sandstones are less than 40 ft thick.

The sequence above the suggested equivalents of the 'First Grit' (Fig. 5), comprises rather varied strata; they are generally 20 to 30 ft thick, but at Ryal consist of about 50 ft of 'shale', 'metal', 'metal stone' and thin sandstone or 'post' bands. In the South Hetton Bore, a 9-in coal was recorded, and limestones 1, $1\frac{1}{2}$, $3\frac{1}{2}$ and 5 ft thick have been proved in the Ryal, Hart, Tudhoe Colliery and Bishop Middleham bores respectively. At about the same horizon in the Middle Stotfold Bore, a 1-ft bed of grey shaly mudstone at a depth of 990 ft 1 in yielded, in addition to carbonized plant fragments, the following: *Planolites ophthalmoides*; *Campylites* [*Sphenothallus*] *sp.*; *Lingula mytilloides*, *Productus carbonarius* de Koninck; cochliodont fish tooth.

The '**Second Grit**', which is over 100 ft thick in the Woodland Bore is readily recognized in the Durham–West Hartlepool district. In the Littletown Lady Alice Pit Bore it is represented by $66\frac{1}{2}$ ft of 'very coarse grey post', and in the Rainton Alexandrina Pit Bore nearby is 63 ft 11 in thick and is made up of 'hard light grey post' 14 ft 14 in, on 'very coarse post' 41 ft, 'fine dark post with shale partings and coal threads near the bottom' 7 ft 7 in, and 'conglomerate' 1 ft at the base. In most other provings within the district, the equivalent sandstone is between 40 and 50 ft thick, though in the Hart and Middle Stotfold bores it is only about 30 ft. At Bishop Middleham, the basal layer consists of 10 ft 4 in of 'white sandstone, very coarse and full of pebbles' overlain by 2 ft of 'dark conglomerate'. The remaining 32 ft comprises 'grey post, with shale bands'.

In the Middle Stotfold Bore, the sandstone considered to be equivalent to the 'Second Grit' is overlain by seatearth-siltstone, which is taken to be the topmost stratum of the Millstone Grit Series. In the older records this seatearth is not recorded, and the boundary is consequently placed at the top of the bed correlated with the 'Second Grit'.

E.A.F.

The base of the Lower Coal Measures is now defined as the base of the lowest of a group of marine bands which overlie the 'Second Grit' in West Durham (Fig. 5). In the Durham–West Hartlepool district, the presumed equivalents of these marine bands lie between about 934 ft and 943 ft in the Middle Stotfold Bore, and have yielded *Planolites?* at 934 ft 2 in, *P. ophthalmoides* at 937 ft 3 in, and fragments of costate Productids between 941 ft 10 in and 942 ft 1 in, and at 942 ft 5 in. Marine fossils have not been recorded at the same horizon in any other boring in this district, but the generally argillaceous sequence which fills the interval between the 'Second' and 'Third' grits can be distinguished in most of the old logs. The sequence includes a coal which is up to 7 in thick in the Bishop Middleham and South Hetton Colliery bores (Fig. 5) and also in the Hart Bore.

Above the coal are sandstones of variable thickness and extent: they are thought to correspond with the 'Third Grit' which in West Durham attains a thickness of 50 or 60 ft. In the Durham–West Hartlepool district, however, a single thick sandstone at this position is exceptional, apart from one of about 40 ft in the Hart Bore, and 30 ft in the Tursdale and Windlestone Colliery bores (Pl. III).

The measures above the 'Third Grit' sequence include ganister in the Rainton Alexandrina Pit and Windlestone Colliery bores. The Littletown Lady Alice Pit Bore passed through 'mild grey post with silica' at this horizon, and in the East Hetton Busty Bore the Gubeon Coal is underlain successively by 1 ft 7 in of ganister and 10 in of hard sandy fireclay. The presence of ganister in these bores supports the correlation with the 'Third Grit', for ganister is otherwise rarely recorded in this district. In Offshore No. 1 Bore the floor of the same coal is a grey-brown rather silty seatearth which passes down into a brown sandy seatearth with sphaerosiderite.

The Gubeon Coal which underlies the Gubeon Marine Band (Pl. III and Fig. 3), has been proved in many bores in the

district, but is everywhere thin. In Offshore No. 1 Bore, it is 18 in thick with a $2\frac{1}{4}$-in dirt band 12 in above the base; in Ten-o'-Clock Barn No. 1 Bore, it is 17 in; in Rainton Alexandrina Pit Bore, 10 in; and in Littletown Lady Alice Pit Bore, 9 in. The correlation is not certain in the Bishop Middleham and Windlestone Colliery bores, where two thin coals are recorded at about this horizon and where the overlying Gubeon Marine Band has not yet been recognized. In the Bishop Middleham Bore, the lower coal, $7\frac{1}{2}$ in thick, is separated from the upper $14\frac{1}{4}$-in coal and bands by about 15 ft of 'grey metal, with post partings'; and in the Windlestone Bore, the lower 11-in coal (with a 4-ft ganister seatearth) is separated from the upper 12-in coal by about 21 ft of 'metal' and 'post' with $1\frac{1}{2}$ ft of 'seggar' at the top.

The Gubeon Marine Band has yielded the following fauna from Offshore No. 1 Bore: foraminifera; *Planolites ophthalmoides*; *Lingula mytilloides*; undetermined Palaeoniscid scales, *Rhizodopsis sp.* In this bore *Lingula* was recorded at intervals through 7 ft of the shales above the coal, but in the East Hetton Busty Bore *Lingula* and sponge spicules were found only in the lowest 4 in of 3 ft 9 in of 'bluestone' (i.e. mudstone) forming the roof of the Gubeon Coal.   E.A.F.

The measures between the Gubeon and Ganister Clay coals (Pl. III) range from about 40 ft to over 70 ft in thickness. At Rainton Alexandrina Pit, $61\frac{1}{2}$ ft of strata separate the Gubeon Coal from 4 ft of 'grey shale' at the horizon of the Ganister Clay Coal, and of this, $56\frac{3}{4}$ ft consists of 'white and grey post' (Fig. 5). In Littletown Lady Alice Pit, $1\frac{3}{4}$ miles to the south, the interval of $46\frac{3}{4}$ ft includes only 24 ft of 'grey post with shale partings'. However, there is no direct relationship between the thickness of the interval and that of the sandstone, for in the East Hetton Busty Bore (Pl. III) where the interval is $55\frac{1}{2}$ ft, there are only $17\frac{1}{4}$ ft of 'tough white micaceous shaly post' at about $7\frac{1}{2}$ ft above the base and two thin bands near the top. Farther to the southeast, in Ten-o'-Clock Barn No. 1 Bore, the strata comprise 44 ft of mainly argillaceous beds, with a few thin sandy bands bearing plant debris. In some borings, this part of the sequence is so variable that correlation is uncertain. In the South Hetton Bore, for instance, the first seam recognizable above the Gubeon horizon is the Brockwell Coal, and two 60-ft sandstones, separated by about 10 ft of ' blue metal with ironstone ', lie between the horizons of the Gubeon Marine Band and the Victoria Coal (Pl. III).

The upper part of the sequence between the Gubeon and Ganister Clay coals in Offshore No. 1 Bore (Pl. III) is:

	ft	in
GANISTER CLAY COAL (at 2095 ft 11 in)		9
Seatearth and seatearth-mudstone, sandy in lower 2 ft ...	3	1
Mudstone, sandy, micaceous, with some rootlets ; plant fragments	9	2
Shale, grey, micaceous, with very thin sandy alternations and some thicker sandstone partings (up to 1 in) ; many worm tubes	3	4
Shale, grey, micaceous ; scattered plant fragments and *Curvirimula sp.*	3	6
Mudstone, grey, sandy, rather shaly with micaceous partings ; plant debris	9	5

The *Curvirimula sp.* from 16 ft below the Ganister Clay Coal is stratigraphically the lowest non-marine lamellibranch yet found in the Coal Measures sequence in this district (see p. 23).   E.A.F., D.B.S.

In south-west Durham, an important refractory seatearth underlies the Ganister Clay Coal, but in the Durham–West Hartlepool district this seatearth is not prominent, though thin fireclays and hard sandstones (sometimes called 'ganister-like ') are recorded. In Fishburn No. 2 Bore, for instance, 3 ft of ' strong ganister-like post' lies below the coal, but the topmost 1 ft probably consists of fireclay. In Ten-o'-Clock Barn No. 1 Bore, 1 ft 11 in of ' very hard grey sandstone ' lies about 5 ft beneath the coal. At other places ganister-like sandstone is present

above the coal, as in the East Hetton Busty Bore, where the section (by G. A. Burnett) is :

	ft	in
Bluestone with iron balls, sandy near base	5	8
Sandstone, tough, micaceous, ganister-like	1	6
Sandy bluestone, with ?shells, very crushed		2
Hard splint COAL (?GANISTER CLAY COAL split)		2
Dark sandy fireclay-shale ...		10
Sandstone, grey, micaceous ...	4	2
GANISTER CLAY COAL		7
Grey fireclay-shale	2	8
Sandy fireclay and iron balls...	2	2
Dark fireclay-shale	1	11

E.A.F.

The **Ganister Clay Coal** is of no economic importance in this district. It is more than 1 ft thick only in Windlestone Colliery Shaft (17 in) and Fishburn No. 2 Bore (15 in). In Murton Blue House Lane Bore it is 12 in thick, and in South Wingate Colliery and Offshore No. 1 bores 9 in thick. In Littletown Lady Alice Pit an upper 6-in coal is separated from a lower 4-in coal by 2 in of ' grey shale '. Elsewhere the coal is either an inch or two thick or it is absent.

The **measures between the Ganister Clay and Marshall Green coals** range from about 30 ft to 90 ft in thickness, averaging about 55 ft. In most of the twenty or so bores which have proved this part of the succession, the strata are predominantly arenaceous, with argillaceous beds at the base and top. An exception is the East Hetton Busty Bore where these measures largely consist of ' bluestone' with many ' iron ribs and balls '. This bore is also exceptional in providing the only record of non-marine lamellibranchs from this part of the succession. In Fishburn No. 2 Bore the roof measures of the Ganister Clay Coal contain *Rhadinichthys sp.* and pyrite-filled burrows ; the lithology is that of a typically marine sediment, but no marine fossils were found. In addition to the crushed shells above the Ganister

Clay Coal split (see above), *Carbonicola* cf. *communis* and fragmentary *Curvirimula sp.* were identified from 9 in of 'black bluestone with ironstone ribs' and the base of the overlying 'sandy bluestone' about 30 ft above the Ganister Clay Coal and 19 ft below the Marshall Green Coal. In Offshore No. 1 Bore, the roof of the Ganister Clay Coal consists of 9 in of 'grey sandy shale' with *Annularia radiata* Brongniart.

Because of insufficient evidence, the correlation of the **Marshall Green Coal** and associated measures is uncertain in some places. In the East Hetton Busty Bore the coal is 6 in thick lying about 43½ ft above the thin coal taken to be a top split of the Ganister Clay Coal. In the Windlestone Colliery Shaft two 8-in coals at the Marshall Green position are separated by 6¾ ft of strata, and though a split has not been proved, it seems reasonable to name these coals Top and Bottom Marshall Green. In Offshore No. 1 Bore, the supposed Bottom Marshall Green consists of an upper 4-in, a middle 1-in and a lower 8-in coal lying in 8 ft 2 in of generally arenaceous measures; the 14-in Top Marshall Green Coal lies 37½ ft higher. The thickest single seam recorded at the Marshall Green position in this district is one of 18½ inches in Littletown Lady Alice Pit, and even this contains a 3½-in 'shale' band 5 in above the base. A few scattered records show coal of about 12 inches in thickness.　　　　　E.A.F.

The measures between the Marshall Green and Victoria coals are again not known in detail, but it is clear that they vary considerably in lithology (Pl. III). Where the supposed Top Marshall Green Coal is found, the measures both below and above it commonly include grey, medium- to coarse-grained feldspathic sandstone bearing pebbles in many places. The immediate roof of the Top Marshall Green is, however, commonly composed of mudstone, ranging in thickness from a few inches to about 15 ft. The base of this bed is thought to be the horizon of the Stobswood Marine Band of the Northumberland Coalfield (see 6-in published sheet NZ 27 NE). Although *Planolites ophthalmoides, Curvirimula sp.* and

fish remains have been found in the Fishburn area, no marine fauna has yet been found in Durham. Non-marine fossils have been found at two higher levels in the East Hetton Busty Bore where the succession (recorded by G. A. Burnett) is:

	ft	in
VICTORIA COAL (at 176 ft 7 in)		4
Sandy shale-fireclay and fireclay-shale　...　...　...	6	4
Bluestone with *Spirorbis sp.*; *Curvirimula trapeziforma?* (Dewar), *Naiadites flexuosus*	1	8
Fireclay　...　...　...	1	9
Blue shaly sandstone, sandy bluestone and sandstone ...	5	1
Sandy bluestone with *Spirorbis sp.*, and *Carbonicola* cf. *pseudorobusta*　...　...	3	9
Black bluestone with *Carbonicola sp.*, *?Anthracosphaerium dawsoni*; fish scales throughout　...　...　...　...	2	10
Dark mudstone, no fossils seen	1	3
Bluestone　...　...　...		11
Sandy bluestone with indeterminate shells　...　...	1	8
Sandy bluestone, sandstone, sandy shale, shaly sandstone and bluish sandstone with 8 in of conglomeratic sandstone 7 ft 2 in above the base　...　...　...　...	35	4
MARSHALL GREEN COAL (at 237 ft 8 in) ...　...　...		6

In Ten-o'-Clock Barn No. 2 Bore, the positions of the two 'mussel'-bands are marked by thin coals, but no fossils were recorded from their argillaceous roofs. Coals up to a few inches thick occur at similar horizons in other scattered localities, but elsewhere this part of the sequence is commonly occupied by sandstone.　　　　E.A.F., D.B.S.

The Victoria Coal is everywhere too thin in this district to hold much prospect of it being worked. The maximum proved thickness is in Ten-o'-Clock Barn No. 1 Bore where it is 22 in, including a ½-in 'dirty rib' 12 in above the base. This record, however, is exceptional, for over most of the western and southern parts of the district the coal is only 3 to 12 in thick. Farther east it is less than

3 in, and in some places absent, its position being marked only by seatearth.

E.A.F.

The measures between the Victoria and Brockwell coals are between about 40 and 85 ft thick and include two 'mussel'-bands, one at the base and the other nearly mid-way up the sequence. The basal band, forming the roof of the Victoria Coal, is persistent throughout the district, though the abundance and preservation of the fauna is variable. In the East Hetton Busty Bore, for instance, preservation is poor and the fauna, together with macrospores, is restricted to the lower part of 5 ft 8 in of 'bluestone', the upper part of which is sandy. Around Fishburn, however, the band is nearly twice as thick, as in Fishburn No. 1 Bore where 9 ft 8 in of shales overlie the Victoria Coal. This band contains : *Planolites ophthalmoides* ; *Carbonicola* aff. *declivis, C. pseudorobusta, C.* cf. *rhindi* [juv.], *C. sp.* cf. *cristagalli, Curvirimula subovata, C. trapeziforma* ; *Carbonita humilis, Geisina arcuata.* Other borings nearby have yielded *Naiadites sp.*

E.A.F., D.B.S.

Similar forms have been obtained from Elwick No. 1, Dalton Nook and Low Throston bores. In Offshore No. 1 Bore *P. ophthalmoides* is restricted to a 9-in band 1 ft 9 in above the base of the bed of shales, where it is intercalated with layers containing 'mussels'. Fish remains including an Acanthodian spine, *Elonichthys sp., Rhadinichthys sp.* and *Rhizodopsis sp.* were found with a similar fauna of lamellibranchs and ostracods in Offshore No. 2 Bore.

The 'mussel'-band in the roof of the Victoria Coal is overlain successively by sandstone and seatearth. In the East Hetton Busty Bore the sandstone, about 21 ft thick, is hard and contains thin bands of 'sandy bluestone' close to the base and top. Around the Fishburn area it generally ranges between 10 and 25 ft in thickness, though exceptionally it is reduced to 2 or 3 ft. The same sandstone can be traced through the Elwick series of bores and is distinguishable in Offshore Nos. 1 and 2 bores, though in the latter it is only 2 ft thick. The overlying seatearth consists of 8 in of 'dark fireclay

with soft bands' in the East Hetton Busty Bore, and is persistent over much of the southern and offshore parts of the district, but is not recorded in the north-western part. It is thought to be the equivalent of a prominent fireclay or fireclay-mudstone which forms the roof of the Victoria Coal in parts of West Durham.

The seatearth is overlain by a bed of mudstone up to 16 ft thick and containing non-marine lamellibranchs in the lower part. This is the **Upper Victoria Shell Band** or '*declivis* Horizon' of Magraw and others (1963, pp. 197-203 ; pl. xix) though it is now clear (p. 24) that the latter term has been misapplied and should be discontinued. The fauna is distinctive in consisting mainly of small poorly preserved *Carbonicola* by means of which the band can be traced over most of the district. A representative faunal list obtained from Fishburn Nos. 1 and 2 bores is as follows: *Carbonicola* cf. *acuta* [small], *C.* cf. *communis, C.* aff. *declivis, Naiadites flexuosus* ; *Carbonita?, Geisina arcuata* ; *Elonichthys sp.* In addition to these *Curvirimula subovata, Carbonicola* cf. *polmontensis* (Brown) and *Pleuroplax sp.* [tooth] were obtained from the band in Offshore No. 1 Bore and *Spirorbis sp.* [large] from No. 2 Bore of the same series.

Between this 'mussel'-band and the Brockwell Coal are up to 85 ft of strata which include a median bed of seatearth. In places this seatearth is missing, as in Elwick No. 1 Bore where the whole of the interval is occupied by 43½ ft of cross-bedded sandstone overlain by 12 ft of interbedded sandstone, siltstone and mudstone. In other places the seatearth is overlain by a coal which is normally thin, though in Offshore No. 1 Bore it is as much as 47 in thick and has been named '**Bottom Brockwell**' (Magraw and others 1963, pl. xix). Although in the Nunstainton No. 1 and Stonyhall bores this seam lies so close beneath the County Brockwell Coal as to form with it a single banded section of up to 107 in, it is in most other places separate from the latter.

Fig. 6 illustrates this separate identity of the 'Bottom Brockwell Seam' in the southern and south-western parts of the district, and shows that the coal worked as

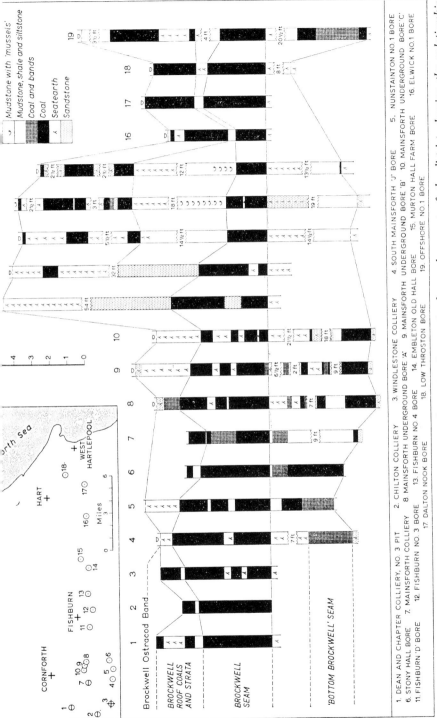

FIG. 6. *Comparative sections of the Brockwell Seam and associated strata in the southern part of the district, showing the relationship of the Brockwell Seam to the 'Bottom Brockwell' Seam and Brockwell Ostracod Band*

Brockwell Seam around Dean and Chapter Colliery and at Chilton Colliery lies at a higher stratigraphical level. Farther east the 'Bottom Brockwell', in six leaves, reaches a thickness of 31 inches in South Mainsforth 'J' Bore [3104 2933], 650 yd W.N.W. of Nunstainton East Farm, where it lies 9 ft 8 in below the Brockwell Coal. In three bores respectively 400 yd S.S.E., 200 yd S.S.W. and 200 yd W.N.W. of Hope House (Mainsforth Underground Bores A, B and C, in Fig 6), it is respectively 21 in, 17 in and 20 in thick lying at intervals below the Brockwell Coal of 13 ft, $22\frac{1}{2}$ ft and $44\frac{1}{2}$ ft. It is probable that the 31-in 'Bottom Brockwell' of the South Mainsforth 'J' Bore includes thin coals which develop in the strata between the County Brockwell and 'Bottom Brockwell' of the Mainsforth Underground bores. In the Bowburn, Shincliffe, Sherburn, Low Grange and Rainton areas, the 'Bottom Brockwell' is represented by a thin coal, which only exceeds 7 inches in a bore [3052 3786] 130 yd S.E. of Bowburn D.C. Pit, where it reaches 16 in.

The Brockwell Seam (Figs. 6 and 7) is the lowest worked in the district, but the workings are restricted to two areas in the south-west and south. The larger of these is around Chilton Colliery and Ferryhill, extending northwards to Croxdale Hall and eastwards to Mainsforth Hall ; the smaller occupies an isolated area south of Fishburn, between Holdforth, Weterton House and Bridge House. In these areas, the coal is generally $3\frac{1}{2}$ to $4\frac{1}{2}$ ft thick, locally increasing to over 5 ft, but reserves of this thickness have now been exhausted and mining is now being extended only at Fishburn Colliery, east-south-eastwards from Bridge House, where the coal is between 2 and 3 ft thick. E.A.F.

The seam may be similarly 2 to 3 ft thick over an area of about 35 square miles east of Easington, south of Cold Hesledon and north of Hart Station, extending seawards at least as far as Offshore Nos. 1 and 2 bores, but in this large area, it has so far been proved only in 7 bores.　　　　D.B.S.

The Brockwell Seam has been worked from Chilton, Dean and Chapter, and East Howle collieries in the south-western area as a good quality coal 40 to 65 in thick,

in which bands are absent or very thin. To the east, the coal becomes too banded for profitable mining, and the line along which workings have terminated for this reason has been traced between Croxdale Hall, Broom Hill and Cookson's Green (Fig. 7). Immediately to the east of this line, the thickness of the banded seam is maintained in a narrow zone, but still farther to the east, the thin leaves of coal die out and in some places only one survives. A typical section through the banded seam at the limit of workings [2801 3768], $2\frac{1}{2}$ miles N.N.W. of East Howle Colliery, is coal 15 in, on band 3 in, coal 5 in, band 7 in, coal 4 in, band 7 in, coal 9 in.

Farther south, the banded zone broadens between Windlestone, Mainsforth and Bishop Middleham and the direction of deterioration is from south to north. Within the zone there are small areas in which the banded coal is more than 24 in thick and one of these has been worked south of Mainsforth Colliery shafts. Near Hope House, three Mainsforth Colliery underground bores (Fig. 6) illustrate the progressive deterioration first by the incoming of a band, and finally by the reduction in thickness of the individual leaves of coal.

The Brockwell Coal attains its maximum unbanded thickness of 63 inches in the southern part of the Fishburn area at a locality [3616 3128], 550 yd E. of Lizards Farm. Towards the southern limit of working, between 450 yd N.W. and 250 yd N.N.W. of Weterton House, its thickness is reduced first to between 38 and 46 in and then to 18 in. The termination of the workings coincides with a line of impoverishment which has been proved eastwards by borings for over 2 miles (Fig. 7), and which has been interpreted as the margin of a washout, though alternatively, since its trend is subparallel to the outcrop on the sub-Permian surface, the absence of the seam farther south may be a result of post-Carboniferous pre-Upper Permian oxidation (see also p. 18). In Fishburn 'D' Bore (Fig. 6), the coal is 66 in thick including an 11-in band 19 in above the base of the seam. There is no record of a band in the log of Fishburn Colliery North Shaft, where the seam is 55 in thick, nor within a limited

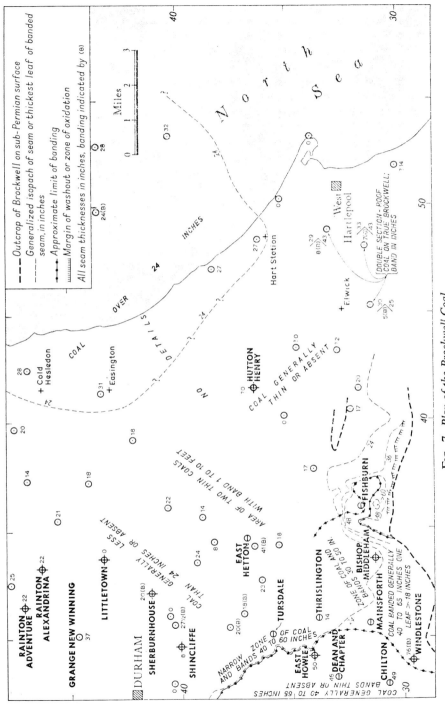

Fig. 7. *Plan of the Brockwell Coal*

area between the colliery shafts and Holdforth, but a thin band is generally included in the seam west and north-west of the shafts. A typical section at a locality [3508 3201], 1150 yd W.N.W. of the North Shaft, is coal 33 in on band 3 in, coal 13 in. At another locality [3387 3352], 300 yd E. of Mahon House, the seam comprises coal 6 in, on band $\frac{1}{4}$ in, coal $4\frac{1}{2}$ in, band $\frac{1}{2}$ in, coal $5\frac{1}{4}$ in, band 2 in, coal $\frac{1}{4}$ in, band $\frac{1}{2}$ in, coal 9 in, splint coal $7\frac{1}{4}$ in. Still farther north, beyond the banded zone, few of these thin coals persist; thus in a bore [3427 3561], 1 mile S.S.W. of East Hetton Colliery, the section reads coal 18 in, on dark grey seatearth-mudstone with pyritic partings 14 in, coal 7 in. A thickness of 41 in of coal with bands, proved in the East Hetton Busty Bore, 1200 yd farther north, is exceptional for this part of the district.

North-east of Fishburn, impoverishment is again evident. The coal is 34 in thick in a bore [3672 3258], 550 yd N.N.E. of Fishburn crossroads, and only 24 in 250 yd farther to N.N.E. Workings were terminated 50 yd still farther to the north where the seam is 20 in thick. In Fishburn No. 1 Bore, 1300 yd N.N.E. of Fishburn crossroads, the coal is 25 in thick, but in Fishburn No. 2 Bore and Fishburn 'B' Bore, farther to the north-east, it is only 16 in and 17 in thick respectively. Seam thicknesses recorded from other bores in this area are as follows: bore [3803 3234] 700 yd W. of Humble Knowle, 24 in; Fishburn No. 4, 27 in; Whin Houses, 23 in; Embleton Old Hall, 28 in (with a 2-in band 15 in above the base); Murton Hall, 20 in.

Over much of the central and western parts of the district, the Brockwell Seam ranges in thickness from 0 to 24 in, though exceptionally in Grange New Winning Bore, north of Carrville, it is said to be 37 in.   E.A.F.

Between Elwick and West Hartlepool (Fig. 7), the coal forms the lower leaf, 20 to 43 in thick, of a composite seam separated by 5 to 8 in from an upper leaf 29 to 40 in thick. This upper leaf is represented farther west by one or more thin coals (Brockwell roof coals of Fig. 6) lying between the Brockwell Seam

and the overlying Brockwell Ostracod Band. In Offshore No. 1 Bore the roof coals consist of coal 32 in, on band 22 in, coal 3 in, and are separated from the County Brockwell by 57 in of strata. The 32-in roof coal is correlated with the Three-Quarter Seam by Magraw and others (1963, pl. xix).

At Dean and Chapter, Chilton and Windlestone collieries there are one or two thin roof coals separated from the worked coal by a thin band. North-eastwards from Dean and Chapter, this band becomes thicker, as in the East Hetton Busty Bore, where the sequence is:

	ft	in
BROCKWELL OSTRACOD BAND:		
bluestone with ironstone ribs and nodules, ostracods and non-marine lamellibranchs in the lower 2 ft	4	8
Sandy shale-fireclay, micaceous and more sandy to base ...	3	4
Coal, rather inferior, shaly	1	2
Sandy fireclay	3	0
Sandy bluestone	5	9
BROCKWELL SEAM (at 94 ft 6 in)		
Coal with bluestone bands	7	
Coal, soft, with thin dirt bands	2	10

Elsewhere the intervals containing the thin coal or coals between the worked seam and the ostracod band may be as little as a few inches or as much as 65 ft. The latter value was recorded in Fishburn 'D' Bore, where the intervening strata comprise 40 ft of coarse gritty micaceous sandstone, succeeded by 18 ft of dark grey micaceous sandy shale with plant debris and 7 ft of grey sandy fireclay with ironstone nodules. In other bores south of Fishburn shafts and near Weterton House, the interval is 40 to 50 ft, largely filled by sandstone, with seatearth and with a coal 1 to 4 in thick at the top. In Fishburn No. 1 Bore near Trimdon, the interval is reduced to $26\frac{1}{4}$ ft of which only 2 ft consists of 'post' and this has shaly partings, but in Fishburn 'B' Bore, the equivalent strata are $61\frac{1}{4}$ ft thick including about $34\frac{1}{2}$ ft of sandstone overlain by about 18 ft of sandy shale and sand-stone with shaly bands. To the south-east of Fishburn shafts, in Fishburn No. 3 Bore (Fig. 6), the thin coal underlying the Ostracod Band is represented by 4 in of black cannelly shale, but an addi-tional 9-in coal is present a little lower in the sequence. The appearance of this additional coal is the first indication of the incoming of the Brockwell roof coals in south-east Durham. In Fishburn No. 4 Bore, farther east, another thin coal is found at a still lower level so that three Brockwell roof coals are present. Still farther to the east, in the Embleton Old Hall Bore, the two lowest are split to form a total of five thin coals in this part of the sequence. The roof of the Brockwell Seam in both these bores con-sists of mudstone with *Curvirimula sp.* In the Murton Hall Farm Bore there is no split and the strata between and below the lower two roof coals are thin-ner than in Fishburn No. 4 Bore. In Elwick No. 1 Bore the band separating the roof coals from the Brockwell Seam is further reduced to only 5 in and the lower two of the roof coals are com-bined into one 30-in seam. Farther east, in the Dalton Nook and Low Throston bores, all the roof coals are united and lie close above the true Brockwell, form-ing with it a single composite seam, 76 to 83 in thick with one 7- or 8-in band. In the Hartlepool Lighthouse Bore the whole of the composite seam is missing, and its position is marked only by a seatearth. The Brockwell is absent also in the North Sands Bore, but the roof coal is 24 in thick. Although these two bores penetrate quite a different type of sequence from that proved in the bores between Embleton and West Hartlepool, a sequence more comparable to the latter is found in Offshore No. 1 Bore (see above).

The Brockwell Ostracod Band, which is a useful and widespread marker band, consists of a few feet of shaly mudstone from which the following composite fauna has been obtained : *Gyrochorte carbonaria* Schleicher, *Spirorbis sp.* [large], *Planolites ophthalmoides* ; *Anthraconaia* aff. *fugax, Carbonicola* cf. *browni,* cf. *C. circinata, C. declivis, C.* cf. *martini* Trueman and Weir, *C. pseudorobusta?,*

C. cf. *robusta, Curvirimula candela, C. subovata, C. trapeziforma* ; *Carbonita humilis, C. pungens, Geisina arcuata* ; and fish remains.

The measures between the Brockwell Ostracod Band and the Three-Quarter Coal vary in thickness and lithology. Where the Ostracod Band forms the roof of the Brockwell Seam, the measures may be as much as 50 ft, but elsewhere the interval is reduced and is locally as little as 10 ft. In the East Hetton Busty Bore, the sequence begins at the base with ' bluestone ' and sandstone 6½ ft, overlain successively by seatearth 2 ft, coal 10 in, plant-bearing bluestone and sand-stone 10½ ft, and hard coaly fire-clay-shale 15 in. In South Mainsforth ' J ' Bore, the succession is as follows :

	ft	in
THREE-QUARTER COAL (at 512 ft 9 in)		10
Carbonaceous shale with coaly partings		3
Seatearth - siltstone, b l a c k, argillaceous, micaceous ...		7
Sandstone, whitish grey, fine-grained, thin and wispy, bedded with dark micaceous partings	5	8
Siltstone, grey, fissile, very argillaceous ; scattered plant fragments ; passage at base	3	9
Mudstone, grey, shaly, with non-marine lamellibranchs ; black, shaly, silty and finely micaceous at base, with fish scales	3	0
Coal, with 1-in carbonaceous shale parting		4
Seatearth-mudstone, grey, with slickensides	1	8
Siltstone, grey, argillaceous, passing down into white, fine-grained wispy-bedded sandstone in repeated 6-in to 12-in rhythms ; scattered rootlets and other plant de-bris throughout	4	6
Mudstone, grey, shaly, silty, with some thin sandy bands at top ; well-preserved plant leaves and stems, abundant towards base ; bottom 3 in packed with coalified material	4	6

	ft	in
Seatearth-mudstone, dark grey, slickensided		4
Siltstone, thick-bedded, scattered rootlets and grey sandy bands at top, increasingly argillaceous below with other plant debris	7	8
BROCKWELL OSTRACOD BAND: Mudstone, grey, shaly, with non-marine lamellibranchs and ostracods; black and very shaly with fish debris in lowest 12 in; carbonaceous and pyritic in bottom 1 in	3	2

The 3-in coaly base of the mudstone 8 ft above the Brockwell Ostracod Band in this bore is equivalent to a 4-in coal lying 20 to 25 ft above the Brockwell Coal at Chilton and Dean and Chapter collieries. The 4-in coal higher in the bore has also been proved at those collieries, and at Dean Pit it is 11 in thick. Farther north-east, at East Howle Colliery, the two coals beween the Brockwell and Three-Quarter coals are closer together (Pl. III), the lower 7-in coal being separated from the upper 4-in coal by only 4 ft 7 in of 'white post' and dark fireclay.

In Tursdale South Shaft, these thin coals are thought to be represented by two bands of 'black stone', the lower 22-in band being separated from the upper 18-in band by 2½ ft of 'grey metal'. There is no evidence that these intermediate coals are splits from the higher Three-Quarter Seam and the records suggest that they come in when the interval between it and the Brockwell Coal exceeds about 40 ft.   E.A.F.

The 4-in coal of the South Mainsforth 'J' Bore can be traced through Mainsforth and Bishop Middleham collieries into the Fishburn area and it attains 20 inches in the Green Lane Cottages Bore [4095 2993]. The fauna in its roof includes : Spirorbis sp. ; Carbonicola pseudo-robusta, Curvirimula candela, C. sub-ovata, Naiadites cf. flexuosus ; and fish remains. Non-marine lamellibranchs have been recorded from the same horizon in

Fishburn Nos. 3 and 4 bores, and farther east, in Embleton Old Hall and Murton Hall bores. Fragmentary 'mussels' are also recorded above a 6-in coal at this horizon in a bore [3442 3853] 650 yd south-west of Dene House Farm, near Cassop Colliery. In Elwick No. 1 Bore, the position of the same coal is marked by roof shales containing macrospores, large fragmentary Curvirimula sp. and indeterminate Palaeoniscid scales. No fossils were found in this part of the sequence in the Dalton Nook Bore, where the interval between a 23-in coal containing a median 9-in band and the Three-Quarter Coal is largely filled by 42½ ft of sandstone. In the Low Throston Bore, this interval amounts to only 7 ft 7 in, and the lower coal is only one inch thick. In the North Sands Bore, it is represented by coal 9½ in, band 2 in, on coal inferior 2 in, lying about 50 ft below the assumed position of the missing Three-Quarter Coal.

                                        E.A.F., D.B.S.

The Three-Quarter Seam (Pl. III) is 17 to 19 in thick at Dean and Chapter Colliery, but only 4 in thick at Chilton Colliery. At East Howle Colliery, it is a 16-in seam overlain by 27 in of 'black stone' upon which rests 9 in of 'fireclay' forming the floor of an unnamed 6-in coal. At Croxdale Thornton Pit in the Wolsingham (26) district, the Three-Quarter Seam comprises 18½ in of coal separated by a 1½-in band from an overlying 4-in seam. A staple-pit near Tursdale South Shaft proved a similar section in which an upper 4-in coal is separated from a lower 12-in coal by a 6-in 'slaty band'. In the nearby shaft, the Three-Quarter Seam is represented by 26 in of 'coal and bands', while in Tudhoe West Pit, just beyond the western margin of the district, it is a banded coal 36 in thick.

Traced into the north-western part of the district, from the adjacent Wolsingham district, the seam undergoes an eastward deterioration. In two bores near Bishop's Grange it is 39 in and 26 in with bands respectively, while other records to the south-south-east are Kepier Florence Pit, 21 in ; Elvet South Engine Pit, 16½ in ; Old Durham Colliery, 24 in ; and

Shincliffe Colliery, 5 in. In the Sherburn and Rainton areas, the Three-Quarter Coal does not exceed 15 in. E.A.F.

In the eastern part of the district, the Three-Quarter Coal is commonly 18 in or less in thickness and locally it is absent. The coal is 14 and 19 in thick respectively in the Dalton Nook and Hartlepool Lighthouse bores, and in the Low Throston Bore it comprises coal 3 in, on band 7 in, coal 10 in. Though only 8 in thick in Offshore No. 1 Bore, the seam amounts to 34 inches in No. 2 Bore. D.B.S.

At Bishop Middleham Colliery and Fishburn No. 4 Bore, the coal is only 1½ in and 3 in thick respectively, and in the Embleton Old Hall, Murton Hall Farm, Elwick No. 1 and North Sands bores, it is absent, its horizon being marked by silty beds or seatearth. It is also absent in a bore [3442 3853] 650 yd S.W. of Dene House Farm, and in the East Hetton Busty Bore it comprises clean coal 8 in, on coaly fireclay 1 in, dirty coal with pyrite rib 3 in. E.A.F.

The roof of the Three-Quarter Coal locally contains ' Estheria ', and is thought to be the equivalent of the Low ' Estheria ' Band of the East Pennine Coalfield (p. 25). It has been proved at several places in the district and is here named the **Fishburn ' Estheria ' Band** (Fig. 3). In Fishburn No. 4 Bore, the band, only 1 ft 10 in thick and overlain by sandstone, contains only ' Estheria ' sp. In Fishburn ' D ' and No. 3 bores, however, where the thin Three-Quarter Coal is overlain respectively by 16 ft 11 in and 18 ft 8 in of shale and shaly mudstone with ironstone ribs and nodules, ' Estheria ' sp. is found near the base and non-marine lamellibranchs above. The presence of ' Estheria ' and a non-marine lamellibranch fauna in the Offshore No. 1 and North Sands bores assists in identifying the position of the Three-Quarter Coal, close below the Busty coals, and the correlation is consequently different from that in Magraw and others (1963, pl. xix). E.A.F.

In Fishburn No. 3 Bore the following were identified: Carbonicola pseudorobusta, Curvirimula subovata ; Carbonita cf. humilis, Geisina arcuata ; and fish remains. Cochlichnus kochi was obtained

3 ft above ' Estheria ' in Hartlepool Lighthouse Bore. The top of this ' mussel '-band in Fishburn ' D ' and No. 3 bores lies only 6 ft 8 in and 22 ft 3 in respectively below the Bottom Busty Seam. In South Mainsforth ' J ' and Fishburn No. 4 bores, by contrast, this interval includes nearly 60 ft of sandstone. This bed reaches a maximum of 94 ft at Croxdale Thornton Pit, just beyond the western margin of the district, but it is about 70 ft thick at East Howle Colliery and ranges from 40 to 70 ft over much of the south-western part of the district (Pl. III). E.A.F.

In an underground bore [3658 3617], 550 yd N.N.W. of Trimdon Grange East Shaft, the sandstone is 70 ft thick extending down, with no trace of the Three-Quarter Coal, to the roof of the thin coal about the Brockwell Ostracod Band.

The Top and Bottom Busty coals (Fig. 8) are normally separate and distinct, and in most places the Bottom Busty has been the more extensively worked. In three areas, however, the two coals form a single, usually banded, seam. Two of the areas are extensions from adjacent districts. Between Hetton le Hole and Murton Moor East Farm, this single seam is currently being worked southwards from Eppleton Colliery in the Sunderland (21) district, where a typical section measured in the workings [3624 4716] is coal 19 in, on band 2 in, coal 3 in, band 6 in, coal 24 in. The composite seam extends, in similarly banded form, around New Hesledon and has been proved in an underground bore [4117 4669] 300 yd W. of Cold Hesledon. D.B.S.

The Top and Bottom Busty coals are again close together in the extreme northwest of the district between Crook Hall and Ford Cottage, and the combined seam has been worked from the west. A bore [2750 4742], 2½ miles N. of Durham Main Colliery, proved top coal 24 in, on band 12 in, bottom coal 27 in. To the east of this bore the coals become thinner and the band increases to over 3 ft. At the limit of workings 750 yd E. of the bore, the section is Top Busty 28 in (coal 21 in, on band 4 in, coal 3 in), band 3 ft 10 in, Bottom Busty 17 in. Farther south, at a locality [2757 4633], 250 yd N.

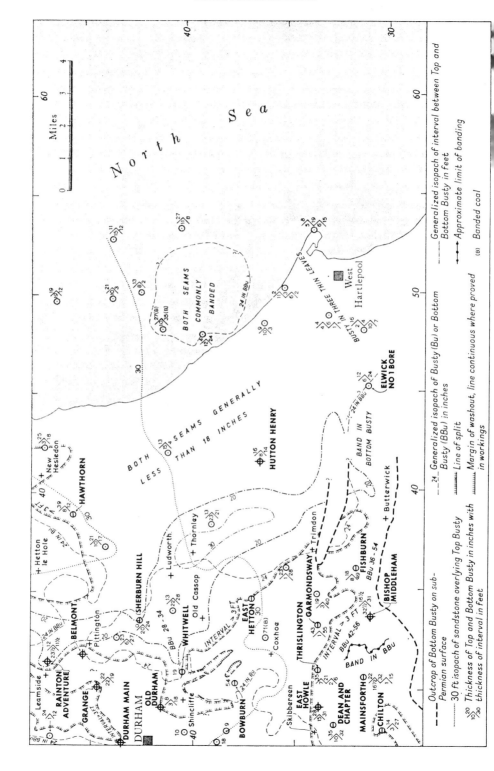

of Red House, the combined seam comprises coal 19 in, on band 14 in, coal 24 in ; but 425 yd E.N.E. of Red House, it is coal 21 in, on band 2 ft 10 in, coal 19 in ; and this is close to the limit of workings. Between two localities [2763 4460 and 2778 4420], ¾ mile and ½ mile N. of Durham Main Colliery, the band increases from 2 to 15 ft, and as both top and bottom coals are there only 19 to 21 in, the workings have not been continued to the east. Immediately north of the colliery, the line of split follows a course generally sub-parallel to the 24-in isopach of the Bottom Busty Seam (Fig. 8). To the east of the colliery, the course of the line of split is more conjectural, but some evidence is provided by the Kepier Florence Pit section in which the whole Busty comprises top coal ¾ in, band 22 in, bottom coal 20 in ; and by the Grange Pit section, in which it is top coal 22 in, strata about 22 ft, bottom coal 19 in.

The main area of combined Busty coals lies in a triangle between Shincliffe Colliery, Trimdon and East Howle. The composite seam has been worked from Bowburn Colliery, where the Downcast (West or New) Pit proved 54 in of unbanded coal. The same thickness was proved in a bore [2986 3639], 640 yd N.E. of Broom Hill. To the north of Bowburn, bands come in near the middle of the seam, and a bore [3003 3921], 190 yd S.E. of Shincliffe Station, proved coal 24 in, on plate band 1¾ in, coal 4 in, seggar-clay band 3 in, coal 25½ in. To the south of Bowburn also, the composite seam becomes banded as shown by the following section in a bore [2911 3659], 550 yd S.S.W. of Tursdale House : good coal 5½ in, on black shale with coal scars 1¼ in, good coal 3½ in, grey metal shale 2¼ in, coal with danty partings 4 in, good coal 6 in, brown band ½ in, good coal 2½ in, rather danty coal 2 in, good coal 22½ in. This is close to the western limit of workings beyond which the single seam is impoverished.

Mine plans show the Busty Seam to be washed out in two areas to the north-east of Bowburn (Fig. 8). One area, about 350 yd N.N.E. of Bowburn Downcast Pit, is about 1300 yd long by 300 yd wide trending in a north-westerly direction. A larger area of 'nip-out' measuring 1800 yd by 400 yd and elongated in the same direction, lies about ¼ mile N.E. of the same pit. At the north-western limit of this area [3067 3938], 800 yd E. of Shincliffe Station, the seam is recorded as diminishing in thickness from 24 in to 4 in over a distance of 60 yd from west to east. Farther north, towards Shincliffe Station and Shincliffe Colliery the thicknesses of the worked Busty seams range as follows: top coal 13 to 24 in, band 12 to 26 in, bottom coal 21 to 23 in. On the western side of this area, the top coal decreases from 24 to 18 in near High Grange, and on the eastern side the bottom coal is reduced to 18 in near Whitwell Beck. It would therefore seem that the Busty split develops at these extremities.

In the Busty workings [2950 3824] 1050 yd W.N.W. of Bowburn Downcast Pit, the 23-in top coal is separated from the 16-in bottom coal by a 4-ft band. At South Grange, nearby, the section is top coal 5 in, band 14 in, bottom coal 31 in ; and the workings are limited to the west by a 'nip-out'. Between 100 and 350 yd N.E. of Shincliffe Colliery the seam comprises top coal 22 to 24 in, band 3 to 8½ ft, bottom coal 19 to 22 in. The increase in thickness of this band and reduction of both coals to less than 2 ft has precluded mining farther north.

To the west and north-west of Bowburn lies an area in which the Busty is impoverished and unworked. Throughout this area, extending from Skibbereen north-north-eastwards through High Croxdale, Shincliffe, Old Durham, Gilesgate Moor and Carrville to Leamside, the Busty Seam consists generally of a single coal less than 18 in thick. Thus, two bores [2806 3832 and 2754 3875], 2 miles and 1½ miles W.N.W. of Bowburn Colliery, proved 9 and 18 in of coal respectively. At Old Durham, however, a 3-ft band separates an upper 8 in from a lower 18-in coal, suggesting that the Busty split inferred at the limits of the Shincliffe workings extends this far north.

The Busty Seam has also been worked north-east of Bowburn in a narrow zone bounded on the west by the Heugh Hall 'nip-out'. The northern boundary is marked by cindered coal. Thus in the workings [3232 3914], 1¼ miles N.E. of Bowburn Upcast Pit, the coal is 33 in thick, but 100 yd farther to the north-north-east, 15 in of cindered coal lies on 9 in of uncindered coal and the seam is overlain by 18 in of 'whin' (dolerite). At the northward end of a 250-yd exploration [3197 3938] within the limit of cindering, the section is cindered top coal 19 in, band 37 in, 'whin' (dolerite) 22 in, cindered bottom coal 24 in. This cindered zone lies on the south side only of the Ludworth Dyke (p. 189) and is 700 to 1000 yd wide in this area lying to the east of the Heugh Hall 'nip-out'. The last section also demonstrates a split in the seam, which widens north-eastwards so that in a bore [3354 3929], 2 miles E.N.E. of Bowburn Colliery, the 39-in Bottom Busty is separated from the Top Busty coals by about 22 ft of strata.

South-east of Bowburn in the area around Coxhoe Hall, bands are present in most sections of the Busty Seam, which ranges in thickness from 44 to 71 in. In the workings 750 yd N.N.E. of the Hall, a top coal, 27 in thick, is separated from a bottom 31-in coal by an 8-in band, but 500 yd N.E. of Low Raisby, the top coal is reduced to 22 in and the bottom coal to 27 in, while the separation is as much as 4½ ft. This and other records between Quarrington Hill and Trimdon indicate the approximate course of the 3-ft isopach on the Busty split and the continuation of the south-easterly trend from the Cassop Moor area (Fig. 8).

The southern boundary of the area of composite Busty Coal lies between Trimdon and Skibbereen. In a bore [2857 3506] ¾ mile N.N.W. of East Howle, it consists of coal 23 in, on 'brown metal mixed with coal' 3 in, coal 9 in, 'white metal' 2 in, 'white post' 16 in, coal 20 in. Another bore [279 341], 475 yd N.E. of Skibbereen, proved Top Busty Coal 30 in, 'grey metal' 5 ft 10 in, Bottom Busty Coal 24 in. At East Howle, the 28-in Top Busty and 31-in Bottom Busty coals are 15 ft apart. Farther east, at Thrislington Florence Pit, the Top Busty is 35 in thick, and the Bottom Busty Seam, 6 ft 9 in

below, consists of coal 21 in, band 28 in, on coal 6 in.

The Bottom Busty Seam has been worked extensively in the south-western part of the district, where it is now largely exhausted. In the Dean and Chapter Colliery shafts, the Bottom Busty is 32 in thick and is separated from the 35-in Top Busty by 5 to 10 ft of strata. At Chilton, the interval is reduced to 3 ft, between a 27-in bottom coal and a 14-in top coal. To the east, a band noted in the Bottom Busty at Thrislington persists to Mainsforth, the shaft section being coal 32 in, on band 11 in, coal 15 in. The Top Busty, 22 in thick, lies 16½ ft higher. Farther east between Thrislington, Mainsforth, Garmondsway and Bishop Middleham, the Bottom Busty is generally 42 to 56 in thick, though it is locally less, as at Bishop Middleham itself where it amounts to only 31 in. At Bishop Middleham the 16½-in Top Busty lies 32½ ft above the Bottom Busty, but to the north this interval decreases and at Garmondsway Pit the whole Busty section is coal 43 in, on band about 12 in, coal 55 in. At an intermediate locality [3374 3314], 400 yd S.S.E. of Mahon House, it is coal 37 in, on band 24 in, coal 53 in. This section evidently lies close to the line of split as do the following: [3445 3286] coal 20 in, on band 4 in, coal 29 in ; [3465 3309] coal 29 in, on band 3½ ft, coal 41 in ; [3495 3330] coal 30 in, on band 6 in, coal 30 in ; [3479 3360] coal 8 in, on band 1 in, coal 7 in, band 6 in, coal 34 in.

The Bottom Busty Seam has been extensively worked at Fishburn Colliery and is still being developed in the Butterwick Moor area. Near the south 'crop' between Weterton House and Butterwick, the interval between the Top and Bottom Busty is 20 to 30 ft. The Bottom Busty around Fishburn shafts is generally 3 to 4½ ft thick, but locally it increases to over 5 ft, as in a bore [3646 3079], 400 yd N.E. of Weterton House, where the whole Busty sequence is Top Busty (coal ½ in, on seatearth-mudstone ½ in, on coal 27 in), black carbonaceous shale 3 in ; sandstone and siltstone 22 ft 9 in ; Bottom Busty (coal 67 in, on seatearth-mudstone 4 in,

Inferior shaly coal 3 in). To the east-north-east between Cowburn and Butterwick Moor, the thickness of the Top Busty is maintained at 20 to 30 in, but the Bottom Busty decreases to about the same range. Concomitant with this deterioration, the Busty split is reduced from over 20 ft to a few inches. Thus, in Fishburn No. 3 Bore, the section is Top Busty (black parroty shale 2 in, coal 18 in, on pyritic splint 1 in), 'post' with sandy shale above and below, and with sandy fireclay-shale at top 22½ ft, Bottom Busty (coal 33 in, on splint and coal 5½ in, coal 2½ in, splint 1 in). About 350 yd S.E. of Bridge House the Bottom Busty is 36 in thick, and in Fishburn No. 4 Bore, 650 yd farther N.N.E., the section is Top Busty Coal 28 in, 'post' and sandy fireclay-shale 7 ft, Bottom Busty Coal 24 in. In workings 200 yd N.E. of this bore there is a composite Busty section of coal 29 in, on band 3 ft, coal 26 in, and at a further locality [3887 3180], 450 yd N.W. of Butterwick Moor, it is coal 26 in, on band 2 ft, coal 27 in. At 100 yd farther to the north-west the band is only 6 in thick and separates top coal 28 in from bottom coal 26 in, and a comparable section [3887 3211], 300 yd S.E. of Humble Knowle, is coal 30 in, on band 6 in, coal 24 in.

To the north-east of Fishburn, a similar decrease in the thickness of both the Bottom Busty Seam and the strata separating it from the Top Busty Seam can be detected. In a bore [3672 3258], 550 yd N.N.E. of Fishburn crossroads, the section reads: Top Busty Coal 18 in, on 'grey metal stone' 28½ ft, 'dark blue metal' 1½ ft, Bottom Busty Coal 42 in ; and at a locality [3498 3258], 250 yd E. of this bore, Top Busty (coal 28 in, on band 1 in, coal 7 in) on 'black shale' 17 in, 'blue metal' 5 ft 10 in, Bottom Busty Coal 38 in. At two other localities [3681 3274 ; 3670 3277], 225 yd N.N.E. and 250 yd N. of the same bore, the separation is 6½ ft and 4 ft respectively, and the Top Busty is 33 in thick. The evidence in the Humble Knowle and Hope House areas suggests that the area of composite Busty Coal north-west of Trimdon might extend into the north-eastern side of the Fishburn Colliery take as far south-east as Cowburn (Fig. 8). E.A.F.

Between Butterwick and Elwick, beyond the present Fishburn workings, bores indicate a split of between 5 and 25 ft in the Busty Seam. Around Whin Houses, West Murton Blue House, Murton Hall and Embleton Old Hall, the Top Busty Coal is 20 to 28 in thick, and the Bottom Busty Seam is generally over 3 ft thick, but tends to include one or more bands. Thus, in the Embleton Old Hall Bore, the bottom seam comprises coal 17 in, on seatearth-mudstone 5 in, coal 20 in ; and in the Murton Hall Farm Bore: coal 15 in, on seatearth-mudstone 23 in, coal 5 in, seatearth-sandstone 10 in, coal 11 in. These records indicate a northwards impoverishment, for the Whin Houses Bore proved 53 in coal and the West Murton Blue House Bore proved 37 in coal. Farther east, in Elwick No. 1 Bore, the bottom seam is only 24 in thick, and still farther east, in the West Hartlepool area, neither leaf is thicker than 19 in. E.A.F., D.B.S.

The Bottom Busty Coal has been widely worked in the Sherburn Hill area where the Top Busty, generally less than 20 in thick, is locally absent. The interval between the two seams is generally 15 to 30 ft, but this decreases rapidly to the west between Belmont and Whitwell. Thus, a bore [3104 4060], 2 miles S.W. of Sherburn Hill Colliery, recorded black stone (Top Busty position) 3 in, 'white post' and 'grey metal stone' 17 ft 10 in, Bottom Busty Coal 19¼ in ; and a second bore [3058 4038], 560 yd farther to the south-west, proved the combined seam to be coal 10 in, on 'black stone with coal threads' 1 in, coal 20 in. Over most of the area between Sherburn Hospital, Pittington, Ludworth and Old Cassop, the Bottom Busty is 2 to 3 ft thick. Near Pittington, bands of unspecified thickness are recorded in a 34-in seam. Along a 950-yd line trending north-west from Littletown Colliery, the workings terminate against the western margin of a 'nip-out', but the width of this is not known. About 450 yd E. of Sherburn Hill Downcast shafts, the Bottom Busty workings stopped where the thickness of the seam decreased to 15 in, and the

24-in isopach can be traced (Fig. 8), thence roughly eastwards to a locality [3524 4198], 650 yd S.S.W. of Haswell Moor and on to Thornley, Trimdon Grange and Trimdon.   E.A.F.

With the exception of the areas in the north, the Busty seams in the whole district east of a line joining Pittington, Ludworth, Thornley and Trimdon are impoverished. Even where the thickness of either seam increases to over 24 in, as in the approximately circular offshore area, 3 miles in diameter, beyond Black Halls Rocks, the coal is generally banded. Over the whole of this south-eastern part of the district both of the Busty seams are commonly less than 18 inches in thickness. Near Thornley, the interval between them increases to over 30 ft; 32 ft is recorded in a bore [3898 3767], 880 yd N.N.W. of Wingate Grange. Over most of this part of the district the seams are separated by sandstone, though 'metal stone' is not uncommonly mentioned in the old records. In the offshore area the interval is 6 to 20 ft, and the coals are both thin and banded.   D.B.S.

In the south-western part of the district **the measures between the Busty or Top Busty Seam and the Bottom Tilley Seam** consist of 20 to 35 ft of strata variously recorded as 'post', 'metal stone' or 'metal'. The interval is increased to 50 ft at Garmondsway Pit but is only 14 to 35 ft farther east in the Fishburn area (Pl. III). In Fishburn No. 3 Bore the Top Busty Coal is overlain by 9 ft of shale with *Anthracosphaerium* cf. *dawsoni,* preserved as internal moulds in ironstone, small *Carbonicola sp.* [juv.] and *Naiadites flexuosus.* Shells have been recorded at the same position in Ten-o'-Clock Barn No. 2 Bore and from borings in the Embleton area, while *Carbonicola rhomboidalis?* was obtained from this position in a bore [3807 3040] in the Butterwick area and from Fishburn 'B' Bore.

Just beyond the western margin of the district, the Bottom Tilley Coal lies 29½ ft above the Busty in Aykley Heads Pit. The interval amounts to 31 ft in a bore [3003 3921], 190 yd S.E. of Shincliffe Station, but is only 13½ ft in Littletown

Lady Alice Pit. Just beyond the northern margin of the district, between Great Lumley and Chilton Moor, the interval is 20 to 35 ft, but over much of the north-central part of the district, between Elemore, Coxhoe and Castle Eden, the Bottom Tilley Seam is absent and a thick sandstone occupies this part of the sequence (Pl. III). This bed reaches a recorded maximum of 106 ft 5 inches in a bore [3312 4313], 550 yd S.S.E. of Hallgarth church, where it consists mainly of medium- and fine-grained pale grey sandstone, generally massive, but 'wispy-bedded' with silty and micaceous bands and partings in the top 14 ft. Siltstone and shale pellets are so abundant in a 3½-ft layer, 25 ft above the base of the bed, as to form a breccio-conglomerate, and they are also scattered through a thickness of 30 ft of the overlying sandstone. Coal scars are recorded at about 10 ft and 50 ft above the base of the bed, and poorly preserved plant debris and rootlets from the top 10 ft. The sandstone is overlain by 6 ft 8 in of grey shaly mudstone with scattered rootlets, the top lying only about 5 ft below the Harvey Coal. Sandstone, 94 ft thick, was proved in two bores [3263 4362 ; 3201 4357], 250 yd W.N.W. and 900 yd W. respectively of Hallgarth church. Between Pittington and Sherburn Hill, and between Hetton le Hill and Shincliffe Colliery, the sandstone is generally over 60 ft thick, and a similar thickness is proved between Shincliffe Station and East Hetton, and around Dene House Farm and Ludworth. The sandstone may be continuous eastwards beyond Haswell Plough, with thicknesses of 65 to 85 ft extending north-eastwards at least as far as Easington and New Hesleden. To the west the sandstone extends to Kepier Florence Pit, where it amounts to 87 ft, and thence through Durham City, Grange Colliery and Low Newton to Rainton Adventure Pit where a thickness of 70 ft is recorded (Pl. III).

The Bottom Tilley Seam, where present, consists generally of one or more thin coals, though one of these reaches a thickness of between two and three feet. At the western margin of the district, in Tudhoe West Pit, the seam section is coal

3 in, on band 24 in, coal 2 in, band 26 in, coal 10 in. In the surrounding area, the seam comprises either a single coal or two parts separated by up to about 8 ft of strata. To the south-east, in the Ferryhill area, the second alternative is typical, but traced from Tudhoe to Mainsforth the coals become closer, thin, and finally die out. Thus, at East Howle Catherine Pit, the Bottom Tilley comprises coal 18 in, on 'dark fireclay' 11 in, coal 3in, 'fireclay' 30 in, 'blue metal' 33 in, 'post' 28 in, coal 5½ in, 'fireclay' 17½ in, coal 13 in; at Dean and Chapter No. 3 Pit, coal 1½ in, on 'black stone' 2½ in, coal 17 in, 'seggar' 26 in, coal 5½ in, 'seggar' 6 in, coal 7 in, 'seggar' 34 in, coal 18 in; at Dean Pit, coal 12 in, on 'hard seggar-clay' 25 in, coal 17 in; at Chilton Colliery coal 6 in, on band 9 in, coal 15 in; at Windlestone Colliery, coal 18 in; and at Mainsforth Colliery, coal absent. E.A.F.

Farther east in the Fishburn area, at Fishburn North Downcast Shaft, the Bottom Tilley is a single seam, 26 in thick, and in a bore [3672 3258], 350 yd S.S.E. of Hope House, it is 29 in thick. The best section in this area is proved in Fishburn No. 3 Bore: upper leaf 30 in, on fireclay 30 in, grey fireclay-shale with ironstone nodules 16 in, rather coarse micaceous sandstone 10 in, lower leaf 23 in. In the Ten-o'-Clock Barn, Embleton and Elwick areas, the Bottom Tilley Seam generally comprises two or three thin coals separated by a few inches to 5 ft of strata. In the Dalton Nook Bore, one of these leaves attains 29 in in thickness, but elsewhere in these areas 20 in is the maximum and 4 to 10 in normal. *Spirorbis sp.* and *Naiadites sp.* [juv.] were obtained from the roof of the Bottom Tilley Seam in the Hartlepools Lighthouse Bore. E.A.F., D.B.S.

The Bottom Tilley Seam is less persistent in the north-western part of the district than in the south-west and south. At Aykley Heads Pit, just beyond the western margin of the district, the lower leaf comprises 18 in of coal, and the upper leaf is absent, though its seatearth lies 14 ft higher. At Durham Main Pit, 400 yd to the north-east, both leaves are present and are 14 in and 3 in thick respectively, but

at Old Durham Colliery Upcast Pit there is only a single Bottom Tilley coal, 20 in thick, and at Shincliffe Colliery and in a boring near Shincliffe Station, the single seam is only 7 and 10 in thick respectively. Further to the east, in a bore at Sherburn Lady Durham Pit, the Bottom Tilley comprises 4 in of 'coal mixed with stone', but is absent in a bore said to be put down about 30 chains due east of this pit. At Littletown Lady Alice Pit, the Bottom Tilley consists of 'coal and black stone' 3 in, on 'grey shale' 5 in, coal 7 in. Just beyond the northern margin of the district, between Great Lumley and Chilton Moor, the seam is represented in borings by one thin coal or two leaves separated by a band up to 26 in thick. E.A.F.

Around Murton, in the offshore area east of Horden, and also to the south of Thornley, the Bottom Tilley is formed by one or more thin or banded coals. The seam is absent over much of the north-central part of the district between Elemore, Coxhoe and Castle Eden, where a thick sandstone occupies this part of the sequence (p. 50). D.B.S., E.A.F.

The Top Tilley Seam is a variable, banded coal which is of economic importance only in the western part of the district between Sherburn Hill, Trimdon, Whitwell and Wingate. Sections near the western margin of the worked area illustrate the degree of lateral variation. In the Sherburn Hill and Sherburn House shafts no coals are recorded between the Busty and Harvey seams, but in a bore [3370 4127], 500 yd S.S.W. of Crime Rigg, the Top Tilley Seam consists of coal 12 in, on band 7 in, coal 17 in; and it is represented by a single 21-in coal in a second bore [3321 4106], 575 yd farther to W.S.W. To the south-east many, though not all, sections are highly banded. Thus, a bore [3402 4089], 200 yd W.S.W. of the western end of Shadforth, proved coal 3 in, on band 2 in, coal 2 in, band 2 in, coal 3 in, band 9 in, coal 10 in, band 1 in, coal 2 in, band ½ in, coal 1 in. By contrast, a bore [3385 4085], only 200 yd farther to west-south-west, proved coal 9 in, on band 8 in, coal 19 in. A zone in which Tilley coals are absent extends through Fatclose House, Strawberry Hill and Old

Cassop, and separates this area of banded coal around Shadforth from the main area of worked Top Tilley Seam to the south and east: it seems to trend parallel to a similar zone through Sherburn House and Sherburn Hill.   E.A.F.

The Top Tilley Seam is of most economic importance between Thornley and Garmondsway, Cassop Moor and South Wingate, where it is 3 to 5 ft thick, including four or more bands, and has been worked from East Hetton, Thornley, Wheatley Hill, Deaf Hill, Wingate Grange, Trimdon Grange, South Wingate, Trimdon, and Hutton Henry collieries. Typical sections are : [3516 3732], 425 yd E.S.E. of Hole House (East Hetton), coal 10 in, on band 3 in, coal 5 in, band 7 in, coal 10 in, band 3 in, coal 11 in, total 49 in ; in a staple [3714 3975], 700 yd E.N.E. of Thornley Colliery, coal 25 in, on band 4 in, coal 16 in, band 28 in, coal 15 in ; [3983 3455], 300 yd W. of White Hurworth, coal 16 in, on band 1 in, coal 14 in, band 2 in, coal 16 in, total 49 in. In some of the workings, part or all of the uppermost leaf of the seam is inferior and is not mined.   D.B.S., E.A.F.

The Top Tilley seems to be united with the Bottom Tilley in a small part of the Wheatley Hill area, but elsewhere in the eastern part of the worked area, the Top Tilley is up to 30 ft above the thin Bottom Tilley. To the north, between South Hetton and Murton collieries, the interval is again small, and at Hawthorn Colliery Shaft, the following section was measured: Top Tilley Seam (coal 18 in, on band 6 in, coal 7½ in, band 6 in, coal 10 in), seatearth and siltstone 9 ft 7½ in, Bottom Tilley Seam 5 in.   D.B.S.

Beyond the southern margin of worked Top Tilley, the seam is absent in the Thrislington area. To the east, a bore [3538 3313], 800 yd S.S.W. of Catley Hill, proved coal 14 in, on ' dark metal ' 3 ft 7 in, on coal 15 in. In Garmondsway Moor Pit, about 1150 yd to the north, the Top Tilley is typically banded: coal 10 in, on band 2 in, coal 10 in, band 6 in, coal 28 in, band 4 in, coal 19 in. In a bore [3038 3313], 800 yd S.S.W. of Catley Hill, the Top Tilley comprises foul coal 5 in, on ' grey metal scared with coal ' 24 in ;

and in another bore [3672 3258], 350 yd S.S.E. of Hope House, it is only 4 in thick.   E.A.F.

In the coastal and offshore areas, additional splitting is characteristic of the Tilley coals and their clear distinction becomes difficult (Pl. III). It is possible that the Top Tilley is absent in this area, but the data are too scattered for correlation to be reliable. Between South Wingate and Hutton Henry collieries, the Top Tilley closely underlies the Harvey, with a thin coal between. The section at Hutton Henry Perseverance Pit reads: Harvey Seam (coal 45 in, on band 1¾ in, coal 20 in) on seggar 34 in, coal 7½ in, seggar 41 in, Top Tilley Seam (coal 18 in, on band 4½ in, coal 11 in, band 2½ in, coal 8½). The intermediate 7½-in coal is thought to represent the Hodge Seam of north-west Durham.   D.B.S.

In the south-western corner of the district, **the measures between the Bottom Tilley and Harvey coals** range in thickness from 50 or 60 ft (as at Tudhoe) to 101½ ft (at Thrislington Jane Pit) and to 123 ft (at Garmondsway Pit). Farther north, at Durham Main Pit, these coals are 85 ft apart and this interval again increases to east and south-east so that at Old Durham Colliery Upcast Pit, at Shincliffe Colliery and in a boring 190 yd S.E. of Shincliffe Station, it is 96, 97 and 101 ft thick respectively. It reaches 112 ft in a bore at Sherburn Lady Durham Pit. In much of this area, the Top Tilley Seam is missing. The interval between the Top Tilley and Harvey coals is commonly between 30 and 50 ft in the western part of the district.   E.A.F.

In other areas where the Top Tilley is present, the range is from 7 to 70 ft corresponding respectively with lithologies consisting mainly of seatearth, as at Hutton Henry, and sandstone as in the undersea areas off Blackhall, Horden and Easington. Sporadic non-marine lamellibranchs are found in a shaly mudstone or shale commonly lying at the base of this sandstone: they include fragmentary *Anthraconaia sp.* and *Naiadites sp.* obtained 35 ft below the Harvey Coal in Eppleton No. 2B Underground Bore [3677 4816]. It is not clear whether this band is more closely associated with an upper leaf of the Bottom Tilley or with the Top Tilley group. Where the thin coal thought to be the equivalent of the Hodge Seam of north-west Durham is present (see p. 15 and below) the strata between it and the Harvey are generally dominantly argillaceous, and in places its seatearth seems to merge with that of the Harvey, the section at Hutton Henry being an example approaching this extreme.   D.B.S., E.A.F.

Along the western border of the district, west of the worked area of Top Tilley, the only coal between the Busty and Harvey seams, with the exception of a sporadic thin Bottom Tilley Seam, is 2 to 6 in thick and lies 14 to 27 ft below the top of the sequence. This thin coal is presumed to be the Hodge Seam of West Durham, because as it is traced westwards the Tilley Coal is found 9 to 30 ft below it. A thin coal representing the Hodge Seam can be recognized around Whitwell and Hallgarth. For instance, it is 5 in thick, 14 ft below the Harvey Seam, in a bore [3058 4038], 420 yd N.W. of Whitwell House, and in another bore [3104 4060], 525 yd N.N.E. of Whitwell House, it is represented by 18 in of 'black partings' 17 ft below the Harvey, while a 6-in coal 49 ft below the Harvey in the same bore can be correlated with the thicker Top Tilley Seam farther east. Similarly, a bore [3189 4407], 850 yd N. of Broomside House, proved a 6-in coal about 17 ft below the Harvey, with no trace of coal or seatearth in the measures immediately below, while in a bore [3201 4357], 350 yd N.N.E. of Broomside House, a seatearth marks the position of the Top Tilley lying 21½ ft below the Harvey and 6½ ft below the Hodge which is represented by coal 2 in, on seatearth-mudstone 1 in, inferior coal 1 in. Fossils have been obtained from between the Hodge and Harvey coals only in Easington No. 5 Bore where they include *Spirorbis sp., Anthracosia* aff. *regularis,* cf. *Carbonicola venusta* Davies and Trueman, and *Naiadites sp.*

The Harvey Seam (Fig. 9) is one of the most important in the district. It has been exploited most extensively in the south-central and south-western areas

97544

where it is generally over 3 ft thick. At Chilton and Hutton Henry, and between Thrislington and Garmondsway Moor, it exceeds 5 ft. At Chilton, however, the seam, 68 in thick, includes an 8-in band 3 in from the top, and a 5-in band 40 in from the base. The lower of these bands persists as far as Fishburn. Thus, the section at Mainsforth is coal 19½ in, on band 4 in, coal 38 in; and at a position [3415 3313], 1160 yd S.E. of Garmondsway Pit, it is coal 16 in, on band 1 in, coal 45 in. At Windlestone, in the south-western corner of the district, the band has thickened to 11 in, the top coal being 14 in and the bottom coal 30 in thick. The Harvey Seam has been worked in the area east of Chilton East House, and at the southern limit of workings [3127 3055] is recorded as being 23 in thick. This probably refers to the Bottom Harvey alone for the evidence of borings suggests that the parting, only inches farther north, is here much thicker. Thus in Nunstainton No. 2 Bore, the 30-in Bottom Harvey Coal is separated from a 12-in Top Harvey Coal by 17¼ ft of strata; and in Windlestone 'E' Bore [2875 2896], just over 1 mile S.S.E. of Chilton Colliery, the Top Harvey Seam (shaly coal 1 in, on black shale ½ in, shaly coal 2 in, seatearth-mudstone 29½ in, coal with pyritous partings 6 in) lies 26 ft above the Bottom Harvey Seam (coal 9 in, on seatearth-mudstone 2 in, coal 4 in, seatearth-mudstone 6 in, coal 8 in). The line of split seems to have a west-south-westerly trend in this area, but there is only scattered information relating to its course to the east. It must, however, lie to the north of a bore [3243 3039], 1050 yd S.S.W. of Bishop Middleham church, where the section is Top Harvey Coal 9 in, strata 24 ft, Bottom Harvey Coal 28 in. The parting must also increase in thickness immediately to the north of Bishop Middleham East Pit where the 37-in Top Harvey is separated by 5½ ft of strata from the 33-in Bottom Harvey.

Farther east, in the Fishburn area, the Harvey Seam is again thick and affected by splitting. At Fishburn North and South shafts, it is 75½ and 78 in thick respectively. In a bore [3552 3251], 1000 yd N.W. of the shafts, it is 55 in

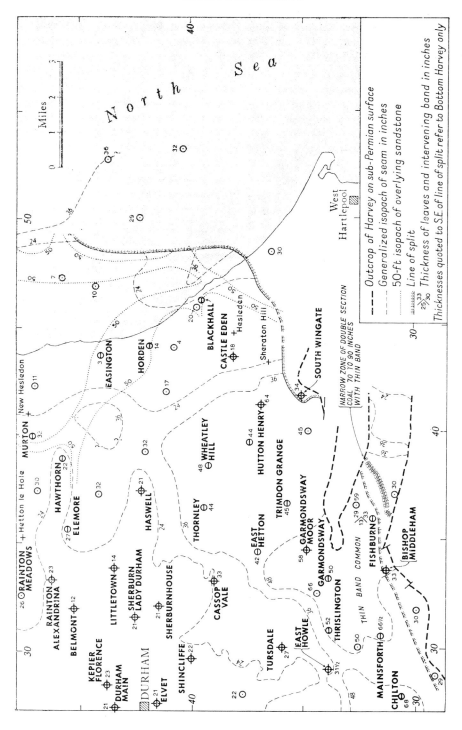

FIG. 9. *Plan of the Harvey Coal*

thick, but at the northern limit of workings [3593 3301], 700 yd N.E. of the bore, a 2-in band intervenes between top coal 14 in and bottom coal 34 in. In other bores [3672 3258 ; 3653 3228], 1000 yd N.N.E. and 650 yd N.N.E. of Fishburn Colliery, unbanded sections of 59 in and 58 in respectively were recorded, but a 12-in band separates top and bottom coals, each 30 in thick, in workings [3536 3170] ½ mile W. of Fishburn South Shaft. At two places [3509 3145 ; 3534 3145], 1200 yd and 950 yd W.S.W. of the same shaft, however, the worked coal is only 26 in and 32 in thick respectively, and it is thought to be Bottom Harvey alone. A 27-in seam worked [3651 3159], 500 yd E.S.E. of the shaft is similarly taken to be Bottom Harvey alone. In the area north of Galley Law, the Bottom Harvey, 28 to 30 in thick, is separated by a band, 4 to 15 in thick, from the Top Harvey, which is 26 to 32 in thick including, locally, a 3- to 6-in band near the top. At a locality [3758 3197], nearly 1 mile E. of Fishburn Colliery, only the Bottom Harvey, 29 in thick, has been worked, but at another place [3697 3218], 500 yd to E.N.E. the section is coal 25 in, on band 27 in, coal 27 in. These records suggest that the line of Harvey split (Fig. 9) traced from Windlestone to Bishop Middleham Colliery continues south of Holdforth and southeast of Fishburn Colliery and thence a short distance north of Galley Law.

To the south-south-east of the major line of split, only the Bottom Harvey Seam is generally worked, but along a narrow strip, about 20 yd wide, the Top and Bottom seams are united, and have been worked together across the Fishburn take (Fig. 9). The reduction in thickness of the parting is abrupt. For instance, it is 21 ft 5 in thick in a bore [3703 3139], 1100 yd E.S.E. of the colliery, but only a few inches thick 125 yd to the south-south-east. Within the strip, the Harvey Seam is over 6 ft thick, with a band ranging from 1 to 18 in. A typical section measured near to the middle of the strip [3731 3131], 1400 yd E.S.E. of the colliery, is coal 49 in, on band 8 in, coal 35 in. To the east of the strip, the Top and Bottom Harvey coals are again separated by up to 20 or 30 ft of strata and the

Bottom Harvey alone is worked, its thickness between Cowburn and Fishburn No. 3 Bore being 28 to 30 in. In the latter bore, the interval is 25 ft and the Top Harvey 33 in, but in Fishburn No. 4 Bore (Plate III) the Top Harvey is absent and its seatearth lies 30 ft above the 33-in Bottom Harvey Seam. E.A.F.

The Bottom Harvey Coal is 42 in thick in workings [3903 3103], 2 miles E.S.E. of Fishburn Colliery, and bores show it to be over 3 ft thick to the north-east and south-east of the workings. However, this area is limited to the north and south by the outcrop of the seam on the base of the Permian, and to the east bores in the Whin Houses area show the Bottom Harvey to be between 2 and 3 ft thick. Still farther east, beyond Embleton Old Hall, the thickness of this seam is generally less than 2 ft, except in the Dalton Nook Bore, where it is 29 in.

E.A.F., D.B.S.

The Harvey split can also be traced in the eastern part of the district. It is proved in the workings south of Hutton Henry, between Catlaw Hall and Sheraton Hill, and in bores between Thorpe Bulmer and the coastline ¾ mile N.W. of the mouth of Crimdon Beck. Beyond the coastline the line of split is thought to turn northwards so as to lie 2½ miles offshore from Easington Colliery (Fig. 9). Seawards of this line, the Bottom Harvey is generally 2 to 3 ft thick and comprises an important reserve. The Top Harvey, up to about 30 ft higher, is thought to deteriorate to the east, for the offshore bores proved only 6 to 12 in of coal at this horizon.

To the north and west of the line of split in the coastal and offshore region and as far inland as a line joining Sheraton Hill and New Hesledon, the single Harvey Seam is generally less than 2 ft in thickness. In the area between the line of split and Blackhall Colliery, however, the seam is over 2 ft thick, increasing to more than 3 ft in a small area offshore (Fig. 9). In the Hesleden and Horden areas, the impoverished area is about 2½ miles wide ; at Easington Colliery, it is about 4 miles wide, and east of New Hesledon it seems to be about 5 miles wide. Beyond about 3 miles east of the coastline in the northern part of the district, the seam is generally 2 to 3½ ft thick, though records of over 3 ft are rare. In the impoverished area as a whole, the Harvey is rarely absent, and typical sections were recorded at Castle Eden Colliery, 18 in, at Horden Colliery, 14 in, and at Easington Colliery, 3 in.   D.B.S.

In the north-central part of the district there is an irregular zone in which the Harvey Seam is more than 2 ft, and in places over 3 ft thick (Fig. 9). In the north-eastern part of this zone, the seam has been worked from Murton Colliery on a restricted scale south and west of the shafts. Farther to the west, around Hetton le Hole and Rainton, the coal is again over 2 ft thick, but is reduced to 23 in at Rainton Alexandrina Pit. At Rainton Adventure Pit, it is 27 in thick, but dirty. Beyond these pits and throughout the north-western part of the district, the Harvey Coal is less than 2 ft thick and is commonly dirty and sulphurous ; locally, as at Old Durham, the seam is absent. This area of impoverishment extends beyond the district to affect the Harvey Seam over much of central Durham.   E.A.F., D.B.S.

The strata between the Harvey Seam and the Harvey Marine Band constitute the highest part of the Lower Coal Measures. Over much of the district, where the Harvey is a single seam, the Harvey Marine Band lies between about 20 and 50 ft higher in the sequence, and the intervening strata generally consist of mudstone, siltstone or sandstone with silty bands (Pl. III). In eastern and north-eastern parts of the district, however, where the interval is filled largely by sandstone, it is over 50 ft, and exceptionally over 70 ft thick. Where the Harvey is split, these measures seem to remain thick, but sandstone is generally absent.   D.B.S., E.A.F.

The roof of the Harvey Seam constitutes an important faunal marker band, distributed widely in the Durham and Northumberland Coalfield. It contains a distinctive fauna of non-marine lamellibranchs and ostracods, and was originally described by Hopkins (1928, p. 6 ; 1929, p. 8) as the Ostracod Band. In order to avoid confusion with other ostracod bands discovered later, however, Armstrong and Price (1954, p. 977) renamed it the **Hopkins Band** (Fig. 3). Hopkins traced the band into this district as far as Bowburn and Tudhoe collieries, but did not detect it in the south-east at Wheatley Hill and Thornley. According to Hopkins (1929, p. 8) the " band, wherever it is found, shows a constant three-fold division, as follows :

(1) An upper division composed of *Carbonicola* and *Naiadites* with occasional *Anthracomya ;*

(2) A middle division composed of numerous *Spirorbis* with a few *Carbonicola* and ostracods ; and

(3) A basal division composed entirely of ostracods " *Beyrichia arcuata* (Bean) ". Hopkins continues : " It should be noted that it is not the presence of the ostracods that makes the band distinctive, but the assemblage and order of deposition of ostracods, annelids and mussels ". This is broadly supported by some borings in the northern, north-western and western parts of the district where the Harvey Seam is directly overlain by a thin black or dark grey shale or shaly mudstone with abundant ostracods and sporadic fish debris, above which is mudstone in which ostracods are less abundant and non-marine lamellibranchs are common. However, so many exceptions to the generalization are found in this district that it is concluded that the Hopkins Band is subject to marked local variations in character and thickness. Thus, in a bore [2895 3817], 1 mile W. of Bowburn, the band consists of a mere 2 in of dark grey shaly mudstone with ' mussels ' and ostracods interposed between the 16-in Harvey Coal and 29 ft of sandstone which underlie the 8-in seatearth-mudstone forming the floor of the Harvey Marine Band. In another bore [2888 3877], ¼ mile farther north, at High Butterby, the argillaceous sequence forming the roof of the Harvey Coal is :

	ft	in
Mudstone, grey, shaly ...		8
Mudstone, dark grey, very shaly, with abundant ostracods and *Spirorbis sp.*		2

ft in

Mudstone, dark grey, shaly, with abundant ' mussels ', including *Naiadites sp.* with *Spirorbis sp.*   3

Mudstone, lighter grey, with sporadic *Naiadites sp.* and ' mussel ' debris ; some clay-ironstone layers, and ferruginous in bottom 1 in   1   2

HARVEY COAL (at 533 ft 5 in)   1   5

In a bore [3312 4313], 550 yd S.S.E. of Hallgarth church, the basal layer of the band contains ' mussel ' fragments and ostracods, with poorly preserved ' mussels ' above. Another bore [3470 4127], 250 yd W.N.W. of Hill House, Shadforth, includes a basal layer, 1 in thick, with fish scales and spines, and above this both ostracods and ' mussels ' are found. Close by, in a bore [3466 4188], 750 yd N.N.W. of Hill House, the basal layer, 2 in thick, consists of black, very shaly mudstone containing coalified comminuted plant fragments, with ostracods and fragmentary *Naiadites* above, and other ' mussels ' in the upper part of the band, the total thickness being here about 11 in. On the extreme northern margin of the district, a bore [3256 4768], 1100 yd W. of East Rainton church, proved the Hopkins Band divided into two, as follows:

ft in

Mudstone, silty   2   5

Mudstone, grey with abundant ' mussels ' associated with ostracods at 1 ft from top ; plant debris below   3   0

Sandstone, white, fine-grained   2   0

Mudstone, grey, silty in parts, thin bands and patches of ironstone ; ' mussels ' near base ; abundant ostracods ...   6   8

HARVEY SEAM (at 223 ft 10 in)   2   2

The Hopkins Band is less persistent to the south, south-east and east of Shadforth. It has been proved in a bore [3445 3764], 400 yd N.E. of Hole House, East Hetton, where it is 2 ft thick, consisting of grey shaly mudstone with ' mussels ' (including *Anthracosia*) at the top, with ostracods and ' mussels ' in the lower half, and with poorly preserved coalified plant

debris in the bottom 3 in. The band here rests on a seatearth-mudstone about 10 ft above the Harvey Seam goaf. In another bore [3442 3853], 650 yd S.W. of Dene House Farm, near Cassop Colliery, the roof of the Harvey Seam cavity consists of about 4 ft of seatearth-mudstone. This is overlain by shaly mudstone with rootlets 2 ft, seatearth-mudstone 11 in, siltstone 1 ft 4 in and ' mussel '-bearing mudstone 16 ft, sandstone 6 ft, seatearth-mudstone 2 ft 4 in and coal 2 in. No ostracods were recorded in this bore, nor in a bore [4016 3532], 1¼ miles E.S.E. of Trimdon Colliery, where the 39-in Harvey Seam is overlain by 1½ ft of mudstone which contained only doubtful fragmentary shells at the base.

To the south and east of the line of Harvey split, the Hopkins Band, where present, lies in the roof of the Top Harvey Coal. A bore [3243 3039], 1050 yd S.S.W. of Bishop Middleham church, proved Top Harvey overlain by 8 in of barren mudstone and the overlying Hopkins Band consists of 8 in of grey shaly mudstone with a few ostracods at the base and ' mussels ' including *Naiadites* above. The barren mudstone contains ' abundant, light buff, flattened small angular silty (? clay-ironstone) inclusions ', which were also found above the Top Harvey in a bore [2875 2896], 650 yd W.S.W. of Standalone, Chilton. In the Fishburn area the roof of the Top Harvey is generally barren, apart from local plant debris.   E.A.F.

In Offshore No. 1 Bore the 6½-in Top Harvey Coal has a dark grey shaly mudstone roof, 2 ft 3½ in thick, containing: *Anthracosia regularis, Naiadites sp.* intermediate between *productus* and *quadratus* associated with *Spirorbis sp.* ; *Geisina arcuata ;* and fish remains including Palaeoniscid scales. Only fragmentary ' mussels ', identified as *Anthracosia?*, were collected from an inch of shaly mudstone forming the roof of the 6-in Top Harvey Coal in Offshore No. 2 Bore.   D.B.S.

The strata between the Hopkins Band and the Harvey Marine Band are variable in this district. The sequence in places includes one or more seatearths, locally carrying thin coals. In some areas, a thin coal underlies the horizon of the marine band, but it is rarely more than a few inches thick. In the eastern and north-eastern parts of the district the sequence includes a thick sandstone which reaches 65 ft in a bore [4773 3981] about 1¼ miles E. of Blackhall Colliery station. Except in the extreme north-east, in the undersea area off Dawdon and Easington collieries, the presence of the thick sandstone has little effect on the thickness of the interval between the Harvey Seam and the Harvey Marine Band, which maintains an average of around 55 ft over most of the eastern part of the district. In the south-western part of the district another thick sandstone is present and reaches 52 ft in Little Chilton Cragg Pit. This sandstone extends from near Croxdale Wood House south-south-eastwards to Mainsforth, where it may unite with the sandstone of the Tudhoe area. Above the sandstones and equivalent strata, is a widespread seatearth which forms the floor of the impersistent coal immediately underlying the Harvey Marine Band and forming the top of the Lower Coal Measures (Pl. III).

E.A.F., D.B.S.

MIDDLE COAL MEASURES

The Harvey Marine Band, at the base of the Middle Coal Measures (Fig. 3 ; Pl. IV), is a black or dark grey micaceous shale or mudstone, rather silty in places, lying between 20 and 70 ft above the Harvey or Top Harvey Seam. It is generally thin and in some places it is absent. The fauna is usually restricted to *Lingula mytilloides* (up to 6 mm in length) and fish scales and teeth, but other fossils recorded from it include sponge spicules, *Orbiculoidea sp.,* and *Hollinella?.*

The band seems to have been first recorded in the Northumberland and Durham Coalfield by G. A. Burnett during examination of a staple pit sunk in 1929-30 at Harton Colliery in the district covered by the Sunderland (21) Sheet. It was subsequently noted in the Bates Sinking in the Tynemouth (Sheet 15) district by Hopkins (1934, p. 186 ; 1935, pp. 5, 9). At both localities the recorded fauna consists of *Lingula mytilloides* alone. Since then, the band has been recognized in numerous bores. In some places it seems to be represented solely by a film of black shale with numerous well-preserved brown *L. mytilloides.* Generally, the *Lingula* are scattered over the bedding plane or planes, but occasionally they are found in a vertical position. *Lingula* was proved in 13 in of very dark grey finely micaceous shale in a bore [3868 3051], 900 yd W.N.W. of Ten-o'-Clock Barn, and in 5 in of shale in a bore [4015 3473], about 200 yd N. of White Hurworth. Trechmann (1941, p. 314) recorded a 4-ft black bed " with well preserved *Lingula mytiloides* and traces of fish bones " at a depth of 1,000 ft in the Thorpe Bulmer Bore, and this is here identified as the Harvey Marine Band.

In the Hartlepool Lighthouse Bore the band contains fragments of *Orbiculoidea sp.* associated with pyritized specimens of *Hollinella?,* and fish remains including scales of *Rhadinichthys sp.* and *Rhizodopsis sp.* The marine band is there composed of 3 in of dark micaceous silty and shaly mudstone, and the fragmentary nature of the fossils suggests that they have been transported and redeposited. Pyritized sponge spicules have been found in 4 in of black slightly silty finely micaceous mudstone in the Dalton Nook Bore, together with *L. mytilloides* and a tooth of *Helodus sp.* In Offshore No. 2 Bore the basal 2 in of a bed of black silty shale at the Harvey Marine Band position contain fish debris, while the overlying 4 in contain juvenile and fragmentary *Lingula sp.,* fish remains including *Rhadinichthys sp.* and pyritized strap-like markings sometimes referred to as ' fucoids '. The fossils again appear to have been transported. In Offshore No. 1 Bore, 2½ in of black shale at the marine band position contains no specifically marine forms, though they have yielded conspicuous fish remains including *Megalichthys?,* an indeterminable Palaeoniscid and *Rhizodopsis sp.* together with *Spirorbis sp.* and *Naiadites?* In many places the marine band is between 4 and

10 in thick, and there are a few records of it exceeding 12 in. One of these is in a bore [4979 4104], 2½ miles E. of Dene Mouth, where abundant *Lingula* lie in 14 in of black shale, overlain by black splintery non-marine shale. E.A.F., D.B.S.

The measures between the Harvey Marine Band and the Hutton Seam (Pl. III) range in thickness from about 75 ft in the Fishburn area to about 150 ft at Ferryhill, where they consist mainly of sandstone. Their average thickness in the district is 90 to 100 ft, which is divided by three relatively persistent seatearths, of which the middle one carries a thin coal here correlated with the Ruler Seam of north-west Durham.

The strata between the marine band and the lowest of these seatearths are generally 30 to 50 ft thick, including near the base one of the thickest and most persistent 'mussel'-bands in the district. The following is a typical section through this sequence as logged in Offshore No. 2 Bore:

	ft	in
Seatearth, sandy, micaceous	1	9
Sandstone, grey, micaceous, with thin bands of silty mudstone and with rootlets	4	6
Mudstone, sandy, with carbonaceous, micaceous partings and carbonized plant debris...	2	6
Mudstone, grey, silty, with a 3-ft bed of siltstone 10 ft above base ; otherwise with 'mussels'	24	0
Shale, black, silty with fragmentary 'mussels'		10

HARVEY MARINE BAND

| Shale, black, silty, with *Lingula sp.* and fish debris   ... | | 2 |

The 'mussel'-band near the base of the sequence is rarely less than 10 ft thick and serves to fix the position of the base of the Middle Coal Measures where the marine band is absent. Towards the base of the bed the 'mussels' are associated with fish debris, while *Spirorbis sp.* is also common and ostracods and '*Planolites*' have been recorded locally.

The band is well developed in Fishburn No. 3 Bore, where 3 ft of mudstone some 35 ft above the Top Harvey Coal contain the following typical fauna : *Spirorbis sp.* ; *Anthraconaia modiolaris, Anthracosia* aff. *aquilina, A. lateralis* (Brown), *A. ovum, A. subrecta, Naiadites* aff. *quadratus* and *N. triangularis.* Species recorded elsewhere in the district include : *Anthraconaia* cf. *curtata, Anthracosia* aff. *phrygiana, A.* cf. *regularis, Anthracosphaerium* cf. *affine, A.* aff. *exiguum, A. turgidum, Naiadites sp.* [marked posterior wing] ; *Carbonita humilis* ; scales of *Rhabdoderma sp., Rhizodopsis sp.,* and of indeterminate Palaeoniscids. Locally, the band is divided into two by barren measures of the order of 10 to 15 ft including, in some areas, a thin seatearth. The shells in the upper part are commonly of large size, as in the South Hetton Nos. 6 and 7 bores from which the following assemblage was obtained : *Spirorbis sp.* ; *Anthraconaia* cf. *curtata, A. sp.* cf. *salteri* (Leitch) [cf. Wright 1938, fig. 5a], *Anthracosia* cf. *ovum, A.* aff. *phrygiana, A. subrecta, Anthracosphaerium exiguum, Naiadites sp.* ; and *Carbonita humilis.* A similar fauna was present in the Blackhall Colliery No. 36 Underground Bore at 226 ft.

The seatearth, which forms the top of this sequence, locally carries a thin coal which reaches 8 in at Croxdale Thornton Pit and 9 in at Aykley Heads Pit just beyond the western margin of the district. Exceptionally, there are two seatearths at this position, as for instance, in Offshore No. 1 Bore where they are separated by 2 in of black shale containing *Anthracosia ovum* and rootlets. The overlying 20 to 30 ft of measures up to the next relatively persistent seatearth (here taken to be the floor of the Ruler Seam) are argillaceous below and sandy or silty above. 'Mussels' are common at the base and include *Anthracosia ovum* and ?*Anthracosphaerium exiguum.* In Offshore No. 1 Bore, there are 'mussels' at two levels within this part of the sequence as follows :

	ft	in
Seatearth, carbonaceous near top		11½

	ft	in
Seatearth, sandy, with thin bands of wispy-bedded sandstone and some turbulent bedding	3	0
Sandstone, grey, fine-grained, micaceous, wispy-bedded and thin-bedded ; base lies on eroded surface, coaly streaks 4 in above base	4	10
Mudstone, grey, silty with fragmentary *Naiadites?* 9 in above base and coalified plant debris near top ...	2	2
Sandstone, grey, fine- to medium - grained, thin - bedded to wispy-bedded with micaceous and carbonaceous partings	6	6
Shale, grey, sandy, micaceous, with sandstone partings ...	1	6
Mudstone, shaly, carbonaceous, cannelly with fish remains near base, a few ironstone bands above and sandy micaceous bands near top ; scattered ' mussels ' including *Anthracosia ovum* and *Naiadites quadratus?* together with *Spirorbis sp.* and plants	4	0

The thin persistent coal tentatively correlated with the **Ruler Seam** of northwest Durham is 20 in thick in a bore near Framwellgate Moor, just beyond the western margin of the sheet, but in the shafts at Framwellgate Moor Colliery it is reduced to between 8 and 10 in. The seam is generally at its thickest in the southern part of the district. In Fishburn Colliery North Shaft, where it has been mistaken for the Bottom Hutton Coal, it comprises coal 6 in on band 2½ in, coal 1½ in, band 2 in, coal 11 in, band 3 in, coarse coal 12 in. In a nearby bore [3672 3258], it is made up of coal 6 in, on band 2 in, coal 12 in, band 1 in, coal 10 in. At Bishop Middleham Colliery to the west, it is 20 in, and at Mainsforth East Pit, still further west, it is 13 in thick. To the east of Fishburn, in the Dropswell and Fishburn No. 4 bores, the seam is represented by black carbonaceous shale, but farther to the

east, banded coal comes in again and the maximum thickness for the seam in the district is recorded in a bore [4091 3100], 280 yd W.S.W. of Embleton Old Hall, where it comprises coal 7 in, on seatearth-mudstone 14 in, coal 11 in, seatearth-mudstone 2 in, coal 12 in. Elsewhere in the district, the Ruler Coal is generally less than 1 ft in thickness.

The measures between the Ruler Seam and the Bottom Hutton are thinnest in the Fishburn area, where they are 15 to 30 ft thick and are not divided by a seatearth. Elsewhere in the district, there is another relatively persistent seatearth with or without a coal up to 9 in thick between the Ruler and the Hutton. At present the relationship between this thin coal and the Ruler and Hutton coals is not clear: it is possible that it unites, in places, with one or other of them. Where all three relatively persistent seatearths can be recognized between the Harvey and Hutton coals the roof of the Ruler, in contrast to the roofs of the other thin coals above and below it, tends to be barren of non-marine lamellibranchs. However, in Shotton Colliery No. 6 Bore [4201 3929], ½ mile S.E. of Shotton village, both the Ruler and the higher thin coal are overlain by ' mussel '-bands.   E.A.F., D.B.S.

The faunal list from the roof of the Ruler in South Hetton No. 6 Bore includes: *Spirorbis sp.* ; *Anthraconaia sp.* of *A. modiolaris* group, large *Anthracosia* shells including *A. ovum*, and *Naiadites sp.* Elsewhere in the district *Anthraconaia* cf. *curtata, Anthracosia sp.* of *phrygiana/subrecta* affinities, *N. quadratus* and *N. triangularis* have been collected.

In Offshore No. 1 Bore this ' mussel '-band is absent, but from a bed of sandy and silty mudstone, 3 to 11½ ft above the seam, Dr. W. G. Chaloner has identified *Neuropteris attenuata* Lindley and Hutton, and *N. tenuifolia* Schlotheim. Above the higher thin coal in the same bore, 4 in of mudstone contain: *Spirorbis sp.* ; *Naiadites* aff. *quadratus* ; and fish remains including a Palaeoniscid scale. The same horizon in Blackhall Colliery No. 33 Underground Bore yielded a fauna similar to that from above the Ruler position: *Spirorbis sp.* ; *Anthracosia* cf. *ovum,*

A. aff. *phrygiana, A. sp.* intermediate between *phrygiana* and *ovum, Naiadites* cf. *triangularis* ; and fish remains. In Easington Colliery E. 3 Bore, this band contained, in addition to *Anthracosia* and *Naiadites,* several ostracods including *Carbonita humilis* and *Hilboldtina sp.*

The Hutton Coal (Fig. 10) is the lowest worked seam of the Middle Coal Measures. To the north of a line between Shincliffe Colliery and Horden, it is a single seam, but to the south of this line, it is represented by the Top and Bottom Hutton seams.   E.A.F., D.B.S.

In the north-western part of the district, the single Hutton Seam is generally a high quality coal, 3 to 5 ft thick, locally including near the base a band which lies lower than the horizon of the main split. West of the outcrop along the buried valley of the River Wear, between Bishop's Grange Farm and Durham city, the seam is 3 to 4 ft thick, and at a locality [2769 4689], 200 yd E. of the farm, the section is coal 48 in, on band 3 in, coal 15 in. A similar section is recorded 150 yd farther south [2768 4675], but the bottom coal is here inferior and only 8 in thick. East of the buried valley, the single Hutton Seam is generally 4 to 5 ft thick, and it reaches 65 in at Grange Colliery where two bores of uncertain position record a 2- to 4-in band 8 to 12 in above the base of the seam. Farther south, at Old Durham Colliery, the section is coal 47 in, on band 5 in, coal 8 in. At Whitwell ' A ' Pit and Shincliffe Colliery Shaft, the seam is 70 to 73 in thick with a $\frac{1}{2}$-in band lying $20\frac{1}{2}$ in from the base. The main split appears in the higher part of the seam in the southernmost workings of the Whitwell and Shincliffe collieries. At Whitwell ' B ' Pit, the section is coal $23\frac{1}{2}$ in, on ' stone ' band 15 in, coal 39 in ; and sections at the southern limit of Shincliffe Colliery workings variously prove coal 25 to 30 in, on band 13 to 39 in, coal 18 to 24 in. The main split can also be traced near the western margin of the district. Thus, at a locality [2721 3884], $1\frac{1}{4}$ miles W.S.W. of Shincliffe Colliery, the section is coal 22 in, on band 30 in, coal 23 in ; and a bore [2754 3875], 375 yd to the E.S.E.,

proved Top Hutton Coal, with ' brass ' near bottom, 15 in, on grey shale $3\frac{1}{2}$ ft, white ' post ' 15 ft 2 in, on Bottom Hutton Seam, consisting of coal 20 in, dark shale 3 in, coarse coal with shale partings and brass $8\frac{1}{2}$ in.   E.A.F.

In the north-central and north-eastern parts of the district, the single Hutton Seam is generally thicker than 5 ft. It is 70 in thick at North Pittington Colliery, $61\frac{1}{4}$ in at Rainton Alexandrina Pit, and $64\frac{1}{4}$ in, including a $\frac{1}{2}$-in band 20 in from the base, at North Hetton Moorsley Winning. The unbanded seam, 5 to 6 ft thick, extends eastwards in a narrow strip from Sherburn Hill to Warren House near Horden and thence $2\frac{1}{4}$ miles offshore. At the southern margin of the strip the major split can again be traced from a band in the upper part of the seam, while at the northern margin a thinner band appears near the base of the seam : both bands are present in parts of the offshore area. Typical sections illustrating the lower, thinner band in the north-eastern part of the district are : Hetton Lyons Pit, coal 56 in, on band 4 in, coal 15 in ; Easington Colliery, coal 48 in, on band 2 in, coal 15 in ; Murton Colliery shafts, coal 60 in, on band 12 in, coal 15 in. Sections which show the upper, principal splitting along the southern margin of the strip are seen in Thornley Hutton workings as follows : [3339 4008] coal 25 in, on band 6 in, coal 39 in ; [3456 4010] coal 24 in, on band 25 in, coal 38 in ; [3424 3990] coal 26 in, on band $3\frac{1}{4}$ ft, coal 37 in. Half-a-mile south-east of Ludworth, the single Hutton is 72 in thick, but 1000 yd S.E. of the colliery the section is coal 25 in, on band 6 in, coal 38 in, while 300 yd farther south the band is 3 ft thick. West of Acre Rigg (west of Horden) the band is less than 3 ft thick extending in a belt about $\frac{1}{4}$ mile wide, but to the east this belt increases to nearly 2 miles wide in the area north of Black Halls Rocks.

                                    D.B.S., E.A.F.

Where, in parts of the offshore area, the Hutton Seam includes both bands (Fig. 10), examples are as follows : [4623 4659] coal 34 in, on band 2 in, coal 19 in, band 2 in, coal 19 in ; [4809 4581] coal 20 in, on band 1 in, coal 24 in, band 27 in,

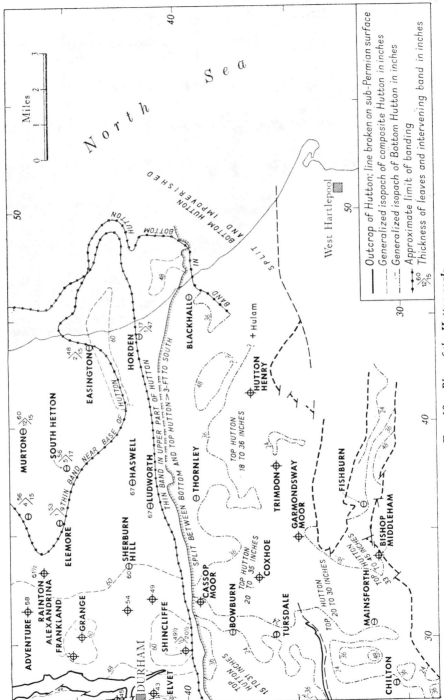

FIG. 10. *Plan of the Hutton coals*

coal 38 in. Farther to the east, just beyond the limit of workings, two bores record only 23 in of coal at the level of the Hutton Seam, and still farther east, Easington No. 9 Bore [4969 4649], 3½ miles E. of Chourdon Point, proved only 15 in. of coal. However, sections near the northern margin of the district about 2 miles E. of the coastline show unbanded coal 77 in thick. Farther south, at a locality [4926 4358], 3½ miles E. of Easington Colliery, the seam includes three bands as follows: coal 28 in, on band 20 in, coal 3 in, band 3 in, coal 3 in, band 28 in, coal 28 in; but 570 yd to the E.N.E., Horden No. 15 Bore [4974 4374] proved only 14 in of coal. Further evidence to suggest that the Hutton is thin or banded beyond the present limit of workings in the offshore area is provided by Offshore No. 2 Bore, which proved coal 6 in, on band 6 in, coal 15 in, at this horizon.   D.B.S.

The Bottom Hutton Seam is generally over 2 ft thick and in two bores [2906 3154 and 4201 3929] near Chilton and Castle Eden, it measures as much as 63 in. Elsewhere, the seam approaches a thickness of 4 ft only in the area east of Fishburn Colliery, between Bridge House, Butterwick and Ten-o'-Clock Barn, where it locally contains a band. In a bore [3679 3123], ½ mile S.W. of the Colliery it is 46 in thick. Farther east, Fishburn No. 4 Bore proved coal 24 in, on band 3 in, coal 21 in, and still farther east, the seam is 27 in thick in the Whin Houses Farm Bore.

The seam is more than 3 ft thick adjacent to the line of split in the area extending between Cassop Moor and the coast, and again in a small area east of High Butterby. It is of comparable thickness between Monk Hesleden and Eden Vale, but is probably impoverished to the south-east, beyond Hulam, being only 30 in thick in a bore [4349 3653], 570 yd W.S.W. of Hulam, and absent in the Thorpe Bulmer Bore.   E.A.F., D.B.S.

In the south-western part of the district the seam is locally about 40 in thick as in bores [3003 3921 and 3048 3923] 1450 yd N.N.W. and 1450 yd N. of Bowburn Colliery respectively. North-west of the colliery, a bore [2888 3877] at High Butterby, proved coal 18 in, on band 3 in, coal 10 in, band 5½ in, coal ½ in. The coal exceeds 3 ft in thickness in workings near the western margin of the district, about 800 yd W. of High Butcher Race, in Chilton 'E' Bore [2825 3036], 620 yd S.E. of Chilton Colliery, and in another bore [2745 2883], ½ mile W. of the inn at Rushyford. It might also be as thick over a limited area between West Close and Ferryhill Reservoir, but information is inadequate both here and in the area to the north and east of Bishop Middleham: the only recent bore [3433 3211], 1150 yd N.E. of Bishop Middleham Colliery, proved 45 in of Bottom Hutton, though at the Colliery itself, it is only 36 in. However, the seam is relatively close to the Permian unconformity in much of this ground.

In the remainder of the district, the Bottom Hutton Seam is generally between 2 and 3 ft thick. West of Bowburn and Broom Hill, it includes one or more thin bands, and is subject to variation over relatively short distances. Thus, at Bowburn Colliery Downcast Pit it comprises 25 in of coal, but in a bore [2978 3796], 680 yd to the west, the section is coal 24 in, on band 1 in, coal 13 in; and in a bore [2929 3787], 560 yd farther W.S.W., it is coal 18½ in, on black shale band 1½ in, coal 13 in. This band is generally 2 in or less in thickness, except in a bore [2806 3832], 1½ miles W.N.W. of Bowburn Colliery, where the section is coal 17¾ in, on 'black stone with coal threads' 4¼ in, coarse bottom coal 3 in. The seam is 18 in thick in a bore [2939 3591], ½ mile W. of Tursdale Colliery, but nearby, in three other borings [2981 3602, 2960 3646 and 2986 3639], it is 54, 21 and 37 in thick respectively.

The seam is less than 2 ft thick (Fig. 10), in a belt trending south-south-west through Cassop Moor Pit (20 in), Bowburn Upcast Pit (22 in) and Tursdale Colliery (17 in). Farther to the south this impoverished belt seems to divide, one branch extending to Chilton (22 in) and the other via East Howle to Chilton Colliery (12 in). The seam is less than 2 ft in a small area between Trimdon Colliery and Hurworth Bryan and in a narrow belt adjacent to its northern outcrop in the area east of Fishburn. In the Dalton Nook Bore, it is probably represented by a 13-in coal at a depth of 767¾ ft.

The strata between the Bottom and Top Hutton seams range from 13 ft in thickness, as at Wingate Grange Colliery, to 43 ft, as proved in two bores [3981 3555 and 4016 3534], 400 yd W. and 240 yd S. of Woodlands Close respectively. Elsewhere the interval is generally between 20 and 40 ft, except in the area of impoverished banded Hutton extending from south-east of Tweddle Black Halls to Offshore No. 1 Bore.   E.A.F., D.B.S.

The lowest part of this sequence, forming the roof of the Bottom Hutton Seam, consists of mudstone containing non-marine lamellibranchs, though these are usually few and fragmentary. In the Dalton Nook Bore, *Spirorbis sp., Anthracosia ovum* and *Naiadites quadratus* were obtained at this horizon. In the two offshore bores, where the Hutton sequence is complicated by further splitting, there are two additional fossiliferous bands. In Offshore No. 1 Bore, the Bottom Hutton is probably represented by an upper 2-in and a lower 13-in coal. The lower coal is overlain by seatearth which carries a mudstone with shell fragments. The roof of the upper leaf is the normal Bottom Hutton shell-bed and contains *Anthraconaia sp.* [juv.], *Anthracosia* cf. *ovum* [pyritized], *A.* aff. *phrygiana, Anthracosphaerium* cf. *turgidum, Naiadites quadratus* and fish debris. The higher additional band occurs immediately beneath the Top Hutton and is represented by a cannelly shale containing coalified plant fragments, together with fish remains, including *Elonichthys sp.* and ostracods. In Offshore No. 2 Bore, the roof of the Bottom Hutton contains only poorly preserved, indeterminable non-marine lamellibranchs, but the upper fossiliferous band contains: *Spirorbis sp.; Naiadites sp.* [fragments]; *Carbonita humilis, C. sp.;* fish remains including *Megalichthys sp.,* indeterminate Palaeoniscid and Platysomid scales, *Rhadinichthys sp.* and *Rhabdoderma sp.* In Deaf Hill No. 4 Bore [4016 3534] and Park Farm Bore [3900 3440], the following were obtained from 25 ft above the

Bottom Hutton: *Spirorbis sp.; Anthraconaia* cf. *pumila* (Salter), *Anthracosia* cf. *ovum, A. phrygiana, A.* cf. *subrecta, ?Anthracosphaerium turgidum, Naiadites quadratus*; and *Carbonita humilis.*

No pattern of distribution can be inferred for the strata above the mudstone roof of the Bottom Hutton. Sandstones are commonly interbedded with the higher mudstones and siltstones, but they are not persistent apart from a bed which occupies most of the interval between the Bottom and Top Hutton immediately south of the major split at Bowburn, Cassop Moor, Cassop, East Hetton, Thornley, Wheatley Hill and Blackhall collieries. This sandstone is exemplified in a bore [2929 3787], 1100 yd S.S.E. of High Butterby, where the Bottom Hutton roof comprises 2½ ft of mudstone with clay-ironstone and ' mussel ' fragments, overlain by 25 ft of light grey, very fine-grained sandstone, with dark micaceous partings in the bottom 3 ft and silty laminae in the top 1 ft, on which rests 2½ ft of light grey seatearth-mudstone forming the floor of the Top Hutton Seam. In a bore [2888 3877], 40 yd E. of High Butterby, the sandstone makes up the whole of the sequence between Bottom and Top Hutton seams.

The Top Hutton Seam is of less economic importance than the Bottom Hutton. Its thickness is very variable, ranging up to 2½ ft, especially between Bowburn, Wheatley Hill and Garmondsway Moor. Where it is thicker it generally includes one or more bands or becomes shaly. Thus, in a bore [3433 3211], ¾ mile N.E. of Bishop Middleham Colliery, it comprises coal 15 in, on band 12 in, coal 12 in, band 2 in, coal 4 in ; and at the colliery itself the 33-in seam includes a 10-in band 4 in above the base. A bore [3142 3004] at Thrundle, 1250 yd E.S.E. of Chilton East House proved the Top Hutton to be divided into two 13-in leaves by 3 ft 4 in of strata. South of Bowburn Colliery, between Hett and Park Hill, it attains a maximum thickness of over 50 in, but is banded as it is also south of Hett, where the seam is over 2 ft thick. Around Croxdale Hall, though free from bands, it is generally less than 18 in. Between Bowburn and Kelloe Law, it is generally 2 to 3 ft thick and has been worked in places, though many of the thicker sections include a band of up to 5 in.   E.A.F.

At Thornley Colliery Shaft, it measures only 11 in, but it has been worked over a small area farther south where it is about 2 ft thick. Other small areas have been worked north and south of Wheatley Hill, where it is only 19 to 21 in thick as compared with a shaft thickness of 24 in. Farther to the east, the coal is thin and variable and only rarely exceeds 2 ft, while in the offshore area, further splitting takes place.   D.B.S.

The measures between the Hutton and Bottom Brass Thill coals range in thickness from less than 8 ft up to perhaps as much as 94 ft, and include several sandstones. The most persistent is found in the north-eastern part of the district where the interval between the two coals is greatest (Pl. IV). This sandstone at Hetton Lyons Pit is 24 ft thick, at Elemore Colliery 33 ft, at Murton Colliery 19 ft, at Haswell Colliery 31 ft, at Horden Colliery 37 ft, at Easington Colliery 19 ft, at Dawdon Colliery 38 ft, at Blackhall No. 36 Underground Bore 71 ft, at Offshore No. 1 Bore 68 ft, and in a bore [4201 3929], ¼ mile S.E. of Shotton, 25 ft. Another such sandstone, generally less than 25 ft thick, lies between the Top Hutton and Bottom Brass Thill west of a line joining Old Cassop and Bishop Middleham, but it is impersistent between Tursdale House and Dean Bank. In the remainder of the district, this part of the sequence consists of mudstone and siltstone with sandy bands, as in a bore [3442 3853] near Cassop Colliery :

	ft	in
BOTTOM BRASS THILL COAL ...		6
Mudstone, grey, silty, with rootlets abundant at top ...	4	0
Siltstone, grey, sandy, with sporadic sandy partings ; scattered plant leaves and stems ; rootlets in upper 7 ft	15	5
Mudstone, grey, shaly, very silty, with scattered plant stems and leaves	3	8
Siltstone, grey, thin-bedded,		

	ft	in
with sandy and micaceous partings	1	2
Mudstone, grey, shaly, silty, with scattered plant stems and leaves and some irony layers	16	11
TOP HUTTON COAL	2	2

Locally a thin bed of shaly mudstone at the base of the sequence contains non-marine lamellibranchs, but the shells are generally sparse and poorly preserved.
                                    E.A.F., D.B.S.

The assemblage from above the Top Hutton is best represented by the collections from 437 to 442 ft in Fishburn No. 4 Bore. The identifications include : *Anthraconaia sp.* [large], *Anthracosia sp.* cf. *caledonica* A. *ovum*, A. aff. *phrygiana*, A. *sp.* intermediate between *ovum* and *phrygiana*, *Naiadites quadratus* ; *Carbonita humilis* ; and fish.

The Brass Thill coals of the Durham district are splits from the single seam of that name in the Sunderland (Sheet 21) and Barnard Castle (Sheet 32) districts to the north and south-west respectively. The primary split is into Bottom and Top Brass Thill coals, of which the Top is generally the thicker and more persistent, while the Bottom Brass Thill is commonly subject to further splitting (Pl. IV). The primary split is smallest in the Cocken area where locally only a thin band separates the two leaves. The seam was worked from Cocken Drift as a single section [2897 4697], 700 yd W.S.W. of Finchale Priory, and consists of coal 15 in, on band 3 in, coal 15 in. Other sections proved in the workings and in bores in the surrounding area demonstrate that the band increases to over 10 ft within ¼ mile, as in a bore [3023 4745], 700 yd E.N.E. of the Priory, where the two coals, each 12 in thick, are separated by 10 ft 5 in of strata. On the left bank of the River Wear the following section [2868 4689], 1025 yd W.S.W. of the Priory, is exposed:

	ft	in
Sandstone, fine, commonly massive	10	0
Mudstone and siltstone, some thin sandstone bands, becoming more sandy to west ...	10	0

	ft	in
Brass Thill Shell-bed: mudstone, shaly, with non-marine lamellibranchs and thin bands and nodules of clay-ironstone	2	6
BRASS THILL SEAM: **coal** 30 in, on fireclay-mudstone 22 in, **coal** 8 to 9 in (at approx. + 50 ft O.D.)	5	0
Grey fireclay	1	6
Fireclay-mudstone, with ironstone nodules	2	0
Sandstone, shaly (to river level)	8	0

The Brass Thill coals are proved in a small area worked from Harbour House Drift, where the band is up to 3½ ft thick, separating the Bottom Brass Thill, 8 to 10 in, from the Top Brass Thill Coal, 29 to 36 in.

Traced to the east of Finchale Priory, the Bottom Brass Thill splits, as in a bore [3092 4707], ¼ mile N.W. of Leamside Station where the section is Top Brass Thill Seam (cannel 5 in, on bright coal 11 in, cannel 7 in), on seatearth-mudstone 4 in, sandstone 10 in, mudstone 3½ ft, Bottom Brass Thill Seam (coal 8 in, on seatearth-mudstone 11 in, coal 6 in). However, farther to the east and throughhout the Rainton area, there are only two thin coals each 7 to 12 in thick and about 1 to 3 ft apart. To the north, beyond Rainton Meadows Pit, in the Sunderland (Sheet 21) district, these coals reunite, and at Rainton Plain Pit, just over ¼ mile N.N.W. of the Meadows Pit, there is a single Brass Thill Coal 40 in thick. Farther to the east, at Houghton Pit, also in the Sunderland district, and at North Hetton Colliery, the Top and Bottom seams are again distinct. In the North Hetton Upcast Shaft, the Bottom Brass Thill comprises two thin coals 7 and 10 in thick, but in the nearby Moorsley Winning these are united in a single 22-in coal.

In the west-central part of the district between Pittington, Sherburn Hill, Bowburn and Kepier Grange collieries, the Brass Thill coals are all thin. The Top Brass Thill reaches a local maximum of 18 in at Kepier Grange, but at Kepier Colliery it is 14 in thick and the Bottom Brass Thill consists of two 4-in leaves of

coal. At Whitwell ' A ' Pit, however, the sequence is Top Brass Thill Seam foul coal 4 in, on ' post ' and ' metal ' 10 ft 10 in, Bottom Brass Thill Seam (coal 9 in, on ' grey metal ' 2 ft 9 in, coal 5 in). At Old Durham Colliery, the Top Coal seems to be represented only by 5 ft of ' black stone ' about 6 ft above the Bottom Seam, here represented by coal 10 in, on ' dark grey metal ' 4 ft, coal 7 in. Farther west, at Elvet South Engine Pit, three coals, comprising, from the top, seams of 10, 11 and 6 in thick respectively, are separated by intervening strata comparable with those of Whitwell ' A ' Pit. One mile to the north-north-west of Elvet, at Aykley Heads Pit just beyond the western margin of the district, there is again a three-fold sequence, but here the Top Brass Thill is split and comprises coal 13 in, on fireclay 5 in, coal 15 in, separated by 4 ft of fireclay from the 6-in Bottom Brass Thill Coal. At Framwellgate Moor Pit, one mile farther north, an unbanded 30 in Top Brass Thill is separated by 3 ft 10½ in of ' grey metal ' from an 8-in Bottom Brass Thill.

In the extreme north-western corner of the district, a 15- to 34-in Top Brass Thill Coal lies up to 6 ft above the 4- to 12-in Bottom Brass Thill Coal. To the southeast of this, the Bottom Brass Thill splits so that a three-fold sequence is recorded at Frankland Furnace and Engine pits. However, local variations are exemplified in a third Frankland Pit which records a 21-in Top Brass Thill and a single 7-in Bottom Brass Thill, while an old bore (Borings and Sinkings 1467) of uncertain site in the Newton Hall Estate proved foul coal 10 in, on ' grey metal ' 6 in, coal 18 in, ' grey metal ' 4¼ ft, coal 4 in, a section closely comparable with that at Aykley Heads Pit.   E.A.F.

In the eastern part of the district, the Top Brass Thill Coal is as much as 3½ ft thick at South Hetton Colliery and 2½ ft in Easington Colliery No. 5 Bore, but has not been worked. The Bottom Brass Thill Coal, though proved to be 20 and 26 in thick respectively at Eppleton Colliery and Horden No. 15 Bore, 3¼ miles E. of Fox Holes, Easington Colliery, is also not worked. In some eastern parts of the district, further splitting takes place to form up to 5 thin seams, some of the lower of which, as in Blackhall No. 36 Bore are locally absent (Pl. IV). D.B.S.

In the south-western part of the district, there is again considerable variation in the Brass Thill coals, but the Bottom Brass Thill is generally less than 1 ft thick where single, and where double, the individual leaves rarely exceed 10 in. South-east of Bowburn, three leaves are present, but the Bottom Brass Thill coals are both thin and the Top Brass Thill is generally less than 12 in. None of these coals was recorded at Coxhoe and East Hetton collieries. South-west of Bowburn, the Top Brass Thill Coal exceptionally reaches 22 inches in a bore [2960 3646], 550 yd N.N.E. of Broom Hill, and south of High Butcher Race a number of bores record the Top Brass Thill as having thicknesses ranging from 9 to 21 in, though the upper leaf of the Bottom coal is only 8 to 10 in and the lower leaf 2 to 8 in thick. Still farther south, at Dean and Chapter Colliery, the Top Brass Thill is only about 18 in thick, but it increases to 27 in farther east in the Cleves Cross No. 2 Bore [2946 3264] and in the Wood Lane Bore [2937 3293]. At Chilton Colliery the Top Brass Thill is 24 in thick, and though one bore [2867 3120] at West Close proved only 1 in of coal at this position, it is consistently 21 to 31 in thick in other bores in the area. To the east of Chilton the Top Brass Thill is variable in thickness, but is generally over 2 ft in the area south of Thrislington Plantation and Highland House, and reaches its local maximum of 34 in in a bore [3234 3105], 550 yd S.W. of Bishop Middleham church. In another bore [3278 3126], 20 yd N. of the church, however, the coal, though comparable in thickness, contains a thin band: coal 11 in, on band 3 in, coal 18 in. To the north around Thrislington and Garmondsway, the thickness decreases, so that at Thrislington Colliery, it is only 22 in, and in the Garmondsway bores, it ranges from 10 to 18 in. E.A.F.

In South Hetton No. 6 and Shotton No. 7 bores, non-marine lamellibranchs were recorded from the roofs of both Bottom and Top Brass Thill Coals. The ' mussel '-band overlying the lower coal

95744

is the less persistent and has yielded the following : *Anthraconaia salteri?, Anthracosia beaniana, A.* cf. *ovum, A.* aff. *phrygiana, A.* aff. *retrotracta* (Wright), *A. sp. nov.* [truncate], *A. sp.* intermediate between *aquilina* and *ovum, Anthracosphaerium* cf. *exiguum, A. turgidum, Naiadites quadratus* ; *Carbonita humilis, C. pungens, Hilboldtina sp.* ; *Rhabdoderma sp.* and *Rhizodopsis sp.* The assemblage from the roof measures of the Top Brass Thill includes : *Anthraconaia sp., Anthracosia beaniana, A. disjuncta, A.* cf. *ovum, A.* aff. *phrygiana, A. retrotracta?, A. sp. nov.* [truncate], *A. sp.* intermediate between *aquilina* and *ovum, Anthracosphaerium* aff. *affine, A. exiguum, A.* aff. *turgidum, Naiadites quadratus* ; *Carbonita humilis* ; and Eurypterid fragment.

The roof of the Top Brass Thill Coal generally consists of shaly mudstone with abundant non-marine lamellibranchs—**the Brass Thill Shell-bed.** This is one of the most persistent ' mussel '-bands of the Coal Measures sequence in the district and resembles the band overlying the Harvey Marine Band in the variety and abundance of its fauna—in marked contrast to the other shell-beds in the lower part of the Middle Coal Measures. However, it rarely exceeds 10 ft in thickness. In the Butterby bores, between Bowburn, High Butterby, Croxdale Wood House and High Croxdale, the shell-bed ranges from 5 to 8½ ft in thickness, and this range may be taken as typical.

In the same bores the Brass Thill Shell-bed is successively overlain by 10 to 15½ ft of mudstone and siltstone, 6 to 10 ft of sandstone and siltstone, and 1 to 3½ ft of seatearth forming the floor of the Durham Low Main Coal. This sequence is fairly typical of the district, though in the area around Durham City the sandstone and siltstone lie immediately above the ' mussel '-band. In south-western parts of the district the interval between the coals is increased and is largely occupied by a thick sandstone. At Chilton Colliery the sandstone, 70 ft thick, extends from the roof of the Top Brass Thill up to a 2½-ft bed of ' fireclay ' forming the seatearth of the Durham Low Main Coal.

F

A small form of *Anthraconaia* compared with *A. pumila* (Salter) was found 16 ft below the Durham Low Main Coal in Shotton No. 6 Underground Bore and at a similar horizon in Blackhall No. 36 Underground Bore.

The Durham Low Main Coal (Fig. 11) has been widely worked in the district. It crops out within the buried valley system of the River Wear north-west of Shincliffe and north-east of Hett, and along the northern side of the Butterknowle Fault in the western part of the district. It is absent from a large part of the southern and south-eastern area of the district as a result of pre-Upper Permian erosion. The average thickness of the seam is 3 ft, but it reaches a maximum of about $7\frac{1}{2}$ ft in Offshore No. 2 Bore where, however, the lower part is banded (see below). Such banding is only locally recorded in this seam. Its distribution is shown in Fig. 11, together with the position and extent of long narrow belts of abnormally thin coal and washouts, which are more abundant in the Durham Low Main than in any other seam in the district.   E.A.F., D.B.S.

The seam is more than 5 ft thick in the north-eastern part of the district in a narrow zone extending eastwards from Cold Hesledon. About $\frac{1}{4}$ mile E.S.E. of Chourdon Point, the thickness recorded on mine plans is 54 in, but this may not include the uppermost part of the seam which is locally inferior and unworked. About $2\frac{1}{2}$ miles E. of Chourdon Point, however, the whole seam becomes thinner as the top inferior coal decreases from 14 out of 72 in to 4 in out of 67 in, and then to 2 in out of 60 in. To the north a thin band appears, as at a locality [4842 4758], 3 miles E. of the coastline, where the seam consists of coal 2 in, on band 5 in, coal 60 in. In most other eastern parts of the district, the seam exceeds 3 ft except in an irregular area extending from Easington through Warren House south-eastwards for over a mile beyond the coast; within it is a smaller, narrow zone where the coal is either less than 2 ft thick, or absent because of washout (Fig. 11). In the offshore area, it is likely that substantial reserves in this seam remain to be won, judging from Offshore No. 1 Bore

which proved 43 in of unbanded coal, and by Offshore No. 2 Bore which proved coal $59\frac{1}{2}$ in, on band 1 in, coal 13 in, seatearth 2 in, coal 1 in, seatearth 10 in, coal 2 in.

Between Murton and The Castle, at Castle Eden, the upper part of the Durham Low Main Coal includes a band which rarely exceeds 18 in, though in South Hetton Colliery Shaft it reaches the exceptional thickness of 7 ft 3 in. Another band occurs low in the seam in an area between Easington Lane and Easington. The two areas overlap between Easington and South Hetton, giving seam sections as follows: coal 2 to 6 in, band 9 to 13 in, coal 16 to 23 in, band 1 to 4 in, coal 16 to 23 in. Farther north, the Durham Low Main Seam lies within 2 ft of the Maudlin Coal, thus forming a compound seam, locally over 8 ft thick which has been worked from Eppleton Colliery just beyond the northern margin of the district. The interval is as little as 4 in near Carr House (see also p. 71).   D.B.S.

West of a line joining South Hetton, Shotton and Castle Eden the Durham Low Main Seam is generally between 2 and 4 ft thick, and is traversed by several belts of thin coal and washouts, some of which have a north-westerly to north-north-westerly trend corresponding with similar belts near the coast (Fig. 11). One of these crosses the northern margin of the district between Hetton le Hole and Carr House and continues as far south as South Hetton.
E.A.F., D.B.S.

A much longer one extends from close to the northern margin of the district, about $\frac{1}{2}$ mile N.W. of Finchale Priory, for over five miles south-eastwards and then turns southwards for more than a further two miles to Cassop Colliery. Sections measured along the margin of this belt show that impoverishment is caused by erosion in some places, but by thinning or banding in others. Although the impoverishment is of more than one kind, its occurrence within a single belt suggests a causal relationship. The northern part of this belt can be seen at outcrop [2873 4788], 1,250 yd

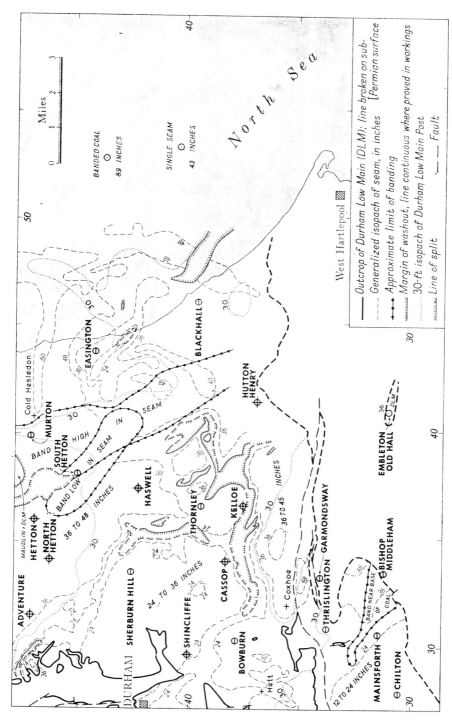

Fig. 11. *Plan of the Durham Low Main Coal*

N.W. of Finchale Priory, in the steep left bank of the Bow Burn, where the Durham Low Main Coal, 12 in thick, is overlain by 19 ft of sandstone. It can be followed downstream for about 75 yd until the coal is cut out by the sandstone close to the confluence of the burn with the River Wear. The coal is absent too, in the right bank of the River Wear gorge for about 400 yd upstream from this confluence as far as an old adit [2864 4746], about 1,100 yd W.N.W. of Finchale Priory. The adit was put in at the level of the seam, though no section is now visible. The belt can be traced south-eastwards first by a bore [2897 4755], about 825 yd N.W. of the Priory, where the seam is absent, and thereafter by mining informa- tion (Fig. 11).   E.A.F.

To the south and east of Thornley Colliery, this belt is linked with other belts of washout, some having a north- westerly trend, while others trend east- wards, so that in places islands of coal are surrounded by washouts.   D.B.S.

The westernmost extremity of the main east–west washout is proved by two stone drifts, respectively 320 and 420 yd long, about ¼ mile E. of Quarrington Hill crossroads, and in a bore [3184 3747], 200 yd S.S.W. of Heugh Hall Row. There is evidence to show that from the westernmost extremity of washout a north-westerly belt of thin coal continues towards Durham City. At Shincliffe Colliery, the section is coal 29 in, on band 7 in, coal 11 in, but the limit of workings lies only 70 yd to the west of the downcast pit and trends generally north-north-west from near Shincliffe Station, where a bore [3003 3921] proved the coal to be only 7 in thick. The limit of workings probably represents the line along which the coal thickness falls below 2 ft. If the north-west trend is continued, this belt of thin coal should pass through Elvet South Engine Pit where the Durham Low Main seems to be represented by an 11-in coal 103 ft 9 in above the Hutton (Dunham and Hopkins 1958, p. 2). The record of an ' old shaft near Durham Grammar School ' (Borings and Sinkings No. 2652), however, shows

the Durham Low Main to be 30 in thick at 75 ft 3 in above the Hutton. Although the precise position of this old shaft is not known, it appears to mark the southern edge of the north-westerly belt of thin coal. The northern margin probably passes south of Crook Hall High Pit [2728 4314], which is thought to be the site of a bore (Borings and Sink- ings No. 927) proving the Durham Low Main Coal to be 30 in thick, and thence close to Aykley Heads Pit where the coal is 26 in thick. Thus, much of the Durham gorge lies within the belt of thin coal. Recent information suggests that the thin coal which underlies the sandstone upon which the Cathedral is built is the Durham Low Main, as proposed by Holmes (1928), and not the Maudlin as shown on the new One-Inch Map, nor the Brass Thill as proposed by Hindson and Hopkins (1947) and by Dunham and Hopkins (1958). The north-westerly belt is joined to the south-east of Shincliffe Colliery by a similar belt of thin coal which extends southwards through Hett. In one bore [2774 3827] sited along this belt the coal is 11½ in thick.

In much of the western part of the district, the Durham Low Main Seam is unbanded, but bands occur locally near impoverished zones. In a bore [3371 3942] at Old Cassop, the section is coal ½ in, on sandstone with large coal scar 8 in, coal 1 in, sandstone 1 in, coal 15 in, seatearth-siltstone 11 in, coal 3 in, and this may indicate proximity to a washout or belt of thin coal extending at least as far as a bore [3264 3952], ¼ mile to the west, where the section is coal 13 in, on band 1 in, coal 5 in. Other banded sections were measured in workings [3487 3562 and 3452 3514], 1¼ miles E. and 1½ miles E.S.E. of Coxhoe, where the seam consists respectively of coal 18 in, on band 8 in, coal 3 in ; and coal 30 in, on band 10 in, coal 3 in. A bore [3164 3584], ½ mile W. of Coxhoe proved coal 8 in, on band 2 in, coal 31 in. The Durham Low Main, like the Busty Seam (p. 48), is cindered over a smaller area to the north of the Ludworth Dyke than it is to the south where it extends at least as far as a bore [3166 3879], 300 yd E. of Cassop Grange, which proved cindered coal 21 in, on band 7 in, cindered coal 5 in.

South of the Butterknowle Fault, the Durham Low Main Seam exceeds 4 ft near Garmondsway and Farnless, and decreases to about 1 ft south-westwards. Near Garmondsway, bores proved coal up to about 5 ft in thickness on the northern limb of the anticline, though it is only 29 in thick in a bore [3426 3453], ¾ mile N.E. of Garmondsway Pit. Some of the sections may reflect minor tectonic thickening, but this is unlikely to account for the thickness of 50 in proved in a bore [3362 3243], nearly 1 mile south of the pit. Still farther south, in a bore [3386 3168], ¼ mile N.N.E. of Bishop Middleham Colliery, the 43-in seam includes a 4-in band 7 in above the base, and this band is also proved in several bores around Bishop Middleham church, reaching its maximum in a bore [3234 3105], 550 yd S.W. of the church where the section is coal 30 in, on band 4 ft 5 in, coal 4 in. The same band is proved in a bore [2996 3275], 1 mile N.N.W. of Mainsforth Colliery: coal 28 in, on band 3 in, coal 5 in. The seam measures 42 inches in a bore [2937 3293], 1¼ miles N.W. of the colliery, but this thickness is locally exceptional for the 24-in isopach of the coal runs south-eastward through the Mainsforth Colliery shafts and the seam consists of only 12 in of shale and coal at Chilton Colliery shafts. A small outlier of the coal, 36 in thick, was proved in the Embleton Old Hall Bore, but it lies too close to the base of the Permian to be of economic value. E.A.F.

The measures between the Durham Low Main and Maudlin seams range in thickness from 4 in near Carr House to about 115 ft at Littletown Engine Pit. Except in some northern, south-western and farthest offshore areas this interval is dominated by thick sandstone–the Low Main Post–which reaches 105 and 104 ft respectively at Hutton Henry Colliery and Littletown Engine Pit. The 30-ft isopach in the sandstone is plotted on Fig. 11 from which it can be seen that there is a direct relationship between most of the washouts affecting the Durham Low Main Coal and the overlying sandstone. An exception is the washout east of Easington

Lane, for close to its southern tip only 5 ft of sandstone is present in the roof of the Durham Low Main Seam at South Hetton Colliery, and to the west of the washout at Hetton Lyons Pit the sandstone is only 11 ft thick, and lies 18 ft above the coal. E.A.F., D.B.S.

Farther west, at Elemore, Rainton Alexandrina, Meadows, Adventure and Resolution pits, the 40 to 70 ft interval between the Low Main and Maudlin seams consists mainly of ' grey metal ' and ' blue stone ' with no thick sandstones. In a bore [3092 4707], 650 yd W. of the Adventure Pit, the Durham Low Main Coal is overlain by a thicker, more sandy sequence, comprising about 60 ft of sandstone in five bands interbedded with siltstone with sandy partings, and overlain by about 20 ft of silty and shaly mudstone topped by the Maudlin seatearth-mudstone. The records suggest that as the succession between the Durham Low Main and Maudlin coals is traced southwards through the area separating Leamside and Elemore Colliery, further bands of sandstone appear, and that these thicken and eventually coalesce into the single thick bed farther south. In places, however, bands of mudstone or siltstone survive. Thus, whereas at Broomside Colliery 89 ft 4 in of ' strong jointy white post ' rest on 3 ft 2 in of ' soft blue metal ' forming the roof of the Durham Low Main, Grange New Winning proved the following contrasting sequence:

	ft	in
?MAUDLIN SEAM horizon ...	—	—
Black metal		3
Dark grey metal	15	0
Strong grey metal, with post girdles in lower 9 ft ...	20	0
Strong grey post	7	0
Grey metal	1	3
Brown freestone with metal partings	61	8
Strong grey metal	7	0
Strong grey post	28	1
Strong grey metal, with black stone or metal 9 in at base	6	9
DURHAM LOW MAIN COAL (at 177 ft 1 in)	3	3

The immediate roof of the Durham Low Main Coal consists locally of black mudstone or shale with crushed, usually indeterminate, non-marine lamellibranchs. The shale is in most places only a fraction of an inch thick, but reaches 2½ in at outcrop [2717 3748], just beyond the western margin of the district. During the resurvey, the stratigraphically higher bed beween the coal and the undulating base of the Low Main Post was examined in the workings south of Sherburn Hill Colliery, where it consists of up to 5 ft of grey silty mudstone and siltstone with sandy partings and abundant clay-ironstone nodules. These argillaceous beds seem to be absent in the areas of true washout in the Durham Low Main Coal, where the overlying Low Main Post presumably descends to or below the floor of the seam.

The Durham Low Main Post crops out along the gorge of the River Wear downstream from Frankland Farm. On the left bank [2881 4698], 875 yd W.S.W. of Finchale Priory, the lowest 20 ft of the post, containing thin coal partings near the base, rest directly on the Durham Low Main Coal (p. 68), and are overlain by boulder clay. On the opposite side of the river between about 800 yd W. and 500 yd W.N.W. of the Priory, up to 40 ft of sandstone are exposed. In the western part of the section up to 10 in of Durham Low Main Coal are seen and are overlain by up to 9 ft of conglomeratic sandstone, the lower part of which contains lenticular masses of sandstone. At Cocken Rock [2896 4727], 700 yd W.N.W. of the Priory, at least 6 ft of breccio-conglomerate is exposed and the fragments consist of masses of sandy siltstone or silty mudstone generally 3 or 4 inches in length, but ranging up to 3 ft long by 1 ft thick. These fragments, which are set in a matrix of clean white sandstone, are generally angular or subangular, but mixed with them are many small round lumps of clay-ironstone or 'boxstone', a few inches in diameter. The coarse detritus is collectively reminiscent of the bed lying between the Durham Low Main Coal and Post in Sherburn Hill Colliery (see above): its imbrication suggests transport from the north-west. To the west, thin beds of sandstone come in in the upper part of the breccio-conglomerate which is succeded by an impersistent massive bed of sandstone up to 2½ ft thick. This is overlain by about 40 ft of cross-bedded, fine-grained sandstone which contains thin layers of small nodules of ironstone, many weathered to ' boxstone '.

These exposures lie along one of the north-westerly belts in which the Durham Low Main Coal is thin. Not all of the attenuation, however, has resulted from erosion at the base of the sandstone. At Littletown Lady Alice Pit, within the same belt, for instance, the sandstone, 87 ft thick, is separated from the 16-in coal by 8½ ft of ' grey metal stone with post girdles ' (Pl. IV).

At the southern margin of the Durham Low Main Post, between Hett and Low Butcher Race, the sandstone splits into separate beds which become thinner and eventually wedge out. To the south-west, though the sandstone reaches 30 ft at Chilton Colliery, it is generally thin or absent, and the interval up to the Maudlin Seam contains mostly siltstone and mudstone. In places, a thin impersistent coal is present below the Maudlin (Fig. 12). In this south-western area and in the outlier penetrated by the Embleton Old Hall Bore, non-marine lamellibranchs have been recorded locally in the roof of the Durham Low Main Coal.   E.A.F.

In some places where the Low Main Post is split or absent ' mussel '-bands, thin coals and seatearths divide the sequence, as in Horden No. 20 Bore (Fig. 12). Less commonly, as in Shotton No. 7 Bore (Fig. 12), a ' mussel '-band with *Naiadites productus* and *Spirorbis sp.* locally overlies a thick Low Main Post. In a bore [3120 3597], 1100 yd E. of Tursdale Colliery, *Carbonita sp.* cf. *scalpellus* was recorded in addition from this position.   D.B.S., E.A.F.

The Maudlin Coal is generally thin and unworkable over most of this district, though in the ground to the north covered by the Sunderland (21) Sheet, it is a consistent seam of good quality with an average thickness of over 4 ft. This order of thickness persists across the northern

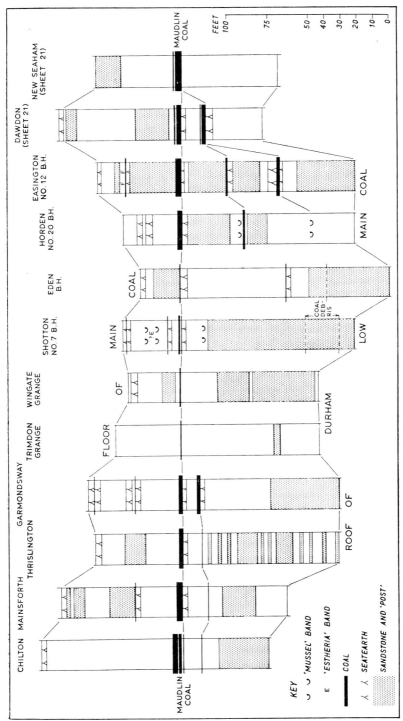

Fig. 12. *Comparative sections of the Maudlin Coal and associated strata, to illustrate variation across the district*

margin of the district and the seam has been worked as far south as Rainton Adventure Pit, Middle Rainton and East Rainton in the north-west, Hetton le Hole and Carr House in the north-centre (where it is locally united with the Durham Low Main Coal, p. 68) and Chourdon Point in the north-east.   D.B.S., E.A.F.

Sections in the Rainton area commonly include a band 1 to 3 ft thick and 12 to 24 in above the base of the seam. The band exceeds $5\frac{1}{2}$ ft just beyond the northern margin of the district at Prior's Close, 1600 yd N.W. of Adventure Pit, and the top coal is heavily banded as follows: coal $2\frac{1}{2}$ in, on band $\frac{1}{4}$ in, coal $4\frac{3}{4}$ in, band $\frac{1}{2}$ in, coal $2\frac{3}{4}$ in, band $\frac{1}{4}$ in, coal 5 in, band $\frac{1}{4}$ in, coal $2\frac{1}{4}$ in, band $5\frac{1}{4}$ in, coal 5 in, band 5 ft 8 in, splint $2\frac{1}{2}$ in, coal 18 in. Traced to the south and south-west of Resolution Pit, the median band thins and eventually disappears, and the whole seam concomitantly becomes too thin to work. Seam thicknesses in this area are as follows: two bores [3133 4676 and 3092 4707], 625 yd W. and 400 yd S.W. of Adventure Pit, $22\frac{1}{2}$ to 24 in, with a median $\frac{1}{2}$-in band; in bores [3023 4745 and 2997 4741], 700 yd E.N.E. and 450 yd N.E. of Finchale Priory, 36 in; at Cocken Pit and in a bore [2937 4745], 425 yd N.W. of Finchale Priory, 30 in. E.A.F.

In broad terms, the deterioration of the Maudlin Coal corresponds to a thickening of the measures between it and the under-lying Durham Low Main Coal (Fig. 12). Thus, between Resolution and Adventure pits, the seam is reduced from 4 to 3 ft corresponding to an increase from 44 to 70 ft of the underlying strata. Similarly the 48 in of ' bad ' Maudlin Coal, 45 ft above the Durham Low Main at Alexandrina Pit, deteriorates further to 6 ft of ' soft black metal stone and coal pipes ' 67 ft above the Durham Low Main at North Hetton No. 2 Upcast Pit, while at the North Hetton Moorsley Winning and Lady Sea-ham Pit, it is absent. It is absent too at Belmont Furnace Pit where its seatearth is separated from the Durham Low Main by 97 ft of ' white post '. Farther south-east, where the ' white post ' is of comparable thickness, the Maudlin is represented by $6\frac{1}{4}$ ft of ' black stone ' at North Pittington

Londonderry Pit, by 3 ft 4 in of ' black stone and sulphur ' at Littletown Engine Pit, and by $3\frac{1}{2}$ ft of ' coal or black stone ' at Littletown Lord Lambton and Lady Alice pits. In places, however, both the Maudlin Coal and Low Main Post are thin as at Trimdon Grange Colliery (Fig. 12), and there is no obvious relationship between the Maudlin coals and Low Main Post in the offshore area. E.A.F., D.B.S.

Southward deterioration of the Maudlin Seam can be traced between Hetton le Hole and Carr House. At Houghton Pit, just beyond the northern margin of the district, the Maudlin is a single coal 61 in thick, but at a locality [3637 4805], 1200 yd N.N.E. of Hetton Colliery, it is repre-sented by coal 25 in, on band 3 in, coal 26 in, band 9 in, coal 6 in. At a locality [3675 4779], 1200 yd N.E. of Hetton Colliery, the section is coal 14 in, on band 5 in, coal 6 in, band 4 in, coal 13 in, and close to the limit of workings [3669 4764], 1000 yd N.E. of the same colliery, it is coal 16 in, on band 9 in, coal 9 in, inferior coal 25 in, band 16 in, coal 18 in. At Hetton Lyons Pit, beyond the limit of workings, two coals survive: they are 14 and 4 in thick, lying 58 and 53 ft respec-tively above the Durham Low Main Seam.

Workings from the Seaham and Dawdon collieries extend southward into the district and in the ground between Hawthorn and Murton collieries the seam is split and only the Top Maudlin is extracted. The split persists into the offshore area, where several bores penetrate a Top Maudlin coal lying up to 50 ft above two relatively persistent coals considered here to be a split Bottom Maudlin. The lower leaf of the Bottom Maudlin is generally 12 to 24 in thick but is as much as 35 inches in Easington No. 20 Bore [4625 4667]: the upper leaf, lying some 10 to 20 ft higher, is commonly 12 to 18 in thick, reaching a maximum of 22 inches in Easington No. 12 Bore (Fig. 12). The Top Maudlin is the most consistent of the three coals, and reaches thicknesses of 48 inches in Horden No. 1 Bore [4657 4206] and 30 inches in Easington No. 12 Bore [4728 4560]. East of Horden and Blackhall Colliery the Top Maudlin deteriorates with the incoming of a median band up to $1\frac{1}{2}$ ft thick.

In the north-central part of the district, between Elemore Hall and Hawthorn, and to the south-east through Shotton Colliery to Oakerside and Blackhall Colliery, the Maudlin coals are absent and correlation between the eastern and western parts of the district is therefore difficult. D.B.S.

In the western part of the district, between Sherburn Hill and Thrislington, there are commonly one or two thin coals 60 to 110 ft above the Durham Low Main Coal. The upper is the more persistent and generally the thicker of the two and is here correlated with the Maudlin Seam of the Rainton area. In the south-western part of the district, beyond Thrislington, this thickens and in a bore [3226 3215], 250 yd N.N.E. of Hope House, it is $5\frac{1}{2}$ ft thick, comprising coal 45 in, on band 5 in, coal 16 in. At 12 ft 8 in below is $1\frac{1}{4}$ ft of black coaly shale, which passes laterally into a thin, relatively persistent coal—the lower of the two coals proved farther north. It is possible that these two coals are the equivalents of the Top and Bottom Maudlin coals of the north-eastern part of the district, but in view of the lack of evidence in the intervening north-central area (see, for instance, Fig. 12) the lower coal in the western area is here regarded as separate and has not been named on the maps.

In a small area north of Mainsforth, the thick Maudlin Coal generally includes more than one band as in a bore [3112 3200], 650 yd N.W. of Mainsforth Hall, which proved coal 28 in, on band 2 in, coal 14 in, band 1 in, coal 4 in, band 1 in, coal 9 in. Farther south-west, at Mainsforth Colliery (Fig. 12), the Maudlin is an unbanded seam $48\frac{1}{2}$ in thick, but still farther south-west as at Chilton Colliery, the seam is again typically banded as follows: coal $26\frac{1}{2}$ in, on band $9\frac{1}{2}$ in, coal $10\frac{1}{2}$ in, band 6 in, coal 8 in. Around Ferryhill, this thick banded seam splits, and in the Dean Pit the section is coal 5 in, on ' seggar-clay ' 5 in, coal 11 in, ' black stone ' 1 ft 8 in, ' grey metal ' 5 ft 3 in, coal $19\frac{1}{2}$ in. The upper split fails nearby and the seatearth is traceable for only a short distance before it too fails. Concurrently, the lower split thins and can be traced into the impoverished Maudlin of the western area. E.A.F.

The strata between the Maudlin Coal and the Main Coal range from about 25 ft in the Eden Bore to about 80 ft at Chilton Colliery (Fig. 12). The sequence can be subdivided into a lower, thinner part topped by an impersistent seatearth and thin coal, and an upper part which includes an important ' mussel '-band either just above the thin coal or below the Main Coal seatearth. The band consists of dark grey or black mudstone or shale containing fish fragments and non-marine lamellibranchs. In some places, as in Offshore Nos. 1 and 2 bores, ' Estheria ' has been found and the band has been named the Blackhall ' Estheria ' Band (Magraw and others 1963, p. 164). The ' Estheria ' are preserved as iridescent films and are associated with fish debris including *Diplodus sp.*, *Megalichthys sp.*, Platysomid scales, *Rhabdoderma sp.*, *Rhadinichthys sp.* and *Rhizodopsis sp.* Other fossils obtained from the band are as follows: megaspores; *Spirorbis sp.*; *Naiadites* cf. *angustus* Trueman and Weir, *N.* cf. *obliquus*, *N. productus*; *Carbonita sp.* cf. *scalpellus*.

The Main Coal (Fig. 13) is thickest in the northern part of the district. It is 6 to 7 ft thick in workings as far south as South Hetton and Low Moorsley and reaches a maximum of 107 in at Rainton Adventure Pit. Between Low Moorsley and Rainton Adventure Pit it contains a band (not plotted in Fig. 13) which is generally only 1 to 2 in thick, though exceptionally as much as 13 in at a locality [3202 4644], $\frac{3}{4}$ mile N. of Belmont Colliery. The band divides the seam into a 15- to 21-in top coal and a 40- to 55-in bottom coal. A thin band lies at a comparable position in the seam east of Murton and Great Coop House, and this thickens south-eastward to more than 3 ft just off shore from Easington Colliery and to 25 ft still farther east in Horden No. 19 Bore [4738 4258], where the Top and Bottom Main coals are 8 and 27 in thick respectively (Pl. IV). The Main Coal has been exhausted in most of this northern part of the district except for a small area around Easington village, where extraction is in progress at the time of writing.

South of a line through South Hetton and Low Moorsley the seam is thinner.

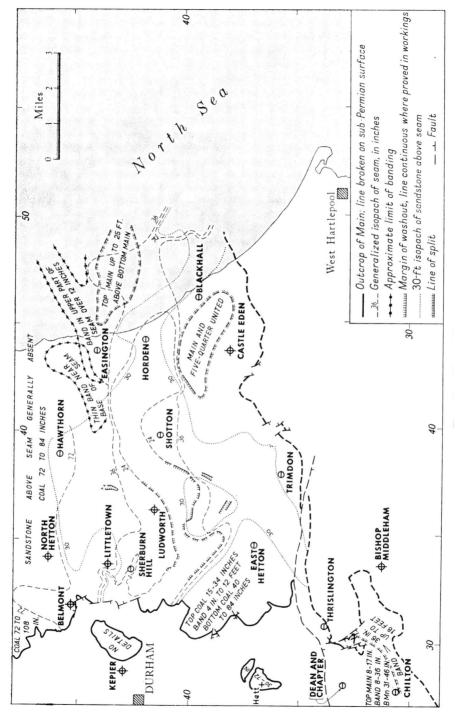

FIG. 13. *Plan of the Main Coal*

Across the central part of the district, west of Horden, is a zone in which it is less than 2 ft thick and locally only 4 in as in a bore [4184 4058], 1000 yd N.E. of Shotton. In the ground to the west of Shotton Colliery it is absent in three bores between Harehill Farm and Low Crow's House and a line of ' washout ' was proved in the workings 2000 yd S.W. of the colliery shafts.   E.A.F., D.B.S.

The seam is absent also farther west, in bores [3450 4019 ; 3400 3966] 1½ and 2 miles S.W. of Ludworth Colliery respectively, where about 70 ft of sandstone were proved above the expected horizon of the coal. A similar absence is recorded in a bore [3361 4063], 1¼ miles S. of Sherburn Hill Colliery. In two other bores [3317 4105 and 3303 4101], about 1 mile S.S.W. of the colliery, the Main Coal is 8 in and 12 in thick respectively, and at Sherburnhouse Colliery it is 15 in. A smaller zone of impoverishment trends in parallel through Sherburn Hill Colliery where coal thicknesses of 6 in and 8 in are proved in the East and West pits respectively. At the limit of workings [3275 4285] on the south side of this zone, the coal is 16 in thick, having decreased from 24 inches only 70 yd to S.S.E.

An outlier of Main Coal has been mapped east of Kepier Pit, in the Carrville–Gilesgate Moor area (Fig. 13) on the basis of rock-head calculations. The area is not known to have been penetrated by bores or shafts and since it is in line with the impoverished belt, the coal, if present at all, is probably thin and close to rockhead.   E.A.F.

The Main Coal thickness also decreases to less than 2 ft in an area east of a line two miles offshore from Dene Mouth and Blackhall Colliery shafts. Two bores

[4981 4104 and 4916 4045], 2¾ miles and 2¼ miles E. of Dene Mouth, record thicknesses of 11 and 16 in respectively, and it may be absent in a third bore [4924 3939] 1½ miles E.N.E. of Blackhall Rocks station. The impoverished area does not extend as far as Offshore No. 1 Bore which proved 34 in of coal at this position. North of the zone of impoverishment, borings and undersea workings in the offshore area have proved a thin band near the base of the lower leaf of the Main Coal (or Bottom Main where the split is more than 3 ft). Below this thin band, the coal is of inferior quality. Between the offshore zone of thin Main Coal and Blackhall Colliery, several borings have penetrated a thin banded coal a short distance below the lower leaf of the seam. It seems possible that this banded coal is so near to the Main Coal as to have been worked with it as a composite seam in parts of Blackhall Colliery workings. Alternatively the banded coal may be the seam underlying the Blackhall ' Estheria ' Band (p. 75).

West of Blackhall Colliery the Main and Five-Quarter coals are so close as to have been worked together from Shotton Colliery in a belt elongated south-eastwards (Figs. 13 and 14). Within this belt the seams are less than 2 ft apart in contrast with over 50 ft in the surrounding area. They are 3 ft apart in Shotton No. 6 Bore [4201 3929], 730 yd S.E. of Shotton Hall, and 7 ft apart at Blackhall Colliery shafts.   D.B.S.

In the remainder of the southern part of the district, south of the central belt of impoverishment, the Main Coal is split into Top Main and Bottom Main. The Bottom Main is generally over 3 ft thick and has been worked under the name ' Main '. Specimen sections are as follows :

		Wheatley Hill	Castle Eden	Trimdon Grange	Mainsforth	West Hetton	East Hetton
		ft in	ft in	ft in	ft in	ft in	ft in
TOP MAIN ...	...	9	1 8	1 3	1 6½	2 4	2 8
Strata	...	3	6	9	4 2	8 0	15 0
BOTTOM MAIN							
Coal	...	⎱				⎛ 1 7	1 3
Band	...	⎰ 3 4	3 3	3 8	3 8	⎨ 5	6
Coal	...	⎰				⎝ 2 8	2 10

The band reaches 24 ft in a bore [3220 3942], 1150 yd N.E. of Cassop Grange, where the Bottom Main comprises coal 3 in, on band 1 in, coal 14 in, band 3 in, coal 33 in, and the Top Main consists of 33 in of coal and bands.

At Coxhoe and Wingate, the Top and Bottom Main coals are united as follows: Wingate 61 in; Coxhoe Bell's Pit, coal 34 in, on band 2 in, coal 48 in; bore [3348 3480], 1075 yd S.S.E. of Coxhoe East House, coal 20 in, on band 15 in, coal 55 in.   E.A.F., D.B.S.

Two small outliers of Main Coal remain at Hett. The southernmost has been explored by opencast drilling which proved a thickness of over 4 ft. North of the railway line, opencast bores have proved pillars of coal, up to 6 ft high, in old workings which were probably entered from a drift [2890 3710] still visible 825 yd N.N.E. of Hett church.
                                                      E.A.F.

Most of the Main Coal has been worked out west of Castle Eden, but to the east, considerable reserves of coal remain, particularly around Hesleden.   D.B.S.

The measures between the Main and Five-Quarter coals range in thickness from less than 2 ft west of Blackhall Colliery to at least 115 ft, including 72 ft of sandstone, in a bore [3361 4063], ¼ mile S.W. of Shadforth. The sandstone exceeds 30 ft in thickness over large areas of the district (Fig. 13), and, where the Five-Quarter Coal is washed out, unites locally with a higher sandstone to form a bed up to 142 ft thick as at South Hetton Colliery Engine Pit (Pl. IV).
                                             E.A.F., D.B.S.

Where sandstone does not form the roof of the Main Coal, argillaceous beds containing rare non-marine lamellibranchs have been recorded. In Easington Colliery No. 12 Bore the coal is overlain by a thin black shale with ' Estheria ', and in other bores nearby, fish debris has been obtained from the same bed. Plant remains are common at this horizon. In Offshore No. 1 Bore, the Main Coal is overlain by 6 ft of mudstone containing many pinnules, mainly of *Neuropteris,* with abundant *Calamites* in the basal 1 ft. Finely comminuted plant debris is also preserved in the overlying 10 ft of mudstone.   D.B.S.

At some places, and especially in the south-western part of the district, a thin coal lies between the Top Main and Five-Quarter coals. It is 8 in thick in East Hetton North Pit, 15 inches in Coxhoe Engine Pit and 21 inches in Mainsforth Colliery (Pl. IV).   E.A.F.

The Five-Quarter Seam (Fig. 14) is 2½ to 6 ft thick and has been worked widely in the district. It is normally unbanded, but impersistent bands are recorded locally and one of these thickens to form a split of as much as 21 ft in Shotton Colliery No. 10 Bore [4067 4125], 1½ miles N.N.W. of Shotton.

In a number of bores at Blackhall and Horden collieries, the seam is composed wholly or partly of thin alternations of coal and carbonaceous shale. Farther west, the Five-Quarter Coal is recorded as being just over 6 ft in bores in the Garmondsway area, near to its outcrop beneath the Permian, but it lies in a narrow zone of steeply dipping strata on the south side of the Butterknowle Fault. The coal is absent between Middle Rainton, Low Moorsley, Easington Lane, South Hetton and Shotton Colliery, where a thick sandstone occupies this part of the sequence (see above).

The measures between the Five-Quarter and Metal coals range in thickness from a few inches to over 80 ft. In the Littletown area, the thickness variation is as follows:

	Sherburn Hill Shafts	Littletown Farm No. 3 Bore [3272 4283]	Littletown Lady Alice Pit	North Pittington Londonderry Pit	Pittington Buddle Pit
METAL SEAM ...	5–6 in	76 in	24 in	61 in (banded)	55 in (banded) 55 in
Band	29–34 in	32 in	18 ft 4 in	21 ft 2½ in	35 ft 6 in
FIVE-QUARTER SEAM	44–51 in	46 in	40 in	45½ in	45 in

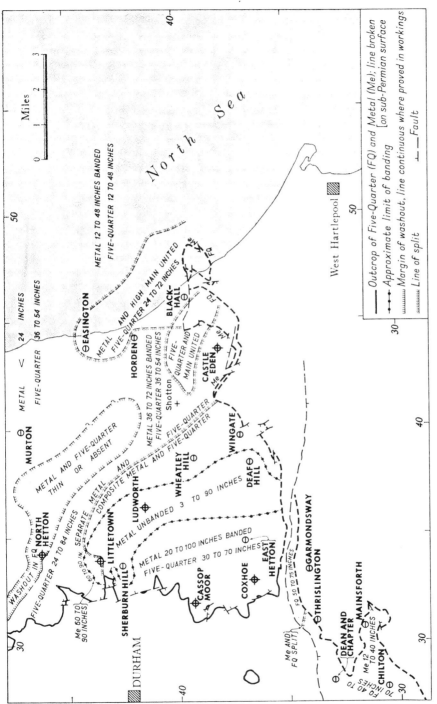

Fig. 14. *Plan of the Five-Quarter and Metal coals*

Where the interval is large, fossils have been obtained from the roof of the Five-Quarter Coal: they include *Naiadites* aff. *alatus* from Offshore No. 1 Bore and plants including *Alethopteris sp.* from Offshore No. 2 Bore. A specimen of *Anthrapalaemon,* associated with megaspores, was obtained from 20 ft above the Five-Quarter Coal in Easington No. 7 Bore. Where the interval is small, as it is in much of the west-central part of the district, the two coals unite to form a composite seam exemplified in a bore [3350 3480], just over ¼ mile N. of Garmondsway Pit, where the **Metal Seam** comprises coal 32 in, on band 28 in, coal 9 in, band 13 in, coal 1 in, band 6 in, coal 15 in, band 3 in, coal 3 in, band 6 in, coal 19 in, separated by a 7-in band from the 77-in Five-Quarter Seam (corrected for high dip to 74 in). In a zone stretching from Littletown to Deaf Hill collieries through Ludworth and Thornley collieries (Fig. 14), the records, which are few and old, show the Metal component of the composite seam to be unbanded and 3 to 7 ft thick, but banding is again recorded farther east at Wheatley Hill and Wingate collieries. Still farther east, where the Metal Seam is separated from the Five-Quarter, banding persists and the seam, up to 6 ft thick, has been worked only in the north-eastern part of Shotton Colliery take. Here, at least, two bands are present, but farther south the number increases, and in Shotton Colliery No. 5 Bore [4149 4028], 600 yd N. of Shotton Hall, seven bands totalling 25 in were proved in a seam section of 68 in. Eastwards from Shotton and Castle Eden the coal seems to thin rapidly, and in the coastal and undersea areas it is generally thin. Exceptionally, as in Offshore No. 1 Bore, it reaches 42 inches in thickness. In the Buddle Pit, the Metal Seam lies approximately halfway between the Five-Quarter and High Main coals.

In a zone between Middle Rainton and Shotton Colliery, the Metal, like the Five-Quarter, is thin or absent (pp. 78 and Fig. 14).

The measures between the Metal and High Main coals are up to 100 ft thick and consist mainly of sandstone in the western part of the district, but in the east, between Easington and Blackhall, the two coals are close together and form a composite seam.   E.A.F., D.B.S.

In Easington No. 7 Bore, the roof measures of the Metal Coal yielded poorly preserved *Anthracosia spp., Naiadites* aff. *alatus* and *N.* cf. *productus* [juv.] ; and at 30 ft higher, a specimen of *Anthracosia* cf. *aquilinoides.* Another ' mussel '-band is locally recorded close below the High Main Seam. In Shotton Colliery No. 6 Bore it contained: *Spirorbis sp.* ; *Anthraconaia sp. nov.* cf. *wardi, Anthracosia sp.* cf. *caledonica, Anthracosphaerium propinquum, Naiadites* cf. *productus* ; *Carbonita humilis* ; and fish remains including *Rhabdoderma sp.*

The High Main Coal (Fig. 15) the uppermost widely worked coal of the Durham Coalfield, is over 5 ft thick in the northern and eastern parts of the district, but it is generally banded and has been mined only from the collieries near the northern margin of the district and on a small scale from Blackhall and Horden collieries. The seam is more than 7 ft thick over an area of several square miles to the north of Haswell and Easington collieries, and is over 8 ft thick at some localities around Rainton and Low Moorsley. In many areas, two or more bands are present of which the highest and most persistent is commonly three to eight inches thick, locally increasing to 18 in around Easington and Easington Colliery. The top leaf of the seam, above this band, is commonly about 10 in thick, increasing to 18 in at Elemore and North Hetton collieries. Other, lower bands in the seam are thin, and are only persistent in the area south of South Hetton and Hawthorn shafts, where they are two or three in number. In the coastal region between Easington and Blackhall, the High Main is joined by the Metal Seam, but the band separating the two cannot always be identified and the High Main, as worked in this area, may locally include part of the Metal Seam. The margin of the united seam (Figs. 14, 15) must therefore be regarded as approximate and it is possible that the area extends considerably farther north-westwards.   D.B.S., E.A.F.

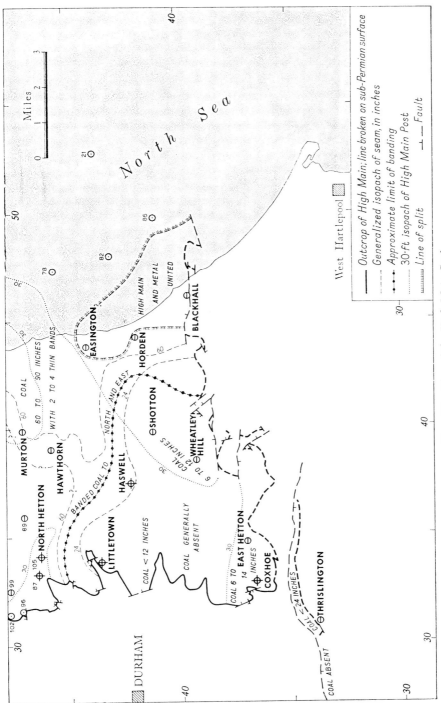

FIG. 15. *Plan of the High Main Coal*

Where present in the south-western part of the district, the seam is unbanded and thinner. It reaches a thickness of only 20 inches in Coxhoe Bridge No. 1 Bore. In the central part of the district, it is impoverished over roughly the same area as that in which the Metal and Five-Quarter coals are close together. Here the High Main Coal is generally less than 18 in thick and is absent in places where the overlying High Main Post joins the sandstones overlying the Five-Quarter and Metal coals (see below).   E.A.F., D.B.S.

The measures between the High Main Coal and the High Main Marine Band range in thickness from 41 ft in Shotton Colliery No. 6 Bore to 102 ft in Murton Shaft (Pl. IV). There is no clearly defined pattern of variation in thickness, but the interval tends to be least in the central part of the district, corresponding with the area in which the High Main Coal is thin or absent, and greatest north of a line joining East Rainton, Hetton le Hole, South-Hetton and New Hesledon.

The roof of the High Main Coal is generally a dark mudstone or shale, from which a sparse fauna of fish fragments and non-marine lamellibranchs has been collected. The following were obtained from Offshore No. 2 Bore: *Anthraconaia?*, *Anthracosia spp.* including *A.* cf. *elegans* (Van der Heide), *Naiadites spp.* In the extreme north-western part of the district, beyond Low Moorsley, Rainton Alexandrina Pit and Rainton Meadows Pit, the succession above the High Main Coal remains dominantly argillaceous, and this is also true of the eastern area beyond Shotton and Murton collieries where the sequence up to the High Main Marine Band consists of siltstone and mudstone with abundant plant debris, and subordinate sandstone. In the intervening ground the siltstone and mudstone gives way to a thick sandstone, the **High Main Post,** which reaches a maximum thickness of over 80 ft at Hawthorn and Hetton Lyons collieries. However, it is only 31 ft at Elemore Colliery, and 27 and 22 ft respectively in the two shafts of North Hetton Colliery.   D.B.S., E.A.F.

At Pittington, the High Main Post is 55 ft thick, and a 40-ft face of cross-bedded sandstone is exposed in a disused quarry [3265 4430], 800 yd S.S.E. of Pittington Station. Farther south, in the Littletown and Sherburn Hill Colliery areas where the High Main Coal is absent, the High Main Post loses its separate identity and becomes vertically continuous with the sandstone overlying the Metal and Five-Quarter coals. South of Sherburn Hill, between Sherburnhouse Colliery and Shadforth, about 70 ft of High Main Post overlies about 4 ft of argillaceous beds forming the roof of a very thin High Main Coal. Between Shadforth and Heugh Hall, however, where the High Main Coal is again absent, a thick sandstone separates the Metal Coal from the horizon of the High Main Marine Band, and this sandstone reaches 105 ft in a bore [3401 3966], 350 yd N.E. of Old Cassop. The High Main Post is absent in bores southwest of Old Quarrington, and in one of these [3185 3747], $\frac{5}{8}$ mile N.W. of Coxhoe Colliery, the High Main Coal is separated from the High Main Marine Band by only 48 ft of strata, which include two thin coals, the uppermost being 14 in thick. At Coxhoe Colliery Engine Pit, the correlative of this 14-in seam is 38 in thick with a median 2-in band, but there is no record of comparable thickness elsewhere at this horizon. In bores near Coxhoe Bridge there is a thin sandstone above the High Main Coal, and the interval between this coal and the High Main Marine Band in Coxhoe Bridge No. 1 Bore is nearly 116 ft, though this figure is not reliable because of faulting and dips up to about 30°. The thin sandstone is overlain by siltstone and seatearth, followed in turn by mudstone which comprises a well-developed 'mussel'-band lying at about 43 ft above the High Main Coal. The fauna is represented by: *Spirorbis sp.*; *Anthracosia acutella, A. atra, A.* cf. *aquilina* Trueman and Weir (*non* J. de C. Sowerby), *A.* cf. *caledonica, A. concinna, Anthracosphaerium radiatum*; *Carbonita sp.*; and indeterminate Platysomid scales. At Thrislington Coke Oven Pit [3043 3405], 400 yd S.S.W. of Cornforth Station, the interval is 89 ft, and no High Main Post is recognisable from the record (Pl. III). However, sandstone certainly comes in to the west, for it has been observed during the resurvey in and near the lane 400 yd W. of

Ferryhill Reservoir. Red micaceous cross-bedded sandstone seems to occupy the greater part of the interval between the Five-Quarter and Ryhope Little coals at this locality, but grey mudstone is present in the ditch by the reservoir approach road less than a quarter of a mile to the east, indicating that the thick sandstone does not continue in this direction. In the area between the reservoir and Thrislington Coke Oven Pit, the Coal Measures outcrop is covered by Permian rocks and drift.

The High Main Post and its lateral equivalents are generally succeeded by a seatearth and by a coal which is usually thin, but exceptionally is 22 in thick at Elemore Colliery Isabella Pit. The overlying **High Main Marine Band** consists of a few inches of black shale containing small specimens of *Lingula mytilloides*. In the eastern part of the district the band is commonly separated from the roof of the coal by up to $6\frac{1}{2}$ ft of mudstone and shale, in which fish fragments have been found near the base, while *Lioestheria vinti,* in places abundant, is distributed throughout, together with sporadic non-marine lamellibranchs. To the west the High Main Marine Band has been found in Heugh Hall No. 1 Bore where 22 ft of black shale and mudstones containing *Lingula* in the bottom 1 ft rest directly on grey sandy seatearth, and in Coxhoe Bridge No. 1 Bore. The marine band is absent in some areas and there are two explanations of this: (i) penecontemporaneous erosion, now indicated by records and exposures showing this part of the succession to be occupied by sandstone, or (ii) non-deposition, for in several recent bores no marine fauna has been identified in spite of careful search. In the second case the position of the marine band can usually be recognized by the associated non-marine fauna which includes: *Spirorbis sp.*; *Anthraconaia spp.* including *A.* aff *librata* and *A. sp. nov.* cf. *wardi* (Hind *pars*), *Anthracosia* cf. *atra, A.* sp. cf. *atra, A. sp.* cf. *fulva* (Davies and Trueman), cf. *A. planitumida* (Trueman), *A.* cf. *simulans* Trueman and Weir, cf. *Anthracosphaerium radiatum* (Wright), *Curvirimula sp., Naiadites alatus, N.* cf. *obliquus*; *N.* cf. *productus, Carbonita sp.* including *C. humilis* and

C. cf. *scalpellus, Geisina sp.*; *Elonichthys sp.,* Palaeoniscid scales, Platysomid scales, *Rhabdoderma sp, Rhadinichthys sp., Rhizodopsis sp., Stemmatias sp.* [tooth].

The measures between the High Main Marine Band and the Ryhope Little Coal, in all but the central part of the district between Coxhoe Bridge, Thrislington and Old Quarrington, generally include two thin coals, each overlain in some areas by shale or mudstone, containing fish debris and non-marine lamellibranchs at base.   D.B.S., E.A.F.

At Cassop 'A' Pit the lower of these two seams is 22 in thick, including a 2-in band 2 in above the base; the upper consists of 21 in of unbanded coal. At North Pittington Colliery Londonderry Pit these two coals are 12 and 20 in thick, and at North Hetton Moorsley Winning they are 15 and 18 in respectively. A surface drift [3246 3860], 450 yd E.N.E. of Hill Top Farm (Cassop cum Quarrington), for which no records have been traced, is thought to have been put into the uppermost of these two seams, here overlain by reddened sandstone which has been worked in a small quarry 175 yd to the west. This sandstone, though virtually absent at Cassop 'A' Pit, thickens westwards and north-westwards along the sides of Cassop Vale, and is exposed in quarries at the tips of the spurs to the north and south of the Vale. At the western end of the northern spur the sandstone is about 50 ft thick, and is seen to be massive and cross-bedded with small quartz pebbles in a quarry [3294 3933], 600 yd W.S.W. of Old Cassop. This quarry provided stone for St. Paul's Church, Quarrington Hill. Near Hill Top Farm, around the western end of the southern spur, about 20 ft of cross-bedded sandstone can be seen, but the original thickness is unknown as a result of pre-Upper Permian erosion. On the southern flank of this spur the sandstone seems to wedge out eastwards.   E.A.F.

In Coxhoe Bridge No. 1 Bore the strata above the High Main Marine Band contain a coal 24 in thick. The supposed correlative of this coal at Thrislington Coke Oven Pit is 26 in thick. At Hawthorn Colliery (Fig. 16) the whole

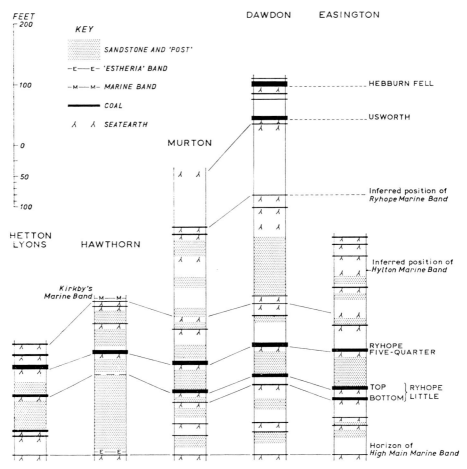

FIG. 16. *Comparative sections of the Middle Coal Measures, above the High Main Marine Band*

of the interval between the High Main Marine Band and the Ryhope Little Coal is filled by sandstone.   E.A.F., D.B.S.

The Ryhope Little Coal lies near the top of the sequence of productive coals of the Durham Coalfield and is worked at Ryhope, 4 miles N. of Murton, as a single seam about 3 ft thick. It is not, however, worked anywhere in the Durham–West Hartlepool district. East of a line joining Murton and Shotton collieries, two coals 10 to 20 ft apart are present in this part of the sequence over wide areas (Fig. 16). Although it is possible that the lower seam is the correlative of a thin coal lying 18½ ft below the Ryhope Little Coal at Ryhope,

there is some evidence that the two coals are splits from the worked seam and they are therefore shown on the maps as the Top and Bottom Ryhope Little coals. The Top Ryhope Little Coal is a consistent seam, 1½ to 2½ ft thick, generally free from bands, and traceable over much of the eastern part of the district. The Bottom Ryhope Little Coal is present in the area east of the line joining Murton and Shotton collieries, and it is commonly about 1 ft thick, though locally it reaches 2 ft. The interval between these two coals is generally occupied by argillaceous strata, and in some places fish debris has been recorded from the roof of the lower coal.   D.B.S.

At North Hetton No. 2 Shaft, the Ryhope Little Coal consists of 3½ ft of coal with bands. There are no records of this seam in the area around East Rainton, about 1 mile to the north, but it is likely to be present, although probably thin or banded. About a mile south-east of Low Moorsley, however, at Elemore George Pit, the Ryhope Little Seam consists of three leaves: the Top Ryhope Little, a 28-in coal, is separated by a band, 3 ft 9 in thick, from the Bottom Ryhope Little comprising coal 21 in, on band 13 in, coal 23½ in. A line of split thus lies between Low Moorsley and Elemore and this is thought to continue north-eastwards to join that assumed between Murton and Ryhope. The southward trend of the line of split is obscure due to the paucity of records and difficulty of interpretation. A bore [3293 4175] in Sherburn Quarry, 1200 yd S.E. of Sherburn crossroads, records a section thought to represent the Ryhope Little Coal: coal 4 in on 3 ft 8 in of seatearth, on coal 2 in. To the south, a 15-in coal at Cassop 'A' Pit has been correlated with this seam. To the west of the district, around Westerton, a 2½- to 3-ft coal now identified as the Ryhope Little Seam has been proved, and on the assumption that this coal extends eastwards, the outcrop has been inserted on the maps north-west of Ferryhill. The line at which impoverishment takes place has not however, been established.

The measures between the Ryhope Little and Ryhope Five-Quarter coals range in thickness from about 35 ft to nearly 70 ft. To the west, at Westerton South Pit, the interval is near the minimum of this range and about half, i.e. 18 ft, consists of 'post', separated from the underlying coal by 2 ft 9 in of 'blue metal'. In the field immediately east of the northern end of the lane between Ferryhill and East Howle, red micaceous sand in the soil is thought to mark the outcrop of the 'post' above the Ryhope Little Coal. In Coxhoe Bridge No. 1 Bore, a seatearth close beneath the base of the Permian strata may belong to the bed underlying the Ryhope Little Coal. Over most of the area lying to the north of the Butterknowle Fault, the strata

between the Ryhope Little and Ryhope Five-Quarter coals are composed consistently of white sandstone, though a variable thickness of argillaceous beds lies above this sandstone in some localities. The interval is generally between about 45 and 60 ft. In most places the sandstone lies directly on the roof of the Top Ryhope Little Coal which is nevertheless rarely affected by washouts. At some localities, a thin shale lies in the roof of the coal and presumably is equivalent to the Little Marine Band or its associated non-marine phase. Near the northern edge of the district, the sandstone was formerly worked in a small quarry [3359 4738], 450 yd S. of East Rainton church, but neither sandstone nor associated strata are now exposed. In this area the sandstone is less constant than in most of the central and eastern parts of the district and is missing from the 30 ft of strata proved above the Ryhope Little Coal at North Hetton No. 2 Shaft, three-quarters of a mile S.S.E. of the quarry near East Rainton. At Elemore George Pit the sandstone is separated from the coal by 16 ft of 'blue metal with ironstone balls'. E.A.F.

The Ryhope Five-Quarter Coal, the highest worked seam in the district, is worked only at Murton Colliery (Fig. 16) near the northern margin, where it is an unbanded coal of good quality between 2½ and 3 ft in thickness. The coal is present in much of the north-eastern sector of the district and is commonly 1½ to 3 ft thick with only slight lateral variations in thickness. The outcrop of the seam has been inferred on the maps of the spur and vale country between Low Moorsley and Quarrington Hill, but no details of its thickness are available. There is a similar lack of information in the narrow zone of steeply dipping strata along the southern side of the Butterknowle Fault, and as no sign of coal was seen at the appropriate horizon during resurvey north of Ferryhill, the outcrop has been omitted from the maps in that area. In the South Pit at Westerton, the Ryhope Five-Quarter Seam, known as the Westerton High Main, has the section coal 22 in, on band 4 in, coal 33 in, and this seam may extend eastward for a short distance crossing the western boundary of the present district.

D.B.S., E.A.F.

The Ryhope Five-Quarter Coal is generally overlain by mudstones. Above the mudstones is a sandstone, up to 28 ft thick, followed by a succession including two or three thin coals, the highest of which underlies **Kirkby's Marine Band.** This band has the characteristic inter-leaved structure found in sections examined in the Sunderland (21) district to the north where non-marine fossils commonly occur between strata containing marine fossils. In a trial borehole at the site of Hawthorn Colliery Shaft, the section is as follows:

	ft	in
Grey silty shale with extensive red staining ; scattered plant fragments	9	0
Dark grey mudstone ...	5	0
Dark grey shaly mudstone with abundant foraminifera, *Planolites ophthalmoides,* and pyrite-filled worm burrows ; scattered small *Lingula* and fish debris near base ...	2	9
Dark grey mudstone with poorly preserved non-marine lamellibranchs, foraminifera, and with pyrite-filled worm burrows in lower part	1	9
Dark grey micaceous shale with infrequent fish scales and very small *Lingula* ...		6
Coal		2

Fossils were identified from this sequence in the borehole and shaft as follows: plant fragments ; foraminifera [abundant] including *Ammonema sp.* and *Tolypammina sp.* ; sponge spicules [pyritized] ; *Planolites ophthalmoides* ; *Lingula* cf. *elongata* [3·0 mm], *L. mytilloides* [5·0 mm] ; *Anthracosia* cf. *atra, Curvirimula*

[rare] ; *Lioestheria sp.* ; fish remains including *Diplodus* tooth ; pyritized strap-like markings (' fucoids '). Kirkby's Marine band has not been identified at other localities, but its position can generally be recognized in older records by the presence of up to 30 ft of argillaceous strata. The overlying sequence includes two more thin coals, the upper being overlain by dark grey mudstone and shale, up to 10 ft thick, marking the position of the **Hylton Marine Band.** This band is known from several recent bores in northern Durham and south Northumberland, but no recent opportunity of examining the horizon in the present district has arisen. The mudstones and shale are generally overlain by a thick sandstone, which reaches a maximum thickness of 100 ft 40 yd N. of the district at Dawdon Shaft (Fig. 16) in the Sunderland district, but thins rapidly southwards. Above it are two thin coals, succeeded by mudstone believed to contain the Ryhope Marine Band, which also has not been proved in this district. The mudstone extends up to a thin coal lying about 10 ft below the Usworth Coal, and this sequence reaches over 120 ft in thickness at Dawdon Colliery (Fig. 16).

The Usworth Coal (Fig. 16) is restricted to the area immediately south of Dawdon Colliery, where the shaft section is 59 in with a median 4-in band. The coal is present offshore, but cannot be regarded as workable anywhere within this district because of its limited extent and thin cover to the base of the Permian.

The Hebburn Fell Coal is even less extensive than the Usworth Coal, and inland it never lies more than 30 ft below the base of the Permian. Although this figure is exceeded offshore, the coal is thought to be too close to the unconformity to be workable. At Dawdon Colliery (Fig. 16), it lies about 48 ft above the Usworth Coal and is 52½ in thick.

D.B.S.

REFERENCES

ANDERSON, W., and DUNHAM, K. C. 1953. Reddened beds in the Coal Measures beneath the Permian of Durham and South Northumberland. *Proc. Yorks. Geol. Soc.,* **29,** 21–32.

ARMSTRONG, G., and PRICE, R. H. 1954. The Coal Measures of North-East Durham. *Trans. Inst. Min. Eng.,* **113,** 973–97. Also discussions in: 1955, *Trans. Inst. Min. Eng.,* **114,** 83–6 and 111–14.

BORINGS AND SINKINGS. 1878–1910. An account of the strata of Northumberland and Durham as proved by borings and sinkings. *North of England Inst. Mining and Mech. Eng.*

CONYBEARE, W. D., and PHILLIPS, W. 1822. *Outlines of the geology of England and Wales.* London.

DAGLISH, J., and FORSTER, G. B. 1864. On the Magnesian Limestone of Durham. *Trans. N. of England Inst. Min. Eng.,* **13,** 205–13.

DEFRISE-GUSSENHOVEN, E., and PASTIELS, A. 1957. Contribution à l'étude biométrique des Lioestheriidae du Westphalien supérieur. *Publ. Assoc. Étud. Paléont. Strat. Houill.,* **31,** 1–71.

DUFF, P. McL. D., and WALTON, E. K. 1962. Statistical basis for cyclothems: a quantitative study of the sedimentary succession in the East Pennine Coalfield. *Sedimentology,* **1,** 235–55.

DUNHAM, K. C. 1948. Geology of the Northern Pennine Orefield. Vol. 1. Tyne to Stainmore. *Mem. Geol. Surv.*

—— and HOPKINS, W. 1958. Geology around the University Towns: Durham area. *Geol. Assoc. Guide,* **15.**

EAGAR, R. M. C. 1956. Additions to the non-marine fauna of the Lower Coal Measures of the North-Midlands Coalfields. *Liv. and Manch. Geol. J.,* **1,** 328–69.

—— 1960. A summary of the results of recent work on the palaeoecology of Carboniferous non-marine lamellibranchs. *C.R. 4me Cong. Strat. Carb. Heerlen,* 1958, **1,** 137–49.

—— 1962. New Upper Carboniferous non-marine lamellibranchs. *Palaeontology,* **5,** 307–39.

EARP, J. R., MAGRAW, D., POOLE, E. G., LAND, D. H., and WHITEMAN, A. J. 1961. Geology of the country around Clitheroe and Nelson. *Mem. Geol. Surv.*

FOWLER, A. 1936. The geology of the country around Rothbury, Amble and Ashington. *Mem. Geol. Surv.*

HICKLING, H. G. A. 1949. The prospects of undersea coalfield extension in the North-East. *Trans. Inst. Min. Eng.,* **109,** 659–74.

HIND, W. 1895. A monograph on *Carbonicola, Anthracomya* and *Naiadites.* Pt. 2, 81–170. *Palaeontogr. Soc.*

HINDSON, G., and HOPKINS, W. 1947. The relationship of the Coal Measures to the 'Wash' between Shincliffe Bridge and Harbourhouse Park, Co. Durham. *Trans. Inst. Mining Eng.,* **107,** 105–17.

HOLMES, A. 1928. The foundations of Durham Castle and the geology of the Wear Gorge. *The Durham University Journal,* March issue, 1–8.

HOPKINS, W. 1928. The distribution of the mussel bands of the Northumberland and Durham Coalfield. *Proc. Univ. Durham Phil. Soc.,* **8,** 1–14.

—— 1929. The distribution and sequence of the non-marine lamellibranchs in the Coal Measures of Northumberland and Durham. *Trans. Inst. Min. Eng.,* **78,** 126–44.

—— 1930. A revision of the Upper Carboniferous non-marine lamellibranchs of Northumberland and Durham and a record of their sequence. *Trans. Inst. Min. Eng.,* **80,** 101–10 ; and discussion, 254–57.

—— 1933. Impoverished areas of the Maudlin and Busty seams in the Durham Coalfield. *Trans. Inst. Min. Eng.,* **85,** 212–22.

—— 1934. *Lingula* horizons in the Coal Measures of Northumberland and Durham. *Geol. Mag.,* **71,** 183–89.

—— 1935. A record of the Upper Carboniferous non-marine lamellibranchs from the Bates Sinking, Blyth, Northumberland. *Trans. Inst. Min. Eng.,* **89,** 48–56.

—— 1954. The coalfields of Northumberland and Durham, *in* Trueman, A. E. *The coalfields of Great Britain.* London.

HOWSE, R. 1857. Notes on the Permian system of Northumberland and Durham. *Ann. Mag. Nat. Hist.,* (2) **19,** 33–52, 304–12, 463–73.

JOHNSON, G. A. L., HODGE, B. L., and FAIRBAIRN, R. A. 1962. The base of the Namurian and of the Millstone Grit in north-eastern England. *Proc. Yorks. Geol. Soc.,* **33,** 341–62.

KELLETT, J. G. 1927. A petrological investigation of the Coal Measures sediments of Durham. *Proc. Univ. Durham Phil. Soc.,* **7,** 208–32.

LUMSDEN, G. I., and CALVER, M. A. 1958. The stratigraphy and palaeontology of the Coal Measures of the Douglas Coalfield, Lanarkshire. *Bull. Geol. Surv, Gt. Brit.,* No. 15, 32–70.

MAGRAW, D. 1960. Coal Measures proved underground in cross measures tunnels at Bradford Colliery, Manchester. Palaeontology by M. A. Calver. *Trans. Inst. Min. Eng.,* **119,** 475–92.

—— 1961. Exploratory boreholes in the central part of the south Lancashire coalfield. *Min. Eng.,* **6,** 432–48.

——, CLARKE, A. M., and SMITH, D. B. 1963. The stratigraphy and structure of part of the south-east Durham coalfield. *Proc. Yorks. Geol. Soc.,* **34,** 153–208.

MILLS, D. A. C., and HULL, J. H. In press. The Geological Survey Borehole at High Kays Lea Farm, Woodland, Co. Durham (1962). *Bull. Geol. Surv. Gt. Brit.*

MYKURA, W. 1967. The Upper Carboniferous rocks of south-west Ayrshire. Palaeontology by M. A. Calver and R. B. Wilson. *Bull. Geol. Surv. Gt. Brit.,* No. 26, 23–98.

SEDGWICK, A. 1829. On the geological relations and internal structure of the Magnesian Limestone, and the lower portions of the New Red Sandstone Series in their range through Nottinghamshire, Derbyshire, Yorkshire and Durham, to the southern extremity of Northumberland. *Trans. Geol. Soc.* (2), **3,** 37–124.

SMITH, E. G., RHYS, G. H., and EDEN, R. A. 1967. The geology of the country between Chesterfield, Matlock and Mansfield. *Mem. Geol. Surv.*

STUBBLEFIELD, C. J., and TROTTER, F. M. 1957. Divisions of the Coal Measures on Geological Survey maps of England and Wales. *Bull. Geol. Surv. Gt. Brit.,* No. 13, 1–5.

TAYLOR, B. J. 1961. The stratigraphy of exploratory boreholes in the West Cumberland Coalfield. Palaeontology by M. A. Calver. *Bull. Geol. Surv. Gt. Brit.,* No. 17, 1–74.

TRECHMANN, C. T. 1941. Borings in the Permian and Coal Measures around Hartlepool. *Proc. Yorks. Geol. Soc.,* **24,** 313–27.

TRUEMAN, A. E. 1946. Stratigraphical problems in the Coal Measures of Europe and North America. *Quart. J. Geol. Soc.,* **102,** xlix–xciii.

—— and WEIR, J. 1946: 1947: 1955. A monograph of British Carboniferous non-marine Lamellibranchia. Pt. 1, 1–8; Pt. 2, 19–44; Pt. 8, 207–42. *Palaeontogr. Soc.*

WEIR, J. 1945. A review of recent work on the Permian non-marine lamelli-branchs and its bearing on the affinities of certain non-marine genera of the Upper Palaeozoic. *Trans. Geol. Soc. Glasgow,* **20,** 291–340.

—— 1960. A monograph of British Carboniferous non-marine Lamellibranchia. Pt. 10, 273–320. *Palaeontogr. Soc.*

WELLS, A. J. 1960. Cyclic sedimentation: a review. *Geol. Mag.,* **97,** 389–403.

WRIGHT, W. B. 1938. The Anthracomyas of the Lancashire Coal Measures and the correlation of the latter with the Coal Measures of Scotland. *Sum. Prog. Geol. Surv.* for 1936, pt. 2, 10–26.

Chapter III

PERMIAN AND TRIAS

INTRODUCTION

CLASSIFICATION AND DISTRIBUTION

SUBDIVISION of the Permian and Triassic rocks was first attempted by Sedgwick (1829) and later successively refined by Howse (1848, 1857), King (1850), Kirkby (1860) and Woolacott (1912). The classification adopted here (Fig. 17) is a modified version of that proposed by Woolacott, in which the base of the Middle Magnesian Limestone was defined as coincident with the base of the reef.[1] In Woolacott's classification, the base of the Middle division could not be recognized directly where reef is absent, and to overcome this difficulty Smith (*in* Magraw and others 1963, p. 172) has redefined the base of the division at the base of a loosely defined group of transitional beds which underlie the reef and can also be recognized in lagoonal and basin areas.

The greatest recorded thickness of Permo-Triassic rocks at any locality in the district is 1,382 ft in the Seaton Carew Bore (Bird 1888, p. 570), near West Hartlepool, but each of the lithological divisions shows marked lateral variation with a general tendency for divisions above the Lower Magnesian Limestone to thicken eastwards. The maximum recorded thickness of Magnesian Limestone is 991 ft in N.C.B. Offshore No. 2 Bore (Magraw and others 1963, pp. 202–3), six miles north of Hartlepool. Over most of the district, the average dip at the base of the Permian rocks is about 125 ft per mile (Pl. XV), but this tilt is partly tectonic and partly primary and higher beds have a somewhat lower average dip.

The distribution of the Permian and Triassic rocks in the district is depicted on Pl. V which shows that no strata higher than the Upper Magnesian Limestone crop out on land north of the West Hartlepool Fault. However, their former presence is indicated by fragments of red and green mudstones of

---

[1] In this chapter the term 'reef' is applied to carbonate rocks which were built up directly or indirectly by sessile living organisms and which formed contemporaneously lithified masses rising from the floor to near the surface of the sea. In addition to autochthonous rock formed directly by lime-fixing organisms such as calcareous algae (reef *sensu stricto*), the term is used here to include other closely related rocks formed from debris of earlier autochthonous rock and the remains of benthonic organisms, which together form perhaps 70 per cent of the complex and are held together by an assortment of organic frame-builders and sediment-binders. The use of the term 'reef' to include stromatolitic rocks formed on the reef-flat is considered justified because, although such rocks are considered by some as organosediments, they are here so intimately associated with other parts of the organic complex that to separate them from it would be impracticable.

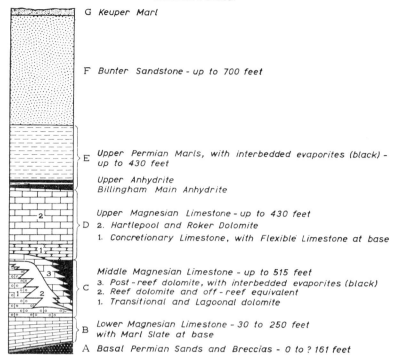

G *Keuper Marl*

F *Bunter Sandstone - up to 700 feet*

E *Upper Permian Marls, with interbedded evaporites (black) - up to 430 feet*

Upper Anhydrite
Billingham Main Anhydrite

Upper Magnesian Limestone - up to 430 feet
D *2. Hartlepool and Roker Dolomite*
1. Concretionary Limestone, with Flexible Limestone at base

Middle Magnesian Limestone - up to 515 feet
C *3. Post - reef dolomite, with interbedded evaporites (black)*
2. Reef dolomite and off - reef equivalent
1. Transitional and Lagoonal dolomite

B *Lower Magnesian Limestone - 30 to 250 feet with Marl Slate at base*

A *Basal Permian Sands and Breccias - 0 to ? 161 feet*

FIG. 17. *Generalized section of the Permo-Triassic Rocks of the Durham–West Hartlepool district*

Upper Permian Marl lithology in many collapse breccias in the Middle and Upper Magnesian Limestone in coastal sections, and the succession is probably complete a few miles out to sea. The Permo-Triassic rocks are largely overlain by drift, especially in the south-eastern part of the district where information is derived mainly from boreholes. Elsewhere most exposures are in narrow belts along the western escarpment, along the coast, and in a number of narrow ravines in the eastern part of the district. Where not overlain by drift the limestones and dolomites give rise to an undulate topography with thin light soils utilized mainly for pasture. Even where buried by drift the underlying rock in many areas exercises a broad topographic control, as for instance, in the case of the drift-covered ridge by which the course of the reef of the Middle Magnesian Limestone is traced between exposures.

CONDITIONS AND HISTORY OF DEPOSITION

Immediately prior to Permian deposition, the gently folded Carboniferous rocks of the district had been eroded to a low-lying plain, reddened at and below the surface by the oxidation of pyrite and siderite *in situ* and by the introduction of red iron oxide along joints and into pore-spaces of sandstones (Anderson and Dunham 1953). The plain, broken by a few scattered hills, had a gentle slope to the east and was bounded by higher ground to the south and possibly also to the west. The small-scale relief of the plain would also appear to have been low judging from the inclination of the plane of the Carboniferous–Permian unconformity in borings and at the few surface exposures.

In the southern part of the district the earliest Permian deposits are breccias and sandstones and widespread patination and the presence of scattered dreikanter suggest that the breccias may have formed part of a stony desert pavement. Many of the fragments are of Carboniferous Limestone, presumably derived from areas a short distance to the south and west. In the northern part of the district (Pl. XV), sands form the characteristic basal deposit. These are mainly aeolian, but are thought to have been partly redistributed subaqueously over wide rock pediments interspersed among the original dunes.

Opinion on the age of the basal deposits is divided. Many writers on the Zechstein of continental Europe, for instance, assign the basal sands and breccias (the Weissliegende and Zechstein Conglomerate respectively) to the Upper Permian. It is here suggested, however, that both these deposits and their equivalents in Durham are more likely to be of Lower Permian age by analogy with the type area of Russia (Nalivkin 1937), where continental beds are assigned to the Kungurian stage of the Lower Permian and the base of the Upper Permian is taken at the base of the marine Kazanian.

Transgression by the Zechstein Sea ended the continental phase, and began a prolonged period of marine conditions. After the transgression, most of east Durham lay in the marginal shelf zone of the Zechstein Sea, and deposits here bear evidence of cyclic sedimentation similar to that which characterizes the Zechstein beds of continental Europe. In England, Sorby (1856), Ramsay (1871) and Green (1872) were the first to recognize that there was an upwards impoverishment in the Magnesian Limestone fauna which could be related to concentration by evaporation, and Bird (1881), Wilson (1881), Lebour (1902), Woolacott (1912) and Trechmann (1913) each reported supporting evidence. Sherlock (1926, p. 59) further supposed that the concentrating water was diluted 'two or three times' and Hollingworth (1942, p. 145) discerned two cycles of 'partial desiccation', followed by a third partial cycle represented by the Upper Anhydrite.

Ideally, the deposits of a complete evaporite cycle should consist of a basal clastic member, followed in lateral and vertical succession by carbonates and sulphates and ending with the precipitation of halite and highly soluble salts of potassium and magnesium. Such cycles are rarely complete in marginal parts of evaporite basins, and this is true of the Permian shelf deposits of Durham where the bulk of the succession is formed by the carbonate phases of three main evaporite cycles corresponding with the main lithological divisions of the Magnesian Limestone. In each of the lower cycles, but especially in the second, the early carbonate deposits contain a relatively rich shelly fauna which becomes impoverished in the later beds. Dolomitization is generally more complete in the later stages of each cycle, and this agrees with observations in equivalent beds in north-east Yorkshire (Dunham 1948, p. 219) where the carbonate directly associated with evaporites is almost exclusively dolomite, some of which may be primary. The margin between shelf and basin in the lowest cycle appears to have lain at about the longitude of West Hartlepool, but in subsequent cycles it followed the outer slope of the reef in the Middle Magnesian Limestone (Pl. V).

The first cycle begins with a fish-bearing bituminous silty carbonate phase, the Marl Slate, followed by up to 250 ft of relatively pure fine-grained carbonate beds which appear, like the Marl Slate, to have been deposited in tranquil water some miles from the shoreline. Sulphate and later phases of this cycle are present in eastern parts of Yorkshire, but in this district are believed to have been relatively minor and have now been removed by solution.

Except in the area south of the West Hartlepool Fault the second cycle is represented in the shelf area entirely by carbonates. The fauna, the rapid lateral variation of lithology, and the widespread oolitic, algal and reef rocks within this cycle suggest deposition in shallow water. The development of a barrier reef near the present coastline defined the margin between shelf and basin more clearly than before, and to the east of this margin deposition of barren carbonates, probably partly chemically precipitated, was followed later in the cycle by precipitation of thick sulphates. The presence south of the West Hartlepool Fault of red clastic deposits of late Middle Magnesian Limestone age suggests a nearly complete infilling of this part of the Zechstein Sea so that by the end of this second evaporite cycle, the depositional surface in Durham and north Yorkshire was almost flat and virtually emergent.

The third cycle commenced with widespread inundation of this surface and the subsequent deposition of the laminated, impure Flexible Limestone which is in many respects similar to the Marl Slate. Higher beds are relatively pure carbonate rocks, which show little lateral variation and, like those of the Lower and Middle Magnesian Limestone, contain very little terrigenous material. Although correlation of the Magnesian Limestone north and south of the West Hartlepool Fault is not yet certain, it appears that the Upper Magnesian Limestone is directly succeeded by the Billingham Main Anhydrite which, east of Billingham (6 miles south of West Hartlepool), is overlain by thick halite.

After the deposition of the Billingham Main Anhydrite the area was covered by red gypsiferous mudstones and siltstones—the Upper Permian Marls—now found in situ only south of the West Hartlepool Fault. These marls contain near their base a relatively thin sulphate bed, the Upper Anhydrite, which is extremely persistent and may be traced at least as far south as Scunthorpe (Lincolnshire). The significance of the Upper Anhydrite within the evaporite sequence is not fully understood, since it is not associated with a well-developed carbonate phase in the Durham district. It may, however, have been deposited in the shelf areas following a relative rise in sea level during a fourth cycle whose carbonate and later phases are restricted to the centre of the basin. Wood (1950, p. 332) notes a westward thinning and increase in the dolomite content of this bed. In the upper parts of the overlying mudstone sequence, sulphate beds become progressively less important and thin beds of red siltstone and sandstone become increasingly prominent. By an increase in the proportion of sandstone the Upper Permian Marls pass gradually upwards by alternation into the Bunter Sandstone, the base of which is almost certainly diachronous in this area. Neither the upper part of the Bunter Sandstone nor the Keuper Marl have been proved in this district, but their presence is inferred from structural evidence.

Earlier writers on the Permo-Triassic have inferred that the climate was arid, and have taken the absence of fossils in the basal deposits to indicate that the land surface was essentially barren. These conclusions are supported by palaeomagnetic evidence (Runcorn 1956, Nairn and Thornley 1961), which suggests that during Permian times, northern England lay within the belt of east-north-easterly trade winds, a direction independently attested by the trend of dunes in the Yellow Sands (Fig. 18) and by aeolian cross-bedding in the Permian Penrith Sandstone in Cumberland (Shotton 1956). At this time a large land mass lay to the east of England and, with easterly winds and high ambient temperatures (Quiring 1954, estimated an average surface temperature of 23°C in the late Permian of Europe), desert conditions are almost axiomatic.

The presence of abundant plant debris in the lower parts of the Lower, Middle and Upper Magnesian limestones suggests that after each transgression, an amelioration of conditions took place around the margins of the Zechstein Sea. It may be supposed that moisture was brought onshore by easterly winds, permitting recolonization of the marginal area by plants with a consequent decrease in the rate of erosion and the formation of a soil cover. Temporary reversions to aridity are suggested by lack or paucity of plant remains in the later stages of each cycle, when deposition took place in water of increasingly high salinity. This does not necessarily imply that temperatures at these times were higher, for it has been shown (d'Ans 1947) that for a given temperature and humidity, evaporation (and therefore subsequent precipitation potential of winds) from concentrated brine is smaller than from water. Depletion of vegetation cover around the margins, presumed to be due partly to this effect and partly to shrinkage of the sea, is reflected by an increased supply of detritus which formed red mudstones and siltstones towards the end of each evaporite cycle. Final infilling of the basin resulted in the shrinkage and disappearance of permanent bodies of open water, and the restoration over the entire area of arid conditions as represented by the barren Upper Permian Marls and Bunter Sandstone.   D.B.S., E.A.F.

CORRELATION WITH YORKSHIRE

In south Durham and north Yorkshire, understanding of the Permian is complicated by intraformational overlap and widespread drift. Borehole evidence (Fowler 1944 ; 1945) shows that the whole succession is condensed and contains non-sequences and red clastic beds indicative of an approach to the limit of marine deposition.

Still farther south, in central Yorkshire, two facies, each showing clear evidence of cyclic deposition, can be distinguished. One of these in the ground to the west of York and Selby is represented by shelf deposits similar to those in Durham. In the other, to the east, sediments are basinal and Stewart (1954) and Lotze (1958) have recognized three main cycles as well as a possible fourth which is incomplete. Correlation between the two facies is difficult, particularly since there is a scarcity of comparative faunal data. The similarity of the shelf deposits of Yorkshire and Durham, however, has been recognized by workers from Sedgwick (1829) onwards. A revised correlation in terms of evaporite cycles is shown in Table 1.

TABLE 1

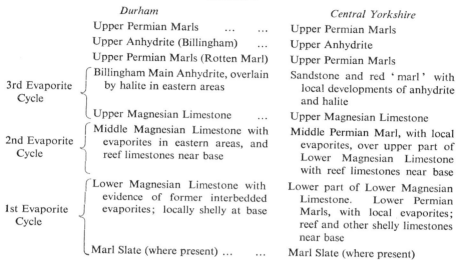

	Durham	Central Yorkshire
	Upper Permian Marls	Upper Permian Marls
	Upper Anhydrite (Billingham) ...	Upper Anhydrite
	Upper Permian Marls (Rotten Marl)	Upper Permian Marls
3rd Evaporite Cycle	Billingham Main Anhydrite, overlain by halite in eastern areas	Sandstone and red 'marl' with local developments of anhydrite and halite
	Upper Magnesian Limestone ...	Upper Magnesian Limestone
2nd Evaporite Cycle	Middle Magnesian Limestone with evaporites in eastern areas, and reef limestones near base	Middle Permian Marl, with local evaporites, over upper part of Lower Magnesian Limestone with reef limestones near base
1st Evaporite Cycle	Lower Magnesian Limestone with evidence of former interbedded evaporites; locally shelly at base	Lower part of Lower Magnesian Limestone. Lower Permian Marls, with local evaporites; reef and other shelly limestones near base
	Marl Slate (where present)	Marl Slate (where present)

The transition from the first to the second evaporite cycle in central Yorkshire is marked by the sporadic occurrence of thin contemporaneous breccias and beds of red and green mudstones (presumably extending westwards over the shelf from the thick evaporites being deposited penecontemporaneously in the basin farther east). The transition appears to coincide with the well-known and widespread lithological change from compact well-bedded dolomitic limestones to granular, minutely cellular, wedge-bedded dolomitic limestones at about the middle of the Lower Magnesian Limestone. Contrary to earlier belief (e.g. Edwards and others 1940, p. 123) field evidence suggests that reef limestones are not randomly scattered vertically throughout the Lower Magnesian Limestone of Yorkshire, but are concentrated a few feet above the base of each of the two lithological (cyclic) subdivisions.   D.B.S.

PALAEONTOLOGY

The fauna of the Magnesian Limestone of East Durham is typical of the marginal shelf areas of the enclosed or nearly enclosed Zechstein Sea. It is comparable as a whole, and in its vertical distribution, with that of the Zechstein beds of Thuringia. Although individuals are locally numerous, the restricted character of the fauna is demonstrated by the small number of species present and the rarity of cephalopods and corals. These facts, together with the absence of fusulinids, rules out direct faunal comparison with the marine Permian deposited outside the Zechstein basin.

The lowest fossiliferous beds are those of the Marl Slate with a fauna almost limited to fish and inarticulate brachiopods and a flora consisting mainly of algae and primitive conifers. In the Lower Magnesian Limestone brachiopods are the most prominent fossils, although foraminifera, ostracods and cryptostome polyzoa are also fairly common. The fauna of the Middle Magnesian Limestone varies with the facies. The reef contains a relatively rich fauna in which brachiopods, lamellibranchs and polyzoa generally dominate, but gastropods, foraminifera and ostracods are not uncommon. Calcareous algae contributed largely to the building of the reef, especially in the upper parts. The beds immediately adjacent to the east and west sides of the reef contain fossils which were either the remains of living assemblages influenced by the proximity of the reef or which were derived from it. Otherwise, the fauna of the lagoonal beds west of the reef consists mainly of lamellibranchs, ostracods and foraminifera. The basinal beds east of the reef are not well known and have yielded little more than brachiopods, foraminifera and ostracods. In the Upper Magnesian Limestone the fauna is more restricted and the only fossils commonly found belong to two lamellibranch genera and algae. However, gastropods, ostracods and other lamellibranchs also occur at some localities.   J.P., D.B.S.

Basal Permian Sands and Breccias

The Basal Permian deposits of the district consist of sands, sandstones and breccias. North and west of a sinuous east-north-easterly line extending from Rushyford to Blackhall Colliery (Pl. XV) the typical deposits are incoherent sands. To the south and east breccias and sandstones predominate. Because of this distribution the Basal Permian deposits at outcrop consist entirely of sands and these are best exposed at Crime Rigg and Sherburn Hill Sand Pits, at Old Quarrington Quarry and in the railway cutting north of Ferryhill Station. The breccias and sandstones do not crop out and are known only from boreholes and shafts. At some localities neither sand nor sandstone and breccia are present and Marl Slate rests directly on Coal Measures.

Basal Permian Sands. The traditional name 'Yellow Sands' was applied to the deposits because, as noted by Sedgwick (1829, p. 65), at outcrop they normally consist of 'yellow incoherent coarse siliceous sand'. In fact, patches of white or light grey sand can also be seen in the large sections now open to view, and below the zone of oxidation grey is the characteristic colour. Studies by Hodge (1932) and Davies and Rees (1944) show that the deposits consist of 85 to 94 per cent of silica, predominantly in the form of quartz and quartzite and with up to 5 per cent of feldspar, commonly microcline. They infer an aeolian origin from the bimodal grain size distribution and from the content of 'frosted', well-rounded 'millet-seed' grains in the coarser fraction. Samples from outcrop contain up to 1 per cent of iron mainly in the form of limonite coating the quartz grains; its distribution is irregular, some of the grains being well coated, others almost clean. Woolacott (1919) suggested that this limonite, which is responsible for the yellow colour at outcrop, resulted from the oxidation of pyrite. During the resurvey, it was noted that while pyrite is absent from the Basal Sands at outcrop, it is abundant in the grey sand at depth; in the latter, limonite seems to be virtually absent. The pyrite may have been formed by reducing waters percolating from the sea bottom when the Marl Slate was being deposited. This would explain why the surface of the sub-Permian

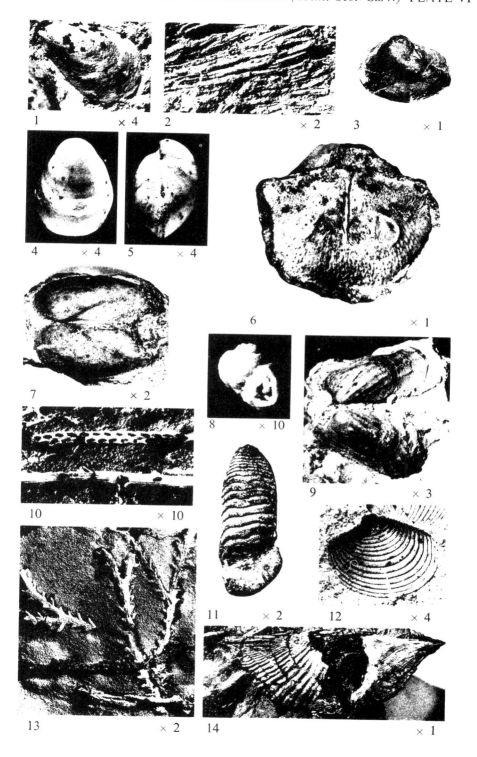

PERMIAN FOSSILS
(For explanation see pp. ix-x)

Coal Measures is similarly grey and commonly pyritiferous, and also why the reddening associated with this surface often begins some inches or feet below it.

Although the Basal Sands are normally incoherent and lack any cementing material other than limonite, some beds –the so-called ' sand rock '– contain a high proportion of interstitial carbonate, generally identified as calcite. Such beds are well exposed in the railway cutting north of Ferryhill Station. The sands normally lying immediately beneath the Marl Slate are also characteristically cemented in this way, and then locally contain the ' small spherical calcareous concretions ' noted by Sedgwick.

From the suite of heavy minerals Hodge (1932) concluded that the sands were derived partly from the Carboniferous sandstones in the immediate vicinity, and party from an area to the north in which sandstones and some granitic and metamorphic rocks were undergoing erosion. In addition to the petrographic evidence of aeolian transport within the Durham district, the bedding too shows features typical of wind-deposited sands. For example the deposit may be divided into distinct lenticular units within which the cross-bedding tends to be in the form of curves with the concave side facing upwards ; curves with the convex side facing upwards are absent.

The Basal Permian deposits are so variable in thickness that a pattern can only be made out where there are a number of closely spaced borings as in the Chilton–Raisby Hill area (Fig. 18). The sands attain a thickness of at least $134\frac{1}{2}$ ft and may be as much as 161 ft according to a possibly unreliable record near Easington Colliery. The amount of variation over small areas can be demonstrated at Sherburn Hill: a bore in the quarry at the western end of the spur proved only 12 ft of sand, but at Crime Rigg (only 1350 yd to the east) about 120 ft are exposed in the sand pit (Pl. VIIA), while at Ludworth Colliery, 2500 yd farther to the east, the thickness has decreased to 30 ft. A similar variation can be traced from 0 ft at Thrisling-ton Colliery to 94 ft near Thrislington Plantation, and back to 5 ft a short distance north of Bishop Middleham.

It was formerly thought (Lebour 1902 ; Hodge 1932) that the sand accumulated in hollows on the eroded surface of the Carboniferous rocks. At the present day, however, aeolian sands often form dunes, particularly in flat featureless country, and if, as seems likely, the Basal Sands were aeolian deposits in this type of environment they also might be expected to include dune forms. In most areas information is too scattered to allow of the location of such forms, but between Raisby Hill and Chilton, a number of relatively closely-spaced bores (Fig. 18) have indicated a ridge of sand which seems to extend west-south-westwards to Chilton Colliery. A secondary spur projects from the area of maximum thickness towards Mainsforth, and it is possible that this is actually more pronounced than shown because the 30 ft of deposits immediately south of the boundary between the sand and breccia facies are tentatively thought to include up to 24 ft of dune sand. The elongated form of this mass of sand is reminiscent of a modern longitudinal or seif dune, examples of which are known to reach a height of up to 300 ft in Egypt. Bagnold (1941) explained the development of seif dunes from isolated barchans (crescentic dunes)

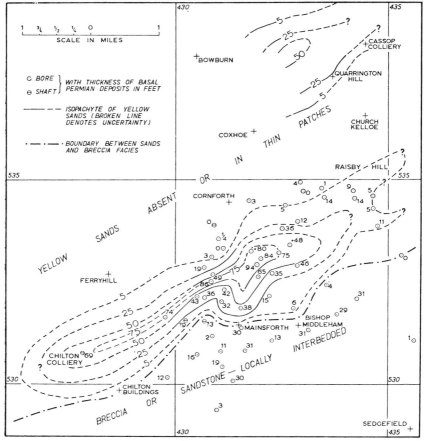

FIG. 18. *Isopachytes in feet of the Basal Permian Sands in the Chilton–Raisby Hill area*

by postulating two alternating wind regimes. His illustration (op. cit., fig. 78d) strikingly resembles the form of the deposit between Raisby Hill and Chilton Colliery, which, according to Bagnold's hypothesis, might have been formed under the influence of a prevailing wind blowing from a point E. 10° N. and an alternating subordinate wind blowing from a point E. 40° N. Such winds by analogy with present-day conditions suggest a tropical latitude and this agrees well with palaeomagnetic evidence (p. 44). It nevertheless seems likely that during the ensuing transgression of the Zechstein Sea, the dune form was modified and that some sand was redeposited subaqueously in interdune areas.

Basal Permian Breccias and Sandstones. The breccias contain rock fragments up to about 3 in long, which are embedded either in a matrix of smaller fragments and a variable proportion of sand, or entirely in sandstone. Although the largest fragments are generally mudstone or sandstone derived from the Coal Measures, by far the greatest number of fragments in many samples consist of Carboniferous Limestone which probably were transported from outcrops in south Durham or north Yorkshire.

Many of the limestone fragments are crinoidal, and separated columnals and portions of stems are common. With the exception of these derived organic remains, most of the fragments are markedly angular or subangular. ' Dreikanter ' pebbles with three well-marked faces and with clean edges have been noted, and these generally have soft leached interiors and hard black skins. Some pebbles in desert areas at the present day exhibit similar characters ; the hard black skin, termed ' desert varnish ', consists of iron and manganese oxides deposited from evaporating moisture drawn to the surface of the pebbles by capillary action. In the sand matrix, the coarse grains are well rounded and generally ' frosted ', and tend to be more abundant in the upper part of the breccia sequence. Much of the sand matrix consists of subangular to angular fine to medium sand and in places this forms beds of sandstone in which breccia fragments are subordinate or absent. Some of these beds contain relatively high proportions of mica and could be mistaken for Coal Measure sandstones. Pyrite is common throughout.

The variation in thickness of the breccia-sandstone facies is pronounced, but is not so great as in the sand facies. The greatest thicknesses proved are $44\frac{1}{4}$ ft in Elwick No. 1 Bore (Magraw and others 1963, pp. 168, 197) and $41\frac{1}{2}$ ft in a borehole a mile south-east of Trimdon Colliery. Apart from these and several records of about 30 ft of breccia with interbedded sandstones in the Bishop Middleham area, the range in thickness is between 0 ft and 20 ft. The information in the Durham district is too scattered for conclusions to be drawn about detailed distribution and thickness variation.

E.A.F., D.B.S.

DETAILS

Basal Permian Sands. In Moorsley Banks, between Coal Bank [345 464] and High Moorsley Quarry [335 456], the sands are about 20 to 30 ft thick, but wedge out in the vicinity of High Moorsley Quarry. Exposures are small and are of typical incoherent yellow sand. Old diggings mark the outcrop along the slopes up to 600 yd north of Elemore Hall [351 442], but the only section now open comprises a few feet of brownish yellow sand in a disused pit 100 yd N.E. of the Hall. A maximum of 20 to 30 ft may be present locally. Up to 2 ft of typical yellow sand were seen at the time of resurvey in a number of small temporary excavations made in the upper slopes of White's Wood, near Littletown Colliery. A section [3425 4332] showing up to 12 ft is accessible in the east bank of a small glacial drainage channel 500 yd E.S.E. of the colliery.

The best exposures in the district are at Sherburn Hill. The original pit [3397 4174] immediately north-east of Crime Rigg Farm is now largely filled

with debris, but a large pit (Pl. VIIA) adjacent to it on the eastern side has a face nearly 300 yd long, with a maximum thickness of Yellow Sands of about 120 ft. Directly beneath the Marl Slate, some cementation has taken place, but the cemented layer rarely exceeds 2 to 3 ft. It is characterized by the small spherical calcareous concretions observed by Sedgwick (1829). Elsewhere, the sands are uncemented, except in some thin layers near the base which have developed iron-pan. In a bore [3293 4175], 1100 yd W. of Crime Rigg Farm, the Yellow Sands were no more than 12 ft thick. In another bore [3361 4063], 1200 yd S.S.W. of Crime Rigg Farm, 19 ft of Yellow Sands were recorded. At Sherburn Hill Sand Pit [344 418], 450 yd E.N.E. of Crime Rigg Farm, over 80 ft of Yellow Sands have been worked and considerable reserves remain. A bore [3473 4059], 500 yd. S.S.E. of Shadforth proved $23\frac{1}{2}$ ft of soft sandstone, interpreted as Yellow Sands, under $84\frac{1}{2}$ ft of drift.

Yellow Sands crop out [335 380] on the northern side of the valley 375 yd E. of Cassop Vale Colliery, but they appear to have wedged out within 300 yd to north-west. They seem to be absent, too, on the southern side of the Vale [3228 3856], 300 yd N.E. of Hill Top Farm, but are certainly present from the Quarrington Hill–Cassop Moor road eastwards to at least as far as Cassop Colliery. Up to 7 ft of typical Yellow Sands are visible in an old quarry [339 383], 1250 yd S.W. of Dene House Farm, where they are im-mediately overlain by the Magnesian Limestone. Since this locality lies on the 500-ft contour and a nearby bore at Cassop ' A ' Pit Shaft is presumed to have cut the base of the Yellow Sands at about 385 ft O.D., their total thickness here-abouts is probably over 100 ft.

At the top of an old quarry [3191 3854], 200 yd N.W. of Hill Top Farm, the Yellow Sands are only about 6 ft thick. Half a mile to east-south-east, at the northern end [3256 3815] of Quarrington Quarry, 50 ft of Yellow Sands were measured between the Marl Slate and light grey Coal Measures mudstone. The thickness appears to decrease along the escarpment towards St. Paul's Church [3347 3792], but in a quarry [3325 3742], 500 yd W. of Quar-rington Hill cross-roads, up to 30 ft of typical Yellow Sands were exposed during the resurvey. Judging from borings at quarries between Quarrington Hill and Coxhoe, the thickness again decreases to-wards the south-west ; and between these quarries and the western end of Raisby Hill Quarry, the Yellow Sands are evidently thin or completely absent. In Raisby Hill Quarry 3 ft of Yellow Sands are exposed in a small cutting [3403 3552] 950 yd E.S.E. of Coxhoe East House.

Yellow Sands up to 6 ft thick are ex-posed on the southern side of the railway cutting [2796 3355], ¼ mile S.E. of Skib-bereen. A bore [2855 3378], 1000 yd E. of Skibbereen, proved 10¼ ft of ' soft and coarse sandstone ' which is interpreted as Yellow Sands. The soil contains abundant grains of Yellow Sands in two patches respectively 900 yd S. and 950 yd S.S.W. of Skibbereen, but there are no sections hereabouts.   E.A.F.

In the Ferryhill Gap at an exposure [3004 3295], 780 yd S.W. of Thrislington Hall, Coal Measures are separated from Marl Slate by fine sand, ½ to 1 in thick. Within this sand, in an almost vertical position, a small indeterminate shell was found. The sands thicken gradually to the south, and 40 ft are exposed in the eastern face of the railway cutting between 350 and 550 yd north of the road bridge at Ferryhill Station. The deposit is largely dune-bedded, but the upper part is either level-bedded or very gently cross-bedded, suggesting aqueous deposition. Small outcrops also occur round the sides of the spur immediately east of this ex-posure.   G.D.G.

Basal Permian Sands have been proved in numerous bores and shafts, and thick-nesses have been inserted on the maps (Pl. XV and Fig. 18).

Basal Permian Breccias and Sandstones. All records are from bores or rarely from shafts ; the following are cited as examples of the thicker sections proved.

In Elwick No. 1 Bore [4531 3117], ¼ mile S. of Elwick, 15 ft 1 in of breccia, containing fragments tending to be roughly flat-lying in the top half and to dip at about 30° in the bottom half, are underlain by 29 ft 2 in of grey and grey-brown sandstone with coarse rounded grains and with subordinate breccia frag-ments and crinoid columnals.   D.B.S.

A bore [3433 3210], 1000 yd E.S.E. of Farnless, proved 30 ft 9 in of Basal Permian deposits as follows :

	ft	in
MARL SLATE	–	–
BASAL PERMIAN DEPOSITS		
Sandstone, medium- to fine-grained with well-rounded grains and conspicuous breccia fragments, mainly of shale and siltstone ...	4	3
Sandstone, whitish, fine, with small flakes of muscovite	4	3
Breccia, with subangular fragments of shale and siltstone, some red, in fine-grained sandstone matrix	5	0
Sandstone, greyish, with fine subangular grains ...	3	3

ft in

Breccia, with subangular
fragments of sandstone
and mudstone, many red,
in fine- to medium-grained
sandstone matrix   ...   10   7
Sandstone, white, fine-
grained, locally tinted pink
near top   3   5

COAL MEASURES: shale (grey)   –   –

A bore [3142 3004] at Thrundle proved
29 ft 8 in of Basal Permian sandstone
and breccia:

ft in

MARL SLATE   ...   .   –   –

BASAL PERMIAN DEPOSITS
Sandstone, grey, coarse- and
medium - grained, with
abundant rounded grains
in upper part; abundant
angular and subangular
fragments of sandstone,
mudstone and shale,
averaging about $\frac{1}{4}$ in   ...   9   2
Breccia, grey, with fragments
(some pinkish) generally
$\frac{1}{2}$ to 1 in   2   0
Sandstone, grey, coarse- and
medium-grained with some
well-rounded fragments ...   1   0
Breccia, with many frag-
ments of Carboniferous
Limestone up to 2 in,
especially at base   1   9
Sandstone, grey, medium-
grained with angular fragments of limestone and of
red pyritic mudstone en-
circled by pyritic rings ...   4   3
Breccia of fragments averag-
ing $\frac{1}{2}$ in: many fragments
are red and the whole rock
has a general pinkish

tinge in which lighter and
darker grey limestone fragments stand out by
contrast   10   7
Sandstone, reddish - brown
and light grey, medium-
to fine-grained with some
coarse grains   11

COAL MEASURES: sandstone
(red and grey)   ...   ...   –   –

A bore [3385 3168], 1000 yd E.N.E. of
Bishop Middleham church, proved
28 ft 9 in of sandstone and breccia:

ft in

MARL SLATE   ...   .   –   –

BASAL PERMIAN DEPOSITS
Sandstone, white; grains
mainly fine and subangular
but coarse and rounded in
some patches; pyritic
inclusions and angular
dark fragments at base ...   2   6
Breccia, pink in part, containing red angular haema-
titized mudstone frag-
ments; many interstitial
subangular sand grains ...   15   0
Sandstone, pink, fine-grained,
with mainly subangular
grains   3   0
Sandstone, pink, fine-grained,
with breccia bands con-
taining angular fragments
of shale and sandy shale   4   0
Sandstone, pink-mottled, fine-
grained, micaceous, with
bands of pink breccia ...   4   0
Breccia, pink   3

COAL MEASURES: shale and
mudstone (red and grey) ...   –   –

E.A.F.

MARL SLATE

The equivalence of the Marl Slate to the Kupferschiefer at the base of the Upper Permian (Thuringian) of north-west Europe was established by Sedgwick (1829, p. 75), the name being an anglicized version of Mergel-schiefer (=Kupferschiefer). It is a silty dolomitic shale or shaly dolomite[1], usually finely laminated and generally somewhat bituminous. At outcrop it is yellowish brown in colour, but in its unweathered state it is a hard compact rock, usually consisting of alternating grey and black laminae. Freshly broken cores commonly have a strong bituminous odour and have a characteristic flecked appearance (Smith *in* Magraw and others 1963, p. 169). The deposit is minutely uneven at the base, especially where resting on breccia (Pl. VIIc), which it locally penetrates. Laminae in the lower part of the Marl Slate conform to irregularities of the base, but are generally smooth and level in the upper part of the deposit. Locally, thin beds of millet-seed sand are present near the base. The upper surface is generally smooth and makes a sharp contact, though in places, especially in south-western parts of the district, thin beds of Marl Slate lithology are interbedded with the lowest beds of Lower Magnesian Limestone. Evidence of contortion by slumping has been noted in several bores.

The Marl Slate reaches a maximum thickness of 18 ft in a bore west of Rushyford, and elsewhere in the area west of Fishburn and Sedgefield it locally exceeds 12 ft (Fig. 19). In most of the district, however, its thickness is 2 ft or less and in bores in the offshore area and at Hartlepool, West Hartlepool and Seaton Carew, the Marl Slate is absent.

The proportions of the main constituents show a fairly wide range from place to place and from bottom to top of the deposit. Aeolian sand grains have been noted in the basal layers both in field exposures and boreholes, and Browell and Kirkby (1866, pp. 209–12) show that whereas the lowest parts of the deposit contain up to about 55 per cent of non-carbonate matter, this proportion generally decreases upwards until the uppermost laminae are composed of almost pure carbonate. This trend is also shown in analyses of specimens of Marl Slate from Blackhall Colliery shaft (Trechmann 1914, p. 249). Proximate calculations of the carbonate fractions show that dolomite is the main carbonate present in all but one of the specimens analyzed by Browell and Kirkby and by Trechmann and this is confirmed by microscopic examination. The non-carbonate fraction, apart from incorporated aeolian sand, consists of angular and subangular quartz grains ranging in grade from silt to fine sand, with mica (mainly muscovite), microcrystalline pyrite, bituminous material, and clay minerals. The pyrite is present mainly in the form of enormous numbers of spheres (Deans 1950, p. 348; Love 1962, p. 354) ranging from 3 to 120μ in diameter and in some cases amounting to over 4 per cent of the total volume of the rock. The spheres are intimately associated with delicate films of organic tissue,

---

[1] Throughout this account the terms 'dolomite' and 'limestone' are used for rocks consisting almost entirely of the minerals dolomite or calcite respectively. 'Slightly dolomitic limestone' and 'slightly calcitic dolomite' are used where no more than 10 per cent of dolomite or calcite are present in otherwise pure limestone or dolomite; and 'dolomitic limestone' and 'calcitic dolomite' are used respectively for rocks containing 10 to 50 per cent of dolomite or calcite.

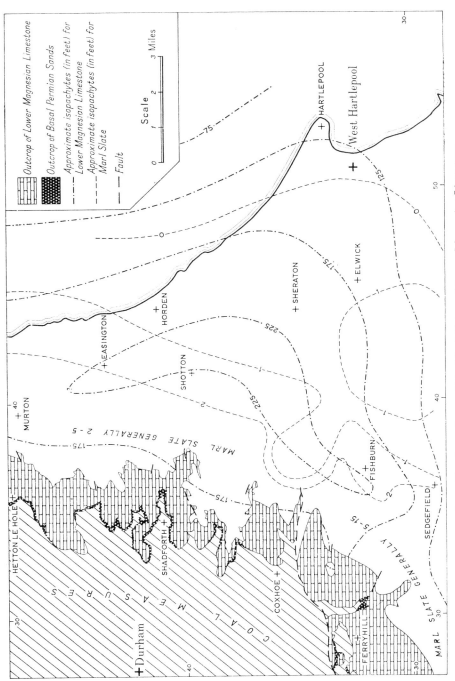

FIG. 19. *Isopachytes of the Marl Slate and Lower Magnesian Limestone*

and these, together with widely distributed black carbonaceous films, account for the high (up to 14 per cent) content of bituminous matter shown in some analyses. Pyrite is also present in megascopic form : form the time of Sedgwick (1829) onwards it has been recorded as coating bedding planes and joints in association with sphalerite and galena. Deans (1950) and Dunham (1960, 1961) postulated a syngenetic origin for the bulk of the metallic sulphides in the Marl Slate, whilst agreeing that subsequently a certain amount of diagenetic redistribution took place. Schouten (1946) and Love (1962) maintained that, in the Kupferschiefer, only the iron sulphide is actually syngenetic and that the remaining metallic sulphides are diagenetic replacements of the primary pyrite.

Deans (1950, p. 349) has demonstrated that the Marl Slate of Durham contains a trace-metal assemblage similar to that in the Kupferschiefer. He points out that while the copper content is normal for sapropel shales, there is a high concentration of lead but little zinc. According to Deans, the Marl Slate carries base-metal concentrations over 100 times greater than normal geochemical background values.

The initial transgression of the Zechstein Sea appears to have been relatively rapid, and resulted in some redistribution of pre-existing aeolian sand (the thin beds of such sand noted above may be evidence of this but are thought more likely to indicate continuing movement of sand around the margins). Subsequent deposition of the Marl Slate took place in a euxinic environment free from bottom scavengers, in water estimated (Love 1962, p. 362) to be 300 to 600 ft deep. The deposition of the Marl Slate is thought (Oelsner 1959, p. 109) to have taken about 17,000 years. Reviews of the conditions of deposition of this deposit are given by Love (1962) and by Hirst and Dunham (1963), who also give much new petrographical and analytical data.

The Marl Slate is well known for its fauna of fossil fish which has been described by Westoll (1941) as including bottom-dwelling elasmobranchs, deep-bodied Platysomids, and large Palaeoniscids. In the past well-preserved specimens were collected from localities such as the Ferryhill Gap, but in recent years few good specimens have been obtained, though scattered scales and spines are recorded from many outcrops and borings. *Lingula credneri* Geinitz is fairly common in the western part of the district and reaches 12 mm in length in Mainsforth No. 5 Bore. About two miles to the south-west, ' *Nautilus freieslebeni* ' Geinitz and ' *Myalina hausmanni* ' Goldfuss have been recorded (Howse 1890) from Middridge Quarry in the north-eastern part of the Barnard Castle (32) Sheet.

Plant remains are more abundant. The railway cutting north of Ferryhill Station is the type locality of ' *Mixoneura huttoniana* ' (King) ; and from other railway cuttings nearby Calvert (1884) has recorded *Ullmannia frumentaria* (Schlotheim) and *Pseudovoltzia liebeana* (Geinitz). The Marl Slate at the western end of Raisby Hill Quarry yielded a branch of *Ullmannia bronni* Göppert nearly 35 cm long (Hickling 1931). A description of plants from the Marl Slate and overlying beds has been given by Stoneley (1958). The assemblage appears to represent the remains of land plants washed into the sea.   E.A.F., D.B.S.

DETAILS

Exposures of Marl Slate are comparatively rare along the escarpment but good sections may be examined at the following localities: cutting at roadside [334 457], High Moorsley; Sherburn Hill Sand Quarries [3435 4175 and 3415 4165]; old quarry [3395 3825] west of Cassop Colliery; Quarrington Quarry [327 379]; Coxhoe Bank Quarries [329 368]; Raisby Hill Quarry [3402 3521]; railway cutting [3084 3230] north of Ferryhill Station; road cutting [285 328] at Ferryhill.   E.A.F.

Marl Slate crops out at several places on the flanks of the Ferryhill Gap. At one locality [3004 3296], 780 yd S.W. of Thrislington Hall, a 10-ft section comprises dark grey to black, laminated, fissile, dolomitic siltstone which contains sporadic fish scales in the lower beds and which becomes more calcareous towards the top where it passes gradually up into rather silty flaggy limestone.   G.D.G.

A thin section (E 21046)[1] of a representative specimen of Marl Slate from a depth of 518 to 519½ ft in Ten-o'-Clock Barn No. 2 Bore [3916 3035] near Fishburn is described by Dr. K. C. Dunham as follows:

" An argillaceous sandy magnesian limestone, dark brownish grey in hand specimen. The rock is composed of crystalline carbonate of average grain-size 0·015 mm, the bulk of which is dolomite with refractive index $\omega = 1·680$. Some calcite, with $\omega = 1·658$ is, however, present, forming groups of crystals coarser than the average. These groups have rounded outlines and may represent organic remains. The bedding of the rock is marked by ramifying layers of a brownish translucent or opaque substance up to 0·05 mm thick, spaced about 0·2 mm apart. A few flakes of mica are associated with these layers. Subangular grains of quartz up to 0·05 mm diameter are scattered through the rock."

This rock has been analysed by Mr. G. A. Sergeant (Lab. No. 1450) as follows (figures in percentages):

SiO_2 21·79, Al_2O_3 8·12, Total Fe as Fe_2O_3 4·88, MgO 10·21, CaO 17·28, Na_2O 0·13, K_2O 1·68, P_2O_5 0·06, TiO_2 0·44, MnO 0·18, BaO 0·07, Cl trace, F trace, LiO_2 trace, SO_3 (in loss-on-ignition residue) 5·00, loss on ignition 30·03* ; total 99·87.

From this analysis percentage mineral composition is calculated to be:

dolomite 46·9, calcite 4·9, siderite 2·1, gypsum 0·6, baryte 0·1, pyrite 4·1, quartz 12·3, muscovite 14·3, kaolinite 6·5, limonite 4·0, organic carbon 4·8.

Dr. Dunham adds : " The richness of the Marl Slate in organic compounds is noteworthy. The dark layers mentioned in the petrographical description evidently consist of bituminous material, mixed with finely divided mica and clay minerals, and with some detrital coarse muscovite. The quantity of the coarse mica, as seen in thin section, is considerably less than 14 per cent ".

The sample was further analysed for metallic trace elements (Deans 1950 p. 348), as follows (figures in parts per million): vanadium 950, lead 500, molybdenum 310, nickel 230, copper 120, cobalt 30, zinc variable but around 20, and chromium 13. Similar figures were obtained by Deans from a specimen of Marl Slate from Blackhall Colliery shaft, and by Dunham (1960, p. 293) from Marl Slate at 546 ft in Elwick No. 1 Bore.

The rock from Elwick No. 1 Bore has been analysed by Pattinson and Stead and the following figures (in percentages) are from Hirst and Dunham (1963):

SiO_2 34·60, Al_2O_3 13·90, FeO 1·00, MgO 1·10, CaO 12·45, Na_2O 0·14, K_2O 1·63, P_2O_5 0·14, TiO_2 0·55, MnO 0·14, SO_3 0·12, BaO 0·18, V_2O_3 0·05, F 0·28,

---

[1] Numbers refer to thin sections in the English Sliced Rock Collection of the Geological Survey and Museum.

* Fe (Approx) 1·71, $H_2O > 105^0C$ (including H_2O from organic H) 5·20, $H_2O < 105°$ 0·86, CO_2 25·28, SO_3 0·33, FeS_2 4·11, Organic C 4·81.

CO_2 10·80, FeS_2 6·45, H_2O+ 4·84, Organic C 11·51, less 0 for F 0·12 ; total 99·76.

A selection of details from representative borehole sections of Marl Slate is given below :

(1) [2745 2882] at Mill Cottages, Windlestone, 1000 yd W. of Rushyford : 18 ft of brownish grey, very hard shale ; black and slightly more fissile in bottom 6 in ; fish scales at 11½ ft above base.

(2) [3361 3242] 175 yd N.E. of Farnless : 14 ft of dark grey, very hard, finely laminated dolomite, with silty partings ; becoming softer towards base with some obscure fish remains.

(3) [3349 3479], 1100 yd S. of Coxhoe East House : 7½ ft of grey, dolomitic, finely laminated shale with finely mottled partings ; hard at top, but softer, black and highly fissile from 2 ft to 2½ ft above base ; fish scales from 1½ ft to 3 ft above base.

(4) [3890 4584] on site of Hawthorn Colliery Shaft : 5 ft 1 in of grey, finely laminated, silty, argillaceous, bituminous, dolomitic limestone with many small cavities in uppermost part and with scattered mica, a few fragments of poorly-preserved plant debris and scattered fish scales.

(5) [3717 3052] 1000 yd S. of Mill House, Fishburn : 2 ft of black, fissile shale, with fine grey mottling on bedding planes and with scattered fish scales, on 3 in of hard grey, siliceous, pyritic limestone.

(6) [4145 3185], at Murton Hall, Embleton ; shale, grey, dolomitic, hard, finely laminated 9 in, on dolomitic limestone 2 in, shale as above, sandy near base 7 in, shale, greenish grey, dolomitic, softer than above, with fish scales 5 in.

E.A.F., D.B.S.

LOWER MAGNESIAN LIMESTONE

The outcrop of the Lower Magnesian Limestone is restricted to a relatively narrow belt (Plate V) along an escarpment, 100 to 200 ft high, between Hetton le Hole and Ferryhill. Farther east these rocks are known only in shafts and boreholes. In some places, where the lower slopes of the escarpment are formed by older strata, only the lower beds of the Lower Magnesian Limestone are found in the quarry sections, and exposures of the upper beds are confined to valleys and to quarries down-dip from the escarpment. Stone from the lower beds was used in the construction of many of the early farms and settlements, but these rocks are now used mainly as refractories for which they are intensively quarried between Quarrington Hill and Mainsforth.

The thickness of the Lower Magnesian Limestone in the district ranges from 40 ft in Offshore No. 1 Bore (4 miles north of Hartlepool) to 245 ft in a bore at Mill Hill, Easington. It may be about 250 ft thick around Dalton Piercy in the southern part of the district also, but the scanty evidence there is ambiguous and is disregarded in Fig. 19 which shows the broad trend of thickness to range from less than 125 ft in the south and east to over 225 ft between Fishburn and Easington.

At outcrop the Lower Magnesian Limestone falls naturally into three main lithological units distinguished by differences in colour, texture, and type and thickness of bedding. The lowest unit comprises 4 to 12 ft of beds which are individually 6 to 12 in thick and which have a granular or finely crystalline texture ; analyses show them to range from dolomite to almost pure limestone. Laminated argillaceous layers are widespread in

A. Basal Permian Sands (Yellow Sands), Crime Rigg Quarry

L 77

Geology between Durham and West Hartlepool (*Mem. Geol. Surv.*)

PLATE VII

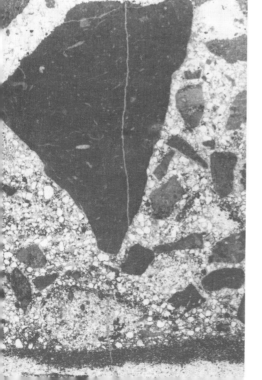

B. Basal Permian Breccia overlying
Coal Measures Sandstone
MLD 1630   × 3·6

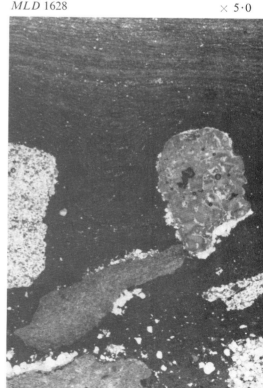

C. Marl Slate overlying Basal
Permian Breccia
MLD 1628   × 5·0

A. Lower Magnesian Limestone, Raisby Hill Quarry   *A* 5250

PLATE VIII   Geology between Durham and West Hartlepool (*Mem. Geol. Surv.*)

B. Lower Magnesian Limestone, Hartlepool Lighthouse Bore × 1·4   *MLD* 803

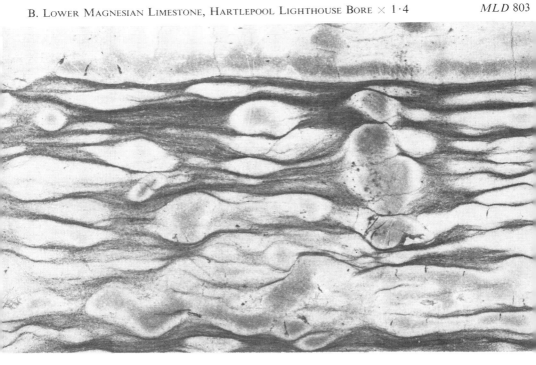

the lower part of these beds in the area between Ferryhill and Kelloe, where there is a passage by alternation from the underlying Marl Slate (p. 102). Small quantities of galena, pyrite and sphalerite are common near the base, and in some places persist for some distance upwards.

The middle division, up to 120 ft thick, is predominantly thin-bedded in layers of 2 to 4 in, grey to buff in colour, and very finely crystalline with grains ranging from mud- to fine silt-grade. Analyses (Table 2) indicate a predominance of dolomite, though samples from Raisby Hill Quarry and the Cotefold Close and Sheraton bores (Woolacott 1919a, b) are partly or wholly calcitic. The bedding, though even and regular in general aspect, is often very uneven and nodular in detail (Plate VIIIB), especially in unweathered sections and borehole cores ; many bedding planes have become stylolitic and bear thin films of argillaceous carbonaceous residue. A nodular structure, first described by Sedgwick (1829, p. 76), is closely associated with grey and buff mottling, in which discrete or interconnected rounded, pale crystalline patches are enveloped in undulant irregularly laminated darker films. Thin sections (E 30183, E 30187–8) show that, apart from a slight concentration of carbonaceous material in the darker bands, the mineral composition of these is the same as that of the pale patches. This is confirmed by chemical analysis (Trechmann 1914, p. 247) of specimens from Blackhall Colliery Shaft. Trechmann (op. cit., p. 255) considered that the mottling reflects a slight segregation or concentration of calcite, and that the nodular structure is due to 'internal bending' brought about by the removal of parts of the more dolomitic portion. An alternative view (Woolacott 1919a, p. 164) is that the nodular bedding planes are due to pressure-solution, and the common association between thin darker bands and bedding planes with stylolites strongly supports this suggestion. In addition to stylolites arranged sub-parallel to the bedding, steeply inclined varieties have been noted in a number of borings. The abundance of stylolites varies greatly from place to place, but there is no apparent relationship between their frequency and the total thickness of beds. In some exposures and bores thin layers of brown leathery clay, often bituminous, occur on non-stylolitic bedding planes low in this middle subdivision and these appear to be original sedimentary deposits.

At slightly higher levels widespread incipient to advanced autobrecciation, accompanied by a complex network of carbonate veinlets, gives rise to a characteristic hackly fracture. Brecciated horizons, first noted by Woolacott (1919a, p. 166), occur towards the top of these beds at Raisby Hill Quarry and in a number of borings in the south-eastern and south-western parts of the district. Woolacott (op. cit., p. 167) suggested that much of the brecciation is contemporaneous, but since gypsum is reported (Woolacott 1919b, p. 456) in geodes at Raisby Hill Quarry and in stratified breccias from boreholes at Embleton and Hartlepool (Smith *in* Magraw and others 1963, pp. 172, 207) it is possible that the breccias are due to collapse following the solution of interbedded sulphate-rich horizons.

The beds of the middle division of the Lower Magnesian Limestone pass laterally in places into lenticular masses of shelly limestone. The most notable in this district is at Raisby Hill Quarry and is 90 ft thick. Trechmann (1914, p. 259 ; 1921, p. 539) suggested that both this mass and another at Thickley Quarry, 3 miles S.W. of Rushyford, in the Barnard Castle (32) district, were of primary limestone which had for some reason escaped dolomitization.

The upper division of the Lower Magnesian Limestone comprises 40 to 100 ft of buff to brown uneven or lenticular beds up to $1\frac{1}{2}$ ft thick. The rock is generally granular in texture and is composed of dolomite grains of silt-grade with small amounts of interstitial calcite. Stylolites and auto-brecciated patches are less common than in the underlying division, and bedding planes are more regular. At Raisby Hill Quarry a group of transitional beds, consisting of interbedded limestones and dolomites, lies at the base of the upper division. Somewhat higher, in boreholes in the Fishburn area and also in Chilton Quarry, Fowler (1943 ; 1956) recorded widespread baryte and fluorite, mainly in veins and cavities, but also in small amounts disseminated throughout the rock.

The general threefold lithological subdivision of the Lower Magnesian Limestone persists eastwards for 5 or 6 miles from the scarp, but is difficult to recognize in boreholes in the Elwick–West Hartlepool area.

D.B.S., E.A.F.

The Lower Magnesian Limestone has a limited fauna, and in many places is almost barren. In all, less than 40 invertebrate species have been recorded from the whole of Durham and some of these were collected outside the limits of the present district. In general, shelly fossils are restricted to the lower 100 ft or so, and are most concentrated in the lowest 20 to 30 ft. Brachiopods, of which the most common are *Strophalosia morrisiana* King and *Horridonia horrida* (J. Sowerby), and crinoid columnals, are virtually confined to the basal beds and to the lenticular masses of shelly limestone. Here, too, are found several species of polyzoa—*Acanthocladia anceps* (Schlotheim), *Batostomella crassa* (Lonsdale) and the cryptostome species described later (p. 181)—not previously recorded from these beds. Higher beds contain only a scanty fauna of lamellibranchs, ostracods and coiled tubular foraminifera. Amphibia and fish are recorded from other parts of Durham (Hancock and Howse 1870, p. 219 ; Howse 1890, p. 247) but are not known in the present district. Plant remains, many of them obscure, are widespread in lower beds of the division.   D.B.S., E.A.F., J.P.

DETAILS

Hetton le Hole to Shadforth. The Lower Magnesian Limestone is overlain by thick drift at Hetton le Hole and the most northerly exposure is in a railway cutting [357 465] 850 yd N. of Elemore Colliery, where 4 ft of cream, very finely crystalline, autobrecciated dolomitic limestone[1] are seen. To the south-west 9 ft of thin-bedded[2] dolomitic limestone are exposed at Coal Bank Quarry [346 464] ; and at

Low Moorsley Quarry [342 462] the section is:

	ft
Dolomitic limestone, brown, thin-bedded, fine-grained ...	20
Limestone, white or light grey, thin-bedded, fine-grained ...	5–6
Limestone, dark brown, thick-bedded and massive, fine-grained	2–6

[1] Cavities, ranging up to 6 in across, are common in almost all outcrops of the Magnesian Limestone. To avoid repetition, they are not mentioned in the ensuing details unless especially rare or abundant.

[2] Throughout this account beds less than about $1\frac{1}{2}$ in thick are described as 'flaggy', beds 2 to 3 in thick as 'thin', beds 4 to 6 in thick as 'medium', and beds averaging or exceeding 6 in as 'thick'. 'Massive' is used to describe rocks in which the beds are over 3 ft thick.

Beds in both these quarries are near the base of the Lower Magnesian Limestone, as are 12 ft of thin-bedded dolomitic limestone exposed in Moorsley Banks Quarry [3362 3591]. The base itself is seen in a roadside section [3342 4569] 500 yd W.S.W. of High Moorsley, where thin-bedded, finely crystalline dolomite is intercalated with Marl Slate. The sequence at High Moorsley Quarry [334 455] is similar to that at Low Moorsley: 42 ft of brown dolomitic limestone are thick-bedded in the lowest 9 ft and thin-bedded above. Haematite and dolomite are abundant on joint faces and in veins on part of the north-west face. The thick-bedded limestone seen at the base of several of the above exposures is over 12 ft thick in an old quarry [3331 4520] 200 yd S. of High Moorsley Quarry, and is overlain by 60 ft of more flaggy dolomitic limestone in Pittington Quarry [332 447]. Thin-bedded dolomite and dolomitic limestone are also seen in the following exposures along the escarpment east and south-east of Pittington Quarry: Warren Quarry [333 445], Cobbler's Quarry [345 449], Hetton le Hill Quarry [348 450], Hetton le Hill Wood Quarry [3500 4495], and in several small old quarries between Elemore Hall [351 442] and Hastings House [3449 4331]. In thin section (E 16136) a specimen from Cobbler's Quarry is seen to be a calcitic dolomite of silt grade, containing a few small angular quartz grains. E.A.F.

East of the escarpment, Lower Magnesian Limestone is exposed in the sides of a glacial drainage channel now occupied by the Coldwell Burn, and at a number of old quarries near Low Haswell. Exposures alongside Coldwell Burn are mainly small and overgrown, and are mostly of thin-bedded, fine-grained, cream, dolomitic limestone. The following is the section in an old quarry [3658 4451] 850 yd N. of High Haswell: dolomitic limestone, cream-buff, massive, hard, semi-porcellanous, autobrecciated, cavernous 10 ft, on dolomite, cream, thin- to thick-bedded, finely granular, with thin bands of semi-porcellanous partly autobrecciated dolomite 10 ft. Autobrecciation is also seen in 7 ft of thin- and medium-bedded sub-porcellanous dolomitic limestone exposed in a small quarry [3656 4463] 130 yd farther north, and again at three localities about 600 yd S.W. of Low Haswell. At one of these, an old quarry [3612 4368], the rock is finely granular dolomite and dolomitic limestone in which large irregular patches are unbedded and autobrecciated, whilst the remainder of the rock shows well-defined bedding including traces of original small-scale cross-bedding. In some softer parts of the granular dolomite, a felted texture (p. 123) is developed as a result of the intersection of large numbers of platy dolomite crystals up to 4 mm across. This texture is taken to indicate a position at or near the top of the Lower Magnesian Limestone, for it is not found at lower horizons. D.B.S.

Scattered exposures between Sherburn, Shadforth, Haswell and Ludworth are all in thin-bedded dolomitic limestone, as follows: old quarry [346 425] at Black Banks; old quarry [352 422] at Limekiln Hill 12 ft; old quarry [335 423] at Sherburn Hill 3 ft; Sherburn Hill Sand Pit [344 418], 0–10 ft, overlying Marl Slate; Crime Rigg Sand Pit [341 416] 12–20 ft (in 1956) overlying Marl Slate; Crime Rigg Quarry [339 419] 10–15 ft; old quarries [345 414] 400 yd E.N.E. of Shadforth Church; and old quarry [3560 4125] at Ludworth Tower. The lowest exposed beds at Sherburn Quarry [328 418] are 12 ft of thick-bedded dolomitic limestone containing ferruginous bands. An old quarry [3575 4125] 100 yd E.S.E. of Ludworth Tower gives the following section:

	ft
Dolomitic limestone, grey-cream, very irregularly bedded, hard, sub-porcellanous, partially or wholly autobrecciated throughout	6–7
Dolomitic limestone, grey-cream, thick-bedded and massive, partially autobrecciated throughout, with many small cavities; slumped bed 1 to 5 ft thick at base	16
Dolomite, cream, thin-bedded (wedge-bedded in part), fairly hard, finely granular, with scattered small cavities	7–10

The base of the middle unit of the section is somewhat uneven and rests on a bed at the western end of the quarry which is 4 ft higher in the sequence than the bed underlying it at the eastern end. The lithology of the lowest unit is like that previously noted near the base of the Lower Magnesian Limestone.

D.B.S., E.A.F.

Shadforth to Garmondsway. The most northerly outcrop between Shadforth and Garmondsway is in a small quarry [3572 4066] 700 yd S. of Ludworth Tower, where 3 ft of grey-cream sub-porcellanous partially autobrecciated dolomitic limestone are exposed. The bedding is generally thin, but extremely uneven. Similar rock is exposed in a group of small quarries in Shadforth Dene, 800 yd farther west. The beds at both localities are about 80 to 100 ft above the Marl Slate. Along the escarpment, the northernmost exposure of any importance is at Running Waters Quarry [334 404], where 30 to 40 ft of thin-bedded dolomitic limestone are visible. Thicker bedding in the lowest part of the section indicates proximity to the base of the division. Thin-bedded dolomitic limestone is 20 ft thick in a quarry [3353 4018] 350 yd N.W. of Cassop Smithy, and forms small outcrops at Strawberry Hill [342 400], and on both sides of a deeply-incised valley [351 399], 1000 yd E. of Strawberry Hill. Farther south exposures are confined almost entirely to the following localities on the scarp: old quarry [3415 3965] 500 yd N.E. of Old Cassop, thin-bedded dolomitic limestone 18 ft; old quarry [3315 3933] 500 yd. W.S.W. of Old Cassop, thin-bedded dolomitic limestone 10 ft; old quarry [3387 3866] 670 yd N.W. of Cassop Colliery church, thick-bedded white ' limestone' 7 ft; old quarry [3464 3870] 350 yd S.W. of Dene House Farm, massive dolomitic limestone 10 ft; old quarries [339 382] 500 yd W. of Cassop Colliery church, thin-bedded dolomitic limestone 25 ft, over thick-bedded dolomitic limestone 11 ft, resting on Basal Permian Sands; three old quarries [329 386] 950 yd E.N.E. of Hill Top Farm, thin-bedded dolomitic limestone spanning the lowest 50 ft of Lower Magnesian Limestone; old quarry [3204 3844] at Hill Top Farm, Old Quarrington, thin-bedded dolomitic limestone 15 ft; and Quarrington Quarry [327 379], thin-bedded dolomitic limestone 12 ft, resting on Marl Slate.

A boring [3323 3815] $\frac{1}{2}$ mile N.W. of Quarrington Hill proved rubbly buff dolomite 36 ft, on bedded dolomite and dolomitic limestone (with autobrecciation between 45 ft and 48 ft from top) 101 ft, on Marl Slate. Analyses show CaO to decrease sharply between depths of 41 ft and 57 ft from the surface, though it remains constant at between 31 and 35 per cent (calcitic dolomite) throughout the remaining 80 ft of beds down to the Marl Slate. Fossils found at depths between 20 and 36 ft may indicate that the highest beds in this bore are Middle Magnesian Limestone.   E.A.F., D.B.S.

Over a mile to the east of the scarp Lower Magnesian Limestone is exposed in the flanks of a system of glacial drainage channels near East Hetton Colliery. Here, two old quarries [3600 3700 and 3592 3710] each show about 30 ft of thin-bedded, cream, finely granular, cavernous dolomitic limestone, overlain in the southernmost quarry by flaggy dolomite of the Middle Magnesian Limestone. A third quarry [3580 3725] contains 20 ft of hard, thin-bedded, cream-brown, sub-porcellanous autobrecciated dolomitic limestone lying 20 to 30 ft lower in the succession, and similar rock is intermittently exposed in the valley sides north and west of Town Kelloe.

D.B.S.

Exposures along the scarp south of Quarrington Hill are dominated by the large quarries at Coxhoe Bank and Raisby Hill. About 130 ft of dolomite, thin-bedded except in the lowest 10 to 18 ft, are exposed in Coxhoe Bank Quarries [325 362]. Thin laminated layers are present near the base of the section. Trechmann (1945, p. 339) records '*Productus horridus*' J. Sowerby [*Horridonia horrida*] 'in plenty' at these quarries.

Analysis (E. G. Radley *in* Thomas and others, 1920, p. 82) of a specimen representative of the lowest beds exposed in Coxhoe Bank Quarries shows that these beds are composed of calcitic dolomite. Beds above, shown by three analyses of representative specimens to be of only slightly calcitic dolomite, are relatively uniform in composition. Thin sections (E 11747–E 11750) show the analysed rocks to be composed of fine-grained aggregates of dolomite crystals (0·03 mm), with a small amount of interstitial calcite in E 11748. Cavities are most common in the highest beds.

Analyses of rocks from closely spaced bores sunk by United Steel Co. Ltd. for the Steetley Company Limited in the ground between Coxhoe Bank Quarry and Kelloe, show a slightly calcitic dolomite down to a level 20 ft above the Marl Slate ; below that level impurities increase sharply.

Raisby Hill Quarries (Pl. VIIIA) yield a complete section through the Lower Magnesian Limestone, mainly in the mile-long north face. A thick lenticular mass of limestone occurs near the base of the division in the central part of the north face, but thins rapidly westwards and is absent in the most westerly faces. The margins of this mass are irregular and diachronous, each component bed passing laterally from limestone to dolomite. Basal beds are exposed in a cutting [3403 3532] in the floor of the central part of the quarry where 10 ft of thin-bedded dolomite or dolomitic limestone overlie 6 ft of dolomitic limestone interbedded with rock of Marl Slate lithology. The following section is exposed [347 352] midway along the north face : thin-bedded dolomitic limestone 80 ft, on alternating beds of limestone and dolomitic limestone 30 ft, on light grey compact limestone 90 ft. The base of this lowest bed probably lies at about, or slightly above, the beds exposed in the cutting. The following have been collected from the lower part of the 90 ft bed of the above section : ‘ *Adhaerentina* ’ *permiana* Paalzow, ‘ *Glomospira* ’ *pusilla* (Geinitz), uniserial chambered foraminifera ; *Acanthocladia anceps* (Pl. VI, fig. 13), *Batostomella sp.*, *Fenestrellina retiformis* (Schlotheim), Polyzoan, *?Gen. et sp. nov.* (Pl. VI, fig. 10) [also from Fishburn No. 4 Bore 330 to 357 ft— see note on p. 181] ; *Dielasma elongatum* (Schlotheim), *Horridonia horrida* [several forms—Eisel's varieties, *hoppeiana, initialis, ?auritula*], *?Neochonetes davidsoni* (Schauroth), *Pterospirifer alatus*

(Schlotheim). (Pl. VI, fig. 14), *Strophalosia morrisiana*, Strophaloslid juv. [*?S. parva* King ; flat, circular, ventral valve with spines] ; *Schizodus rotundatus* (Brown), *Streblochondria pusilla* (Schlotheim) ; nautiloid septa ; ostracods.   E.A.F. At the eastern end of the main face the section [3516 3509] is:

MIDDLE MAGNESIAN LIMESTONE   ft
Dolomitic limestone, cream, flaggy, granular and pisolitic ; conformable base   20

LOWER MAGNESIAN LIMESTONE
Calcitic dolomite, cream-brown, finely granular, thick-bedded with very uneven bedding planes, extensively semi-autobrecciated   35
Calcitic dolomite, cream-brown, finely granular, thick-bedded (fairly regularly), cavernous, extensively semi-auto-brecciated in top 8 ft   ...   15
Calcitic dolomite, cream-brown, finely granular, thin-bedded   3
Dolomitic limestone, grey and brown, sub-porcellanous, with abundant calcite-lined veins and cavities, and very irregular bedding (undulate and of variable thickness)   6
Thin alternations of hard grey sub-porcellanous limestone and cream-brown dolomitic limestone in ratios respectively of 4:1 ; the bedding is very uneven, and the grey limestone beds are commonly broken up into isolated lenses ; upper parts of this bed have collapsed in the east face of the quarry due to solution of lower parts of the bed ...   25
Thin alternations in ratio of about 3:1 respectively of unevenly laminated, hard, cream, sub-porcellanous dolomitic limestone and similar rock containing on selected bedding planes large numbers of irregular rods $\frac{1}{8}$ to $\frac{3}{8}$ in diameter, of grey limestone ; fossils are common throughout (see above list)   ...   .   35+

The lowest 35-ft unit of this section corresponds with the upper part of the lowest (90 ft) member of the previous section. Analysis of a specimen from near the base of this bed (Trechmann 1914, p. 248) shows it to be a slightly dolomitic limestone (calculated calcite 95·27 per cent, dolomite 3·20 per cent), whilst analysis (Thomas and others 1920, p. 82) of a specimen from about the middle of the sequence shows it to be calcitic dolomite ($CaCO_3$ 56·12 per cent, $MgCO_3$ 38·16 per cent). Similar analyses are given by Woolacott (1919a, p. 167), who also records (1919b, p. 456) gypsum crystals from cavities in this quarry. It is possible that the cavernous, autobrecciated beds noted some 50 ft above the base of the sequence represent collapse caused by the removal of earlier interbedded sulphates. It is notable that whereas the lower 55 ft of beds in the east of the quarry are even and of uniform dip, those above this level are slightly undulate (see Plate VIIIA). The uppermost 10 to 20 ft of beds are probably of Middle Magnesian Limestone age.

D.B.S.

A number of small exposures between the eastern end of Raisby Hill Quarry and the railway line are all in highly brecciated dolomite and dolomitic limestone, and malachite and chalcopyrite occur in the most southerly of these. The breccia lies close to the position of the Butterknowle Fault. G.D.G.

Further small exposures of thin-bedded dolomitic limestone are seen in one old quarry [3292 3585] near Coxhoe Hall and another [3460 3610] at Low Raisby. Thin-bedded dolomitic limestone is also seen overlying massive beds of similar composition in an old quarry [3257 3579], 550 yd W. of the ruins of Coxhoe Hall.

E.A.F.

South of Garmondsway. Raisby Hill Quarries are bounded on the south side by the Butterknowle Fault, beyond which the outcrop of the Lower Magnesian Limestone is shifted westwards. Although its thickness decreases to less than 150 ft in places, the Lower Magnesian Limestone outcrop is over 2 miles wide around Ferryhill (Plate V ; Fig. 19),

and this reflects topography and an anticlinal structure. The most northerly exposure is an old quarry [330 348], 1350 yd S. of Coxhoe Hall, which shows 20 ft of cream fine-grained or porcellanous thin-bedded dolomite lying a short distance below the top of the division. A more complete section is seen at West Cornforth Quarry [318 345] where 80 ft of dolomite are exposed. Lower beds here are mainly thin-bedded, fine-grained or porcellanous dolomitic limestone. whilst upper beds are paler, thick-bedded, softer and more dolomitic. A thin section from the latter (E 21138) is described by Dr. K. C. Dunham as a ' yellow fine-grained rock, composed of crypto-crystalline dolomite traversed by thin veinlets of coarser carbonate '. A borehole in the floor of the quarry proved a further 63 ft of thin-bedded dolomitic limestone on 10 ft of slightly dolomitic, finely crystalline massive limestone overlying Marl Slate. The only exposure around Cornforth is an old quarry [309 341], 800 yd N.N.W. of Thrislington Hall, where 15 ft of thin-bedded dolomitic limestone show extensive incipient autobrecciation. South of Cornforth, beds high in the division are exposed in an old quarry [318 324] in Thrislington Plantation, where soft fine-grained cream dolomite overlies hard grey-buff finely crystalline dolomitic limestone. Adjacent boreholes show that the Marl Slate lies at a depth of about 130 ft hereabouts. Finely crystalline dolomitic limestone, 15 ft thick, is seen at a second quarry [316 328] nearby. In both quarries the bedding is of uneven thickness and is undulate, as it is at the same horizon at Raisby Hill Quarries (p. 111). South-west of Cornforth there are old quarries on both sides of the Ferryhill Gap. To the west, in Ferryhill Cliff Quarry [300 330], thin-bedded dolomitic limestone with argillaceous partings is evenly laminated in the lower part and passes down to Marl Slate. Baryte and galena occur at several places in this quarry and are also present in a group of small exposures [3000 3321] 200 yd farther north. Loose fragments of flaggy dolomitic limestone from the latter yielded ' *Ammodiscus* ' *roessleri* (Schmid). Some 10 ft of thin-bedded contorted and brecciated dolomitic limestone containing plant debris are exposed 40 yd farther north. An old quarry [334 329] on the east side of Ferryhill Gap shows 32 ft of thin-bedded cavernous dolomitic limestone.

Of several exposures in the lowest beds of the Lower Magnesian Limestone near Ferryhill Station, the best are at Mainsforth Lime Works Quarry [305 321], Rudd Hill Quarry [302 321] and in a railway cutting [303 323]. At each of these localities about 30 ft of thin-bedded fine-grained and sub-porcellanous dolomitic limestone are seen. The underlying Marl Slate and Basal Permian Sands are exposed on the east face of the cutting. Farther east, two quarries [308 329 and 309 328], 700 yd S.S.E. of Thrislington Hall, have recently been opened in an area where boreholes have proved about 100 ft of thin-bedded fine-grained and porcellanous cream dolomitic limestone. Poikilitic calcite was noted in some of the early faces. South of Ferryhill Station excellent sections about 80 ft high are afforded at Chilton Quarry [300 314] where the lower beds are hard thin-bedded grey dolomitic limestones and the upper are softer grey-brown calcareous dolomites containing hard grey ?calcareous nodules. In thin sections (E 20615–20) from this quarry, both slightly dolomitic limestone and slightly calcitic dolomite can be seen, often in the same specimen and separated by a sharp junction. The occurrence of baryte, fluorite and sphalerite at this locality has been discussed at length by Fowler (1957, pp. 258–60). A borehole [3030 3091], 450 yd S. of the quarry, proved 130 to 150 ft of Lower Magnesian Limestone, its top coinciding approximately with the top of the sequence exposed in Chilton Quarry.   G.D.G.

In the south-western part of the district dolomitic limestones at or near the base of the Lower Magnesian Limestone are seen at the following localities: old quarries [2945 3357] 350 yd S.S.W. of East Howle Post Office, thin-bedded 12 ft on thick-bedded 2 ft; railway cutting [279 336] 350 yd S.S.E. of Skibbereen, thin-bedded 2 ft on Marl Slate 9 ft; Pickering Quarry [274 229], thin-bedded 6 ft; old quarry [2751 3276] 300 yd W. of Low Hill House, Dean Bank, thin-bedded 11 ft over thick-bedded 7 ft; and road-cutting [285 327] Ferryhill, thin-bedded with some finely laminated bands towards base 34 ft.   E.A.F.

East of the outcrop. East of Hetton le Hole and Shadforth, in the northern part of the district, most of the shaft and borehole records are old and unreliable, but Hawthorn Colliery Shaft and Mill Hill Bore, Easington can be taken as representative of more modern sinkings.

The shaft section shows:

LOWER MAGNESIAN LIMESTONE	ft
Dolomite, cream to pale grey, hard, fine-grained to porcellanous, with many cavities and extensive autobrecciation ...	67
Dolomite, cream, buff and brown, finely granular, generally well-bedded and with thin laminated beds at intervals ; harder and finer grained in lowest 48 ft	115
Dolomite and dolomitic limestone, grey and grey-brown, thin-bedded, hard, finely crystalline, with several thin laminated beds and films of dark grey bituminous clay	18
MARL SLATE	—

The bore proved :

LOWER MAGNESIAN LIMESTONE	
Dolomite, buff-brown, hard, finely crystalline to sub-porcellanous, extensively autobrecciated, with beds of soft granular dolomite	50
Dolomite, cream, finely crystalline, saccharoidal	24
Dolomite and dolomitic limestone, mottled grey and buff with extensive incipient autobrecciation and many cavities	78½
Dolomite, cream-buff, fine- to medium-grained, saccharoidal, thick-bedded and massive, with thin knobbly beds of grey fine-grained calcareous dolomite	67½
Dolomite, cream-buff, fine-grained, saccharoidal, flaggy, with films of brown calcareous clay on knobbly bedding planes	15
Dolomite, as above, but predominantly massive	10
MARL SLATE	—

Details of the Lower Magnesian Limestone of Offshore Nos. 1 and 2 bores are given by Magraw and others (1963). The uppermost 48 ft of the division are exposed [4268 4436] in a cross-measures drift leading down from the western end of a tunnel at Easington Colliery, and consist of thin-bedded, fine-grained and sub-porcellanous dolomitic limestone. There is partial autobrecciation near the top of this sequence, which is conformably overlain by granular dolomite of Middle Magnesian Limestone age.

East of Shadforth and Kelloe the Lower Magnesian Limestone has been cut in a number of bores and shafts, but it has been examined in detail only at Blackhall Colliery South Shaft where Trechmann (1913, p. 213) recorded it as 240 ft thick. The uppermost beds of the division are described as ' yellow, bedded friable limestones' containing plant and polyzoan debris. Under the classification used here, however, these are referred to the Middle Magnesian Limestone. Trechmann (1914, p. 247) gave 5 analyses from the ' Lower Magnesian Limestone ' at Blackhall Colliery ; all are of slightly calcitic dolomite. The lowest bed analysed, a hard dark grey rock with carbonaceous partings, contains 4·52 per cent $CaCO_3$ in excess of the molecular proportions of dolomite, and this calcite is visible in thin sections of the rock. Although most of the records of the remaining bores and shafts east of the outcrop area are generalized and unreliable, many show a twofold sub-division into yellow and brown ' limestone ' overlying harder grey, dark grey or blue ' limestone ', the latter often 30 to 40 ft thick.   D.B.S.

South of the Butterknowle Fault modern borings are more numerous, particularly in the Ferryhill–Embleton area. Records between Ferryhill and Bishop Middleham indicate great lateral variations of lithology, but prove a widespread general triple subdivision of the Lower Magnesian Limestone similar to that noted at outcrop (pp. 106-8). A generalized section is:

	ft
Dolomite, yellow to buff, thick - bedded, calcitic, finely granular or finely crystalline	40 to 70
Dolomite, buff to brown in uppermost 50 to 60 ft, often grey in lowest 25 to 50 ft ; thin-bedded, slightly calcitic, very finely crystalline or sub-porcellanous, widely auto-brecciated	95 to 120
Dolomitic limestone or calcitic dolomite, cream to grey, thick-bedded, finely granular, locally argillaceous and / or laminated in lower part	4 to 12

It is apparent from analyses (abstracted by kind permission of the Steetley Company Limited) that although Woolacott (1919a, p. 166) recorded 100 ft of ' highly calcareous rock' from equivalent beds in the Cotefold Close Bore, 2 miles N.N.E. of Embleton, there is no local thick development of calcitic rock low in the sequence as there is at Raisby Hill Quarry. On the contrary, some of the higher beds of the division are fractionally more calcitic than the lower. In the lowest of the three units there is a downward increase to up to 10 per cent of non-carbonate matter. The middle and upper units, however, consist mainly of slightly calcitic dolomite with between 1 and 2 per cent of non-carbonate matter.

Where mottling is well developed in the middle unit, bedding surfaces are markedly knobbly or even semi-stylolitic and bear thin black films of carbonaceous calcareous clay. Thin sections (E 27634–9) of representative specimens from depths between 109 and 131 ft (4 to 26 ft above the base of the division) in Mainsforth Low Main Series No. 9 Bore [3202 3305] are described by Dr. R. Dearnley as ' dolomite siltstone, consisting almost entirely of a mosaic of silt-grade dolomite grains with scattered coarser patches'. Collapse-breccias were seen 108 ft above the Marl Slate in a bore [3141 3230] 900 yd N. of Mainsforth.

Fossils are rare and are found mainly in the lowest 20 to 30 ft of the division. They include *Horridonia horrida* and *Lingula credneri* from Mainsforth No. 2 (1955) Bore [3112 3200], and small

?brachiopods from Mainsforth No. 4 Bore [3227 3215]. A short distance south of the district, in a borehole [2791 2830] near Rushyford, brachiopods including *Lingula sp.* and foraminifera occur in the lowest beds of the division, and foraminifera, polyzoa and plant debris are recorded about 50 to 65 ft higher. Exceptionally, ostracods are found 115 ft above the base of the division in Mainsforth No. 7 Bore [3282 3186].

Boreholes east of Bishop Middleham are less closely spaced than those to the west, and many were cored only in the lower part of the Lower Magnesian Limestone. In this area the division reaches a maximum thickness of over 225 ft, but thins to less than 125 ft near the southern margin of the district (Fig. 19). The triple lithological subdivision recognized west of Bishop Middleham persists throughout the Fishburn–Embleton area, though the lowest member locally becomes progressively more difficult to recognize towards the east. The middle member of the sequence is characterized by pronounced mottling, associated with stylolites and knobbly (semi-stylolitic) bedding planes—features previously described in the ground east of Embleton (Woolacott 1919a, p. 164; Smith *in* Magraw and others 1963, p. 171). The lower part of the middle member of the sequence tends to be calcareous as at Raisby Hill Quarry (p. 111), and this is illustrated in Table 2 (p. 116) where an analysis (Lab. No. 1451) from Ten-o'-Clock Barn No. 2 Bore shows only 0·21 per cent MgO to contrast with approximately 18 to 20 per cent measured in other beds in the bore.

Thin sections of the analysed rocks are described by Dr. K. C. Dunham as follows:

E 21051. Yellow fine-grained calcitic dolomite, composed of dolomite rhombs (average 0·06 mm) with calcite cement.

E 21050. Yellow fine-grained dolomite, composed of dolomite grains (average 0·02 mm). Pore spaces (empty) reach 0·8 × 0·3 mm.

E 21049. Grey fine-grained dolomite, composed of dolomite grains (average 0·03 mm) and a little calcite. Small pore spaces.

E 21048. Light grey dolomite composed of rhombic dolomite grains (average 0·03 mm) with many small empty pores and some pyrite.

E 21047. Grey fine-grained calcite mudstone, composed of even-grained calcite grains averaging 0·01 to 0·015 mm.

Quartz is present in small quantities in slides E 21047–50. It is also seen in thin sections (E 19806–7, E 19809–10) of fine-grained compact limestone from equivalent beds 7, 11½, 17½ and 24¾ ft respectively above the base of the division in the Whin Houses Bore. Minerals such as baryte, fluorite, pyrite, chalcopyrite and sphalerite recorded from the higher beds of the Lower Magnesian Limestone in the Ten-o-'Clock Barn, Whin Houses and other bores in the area (Fowler 1943, pp. 41–51; 1957, pp. 251–65), have been found also in many of the more recent bores.

Fossils collected from Ten-o'-Clock Barn No. 1, Fishburn No. 3, Fishburn No. 4, Fishburn 'D', and the Whin Houses bores are listed on pp. 173-6. Most of them are from the lower 50 ft of the Lower Magnesian Limestone, but in Fishburn 'D' Bore, where the division is about 190 ft thick, fossils (chiefly lamellibranchs and foraminifera) extend upwards to within 70 ft from the top.

Boreholes east of Embleton show a continuation of the general lithological sequence noted above, but the lowest unit becomes distinguishable only with difficulty from the middle unit of the division, and in the extreme east of the district even the middle and upper units tend to lose their separate identities. Details of the Lower Magnesian Limestone in these boreholes are given by Magraw and others (1963). D.B.S., E.A.F.

TABLE 2

ANALYSES OF CARBONATE ROCKS FROM TEN-O'-CLOCK BARN No. 2 BORE

	378	452	452½–457	488–502½	516
Depth (in ft)	378	452	452½–457	488–502½	516
Lab. No.	1455	1454	1452	1451	1453
Slice No.	E 21051	E 21050	E 21048	E 21047	E 21049
SiO_2	0·15	1·89	2·69	0·80	3·71
Al_2O_3	0·13	0·40	0·59	0·22	1·11
Fe_2O_3	2·63	0·71	0·08	Tr	Tr
FeO	NIL	*0·95	*1·30	*0·17	*1·48
MgO	17·81	19·60	19·45	0·21	18·27
CaO	32·28	30·37	29·16	54·79	29·83
Na_2O	0·03	0·04	0·05	0·03	0·05
K_2O	NIL	0·08	0·08	0·03	0·20
$H_2O > 105°C$	0·68	0·45	0·41	0·16	0·53
$H_2O < 105°C$	0·16	0·15	0·13	0·06	0·12
TiO_2	NIL	0·02	0·02	0·01	0·05
P_2O_5	0·02	0·02	NIL	0·01	0·03
MnO	0·18	0·13	0·15	0·10	0·18
CO_2	44·54	45·23	44·52	43·29	43·82
SO_3	0·27	0·04	0·13	0·07	0·07
Cl	Tr	Tr	Tr	Tr	Tr
F	Tr	Tr	Tr	Tr	Tr
FeS_2	0·12	0·09	1·22	0·14	0·51
BaO		NIL	Tr	Tr	Tr
C	0·04	0·05	0·14	0·09	0·15
Ba calculated as $BaSO_4$ ...	1·72				
Total	100·76	100·22	100·12	100·18	100·11
Calculated mineral composition					
Dolomite	81·9	90·2	89·4	1·0	84·0
Calcite	12·7	5·1	3·3	97·4	7·6
Siderite	0·0	0·6	1·2	0·1?	1·5
Gypsum	0·5	0·1	0·2	0·1	0·1
Baryte	1·7	0·0	Tr	Tr	Tr
Pyrite	0·1	0·1	1·2	0·1	0·5
Quartz	0·0	1·4	2·0	0·6	2·4
Muscovite	} 0·3	1·0	1·5	{ 0·0 / 0·4	} 3·0
Kaolinite					
Limonite	3·3	1·2	0·3	Tr	Tr
Organic carbon	Tr	0·1	0·1	0·1	0·2

Analyst G. A. Sergeant

* Figure approximate only, due to presence of organic matter.

The base of the Lower Magnesian Limestone in this bore is at 518 ft.

MIDDLE MAGNESIAN LIMESTONE

Middle Magnesian Limestone beds occupy most of the outcrop of Permo-Triassic formations north of the West Hartlepool Fault (Plate V), a circumstance arising mainly from a combination of a low easterly dip and the gradual easterly fall of the land surface. The division as a whole is interpreted as comprising the deposits of a single major evaporite cycle, although chemical precipitation probably commenced in Durham only at a fairly late stage.

The earliest deposits of the Middle Magnesian Limestone are a relatively thin group of fine-grained calcitic dolomites, here referred to as transitional beds (p. 90, Fig. 17). The junction of these beds with the underlying Lower Magnesian Limestone is usually indistinct, indicating a gradual and irregular dilution of the highly saline water in which the uppermost part of the Lower Magnesian Limestone is thought to have been deposited.   D.B.S.

The earliest transitional beds are barren, but with increasing dilution a limited shelly fauna developed in central and western parts of the district. This fauna is characterized by *Astartella vallisneriana* (King) [Pl. VI, fig. 12], and includes also species of *Bakevellia, Permophorus* and *Schizodus,* as well as foraminifera and ostracods. The later transitional beds, deposited under nearly normal marine conditions, contain a more varied fauna which includes, in addition to the above, a number of species not previously recorded from this part of the sequence. These include the brachiopods *Horridonia horrida* and *Streptorhynchus pelargonatus* (Schlotheim). Also present, mainly as isolated randomly oriented tubes, is the alga *Tubulites permianus* (King), hitherto regarded as an index fossil of the Upper Magnesian Limestone.

D.B.S., J.P.

Perhaps under the influence of longshore currents, shells accumulated in the eastern part of the district in sufficient numbers to form an elongate bank (reef ' A ' of Trechmann 1925, p. 139), which subsequently became the locus of reef formation. The fauna of this shell-bank is notable for the abundance of such brachiopods as *H. horrida, Orthothrix excavata* Geinitz, *Dielasma elongatum* and *Stenoscisma schlotheimi* (von Buch), lamellibranchs including *Bakevellia antiqua* (Münster) and *B. ceratophaga* (Schlotheim) ; and the presence of cryptostome polyzoa such as *Acanthocladia anceps, Fenestrellina retiformis, Synocladia virgulacea* (Phillips) and *Thamniscus dubius* (Schlotheim). Here too are found rarer fossils, including reef-sponges, echinoids, the lamellibranch *Solemya biarmica* de Verneuil and the gastropod *Macrochilina symmetrica* (King).

The formation of the shell-bank and superincumbent reef brought about a fundamental change in depositional environments on their western and eastern sides, so that three distinct and contrasting facies—lagoonal, reef and basinal—can be recognized above the transitional beds (Fig. 20). These facies are not recognized in the highest beds of the Middle Magnesian Limestone which are basinal dolomites deposited over the reef-flat after the latter had become submerged.

On the protected, western lagoonal side of the reef, accumulation in shallow water of a varied sequence of granular oolitic and pisolitic carbonates, now dolomitized, lagged little behind the rapid upward growth of the reef. This is

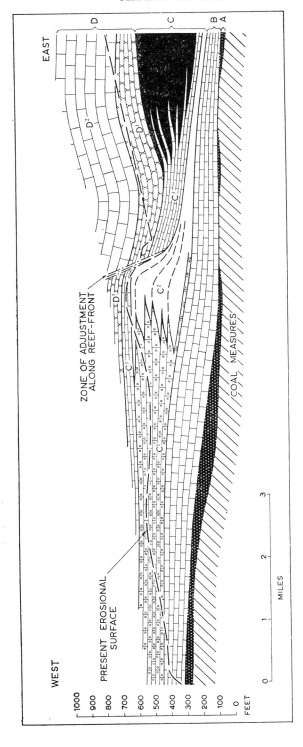

FIG. 20. *Generalized horizontal section showing the relationships of the lithological divisions of the Magnesian Limestone*

indicated by the widespread shallow-water depositional features, especially current-bedding, in most of the lagoonal deposits, and by the interdigitation of these sediments with the uppermost, largely algal, deposits of the reef. In contrast, deposition in deep water on the eastern (basinal) side of the reef was comparatively slow except in the immediate vicinity of the reef front, where fore-reef talus forms a wedge-shaped apron. This relatively slow accumulation can be inferred in a series of tunnels (p. 136, Fig. 21) driven through the Middle Magnesian Limestone at Easington Colliery, where detrital off-reef beds are only 30 to 50 ft thick as compared with over 250 ft of equivalent reef rock some 200 to 400 yd to the west. Corresponding lagoonal deposits probably exceeded 200 ft in thickness.

During the early stages of reef formation, growth was predominantly upward, and the lower parts of the reef are composed of massive biohermal dolomitic limestones and dolomites containing a prolific fauna of stenohaline brachiopods, lamellibranchs and polyzoa, together with scattered poorly preserved stromatolitic (algal) incrustations. The lower part of the reef appears to have been built mainly by cryptostome polyzoa, but it is thought (Smith 1958) that marine calcareous algae also contributed. As the reef was built upwards shallower water encouraged more rapid growth of calcareous algae, and at the same time increasing salinity first led to stunting, and then to gradual extinction of many of the earlier faunal elements (Trechmann 1913, p. 203). Consequently the uppermost parts of the reef are almost wholly of algal origin and contain a wide variety of growth-forms (Smith 1957), of which good examples are seen in Hesleden Dene and in Whelly Hill Quarry, Hart. Growth-forms similar to some of these are found today in Shark Bay, Western Australia (Logan 1961), where they are being formed by fixation of calcareous sediment in the intertidal zone of hypersaline lagoons.

D.B.S.

During the process of extinction of the shelly fauna, the rare forms (p. 117) found in the initial shell-bank were the first to die out. Amongst brachiopods, species such as *Pterospirifer alatus* and *Cleiothyridina pectinifera* (J. de C. Sowerby) were the earliest casualties, followed by *Horridonia horrida, Stenoscisma schlotheimi* and *Orthothrix excavata*. The most adaptable was *Dielasma* which survived almost until the end of reef-formation. Lamellibranchs, less abundant than brachiopods at first, became relatively more prominent later, although the number of species was reduced. The more durable forms include *Bakevellia antiqua, B. ceratophaga, Pseudomonotis speluncaria* (Schlotheim) and *Parallelodon striatus* (Schlotheim). The number of gastropod species also became reduced, although in the later stages of reef-growth the remaining species achieved considerable prominence and locally dominated the fauna.

J.P.

In addition to the vertical changes noted above, there is marked lateral variation in the faunal assemblages of the reef. The rocks deposited near the lagoonal margin of the reef, for instance, contain a fauna dominated by gastropods and lamellibranchs ; they are best seen farther north, in the Sunderland (21) district. Those formed at the reef-crest also contain many lamellibranchs, but they are associated with pedunculate brachiopods, with only rare gastropods and retiform polyzoa. Faunas of the reef-slope, by contrast, are dominated by retiform polyzoa, crinoids, lamellibranchs and larger brachiopods. The distribution of these assemblages appears to be comparable with that found in the Capitan Reef of the Guadalupe Mountains (Newell and others 1953, fig. 79).

The varied growth forms of stromatolites also reflect different environments. Those formed on the upper part of the reef-slope and at the reef-crest are large and complex, those formed on the reef-flat are generally small and uniform, whilst those formed at the lagoonal margin are of intermediate size and complexity and are similar to modern lagoonal and bay-head forms. Stromatolites on the reef-slope form steep sinuous sheets which persist to at least 40 ft below the reef-crest. At lower levels they are progressively less prominent and are almost unrepresented in the talus at the foot of the reef-slope. The precise conditions of formation of a group of large growth forms on the reef-flat in Hesleden Dene is in some doubt, as there appear to be no modern analogues, but it seems possible that they formed in a high energy zone on a newly submerged reef platform. Oncolites are found mainly in the lagoonal deposits near the reef, but also form lenticles within the reef-flat deposits.

The vertical distribution of faunas in the lagoonal environment reflects the same increasingly unfavourable conditions as those of the reef. Forms found in the upper part of the transitional beds at the base of the Middle Magnesian Limestone survive in the overlying lagoonal beds, but show a progressive upwards impoverishment, and are virtually extinct about 80 ft above the base of the division. Exceptions are found only near the reef and these seem to be derived from the reef itself, for they are found nowhere else in the lagoonal beds. The presence in the lagoon of only euryhaline forms and the paucity of stenohaline brachiopods, polyzoa and crinoids, points to a less hospitable (metahaline to hypersaline) environment than that found in corresponding parts of the reef, and the evidence suggests that for a considerable period after the indigenous lagoon fauna became extinct a restricted shelly fauna continued to survive on the reef. However, throughout the period of reef-growth, quantities of derived bioclastic material accumulated in the lagoonal deposits adjacent to the back-reef, forming a narrow belt of coquinas and detrital calcarenites which thin rapidly westwards. Deposits of this type are well exposed in Hesleden Dene and also occur in the Mill Hill Bore, Easington. The middle and later periods of reef-growth are represented in the lagoon predominantly by oolitic rocks, which contain abundant compound pisoliths (oncolites) concentrated chiefly in a belt a few hundred yards wide behind the reef. Similar pisoliths are reported (Newell and others 1953, p. 120) from lagoonal beds adjacent to a reef in the Carlsbad formation of New Mexico and western Texas where they are believed (Johnston 1942, p. 213) to be of algal origin.

The well-defined margins of the preliminary shell-bank and early reef were partially modified as the lagoon filled with sediments, and it is clear that reef growth at times extended both backwards over lagoonal sediments and forwards over talus derived from erosion of early reef-rock. As a result of the differing rates of sedimentation behind and in front of the reef, a sharply asymmetrical form was produced, with back-reef and lagoonal deposits of low surface relief being separated from a steeply-plunging reef-front by a reef-top platform up to a mile wide (Fig. 20).

Interdigitation at the margins of the reef, particularly on the lagoonal side, suggests that reef formation may have been intermittent and that the reef might more properly be regarded as a composite structure built up of a succession of reefs formed, in general, at much the same place. In bore-holes in Hesleden Dene, however, the superimposition of basinal and lagoonal beds suggests that here the reef-front during the late stages of reef growth may have lain west of its earlier position. The course of the reef-front (Pl. V), as inferred during the resurvey, is typical of many present-day reefs, with wide embayments, and in some places, prominent spurs projecting seawards. It is not known whether the reef is a continuous mass or is divided wholly or partially by channels of varying depth and width, but the form of the outcrops from Horden northwards suggests that at least the higher parts of the reef sequence are locally absent. A deep debris-filled cleft in reef-rock at Fox Cover Quarry, Dawdon, is interpreted as a former reef-channel.

Available evidence suggests that reef formation may have continued longer in central and southern parts of the district than in the north. For example, algal rocks of reef-facies appear to persist to a higher horizon in the south, and only a few miles north of the district Burton (1911, p. 301) records wind-blown sand grains in reef-limestones at Tunstall Hills, near Sunderland. This may have been due to a gentle southerly tilting of the whole area, bringing about a gradual emergence of the northern parts of the reef and continuing subsidence of central and southern parts. With subsidence, reef-growth continued until the tolerance of calcareous algae was exceeded, either by increasing salinity or by descent of the reef-top below the light compensation level. The uniformity of the bedded algal reef-rocks south of Castle Eden indicates formation on a wide slightly emergent platform rather than on an exposed reef-flat, however. Eventually increasing salinity and too-rapid subsidence may have operated concurrently in bringing reef-formation to an end. Subsequent temporary returns to more favourable conditions are indicated by algal rocks interbedded with post-reef granular carbonates overlying the reef at a number of eastern and southern localities.

The extent of the reef southwards from West Hartlepool is uncertain. However, indirect evidence yielded by a gravity survey (Bullerwell 1961, pp. 41, 60), and by borings, suggest that the reef may continue as far south as Seaton Carew where, as suggested by Trechmann (1942, p. 320), it seems likely that the reef has a more subdued relief than farther north and that it interdigitates with normal carbonate sediments around its southern margin. South and west of Seaton Carew, the Middle Magnesian Limestone was probably laid down in semi-open shelf conditions rather than in a lagoonal environment, and this would account for the difficulty experienced in distinguishing lower beds of the division from the shelf deposits of the Lower Magnesian Limestone in boreholes at Throston and Brierton. Similar difficulty is experienced east of the reef in more northerly parts of the district.

During the formation of the reef, contemporaneous erosion gave rise to wedge-shaped talus fans at the foot of the reef-front. These were progressively buried as the reef advanced seawards so that the detrital limestones and talus fans now exposed in front of the reef are mainly derived from erosion of the latest reef-growth. Excellent examples of reef talus are exposed at Black Halls Rocks (Pl. XII) and in Crimdon Beck. No examples of detritus fans of early reef-growth are seen at the surface in the district, but shelly dolomite of this facies is present in underground sections at Easington Colliery and in boreholes at Hesleden Dene and Hartlepool. In the boreholes at Hartlepool these beds contain fragments of black fossiliferous siltstone (see p. 146), a rock-type found nowhere else in the Permian of Durham.   D.B.S.

The fauna of the basinal beds of equivalent age to the reef changes with increasing distance from the foot of the reef slope. Near the latter, as seen in the underground sections at Easington Colliery, the fauna is closely comparable with that of the contemporaneous reef from which is is presumably derived. Farther away, as in the boreholes at Hartlepool, species found in the reef are common but there are also uniserial, chambered foraminifera and fish debris which are rare or unknown in the reef. The indigenous fauna of the basinal beds is known only from Offshore No. 1 Bore, 3 miles from the reef; here Chonetids and coiled, tubular and uniserial, chambered foraminifera are the commonest forms.   J.P.,   D.B.S.

It is inferred from the high rate of reef-growth compared with that of accumulation of off-reef material that the front (eastern) face of the reef was ultimately over 200 ft high with a slope of between 30° and 85°. It effectively limited subsequent major sediment deposition to the basinal areas to the east until over 200 ft of sediments had accumulated. Because of the lack of features common to both reef and basin environments, it is not yet possible to equate any basinal bed with the end of reef-growth. However, the virtual absence of a shelly fauna in the highest parts of the reef implies a high salinity and it is thus possible that chemical precipitation of carbonates or even sulphates in adjacent deep water was synchronous with the latest stages of reef growth (Hickling and Holmes 1931, p. 255). The post-reef beds in the basin area comprise fine-grained barren dolomites succeeded by a thick series of evaporites in which interbedded carbonates and sulphates deposited alongside the reef wall pass eastwards into thick anhydrite with subordinate dolomite. Immediately east of the reef these beds are represented by 70 to 140 ft of collapse-brecciated granular and oolitic dolomites. The former presence of anhydrite in this group of strata was inferred by Trechmann (1913, p. 195) from the evidence of the collapse-breccias and of widely disseminated traces of sulphate, and strong supporting evidence comes from the position of the base of the Upper Magnesian Limestone which in places appears to be over 100 ft lower than when originally deposited. The western edge of the collapse-brecciated beds is marked by an adjustment zone of faults and mechanically brecciated rocks which lies immediately east of the reef front (Fig. 20). Borehole evidence (Fig. 22) shows that the zone of interbedded carbonates and sulphates is narrow or absent at West Hartlepool, but appears to widen to 3 to 5 miles in the northern part of the district and in adjacent parts of the Sunderland district. Farther east (Magraw and others 1963) boreholes at Hartlepool and in the offshore area show that the horizon of these beds is occupied by the Hartlepool Anhydrite which is 500 ft thick in N.C.B. Offshore No. 2 Bore. Fowler (1944, p. 205) has suggested that the anhydrite has been formed mainly by replacement of dolomite and there is evidence of such replacement at many localities. It is believed, however, that this process is of relatively minor importance and that on the basis of form and relationship the anhydrite should be regarded as a largely primary deposit. Geophysical evidence (Fig. 23) coupled with boreholes at Hartlepool show that the lateral transition from the predominantly carbonate to the predominantly sulphate phase is fairly sharp, and at Hartlepool the transitional zone is clearly traceable by a marked gravity gradient. A comparable gradient underlies Seaton Carew, and may indicate that the anhydrite which is 35 ft thick and lies at the top of these beds in the Seaton Carew Bore thickens rapidly eastwards.

Whilst anhydrite was being deposited in the eastern part of the district, sediments on the submerged reef-platform were largely of granular and oolitic carbonates which are indistinguishable from the highest lagoonal beds to the west of the reef. In the southern part of the district the lagoon appears to have been wider and deeper than farther north, and to have opened out into a shelf area where sulphates were deposited in alternation with carbonates. These sulphates extend from adjacent parts of the Stockton (33) district into the southern margin of this district and may represent an extension round the southern end of the reef of parts of the thick anhydrite laid down in the basin. In this southern area anhydrite beds are especially common at the top of the Middle Magnesian Limestone, where, west of Billingham, they are interbedded with red mudstone (Hollingworth 1942) and are together thought to correspond with the Middle Permian Marls of Yorkshire (p. 95). There is no evidence that sulphates ever formed primary deposits in the more restricted northern parts of the lagoon, although abundant cavities testify to the former presence of disseminated sulphates (Guy 1911). Chemical precipitation of both carbonates and sulphates was eventually halted by an influx of fresher water under which the widely distributed impure Flexible Limestone at the base of the Upper Magnesian Limestone was laid down.

D.B.S.

DETAILS

In the following account, lagoonal, reef and basin facies are treated separately, but details of the transitional group are included with those of the overlying facies for these beds are generally thin and the junctions at top and bottom are gradational. Details of Middle Magnesian Limestone beds overlying the reef are included in the section dealing with the basin facies. Where possible, faunal lists are included in the text, but some longer lists from exposures and from boreholes are given on pp. 169–72.

Lagoonal Facies

Beds of lagoonal facies occupy most of the outcrop of the Middle Magnesian Limestone. They consist of a thick series of granular, oolitic and pisolitic carbonate rocks which are almost universally dolomitized. In most of these rocks the dolomite forms rhombs of silt- to sand-grade, but is widely distributed irregular patches the dolomite has recrystallized into platy crystals up to 5 mm across arranged in sheath-like and semi-radial aggregates with a loosely cemented matrix of smaller rhombs. This recrystallization texture, which will henceforth be referred to as ' felted ', is virtually confined to the lagoonal beds of the Middle Magnesian Limestone.

Murton. Two small outcrops occur at New Hesledon, ½ mile E. of Murton Colliery shafts. The more northerly [4095 4772] is beside the A.19 (Murton to Easington) road, 350 yd N.N.W. of the Water Works, and comprises about 20 ft of thin- and medium-bedded cream-white finely granular calcitic dolomite in units alternately poorly laminated and finely brecciated, most of the breccia fragments being apparently structureless. The other exposure [4103 4711] is in an old quarry at the north end of the Water Works ; it consists of about 20 ft of thin- and medium-bedded cream sac-charoidal calcitic dolomite, with a wedge of brecciated dolomitic limestone near the western edge of the quarry. Out-crops [4125 4648 and 4140 4636] of similar lithology flank the A.19 road 520 yd S.S.E. of the Water Works, and 200 yd farther S.E. in the eastern side of a glacial drainage channel. Poorly de-veloped irregular lamination is common, and beds and lenses of breccia occur throughout. All these outcrops lie close to the eastern limit of the lagoonal facies, and some of the laminated beds may be partly algal in origin.

South Hetton. Three small exposures occur near South Hetton Station. In an old quarry [3765 4524], 190 yd W.N.W. of the station, about 6 ft of gravel over-lie soft cream finely granular vuggy dolomite, which contains a thin pisolitic bed near the base. A second old quarry [3759 4507], 270 yd W.S.W. of the station, contains about 14 ft of cream, thin-bedded friable finely granular dolo-mite and dolomitic limestone, overlain by about 8 ft of gravel. The dolomite and limestone of the second quarry are prob-ably younger than the dolomite of the first and are composed largely of sand-grade dolomite rhombs commonly strongly cemented into sub-spherical aggregates ; irregular patches with a strongly 'felted' texture are common. At the south-eastern end of this quarry a zone of harder, more calcareous rock is slightly brecciated and has numerous cracks and widened joints. The third exposure [3744 4515] lies on the north side of the channel, 420 yd W. of the station, and contains about 30 ft of highly

weathered soft cream granular dolomite similar to that in the quarry first described.

High Haswell. The main exposures around High Haswell are in the sides of the road north-east of the village and on the hillside 500 yd to the south. The roadside section [368 438] includes about 30 ft of hard thin-bedded cream-grey fine-grained dolomitic limestone, which contains patches of saccharoidal dolomite and scattered traces of small concentric structures which may be recrystallized ooliths. The hillside section [366 434] consists of a number of small outcrops in which about 40 ft of massive cream finely granular dolomite are exposed. Beds at several of these outcrops are highly recrystallized in parts and ex-cellent examples of coarsely saccharoidal (aggregated) and 'felted' dolomite occur. In some cases the matrix of fine dolomite rhombs has been removed by subaerial weathering and only the frame-work of platy dolomite crystals remains.

Haswell to Pesspool Hall. In road and railway cuttings and an old quarry at the north end of Haswell village, a total of about 30 ft of bedded calcitic dolo-mite are visible. The rock is thinly but irregularly bedded and many layers are lenticular ; it is cream in colour, and of granular texture, varying from silt to coarse sand in grade. Cavities occur throughout, and vague traces of cross-bed-ding are seen in places. About 15 ft of beds of similar lithology are exposed in an old quarry [3794 4307], 400 yd S.S.E. of Haswell Station, though parts of the beds are harder and more calcitic than at the previous locality. Abut 7 ft of calcitic dolomite with both saccharoidal (sand grade) and 'felted' textures are exposed in the sides of a large swallow hole [3815 4319] about 80 yd S. of Pess-pool Hall.

At Tuthill Quarry [390 423], 1 mile S.E. of Haswell Station, about 130 ft of beds are seen. The uppermost 100 ft consist of a varied sequence of calcitic dolomites ranging from mud to sand in grade, in which an original oolitic texture is preserved in scattered less-altered patches. Cross-bedding is well exhibited in these patches and is also recognizable in many beds which now have a granular texture and are composed almost entirely of dolomite rhombs. Differential cementation into sub-spherical aggregates of dolomite rhombs is common, as at South Hetton and elsewhere in the lagoonal facies, and ' felted ' texture (E 29448) is locally well developed. A bed of oolitic dolomite of silt-grade, containing many compound ooliths and pisoliths, occurs near the top of the section ; small amounts of free calcite have been noted. Samples of oolitic and granular rock from this quarry were analysed (Trechmann 1914, p. 246) and were found to be of dolomite with only 2 to 4 per cent $CaCO_3$ in excess of that required by the molecular proportions of pure dolomite.

The lowest part of the quarry is in bedded cream finely granular dolomite, containing no traces of cross-bedding nor of ooliths. About 10 ft of carbonate silts and sands, cross-bedded but non-oolitic, form a transitional group between the upper 100 ft and the lower 20 ft of beds. A number of indeterminate shells were found in this lower 20 ft during the resurvey ; they are taken to indicate an horizon near the base of the division. Trechmann (1914, p. 246 ; 1945, p. 352) also recorded the presence of fossils at this locality but he named only ' *Astarte vallisneriana* '.

A feature of the beds at and near Tuthill Quarry is a marked solution-widening of some joints and the formation of large solution cavities in a prominent faulted zone. Several such cavities occur at the south end of the quarry, and other may be located beneath large sink holes in adjacent fields.

Shotton Colliery. Shotton Quarry [399 418] 1050 yd N.E. of Shotton Colliery Station, contains about 25 ft of thin-bedded cream and grey oolitic calcareous dolomite of silt-grade with a few poorly preserved lamellibranchs. Cross-bedding is conspicuous in some beds, but no preferred orientation of foreset laminae is apparent. One bed in the middle of the quarry face contains large numbers of compound ooliths and pisoliths, mostly of irregular shape and of apparently random distribution and orientation. This rock closely resembles

that at South Hetton and near the top of Tuthill Quarry. In thin sections (E 29438–9) the ooliths are seen to be largely replaced by fine-grained dolomite, the original texture being almost completely obliterated. Small amounts of free calcite have also been determined.

Similar rocks are exposed in the sides of a small glacial drainage channel some 300 yd N. of the quarry, but much of their original structure has been obliterated by the growth of aggregated coarsely crystalline dolomite. Sections at the base of the railway cutting north and south of Shotton Colliery Station show a few feet of thin- and medium-bedded cream finely granular and granular calcitic dolomite similar to that found 20 to 30 ft above the base of Tuthill Quarry.

The most complete section of these beds in the Shotton area is in the Mill Hill Bore, Easington ; starting beneath reef limestone, it comprises :

	ft.
Dolomite, cream, soft, bioclastic, with many gastropods	9
Dolomite, buff and grey, soft, granular, with small gastropods and lamellibranchs	30
Dolomite, cream, granular, interbedded with very soft cream oolitic and pisolitic dolomite with scattered lamellibranchs ...	45
Dolomite, cream, finely saccharoidal, with scattered shells and foraminifera in upper 7 ft ...	78
Dolomite, grey and cream, hard, fine-grained, with scattered moulds of foraminifera and *Astartella vallisneriana*	23

The uppermost 39 ft of these beds show close lithological and faunal affinities with reef dolomites located immediately to the east. The lowest 23 ft represent the transitional group at the base of the Middle Magnesian Limestone.

Ludworth. Middle Magnesian Limestone is poorly exposed near Ludworth in small quarries [3709 4140 and 3709 4098], respectively 400 yd W.N.W. and 500 yd S.W. of Harehill Farm. Both are largely overgrown, and contain cream and grey, thin- and medium-bedded granular un-fossiliferous calcitic dolomite and dolo-mitic limestone.

Wheatley Hill. Dolomite with a marked ' felted ' texture occurs in the sides of a sink-hole [3768 4008] in an alluvial flat 780 yd W. of Low Crow's House. The dolomite is thin-bedded, except near a zone of movement planes where it has a saccharoidal texture and is harder and more massive. In a railway cutting [3844 3991] about 100 yd S. of Low Crow's House, about 8 ft of thin-bedded cream saccharoidal (sand-grade) calcitic dolomite are present. An old quarry 40 yd S.E. of Low Crow's House is now filled, but rock obtained from it and built into local walls is of similar lithology to that in the cutting. Another small old quarry [3802 3942] immediately north of Wheatley Hill is in about 9 ft of soft cream thin-bedded calcitic dolomite. Most of the rock is oolitic and unfossiliferous, but a thin non-oolitic bed about the middle of the section has yielded: *Tubu-lites permianus ; ' Adhaerentina ' permiana, ' Ammodiscus '?, ' Glomospira ' gordialis* (Jones and Parker), ' *G* '. *milioloides* (Jones, Parker and Kirkby), ' *G.* ' *pusilla ; Astartella sp., Bakevellia antiqua, Permo-phorus sp. ; ostracods.* The section also includes an 8-in bed containing ooliths and pisoliths similar to those described from Shotton, Haswell and South Hetton. In thin section (E 29440) this rock is seen to be largely recrystallized and to contain dolomite both as rhombs and as small platy crystals in the cores of ooliths and in the matrix. The origin of the compound ooliths is uncertain, but it appears likely that the platy dolomite crystals represent the early stages of the development of the ' felted ' texture noted earlier.

Castle Eden. Rocks of lagoonal facies crop out intermittently in Castle Eden Dene from about 450 yd E. of the main Easington–Castle Eden road to Ivy Rock, midway between Dene House and Dene Leazes, where they are overlain by massive dolomites of reef facies. The lowest exposed beds are about 125 ft above the base of the division, so that most sections are stratigraphically higher than those at Wingate and Haswell. These lower beds

constitute a uniform, relatively evenly bedded series of slightly calcitic dolomites, cream to grey in colour and generally finely saccharoidal (mainly silt-grade) in texture. Small flattened cavities are widespread. The dolomite at many of the outcrops is highly altered and most original sedimentary structures have been obliterated, but traces of current-bedding are common, especially in coarser-grained intercalations. In thin section (Trech-mann 1914, p. 246) these beds are seen to be composed of loosely packed sub-hedral and anhedral dolomite grains, with scattered traces of calcite. Two analyses (loc. cit.) indicated an almost pure dolo-mite, having calculated calcite percentages of only 0·21 and 1·65 respectively ; in-soluble residues of 0·16 and 0·05 per cent are typical of Middle Magnesian Lime-stone lagoonal beds, all of which are remarkably free from detrital inorganic material. These beds are particularly well exposed in the valley sides near Gunner's Pool [419 388] and in the Trossachs [417 388], a spectacular gorge excavated along a fault plane.

Despite recrystallization and brecciu-tion of the rock, moulds of small lamelli-branchs, and a number of species of foraminifera, gastropods and ostracods are common in some beds. The most charac-teristic lagoonal form, *Astartella vallis-neriana,* is plentiful (Pl. VI, fig. 12), associated with *Schizodus sp.* and *Permo-phorus costatus* (Brown). Small tubular rod-like fossils, thought to be *Tubulites permianus,* have also been noted. Trech-mann (1964, pp. 215-6) further record *Bakevellia ceratophaga,* and species of ' *Serpula* ', *Spirorbis* and ' *Vermilia* ' from this locality. A total thickness of about 120 ft of these beds is exposed in the Dene, with a gentle overall easterly dip which carries them below the valley floor between Ivy Rock [435 396] and Craggy Bank [438 398]. They are overlain by 20 to 25 ft of irregularly bedded hard cream-grey lagoonal dolomite, which has a sub-porcellanous appearance when fresh, but which weathers to show its con-stituent sand-grade dolomite rhombs, with some ooliths and highly comminuted shell-debris. Where these beds first appear, at the top of Seven Chambers [431 394], access is difficult, but it becomes easier eastward down towards Ivy Rock. The section there is:

	ft.
Dolomite, buff, massive, shelly (reef)	50+
Dolomite, cream, irregularly-bedded, shelly, calcarenite matrix	5
Dolomite, as above but with nodular weathered surfaces	1½
Dolomite, cream, shelly, oolitic, with irregularly disposed pisoliths	6
Dolomite, cream, hard, thin- and medium-bedded, cavernous; finely comminuted shell-debris ...	1
Dolomite, cream, fine-grained, bioclastic; in two beds ...	3½+

The oolitic bed has a markedly nodular-weathered surface, and is largely recrystallized; the pisoliths near the base are coarse, ranging up to 7 mm, and are similar to those of suspected algal origin in Gilleylaw Quarry, near Sunderland (Smith 1958, p. 75).

Wherever the base of the reef dolomite is seen in Castle Eden Dene it forms an irregular plane transgressing 6 to 8 ft of underlying beds. Lateral passage into reef-rock probably takes place shortly to the east for there are no beds of lagoonal facies in strata of the same age penetrated by Blackhall Colliery Shafts (Trechmann 1913, p. 213), 1½ miles E. of Ivy Rock.

Thornley to Kelloe. In the area between Thornley and Town Kelloe, drift is predominantly thin, and exposures collectively include more than 100 ft of the sequence already described at Tuthill Quarry, Haswell. The most northerly, at Gore Hill [361 400], are of massive cream fine-grained calcitic dolomite. Some 12 ft of thin-bedded cream soft granular calcareous dolomite are exposed in a small quarry [3567 3981], 650 yd W. of Gore Hall, and flaggy beds of similar lithology crop out on a hillside about 130 yd S.S.W. of the quarry. Further beds of this kind are 12 ft thick in a cutting on the disused Cassop Wagonway, 500 yd E.N.E. of Halfway House Inn, and 6 ft thick in a field [3588 3935], 125 yd N. of the cutting. Shallow cuttings along the main Durham–Hartlepool road near to Halfway House Inn contain soft flaggy calcitic dolomite with a well-developed though somewhat patchy 'felted' texture, and similar rock is found in numerous fragments in the soil and in scattered outcrops south and south-west of the Inn. In an old quarry [3544 3827], 920 yd S. of the Inn the 'felted' rock is 12 ft thick: it is also seen at natural exposures [3561 3838 and 3565 3828] respectively 220 yd N.E. and 230 yd E. of the quarry.

Most of the good exposures in the Thornley area are in the sides of glacial drainage channels (p. 222). Of the several exposures of thin-bedded granular and 'felted' calcitic dolomite in the main valley the biggest is a 30-ft sequence [3647 3894], 810 yd N.N.E. of Thornley Hall. Unfortunately most of the of the other exposures are now covered by colliery waste, but 12 ft of soft cream 'felted' dolomite are still to be seen [3656 3872] 270 yd S. of the above locality. A small exposure [3689 3859] on the south side of a tributary valley, about 50 yd S. of White House, contains 10 ft of thin-bedded soft cream calcitic dolomite, in which traces of fore-set cross-bedding laminae dip towards the east. This rock is completely recrystallized and granular with patchy 'felting'. Similar beds are also exposed in an old quarry [3682 3842], 260 yd S.S.W. of White House, where the section is:

	ft
Dolomite, cream, thin-bedded, calcitic, with a 'felted' texture; cross-bedded	8 +
Alternations (¼ to ½ in) of pisolitic and silt- to sand-grade cream calcitic dolomite; partly 'felted'	5
Dolomite, creamy-white, cross-bedded, calcitic, soft, granular (sand-grade)	1½+

All three lithological units of this section are highly weathered and contain pockets of soft cream-white powdery dolomite. Similar pockets are seen in thin-bedded, cross-laminated dolomite in a group of quarries about 600 yd E.S.E. and 350 yd

E.N.E. of Thornley Hall where the texture range from finely saccharoidal to coarsely 'felted' and where parts of the rock adjacent to open joints and movement planes are harder, more massive and more calcitic.

Further exposures of thin-bedded, cream, granular or 'felted' calcitic dolomite, commonly with traces of cross-bedding, extend along the glacial drainage channel south of Thornley Hall. Original textures are largely obliterated by recrystallization, but beds of oolite and pisolite are preserved [3633 3776] 500 yd N.N.W. of Kelloe Law. Oolite lying close to the base of the subdivision and containing some compound ooliths and pisoliths crops out high on the valley side [3585 3739] 760 yd W. of Kelloe Law and at the top of an old quarry [3602 3702] 750 yd W.S.W. of Kelloe Law.

Old Wingate. The Middle Magnesian Limestone has been extensively quarried around Old Wingate. In all there are over two miles of quarry face ranging up to 60 ft high and spanning a sequence of about 90 ft of beds just above the base of the subdivision. Most of the typical lithologies are represented, but undoubted ooliths are rare and no pisoliths were noted. Granular textures predominate, but 'felting' is common and dedolomitization adjacent to widened joints and to a prominent fault system has produced zones of hard massive fine-grained calcareous rock. The southernmost quarry contains, near its base, thin beds of fairly hard cream-brown fine-grained dolomitic limestone which contain scattered traces of small-scale cross-bedding and appear to be less altered than others in the vicinity. Seen in thin section (E 3439) this rock is composed of silt-grade calcite with patchy recrystallization of dolomite.

A mile farther east, an isolated quarry [3933 3789] contains about 20 ft of thin-bedded saccharoidal (silt- to fine-sand grade) calcitic dolomite with small-scale cross-bedding, and small flattened cavities (?shell-casts) throughout. There are local dedolomitized zones and many irregular patches of powdery fine-grained dolomite from which the calcite cement has been leached.

Hesleden. Beds of lagoonal facies, represented by 20 ft to 30 ft of thin- and medium-bedded soft cream-grey shelly dolomitic calcarenite, lie between dolomites of reef-facies at several places in Hesleden Dene, west of Monk Hesleden. Excellent natural exposures occur on the north side of the stream [4358 3783], 700 yd W. of Hesleden Station, at Hairy Man's Hole [4416 3776], 150 yd S. of the station, and at several points between this locality and Jack Rock [4463 3762], 450 yd E.S.E. of the station, where reef-facies limestones overlying the lagoonal beds dip south-eastwards beneath the stream. A gradual and very irregular easterly thinning is apparent from the above sections, and these beds are thin or absent in the core of an anticline 500 yd S.W. of Fillpoke, 1½ miles farther east. They appear to be about 9 ft thick in a borehole [4672 3696], 620 yd S. of Benridge. The irregularity of the eastwards thinning is due to the presence of prominent domes up to 20 ft high (Pl. XA) on the upper surface of the underlying reef limestones. Cliffs [4416 3776] at the side of Hairy Man's Hole provide the most complete single section, although the lowest beds are absent:

	ft
Dolomite, cream, finely granular alga-laminated (reef facies)	30 +
Dolomite, cream, granular, undulating thin bedding ...	3½
Dolomite, cream, granular, irregularly interbedded with harder buff-brown dolomitic limestone ; comminuted shell-debris throughout	18 +

The lowest member of this section is essentially a shelly calcarenite of detrital origin, and though individual fossils are common they are stunted and restricted to a few species of euryhaline gastropods and lamellibranchs. Those collected during the resurvey include: *Cyclobathmus? permianus* (King), *Straparollus permianus* (King), *?Naticopsis leibnitziana* (King), *N. minima* (Brown), *Omphalotrochus helicinus* (Schlotheim), *O. taylorianus* King), *O. thomsonianus* (King), *Pleurotomaria?*, *Strobeus?* ; *Bakevellia antiqua*, *Cardiomorpha modioliformis* King [Pl. VI, fig. 9] ; coprolites. Trechmann (1913, pp. 215–6 ; 1925, p. 136) records the following from this part of Hesleden Dene, but does not specify the horizon, so that some forms (especially the brachiopods) may have been collected from autochthonous reef rock : " *Epithyris elongata* (Schlotheim), *E. sufflata* (Schlotheim), *Strophalosia goldfussi* (Münster) ; *Macrocheilus sp.*, *Natica leibnitziana* King, *?N. minima* Brown, *Pleurotomaria antrina* (Schlotheim), *P. tunstallensis* King, *?Rissoa gibsoni* Brown, *?R. leighi* Brown, *R. obtusa* Brown, *Turbo helicinus* (Schlotheim) *T. mancuniensis* Brown, *T. permianus* King, *T. taylorianus* King, *T. thomsonianus* King ; *B. antiqua*, *B. ceratophaga*, *?B. tumida* King, *Pleurophorus costatus* (Brown), *?Schizodus truncatus* King, *S. rotundatus* ; *Peripetoceras freieslebeni Geinitz ;* ostracods ". Lagoonal beds below those of Hairy Man's Hole are seen near the base of cliffs [4448 3773], 350 yd farther east on the north side of the stream, where they fill hollows between domes and where their average thickness had decreased to about 11 ft.

The following fossils were collected from beds corresponding approximately to those of the Hairy Man's Hole section in a bore [457 361], situated about 330 yd N.W. of Thorpe Bulmer Farm: *Tubulites?* ; ' *Adhaerentina* ' *permiana*, ' *Ammodiscus* ' *roessleri* (Schmid) ; *Neochonetes? davidsoni* ; *Astartella vallisneriana, Bakevellia antiqua, Schizodus sp.*

Trimdon. Lagoonal beds crop out at many localities in the area around Trimdon Grange, Trimdon Colliery and Trimdon, and have been proved in a number of boreholes farther east. At outcrop these beds are as texturally variable as they are in the Thornley–Kelloe area, but in boreholes they are composed mainly of finely cross-bedded carbonate rocks of sand-grade and oolites, generally approaching dolomite in overall composition.

About 40 ft of thin-bedded, granular and finely granular calcitic dolomite are exposed in two old quarries [368 361] immediately north of Trimdon Grange, where bedding is thin, uneven and lenticular. Small scale cross-bedding has foreset laminae dipping generally at less than 15° without noticeable preferred orientation. Recrystallization has produced a saccharoidal texture in most beds with grain size ranging from silt- to sand-grade. In another old quarry [3735 3613], some 500 yd farther east, about 12 ft of similar strata include thin oolitic beds and have incipient ' felted ' texture. Cross-bedded, cream, granular and oolitic dolomite is exposed also in old quarries [383 359 and 383 358] 1250 and 1050 yd N. of Park House, Trimdon Colliery. At the south side of the latter quarry the oolite is coarse, and some individual bodies are of pisolith grade (>2 mm). The most westerly exposure of lagoonal beds is at Raisby Hill Quarry [340 350], where they form the uppermost 20 ft of the section (p. 111). A thin section (E 29452) of flaggy rock from a small exposure [3523 3505] at the eastern end of this quarry is described by Dr. R. Dearnley as " a limestone of silt-grade consisting of a relatively homogeneous mass of calcite and dolomite grains (0·025 mm) ". This rock appears to be an altered oolite. D.B.S.

About 25 ft of dolomite are exposed in a quarry [361 353], 600 yd W.N.W. of The Grange ; it has a predominantly ' felted ' texture, but is otherwise like the rock north of Trimdon Grange. Similar ' felted ' dolomite is 15 ft thick in a railway cutting [364 353] nearby, where foreset cross-bedding laminae are preferentially orientated eastwards. Less highly recrystallized finely granular dolomite with small-scale cross-bedding is exposed in a small quarry [3704 3506] 430 yd E. of The Grange. About 60 ft of cream-buff calcitic dolomite, largely recrystallized, are seen in Grange Quarry [363 346], 600 yd S.W. of The Grange, where cross-bedding and other structures are indicative of an originally granular deposit. In thin section (E 29927) this rock is seen to consist of a porous aggregate of silt-grade rounded and subhedral dolomite crystals (0·02 mm) with small quantities of interstitial calcite. Dips of up to 20° at the north end of Grange Quarry are probably related to the proximity of the Butterknowle Fault, some 200 yd to the north.

About 25 ft of yellow-buff granular calcitic dolomite, mostly thinly cross-bedded, is seen in a small quarry

[3600 3392] 800 yd W.S.W. of Trimdon Cross Roads. This rock is lustre-mottled in hand specimen and appears to be undergoing dedolomitization. An unusual rock (E 29926) occurs in patches on the west side of the quarry; it consists predominantly of an aggregate of equigranular, rounded and subhedral calcite grains of silt- and fine sand-grade, with a poikilitic calcite matrix and with large calcite porphyroblasts (up to $\frac{1}{4}$ in diameter). Cross-bedding and wedge-bedding are ubiquitous in Harap Quarry [3554 3351], 1450 yd S.W. of Trimdon Cross Roads, where 30 ft of granular, oolitic and pisolitic calcareous dolomite and dolomite limestone are to be seen. Some beds contain a mixture of pisoliths and ooliths, many of the former being very irregular. In thin section (E 29924) the shells of ooliths and pisoliths are seen to be almost pure fine-grained calcite while the interiors are either empty or contain sparry calcite, which also fills most of the interstices. Some ooliths and pisoliths are seen also in a 35-ft section of mainly thin-bedded fine-grained rock exposed in an old quarry [361 336] 600 yd E. of Harap Quarry.

A number of the old quarries in the ridge west of Trimdon contain calcitic dolomites similar to those at Harap Quarry. Massive granular dolomite, 12 ft thick, is exposed at the Hare and Hounds cross-roads [3323 3401], and a similar thickness is seen in an old quarry [3397 3395], 750 yd farther east, where the dolomite is extensively recrystallized, and where many pisoliths have been removed by solution. Garmondsway Moor Quarry [349 333], 900 yd S.E. of the cross-roads, contains 35 ft of well-bedded calcareous dolomite, parts of which have a coarse saccharoidal texture. About 12 ft of fine-grained dolomite are exposed in a small quarry [3473 3331] 900 yd farther east where thin interbedded layers contain excellent examples of a mixed oolith/pisolith assemblage, with many multiple ooliths and with a matrix of sparry calcite. These and all other beds exposed west of Trimdon are near the base of the Middle Magnesian Limestone.

East of Trimdon, an old quarry [3772 3336], 800 yd W. of Trimdon East House, contains 15 ft of massive silt-grade saccharoidal calcitic dolomite. Sections measured in Cope Hole Quarries, 200 to 400 yd W. of Trimdon East House, include up to 35 ft of compact to porous granular limestone, with cross-bedding and with scattered ooliths. In thin section (E 29925) it can be seen that the rock consists of about 60 per cent uniformly sized ooliths and pseudo-ooliths with cores and matrix of sparry calcite. Farther east a number of boreholes through thick drift prove the eastward continuation of thick predominantly oolitic dolomites succeeded by predominantly granular non-oolitic beds. Approximately 240 ft of the lagoonal Middle Magnesian Limestone were penetrated in the Cotefold Close Bore, 430 yd S.S.E. of Pudding Poke, and 270 ft were met in the Fifty Rigs Plantation Bore, 1070 yd N. of Sheraton Grange (Woolacott 1919a, p. 165). Lamellibranchs including *Bakevellia ceratophaga* are recorded by Trechmann (in Woolacott 1919a, p. 164) from the lower part of the subdivision in the latter bore: recently they have also been found in equivalent beds in a water borehole, 100 yd W.N.W. of the Cotefold Close Bore. The transitional beds at the base of the division in the water bore are 55 ft thick, and consist of hard fine-grained dolomite and dolomitic limestone.

Mainsforth to Bishop Middleham. A belt of disused quarries, trending north-north-westwards for a distance of about 2000 yd from the east end of Bishop Middleham, is traversed by a fault-zone adjacent to which the beds are hardened and dedolomitized, giving rise to a prominent ridge. About 60 to 70 ft of beds similar to those west of Trimdon are exposed in the quarries. They are predominantly well bedded, granular dolomitic limestones and calcitic dolomites, with abundant traces of cross-bedding and with prominent beds and lenses of coarse oolite or pisolite. As at Trimdon, Haswell and elsewhere, some pisoliths are of bizarre form, and occur in a matrix of ooliths, but unlike those of Haswell and Wheatley Hill they are found in layers displaying current-bedding. Pockets of incoherent silt-grade dolomite and patches of dolomite with 'felted' textures are scattered throughout the quarries. The lowest beds in these quarries lie in the transitional zone at the base of the division.

Beds from the same part of the succession as the quarries are also seen in a number of smaller exposures near Bishop Middleham: old quarry [3447 3200], 560 yd N.N.E. of East House, 6 ft; west bank of stream [3480 3137], 630 yd E.S.E. of East House, 12 ft; valley side [3334 3154], 950 yd W. of East House, 18 ft; excavation [3334 3003], 1920 yd S.S.W. of East House, 12 ft; excavations [3345 3013], 1770 yd S.S.W. of East House, 6 ft; valley side [3275 3035], 2090 yd S.W. of East House, 5 ft; and old quarry [327 318], 650 yd E. of Hope House, 15 ft. A borehole [3282 3186] in the floor of the last-named quarry proved 37 ft of dolomite and dolomitic limestone before reaching Lower Magnesian Limestone. The uppermost $7\frac{1}{2}$ ft of this 37 ft were of lagoonal facies, the rest falling into the transitional beds. Another borehole [3030 3091], 390 yd N.W. of Chilton East House, penetrated 60 ft of Middle Magnesian Limestone, including 30 ft of lagoonal facies. Lagoonal beds are also proved in a number of boreholes in the Rushyford–Sedgefield area.

Fishburn, Sedgefield and Embleton. In this area, where exposures are uncommon, lithologies are similar to those described from the Trimdon area, though oolitic rocks appear to be less common. About 25 ft of hard well-bedded and fine-grained dolomitic limestone are exposed in an old quarry [367 326], 350 yd S.S.E. of Hope House, and traces of cross-bedding are preserved at the west end of the quarry. Cross-bedding is also apparent in 12 ft of yellow, hard, massive, and extensively brecciated limonitic limestone exposed in a railway cutting [3620 3182] east of Fishburn Colliery. A similar thickness of massive limonitic calcite rock of silt-grade is exposed in another cutting [3578 3163], 460 yd N.N.E. of Lizards. The most southerly exposure in this area is in an old quarry [350 309], 730 yd W.S.W. of Lizards, where over 25 ft of yellow thin-bedded compact dolomitic limestone occur.   G.D.G., D.B.S.

97544

The Middle Magnesian Limestone has been completely cored only in a few of the coalfield boreholes drilled in the area east of Fishburn and Sedgefield. The sequence is similar to that met in boreholes east of the Trimdons, consisting of up to 210 ft of essentially granular barren dolomites overlying 100 to 300 ft of oolitic, pseudo-oolitic and pisolitic dolomites which are shelly towards the base. In a borehole [4157 3054] at Tinkler's Gill, 910 yd N.N.W. of Embleton Church, the lower 50 ft of the 210-ft division yielded numerous lamellibranchs including *Astartella vallisneriana*, *Schizodus obscurus* (J. Sowerby) and species of *Bakevellia*. Transitional beds in this bore appear to be only about 2 ft thick. From several boreholes between Sedgefield and Embleton, Mr. A. Fowler has obtained the most complete fauna yet collected from the lagoonal beds of the Middle Magnesian Limestone. The faunal list (quoted in full on pp. 173-6) includes brachiopods such as *Horridonia*, *Neochonetes?* and *Strophalosia*, together with polyzoa previously thought to be confined to rocks of reef-facies, and *Tubulites permianus*. A specimen of rock from a depth of 300 ft (about 30 ft above the base of the division) from Ten-o'-Clock Barn No. 2 Bore [3916 3035], Fishburn, has been analysed by G. A. Sergeant (Lab. No. 1456), with the following result (figures in percentages): SiO_2 0·32, Al_2O_3 0·16, Fe_2O_3 4·27, FeO nil, MgO 16·04, CaO 33·84, Na_2O 0·02, K_2O 0·01, $H_2O+105°$ 0·90, $H_2O-105°$ 0·25, TiO_2 nil, P_2O_5 0·01, MnO 0·30, CO_2 43·87, SO_3 0·12, Cl trace, F trace, FeS_2 0·07, C 0·02, Ba calculated as $BaSO_4$ 0·47, total 100·67. A suggested mineral composition for this rock is dolomite 73·7 per cent, calcite 20·0 per cent and limonite 5·4 per cent. In thin section (E 21052) it is seen to be composed of pseudo-ooliths which have outer rims of heavily pigmented dolomite and cores of sparry calcite; the matrix is of fine-grained dolomite with abundant limonite.   D.B.S.

Elwick, Naisberry and West Hartlepool. Thick drift covers most of this area, but Permian beds of lagoonal facies crop out near Naisberry at two localities. One is an old quarry [477 333], 650 yd S.E. of

Naisberry Farm, where there are about 20 ft of cross-bedded granular and oolitic dolomitic limestone. Thin sections (E 29917-20) from these beds are of highly recrystallized oolitic and pseudo-oolitic calcarenite, with a matrix of sparry calcite, which also forms the cores of many ooliths. Some non-oolitic beds are of crystalline silt-grade limestone (sometimes dolomitic), with a matrix of poikilitic calcite. About 15 ft of bedded dolomitic limestone are exposed in the other quarry [4749 3330], 430 yd S.E. of Naisberry Farm.   G.D.G., D.B.S.

Water boreholes, 610 yd W. of Naisberry Farm, penetrated an interbedded sequence of calcitic dolomites of reef and lagoonal facies, representing a zone of interdigitation on the landward side of the reef. Upper strata in these boreholes are comparable with those exposed in Hesleden Dene, with alga-laminated beds overlying about 12 ft of bedded granular and oolitic dolomite of lagoonal facies containing debris of stunted shells (mostly lamellibranchs and gastropods). These beds are underlain by about 85 ft of bedded and massive shelly and polyzoonal dolomite of reef facies, which pass down by interdigitation into 155 ft of cross-bedded granular, oolitic and pisolitic dolomites similar to those cropping out in the Trimdon area. About 45 ft of transitional beds lie at the base of the division at this locality. Six boreholes

around Dalton Piercy waterworks [465 315] penetrated a maximum thickness of 300 ft of predominantly oolitic dolomite, in which a fragment of *Astartella vallisneriana* was noted 30 ft above the base of the division. This form together with *Streptorhynchus pelargonatus,* species of *Permophorus, Bakevellia* (especially *B. ceratophaga*) and *Schizodus,* is found also at a similar horizon in the Dalton Nook Plantation Bore one mile farther east, and is recorded in Elwick No. 1 Bore, 995 yd S. of Elwick Church. A somewhat richer fauna, including lamellibranchs, polyzoa and foraminifera, was met in the Middle Stotfold Bore.

In the partially cored Dalton Nook Plantation and Low Throston bores and also in a bore [4473 5276] 2 miles S.S.E. of Dalton Piercy there appears to be no clear lithological break at the base of the Middle Magnesian Limestone. In the bore south-south-east of Dalton Piercy the upper part of the lagoonal dolomites contains several beds of anhydrite which may be extensions from the thick Hartlepool Anhydrite, the main mass of which lies in the basin 2 to 3 miles to the east. Farther south, anhydrite becomes increasingly abundant and in a bore at Norton, 4 miles south of Embleton, in the Stockton (33) district, it forms the bulk of the lagoonal deposits, together with subordinate beds of dolomite and red mudstone.

                                                    D.B.S.

Reef facies

Dolomites and dolomitic limestones of reef-facies crop out in a sinuous belt (Pl. V) extending south-south-eastwards from near Murton to West Hartlepool, although for much of this distance drift is thick and the reef is known only from boreholes and scattered exposures in deep quarries and ravines. In the West Hartlepool area the reef is overlapped by basinal beds of the Middle Magnesian Limestone and by higher strata.

Murton to Cold Hesledon. The most northerly outcrops of reef-limestone in this area are found in a cutting [408 477] on the Murton to Easington road, 800 to 1000 yd N.N.W. of the Water Works, New Hesledon, where *Acanthocladia anceps, Fenestrellina retiformis, Stenoscisma schlotheimi, ?Strophalosia morrisiana, Bakevellia ceratophaga* and *Pseudomonotis speluncaria* were collected from hard

massive calcitic dolomite. Similar rock is exposed [4082 4767] slightly farther up the hill side in an old quarry and yielded: ' *Adhaerentina* ' *permiana ; Acanthocladia?, Batostomella crassa, Fenestrellina retiformis ; Dielasma elongatum, D. elongatum sufflatum* (Schlotheim), *Orthothrix excavata ; Bakevellia antiqua, B. ceratophaga, Pseudomonotis speluncaria ; Peripetoceras freieslebeni ;* and ostracods.

Stromatolitic structures occur throughout the rock in the cutting and the quarries, and a number of extensive steeply inclined stromatolite sheets are also present. Much of the shell debris is highly comminuted.

At the time of the resurvey a rich fauna of polyzoa, brachiopods, gastropods and lamellibranchs was collected from Fox Cover Quarry [426 478], 700 yd N.W. of Hesledon East House, where poorly bedded reef-dolomites, 80 ft thick, are now obscured by urban waste. Fossils identified include: algal filaments ; 'Adhaerentina' permiana ; Eudea tuberculata King ; crinoid stems ; Acanthocladia anceps, Batostomella crassa, Fenestrellina retiformis, Synocladia virgulacea, Thamniscus dubius ; Dielasma elongatum, D. elongatum sufflatum, Horridonia horrida, Orthothrix excavata, Spiriferellina cristata (Schlotheim) multiplicata (King), Stenoscisma globulina (Phillips), S. schlotheimi, Strophalosia morrisiana ; Naticopsis cf. leibnitziana (King), gastropod with three whorls in low-angled spire and numerous fine spiral striae ; Bakevellia antiqua, B. aff. antiqua [beak almost in centre of hinge line], Palaeolima?, Parallelodon striatus, Permophorus costatus, Pseudomonotis speluncaria, Schizodus? ; ostracods. The reef-rock in this quarry appears to be naturally disposed in a number of large, mutually interfering, dome-shaped bodies, some of which are over 60 ft high and as much in diameter. Dips on the flanks of such bodies measure up to 85°, and appear to be primary. Spaces between the domes are filled by rubbly reef-detritus, consisting largely of shell-debris, with local coarse breccias of penecontemporaneously eroded reef-rock.

An analysis of a specimen of fossiliferous rock from this quarry (Woolacott 1912, p. 264) shows it to contain 58·02 per cent $CaCO_3$ and 38·33 per cent $MgCO_3$. At the south-east corner of the quarry a 50-ft cleft filled with carbonate sands and breccias of derived reef material appears to be an infilled reef-channel.

Dolomitized limestones of reef facies crop out in a railway cutting [418 473], 1350 yd W. of Hesledon East House, where three distinct sub-facies are recognized. The first is a massive shelly reefrock similar to that in Fox Cover Quarry: it is exposed at the north-eastern end of the cutting and contains: algal pellets and filaments ; Acanthocladia anceps, Fenestrellina retiformis ; Dielasma elongatum ; indet. gastropods ; Bakevellia ceratophaga, Pseudomonotis speluncaria ; and ostracods. The second facies is represented by roughly horizontal and southwestwards dipping bioclastic dolomite, with interbedded alga-laminated sediments and lenses of granular, oolitic and pisolitic dolomite. These deposits yielded Acanthocladia anceps, Dielama sp., indet. gastropod and Parallelodon striatus. The third facies, exposed at the southwestern end of the cutting, is a dolomite composed of a confused mass of mutually interfering small (up to 3 ft diameter) irregular algal colonies (Smith 1958, pl. vib) with interstitial pockets of contemporaneous breccias composed of alga-laminated fragments and algal? pellets. A sparse shelly fauna consisting chiefly of small gastropods and rolled crinoid columnals occurs in parts of this rock, which appears to represent a local backreef deposit. In a lane [4136 4740], 500 yd W. of the railway cutting, temporary exposures through 4 ft of unbedded cream reef-dolomite contained a rich fauna of polyzoa, brachiopods, gastropods and lamellibranchs. Deep subsidence cracks [4220 4735] some 400 yd E. of the cutting reveal dolomite composed entirely of an assemblage of small (less than 1 in) concentrically laminated algal bodies. The remaining exposure of reef-facies in this area comprises 8 ft of finely laminated crystalline dolomitic limestone in a small quarry [4148 4695], 450 yd S.W. of the railway cutting. The laminae are broadly undulant and in part closely crenulated, and appear to be of algal origin.

East of Cold Hesledon, boreholes have proved a maximum thickness of over 100 ft of massive reef-facies dolomite containing a limited fauna of small brachiopods (chiefly Dielasma), gastropods and small lamellibranchs with a few retiform and straggling polyzoa. Fossils identified from this bed in three of the holes include: algal filaments ; ?Acanthocladia anceps, Batostomella crassa, Fenestrellina *retiformis*; *Dielasma elongatum, Ortho-thrix* cf. *lewisiana* (de Koninck), *Spiriferel-lina cristata multiplicata, Strophalosia morrisiana*; *Naticopsis leibnitziana*; *?Pseudomonotis speluncaria*; and ostra-cods. In two of the holes [4206 4607; 4249 4656] shelly reef-limestone was over-lain by irregularly laminated dolomite in which algal structures have been recog-nized in thin section (E 26238 and E 26239). It is probable that in other bores of this group original alga-lamina-tion has been obscured by recrystalliza-tion. In five of the bores the shelly reef-rock is overlain by granular and oolitic bedded dolomite which is interbedded with alga-laminated dolomite in the lower part. These interbedded deposits form part of the lagoonal sequence, but beds above the uppermost algal intercalations are regarded as lagoonward extensions of post-reef basinal deposits.

Hawthorn. Dolomite of reef facies crops out in Hawthorn Quarry [437 463], and in the eastern part of the valley of Haw-thorn Burn. The former is the only quarry still operating near the Durham coast, and is cut into about 80 ft of reef and superincumbent granular calcitic dolo-mite. Reef-facies dolomite forms the lower 30 to 35 ft of the main quarry face, and is a hard, irregularly thick-bedded or massive rock, with scattered poorly preserved fossils, generally simi-lar to those seen in the Cold Hesledon boreholes. They include: ' *Adhaerentina*' *permiana*, ' *Glomospira*' *gordialis*; *Acan-thocladia anceps, Batostomella crassa, Thamniscus dubius*; *Dielasma elongatum, D. elongatum sufflatum* [Pl. VI, figs. 4, 5], *Orthothrix* cf. *excavata*; *Stenoscisma schlotheimi, Strophalosia morrisiana, Nati-copsis sp., Omphalotrochus thom-sonianus, Straparollus sp.*; *Bakevellia antiqua, B. ceratophaga, Liebea squamosa* (J. de C. Sowerby), *Permophorus sp., Pseudomonotis speluncaria*; and ostracods. The uppermost reef-limestones are irre-gularly alga-laminated, the laminae at some exposures showing marked small-scale asymmetric crenulations resembling current ripple-marks. The approximately horizontal bedding in the uppermost part of the reef in the main quarry is inter-rupted in a small neighbouring quarry [4366 4642], now filled, by a number of

prominent dome structures. The largest of these is about 50 ft in diameter and 20 ft high, and has a massive apparently structureless core overlain by alga-laminated deposits similar to those in the main quarry. Dips on the flanks of this and other domes reach 70 to 80°, and are thought to be primary. The top of the reef persists as a relatively flat plane throughout the small quarry and all but the eastern end of the main quarry, where the outer shoulder of the reef is seen trending north-north-west-wards and where bedded reef-flat dolo-mites grade laterally into steeply-dipping massive beds of the reef-front. The north face of the main quarry is now largely obscured, but formerly individual hori-zontal beds deposited on the reef-flat could be seen to pass over the reef-crest and thicken rapidly (Pl. IX) before dipping at angles of up to 80° down the seaward side of the reef. A fault-zone, reversed at the surface but normal in underlying coal workings, lies parallel with the reef-crest and truncates the most easterly beds of the reef-front (equivalent to the highest beds of the reef-flat). Analyses of the reef-rock, though variable, show it to be a calcitic dolomite, with about 28 per cent $MgCO_3$, but evidence of extensive dedolomitization was noted.

A short distance to the south, dolomitic reef-limestones are continuously exposed for about 300 yd in Hawthorn Burn [436 458]. Pockets in the limestones con-tain varied and abundant faunas com-parable in age with that obtained from off-reef beds at Chourdon Point (p. 145), but older than that found in Hawthorn Quarry. The easternmost exposure is in the south bank of the stream, and takes the form of a number of small (3 ft to 6 ft high) mutually interfering domes each composed of a complex of hard smaller rounded masses of dense reef-rock up to about 9 in across. These masses appear to be piled haphazardly, and may represent contemporaneously eroded material accumulated on the reef-front slope. Each mass is enveloped by laminated stromatolitic dolomite, and irregular sheets of similar material ramify throughout and around the domes. Pockets of detritus, some with much shell material, occur between the domes.

Geology between Durham and West Hartlepool (*Mem. Geol. Surv.*)

PLATE IX

L 583

REEF DOLOMITE OF THE MIDDLE MAGNESIAN LIMESTONE, SHOWING BEDDED ROCKS OF REEF-FLAT SUB-FACIES (ON LEFT) GRADING INTO STEEPLY-DIPPING MASSIVE ROCK OF THE REEF-FRONT. THE SECTION SHOWN IS ABOUT 35 FT HIGH: NORTH FACE OF HAWTHORN QUARRY, HAWTHORN

A. COMPLEX ALGAL DOME, MIDDLE MAGNESIAN LIMESTONE, HESLEDEN DENE   *L* 145

PLATE X   Geology between Durham and West Hartlepool (*Mem. Geol. Surv.*)

B. ALGAL COLONY (LOWER RIGHT) WITH STROMATOLITIC OVERGROWTHS, YODEN QUARRY—
   NATURAL SIZE   *MLD* 1625

This type of lithology continues upstream, but bedding gradually becomes more distinct and regular until, at the western end of the outcrop, the reef-rock is relatively evenly bedded, has only a few low domes and contains a more varied and abundant fauna. Here the following were collected from an outcrop [4338 4575] on the south bank of the Burn, 890 yd upstream from the mouth: '*Adhaerentina*' *permiana*, '*Glomospira*' *pusilla*; *Acanthocladia anceps*, *Thamniscus dubius*; *Dielasma elongatum*, *Stenoscisma?* [juv.]; *Bakevellia antiqua*, *B. ceratophaga*; ostracods. The proportion of stromatolitic material remains roughly constant, at 30 to 50 per cent, to about 700 yd upstream from the railway bridge, but thereafter decreases progressively westwards. It is still, however, an important constituent of the rock at the western end of the outcrop.

Easington Colliery. The steep outer face of the reef runs southwards from Hawthorn Quarry and, except around Shippersea Bay, lies a short distance west of the railway line. The hard dolomitic limestone and dolomite of the reef-wall give rise to a marked topographical feature, and have been quarried on a small scale at several localities between Hawthorn Burn and Easington Colliery. The most northerly exposures [4381 4579] are in and around a footpath on a wooded slope, 450 to 490 yd N.N.W. of Beacon House, where a few feet of hard dolomite contain a fauna characteristic of a fairly late stage of the period of reef growth, and include: '*Adhaerentina*' *permiana*, '*Glomospira*' *gordialis*; *Acanthocladia anceps*, *Fenestrellina retiformis*, *Thamniscus dubius*; *D. elongatum*, especially *D. elongatum sufflatum*; indet. gastropods; *Bakevellia* cf. *antiqua*, *Pseudomonotis speluncaria*; and ostracods. Shells and shell debris make up a high proportion of the rock, but the number of species is relatively small. Some shells have thin concentrically laminated stromatolitic envelopes. Stromatolitic material in the form of closely packed concentrically laminated bodies forms most of the hard dolomite exposed in a small excavation [4395 4564], 180 yd N. of Beacon House, and several larger sheets

of laminated stromatolitic dolomite similar to those in the bed of Hawthorn Burn also occur. Some polyzoan and shelly material was also noted. Similar dolomite from the reef-flat sub-facies, near the top of the reef and about 150 yd west of the reef-crest, is found in a small overgrown quarry [4407 4539], 50 yd E. of Beacon House. Specimens from this quarry contain the remains of algal colonies resembling *Girvanella*, *Ortonella* and *Bevocastria* (F. W. Anderson *in* Smith 1958, p. 81, pl. viii). Many excellent examples of stromatolitic structures from the same locality are seen in the walls of Beacon House and around adjacent fields.

Massive steeply dipping beds of the outer reef-slope, equivalent to those 50 yd E. of Beacon House, form 80-ft faces on the west side of the Beacon Hill railway cutting. As at the eastern side of Hawthorn Quarry, primary dips commonly reach 70°, and the beds are arranged in a series of massive, confluent, asymmetrical domes in which the main convexity is eastwards. Many domes are defined by extensive thin steeply dipping sheets of laminated dolomite of algal origin. Towards the base of the cutting these become progressively less common and the proportion of detrital shelly reef-talus increases. Small pedunculate brachiopods are common throughout the cutting, but deeper water forms such as crinoid columnals and retiform polyzoa are abundant only at lower levels. From these lower levels the following have been identified: '*Glomospira*' *milioloides*, '*G.*' *pusilla*; crinoid stems; *Miocidaris keyserlingi* (Geinitz); *Acanthocladia anceps*, *Batostomella crassa*, *Fenestrellina retiformis*, *Thamniscus dubius*; *Dielasma elongatum*, *Orthothrix excavata*, *Spiriferellina cristata multiplicata*, *Stenoscisma schlotheimi*, *Strophalosia morrisiana*; *Omphalotrochus sp.*; *Bakevellia antiqua* [juv.], *B. aff. ceratophaga*, *Parallelodon striatus*, *Pseudomonotis speluncaria*, *Streblochondria pusilla*; and ostracods. Because the original dip of the beds exceeds the slope of the exposure, beds near the base of the Beacon Hill cutting are younger than those at the top. The high dip clearly continues in depth, for 100 ft below the floor of the cutting and 50 yd to the east, bedded brecciated reef-top alga-laminated dolomites are seen at the foot of the cliffs at the western end of Shippersea Bay. These brecciated beds are overlain by an interbedded sequence of crossbedded dolomitic oolites and alga-laminated dolomites, the highest of which is conspicuously laminated and locally contains complex algal growths or is broken down into a mass of large stromatolitic cobbles similar to those of Hawthorn Burn (p. 134). Because of the high degree of recrystallization, brecciation and small-scale faulting, these highest reef-beds can be traced north of Beacon Point for only about 250 yards. Southwards, though still highly recrystallized and brecciated, they can be traced for about half a mile until they pass beneath beach-level some 120 yd S. of Shot Rock. From the evidence of the exposures described above, the reef-limestone must here exceed 250 ft in thickness. A borehole [4405 4542], 80 yd E.N.E. of Beacon House, penetrated dolomitic reef-limestone to 202 ft without reaching the base.

About 15 ft of irregularly bedded, largely algal dolomite of reef-flat sub-facies is exposed in an old quarry [4371 4504], 450 yd S.W. of Beacon House. Small steep-sided domes, up to 4 ft diameter, break the roughly level bedding, and some

thin beds and lenses contain algal pisolitic pellets similar to those described from Silksworth (Smith 1958, pl. vi). A limited fauna of small polyzoa, pedunculate brachiopods and small lamellibranchs occurs in irregular pockets between the domes. Fossils identified from here are *Acanthocladia anceps* and *Bakevellia ceratophaga*. About 15 ft of similar bedded reef-limestones are exposed in an old quarry [4390 4473], 300 yd W. of White Lea. Small gastropods (*Omphalotrochus sp.*) are additional to the fauna noted above, but algal pellets appear to be absent.

The most complete exposures of reef-limestone are in the sides of tunnels driven east and west from Easington Colliery shafts and of an inclined adit linking these to the surface (Fig. 21). For most of its length the tunnel west of the shafts penetrates gently undulating, sparsely fossiliferous, bedded, soft cream dolomite belonging to the transitional beds at the base of Middle Magnesian Limestone. These are about 50 ft thick and show a gradual upwards enrichment in fossil content before passing into a cream friable and highly altered dolomite which forms the sides of the tunnel from 240 to 25 yd west of the West shaft. Least

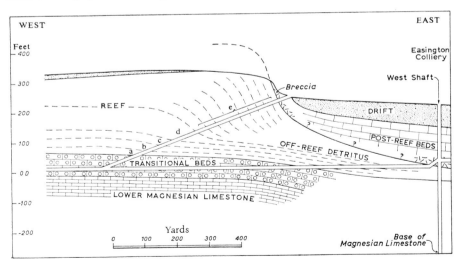

FIG. 21. *Diagram showing the relationship of reef and off-reef facies of the Middle Magnesian Limestone, as proved in underground drivages at Easington Colliery*

altered parts of this rock, which appears to be 30 to 50 ft thick, are composed of shelly reef-talus with an easterly dip of 5° to 10°. Shells, amongst which *Horridonia horrida* is especially common, are most numerous near the base and are progressively less abundant in higher parts. A full faunal list is given on pp. 169-70. The reef-talus is overlain, immediately west of the West shaft, by collapse-brecciated post-reef dolomite.

The coquina at the base of the reef is particularly well exposed at about 65 ft O.D. (a on Fig. 21) in the inclined adit, where it is a poorly bedded, buff, friable dolomite containing an abundant and varied fauna of crinoids, polyzoa, brachiopods, gastropods and lamellibranchs. The dominant polyzoans are cryptostomatous, and the most common brachiopods are species of *Horridonia, Stenoscisma* and Spiriferids (see list, p. 169). Successively younger parts of the reef are exposed towards the mouth of the adit. Above the coquina there is a thinly bedded, bioclastic, dolomitic calcarenite (b on Fig. 21) in which few shells are preserved intact, and this is succeeded (c on Fig. 21) by vaguely-bedded friable dolomite similar to the coquina (a) with the exception that Productid and Spiriferid brachiopods are less common. At a still higher level (d on Fig. 21) the tunnel passes through unbedded hard cream-brown dolomite with a fairly rich fauna dominated by retiform and straggling polyzoa, small brachiopods (especially species of *Dielasma* and *Stenoscisma*) and lamellibranchs (especially *Pseudomonotis speluncaria*). This fauna is closely comparable with that found in Fox Cover Quarry, New Hesledon, and in the lowest part of Hawthorn Quarry. From there to the surface the rock remains generally hard, and the fauna differs only in consisting of fewer and smaller individuals. The main feature of this section (e) of the adit is that the rock is divided into panels 5 to 30 ft long by sinuous, steeply dipping (usually about 80°), thin, laminated sheets of algal origin similar to those noted in the railway cutting east of Beacon House and in Hawthorn Quarry. In thin section (E 29445) one of these sheets is seen to be a cal-

citic dolomite of silt-grade with a poikilitic calcite matrix and obscure traces of indeterminate organic structure. The dip of the laminated sheets is maintained at 80° to 85° until, about 27 yd from the adit mouth, it is reduced to about 45°. In the last 27 yd of the adit parts of the massive dolomite are collapse-brecciated and fragments of soft red and green marl are included.

Brecciated rocks are seen also in an old quarry [4355 4417] on the hillside immediately above the adit, where about 60 ft of limestone and dolomitic limestone dip eastwards at about 60°. Although this high original dip is characteristic of beds of the reef-front in this area, this rock cannot with certainty be recognized as part of the reef because advanced recrystallization and brecciation have obliterated all but major structures.

Easington. Dolomite of reef facies (faunal lists, pp. 170-1) lies between barren post-reef beds and lagoonal dolomite in the Mill Hill Bore, Easington. From the high proportion of alga-laminated material, the limited and generally stunted fauna and the presence of beds of bioclastic and oolitic dolomite, it is concluded that the reef-rock was formed during the last stages of reef formation near the edge of the lagoon. The section is:

	Thick-		Depth	
	ness			
	ft	in	ft	in
DRIFT AND POST-REEF BEDS	48	6	48	6

Dolomite, buff, granular
   and oolitic, with thin
   undulating beds of
   alga-laminated dolo-
   mite at 3-in to 9-in
   intervals. Many of
   these grade down into
   beds of algal pellets
   and these into alg-
   oolites, some ooliths
   of which are multiple.
   Small gastropods and
   lamellibranchs, many
   enclosed in stromato-
   litic envelopes, are
   common at 53 to 54 ft   7   0   55   6

	Thick-ness		Depth	
	ft	in	ft	in

Dolomite, buff; formed of a coarse aggregate of irregular isolated or attached concentrically-laminated algal growth forms up to 6 in across in a bioclastic matrix containing small gastropods and lamellibranchs and many sinuous alga-laminated sheets and lenses 1 0 56 6

Dolomite, cream-buff, composed of a complex of isolated or attached concentrically laminated algal growth-forms in a matrix of algal pisoliths, algoolites and bioclastic material. The latter includes a few small gastropods and abundant stromatolite debris ... 4 3 60 9

Dolomite, cream-buff, consisting mainly of undulating alga-laminated beds separated by pockets, lenses and beds of algal pisoliths and contemporaneous stromatolite debris. Primary dips reach 30° around a large algal dome near base 5 9 66 6

Dolomite, cream-buff, consisting of a densely intergrown mass of concentrically laminated small algal nodules (many contemporaneously brecciated) containing many thin sinuous alga-laminated sheets and scattered small algal domes. Some shell debris ... 2 6 69 0

	Thick-ness		Depth	
	ft	in	ft	in

Dolomite, cream-buff, crystalline, containing abundant gastropods and lamellibranchs and scattered stromatolite debris ... 2 3 71 3

Dolomite, brown-buff and cream-buff, massive, mainly granular at top but with an increasingly high proportion of stromatolite debris towards base. Upper parts of the bed contain an abundant fauna, many elements of which are enclosed in thin stromatolitic envelopes ... 3 6 74 9

Dolomite, cream-buff, massive, granular to coarsely granular, composed mainly of fine bioclastic material with many shelly fossils, widely scattered traces of alga-lamination, and two thin alga-laminated beds near top. Below 75 ft the dolomite is generally softer and finer grained, and its fauna consists mainly of lamellibranchs and small gastropods. Gradational base at about 81 ft 6 3 81 0

LAGOONAL DOLOMITE (p. 125)

Three old quarries are cut into reef-limestone on the east side of Townfield Hill, Easington Colliery, midway between the Cemetery and Paradise. The most northerly, Townfield Quarry [4343 4380], contains about 25 ft of hard, horizontal, irregularly-bedded alga-laminated dolomite similar to the uppermost part of the reef exposed in Hawthorn Quarry, with the exception that none of the domes exceeds 2 to 3 ft in diameter. Several lenses and pockets of pene-contemporaneously brecciated algal sheets and pisoliths were noted in the south-west corner and pockets containing many lamellibranchs and gastropods occur throughout the quarry. Fossils identified from here are: cryptostome polyzoa ; *Dielasma elongatum sufflatum, Orthothrix sp., Stenoscisma schlotheimi* ; *Omphalotrochus taylorianus, O. thomsonianus, Strobeus geinitzianus* (King) ; *Bakevellia ceratophaga, Liebea squamosa, Parallelodon striatus* [Pl. VI, fig. 7], *Pseudomonotis speluncaria, Schizodus truncatus*. Some 15 ft of similar dolomite in a small quarry [4356 4356] at the south end of the hill appear to be slightly higher in the reef-flat sequence than the beds in Townfield Quarry from which they differ in containing a higher proportion of crenulated alga-laminated material and fewer fossils. The latter include: cryptostome polyzoa including *Fenestrellina retiformis* ; *Orthothrix excavata* ; *Bakevellia antiqua, B. ceratophaga*. In the third old quarry [4355 4373], midway between the other two, similar flat-lying sheets of sinuous laminated, fine-grained reef-dolomite is seen to pass over the reef-crest, becoming thicker and steepening to 70° at the quarry floor. Roughly bedded, brecciated and recrystallized dolomitic limestone dips eastwards at about 40° off the steeper reef-face. Rocks in this quarry have yielded: *Batostomella crassa, Fenestrellina retiformis, Thamniscus dubius* ; *D. elongatum sufflatum, Orthothrix excavata* ; *Bakevellia* cf. *antiqua, Liebea squamosa* ; ostracods. Post-reef oolitic dolomite crops out at a slightly higher level, 250 yd S.W. of the middle quarry.

Horden to Castle Eden. The reef forms a ridge which can be traced southwards from Townfield Quarry almost to Castle Eden Burn, but it is exposed only at two localities. One [4315 4176] is on the top of the ridge on the site of the Saxon village of Yoden, now Peterlee, where a largely overgrown quarry contains irregularly laminated algal dolomite of reef-flat sub-facies. There is a wide variety of minor algal growth forms here and several examples of contemporaneous breccias (Pl. XB) composed of fragments of algal material which have been welded together by later algal growth. No shells

were noted in this quarry. The other exposure [4354 4172], on the eastern slope of the ridge, resembles the section in the middle quarry of Townfield Hill, and comprises level thinly bedded, largely algal dolomite of reef-flat type passing abruptly into massive, steeply dipping dolomites of the reef-front. Parts of this rock are abundantly fossiliferous, and have yielded: '*Adhaerentina*' *permiana*, '*Glomospira*' *sp.* ; *Acanthocladia anceps, Batostomella crassa*, ?Polyzoan *?Gen, et sp. nov.* [see note on p. 181] ; *Dielasma elongatum* ; *Naticopsis sp., Pleurotomaria?* ; *Bakevellia antiqua, B. ceratophaga, Liebea squamosa, Pseudomonotis speluncaria* ; ostracods. A thin section (E 29512) of a specimen from this quarry is a calcitic dolomite consisting mainly of dolomite grains (0·10 mm) together with very finely recrystallized areas replacing foraminifera, shell-fragments and pseudo-ooliths (? calcispheres). An old quarry [4347 4127], 620 yd E. of Little Eden, is now filled, but in 1947, Mr. G. D. Mockler saw 12 ft of mainly hard, compact, yellow, oolitic limestone with scattered shells. The adjacent farmhouse is built partly of dolomitic algal limestone similar to that in the quarry at Yoden. Cream dolomite with a sparse fauna corresponding to a position high in the reef sequence was temporarily exposed [4363 4054], 1050 yd S.E. of Little Eden.

Dolomites of reef facies overlie lagoonal beds (p. 126-7) in the valley of Castle Eden Burn, dipping eastwards from Seven Chambers [431 394], where they are high and inaccessible. Throughout the Seven Chambers sector the broadly undulate base of the reef dips gently eastwards and the rock is irregularly bedded in thick, massive units. In the cliffs at White Rock [4350 3963] 6 ft of thin- and medium-bedded dolomite separate a lower 15 ft and an upper 6 ft of thickly-bedded rock. All three units are composed of dense, cream-grey, finely crystalline dolomite containing much comminuted shell-debris and a sparse fauna of polyzoa, gastropods and lamellibranchs (especially *Schizodus* and *Permophorus*). At Ivy Rock [435 396] most of the cliff section is composed of cream porous reef-dolomite, containing a restricted fauna of lamellibranchs and gastropods in a crystalline matrix with much comminuted shell debris: a list of fossils collected at this locality is given by Trechmann (1913, pp. 215–6). The lowest 40 ft of the section are massive or very vaguely bedded, and locally brecciated, but the uppermost 10 ft are thickly and irregularly bedded, and where better exposed, some 100 yd E.N.E. of Ivy Rock, are seen to contain low irregular domes, up to 2 ft high and 9 ft in diameter, of the type previously described in the Easington Colliery area (p. 138). The domes can be traced eastwards for a short distance, but most of the cliff section as far as Craggy Bank [4376 3982] is formed by the lateral equivalent of the main part of the Ivy Rock section—cream, porous dolomite which becomes harder, more massive and, possibly, thicker. At Craggy Bank the fauna, which is dominated by small gastropods and lamellibranchs with subordinate polyzoa and small brachiopods, includes: *Acanthocladia?, Fenestrellina retiformis, Thamniscus?* ; *Orthothrix excavata, Stenoscisma* sp. [juv.] ; *Omphalotrochus* sp., *Pleurotomaria linkiana* King, a turreted gastropod ; *Bakevellia sp.* [juv.], *Parallelodon striatus, Pseudomonotis speluncaria.* The fauna becomes more varied eastwards and at a small exposure [4387 3985], 120 yd E.N.E. of Craggy Bank the reef-rock is composed largely of shells and shell-debris, and has yielded: coiled tubular foraminifera ; *Batostomella crassa, Fenestrellina retiformis, Synocladia virgulacea* ; *Dielasma sp., Orthothrix spp.* [including *O. excavata* but there is a great variety of forms, the broader ones approaching *O. lewisiana*] ; gastropods including *?Naticopsis minima* ; *Bakevellia antiqua, Pseudomonotis speluncaria* ; ostracods. This, the most easterly exposure of reef facies in the Dene, probably lies close behind the reef-front. The highest parts of the reef are not exposed in the Dene, but are seen in a number of small diggings and an old quarry in a wood, 300 yd N.N.W. of Dene Leazes. Here, horizontal, thinly bedded alga-laminated dolomite lies about 200 ft higher than the lowest reef-rocks exposed at Seven Chambers, 350 yd to the north-west. Similar algal beds are exposed in a railway cutting [455 400], N.W.

of Blackhall Colliery Station, where they are arranged in a series of flat-topped domes and are overlain by an interbedded sequence of algal and partly oolitic non-algal dolomites.

Hesleden. Dolomites of reef facies crop out extensively in the sides of the valley south of Hesleden and its eastward extension, Crimdon Beck. They are mostly well bedded and are probably largely or wholly younger than those exposed in Castle Eden Dene. The westernmost outcrops are cliffs [435 378] on both sides of the valley, 780 yd W. of Hesleden Station, where the section is:

	ft
Dolomite, cream, thinly bedded to massive, finely granular, largely alga-laminated ...	20
Dolomite, cream-buff, thin- to medium-bedded, granular, with an abundant fauna dominated by gastropods and lamellibranchs	c.25
Dolomite, buff-brown, irregularly bedded, with a scanty fauna, chiefly of gastropods and lamellibranchs	5

The lowest member of the section is poorly exposed, but appears to be mainly of algal reef-dolomite with markedly undulate bedding. The overlying bed (p. 128) is of lagoonal facies. The uppermost member is of dolomite in which algal laminae are flat, slightly undulate or crinkled. ' Cryptozoon '-type algal domes up to 3 ft diameter are particularly well exposed (Pl. XIB) in low cliffs [4350 3784] on the south side of the stream and form complexes at several levels. Thin beds of granular non-algal dolomite are present in some exposures.

This sequence can be traced downstream for a distance of about 1550 yd before passing below stream level, the maximum exposed thicknesses of the three units being 40 ft, 25 ft and 20 ft respectively. Over this distance, the uppermost unit changes progressively eastwards, as follows: the non-algal, granular dolomites interbedded with the laminated algal beds die out beyond a position 400 yd S.W. of Hesleden Station ; the ' Cryptozoon '-type domes gradually diminish in size and complexity and eventually die out south of Hesleden; and the beds as a whole become more massive and homogeneous. In several places south and south-east of Hesleden these beds form prominent cliffs, and are finely laminated throughout, individual laminae being closely crenulated and exhibiting a wide variety of small-scale algal growth forms. It has not been established whether the laminated deposits are stromatolite-fixed granular sediments or are directly deposited by calcareous algae. They are regarded here as being of reef facies since lithification was clearly contemporaneous, although it is recognized that the reef-flat at the time of their deposition may have been only slightly submerged. No shelly fossils have been found in these beds.

The lowest member of the sequence is poorly exposed at several places near the base of the cliffs, but is well seen [4447 3773] on the north side of the stream, 400 yd E.S.E. of Hesleden Station. Here it is composed of thin- and medium-bedded cream granular dolomite, arranged in a series of compound domes which appear to originate from a single stratum at stream-level and rise to heights ranging up to about 20 ft. (Pl. XA). Fine alga-lamination is preserved in many of the least altered component 'beds'. Alga-laminated rocks are also interbedded with clastic shelly dolomite in depressions between the domes, which are spaced at roughly 25-foot intervals. The fauna is restricted to stunted lamellibranchs (especially *Pseudomonotis speluncaria* and *Bakevellia* spp.), small gastropods (including *Naticopsis minima* Brown [Pl. VI, fig. 8] and euryhaline brachiopods (especially *Dielasma*), and these forms also occur in small irregular pockets within the domes.

The top of the reef dips eastwards beneath interbedded algal and inorganic beds about 200 yd E.S.E. of the above exposure, and reappears in an anticline 1½ miles downstream. The following generalized section was measured in a narrow gorge [4715 3705], 500 yd N. of Middlethorpe:

	ft
Dolomite, buff, thin-bedded, finely granular, alga-laminated throughout	36
Conglomerate, composed of a mass of rounded blocks of crystalline shelly dolomite ...	15
Dolomite, cream, fine-grained irregularly thin- and medium-bedded, with many irregular dome structures up to 3 ft high	?15

The section is directly comparable with that near Hesleden (p. 140), the middle member taking the place of the lagoonal shelly calcarenite. The domes in the lowest unit are most common in the lowest 10 to 12 ft; the uppermost 3 to 5 ft are more evenly bedded with fewer and flatter dome-structures, but with several rounded blocks of dense shelly dolomite which appear to be in or near the position of formation. The blocks forming the overlying conglomerate are of similar rock, but they seem to have been rolled and redeposited as a boulder bed at the reef shoulder. They consist of rounded cobbles and boulders up to about 1½ ft in diameter, each of which has a stromatolitic envelope up to ⅜ in thick. Interstices between the cobbles are partially filled with similar stromatolitic material, often contemporaneously brecciated, which in some cases encloses fossils and fossil debris including: '*Glomospira*' *pusilla*; cryptostome polyzoa; *Dielasma elongatum*; *Omphalotrochus sp.*; *Bakevellia antiqua*, *B. ceratophaga*, *Pseudomonotis speluncaria*; *Peripetoceras freieslebeni*; ostracods. Traced south-eastwards for 100 yd within the gorge the conglomerate thins from 15 to 8 ft. It is thinner also towards the south-west where, at the foot of the railway embankment [4704 3705], it is no more than 2 ft thick. The lamination in the uppermost unit is finely crenulate throughout, but the beds themselves contain many minor domes similar to those near Hesleden. These domes are particularly well seen at a horizon about 18 ft from the top of the section and at the base where there is a complex of structures up to 5 ft high and 8 ft in diameter.

In Hesleden Dene No. 1 Bore [4672 3697] sunk in the valley floor west of the railway embankment, Trechmann (1942, p. 323) records drift on bedded dolomite to a depth of 58 ft and reef-limestones from 58 ft to the final depth of 230 ft. It is difficult to reconcile the sequences described above and on p. 140 either with the record given by Trechmann, or with samples collected from the cores by Mr. A. Fowler, but it seems probable that the fossiliferous bed (faunal list, p. 172) between 58 and 67 ft in the bore is the equivalent of the conglomerate in the section east of the railway embankment. Two other Hesleden Dene bores are mentioned by Trechmann (1954, pp. 203–4) and Fowler (1956, pp. 255–7). One [4660 3703], 160 yd W.N.W. of No. 1 Bore, penetrated cream granular bedded dolomite to a depth of about 190 ft, bedded and massive grey bioclastic calcarenite (faunal lists p. 173) with interbedded shaly siltstone bands to 231 ft, and granular and pisolitic beds to bottom of bore at 252½ ft. The other [4654 3698], 200 yd W. of No. 1 Bore, cut grey bioclastic calcarenite between 232 and 272 ft and pisolitic beds from 272 ft to the bottom of the bore at 298 ft. A shelly assemblage (list, p. 172) remarkable for the paucity of polyzoan debris was collected by Mr. W. Anderson from between 246 ft and 257 ft. Correlation with adjacent surface exposures is again obscure, but it is probable that the place of true reef-limestone is taken in both bores by bioclastic carbonates of off-reef basinal facies. According to Trechmann (1942, pp. 313, 323) no reef-limestone was encountered at its expected position in the Thorpe Bulmer Bore, 1400 yd to the south-west.

Black Halls Rocks. Bedded dolomites of reef facies are exposed for over 1500 yd in coastal cliff sections at Black Halls Rocks where the succession, resembling the upper part of the reef sequence in the Hesleden area, is as follows:

	ft
POST-REEF BEDS	
Dolomite, oolitic and granular	c. 70
REEF-BEDS	
Dolomite, cream and buff, thinly bedded, finely alga-laminated	c. 40
Conglomerate composed of a mass of rounded blocks of crystalline shelly dolomite, with interbedded laminated dolomite ...	4 to 15
Dolomite, cream, irregularly bedded, partly alga-laminated, locally with prominent algal dome structures ...	up to c. 15

The reef dolomites are gently anticlinal, but the limits of outcrop are defined by normal faults which appear to be parallel with the original reef-front. The southern end [4763 3827] of the outcrop lies on the coast 780 yd E. of Black Halls, and is also seen in Cross Gill. In each case the junction with later conformable beds is steep (25° to 70°) and indicates the position of the reef-shoulder, which here trends about west to east. From Cross Gill the reef sequence rises gradually northward, the lowest horizons occurring in the core of the anticline on the foreshore at Black Halls Rocks. The northern limb of the anticline is less clearly seen, but reef rocks can be traced with difficulty through scattered poor exposures to the Blackhall Fault which cuts the cliffs [4683 3939], 550 yd N. of Blackhall Rocks station. The lowest unit of this section is exposed only intermittently beneath the conglomerate and is best seen on the north-facing cliff [4731 3897], 560 yd E. 10° N. of Blackhall Rocks Station, where it is composed partly of contemporaneously brecciated and rolled fragments of small algal domes and stromatolitic debris (Pl. XII).

The conglomerate crops out for about 400 yd in the core of the anticline and reaches a maximum thickness of 15 ft near the easternmost tip of Black Halls Rocks. North and south from there parts of the bed pass gradually into bedded alga-laminated dolomite containing rolled cobbles and similar dense masses which appear to be in their original position of formation. Most of the individual cobbles and boulders of the conglomerate are 6 to 12 in across. Their core consists of a dense mass of fine equigranular dolomite anhedra with abundant, extremely irregular, semi-concentric layers of more turbid finer dolomite crystals, set in a matrix of sparry calcite. Small shells and shell fragments form a minor constituent of most cobbles, especially in less dense outer zones composed of confused stromatolitic material. A hard, smooth finely crystalline stromatolitic envelope of very finely

A. Cryptozoon-type algal colonies, Whelly Hill Quarry   *L* 74

Geology between Durham and West Hartlepool (*Mem. Geol. Surv.*)   PLATE XI

B. Cryptozoon-type algal colonies, Hesleden Dene   *L* 143

A. Breccio-conglomerate of reef dolomite above stromatolite domes, Black Halls Rocks

PLATE XII   Geology between Durham and West Hartlepool (*Mem. Geol. Surv.*)

B. Breccio-conglomerate interbedded with alga-laminated dolomite, Black Halls Rocks

laminated dolomite encrusts each shell and cobble, and similar material forms sinuous sheets in the matrix. Some of these sheets are penecontemporaneously brecciated and have a second-generation stromatolitic overgrowth. Other interstitial material includes pisoliths (also probably organic), fine detritus and shelly material. Apart from stromatolites, fossils in these beds are: *Acanthocladia anceps*; *Dielasma elongatum sufflatum*; gastropods including *Omphalotrochus sp.*; *Bakevellia antiqua, B. ceratophaga, Liebea squamosa, Pseudomonotis speluncaria, Schizodus truncatus*; *Peripetoceras freieslebeni*. Faunal lists for the conglomerate bed are also given by Trechmann (1913, pp. 215–6; 1925, p. 136).

The upper member of the reef sequence is similar in most respects to its correlative in the Hesleden area (p. 140). The laminae form widespread crenulations of up to 3-in wave-length and $1\frac{1}{2}$-in amplitude, and these are particularly well preserved in a massive bed immediately overlying the conglomerate. Counts made at several localities within this bed give an average frequency of about 110 laminae per inch, and other beds of the algal sequence contain 90 to 120 laminae per inch. Most of the beds incorporate low domes somewhere along their exposure, but the most striking feature is the occurrence at several horizons of broad flat-topped domes. In cliff sections most of the laminated beds are parallel and dip at low angles, but the even bedding is interrupted at intervals by steep downfolds in which some or all the beds in a group up to 4 ft thick dip sharply down to a common basal surface. Higher beds in such a group have a progressively shallower downfold, and the irregularities are usually evened out before the common base of the next generation of domes.

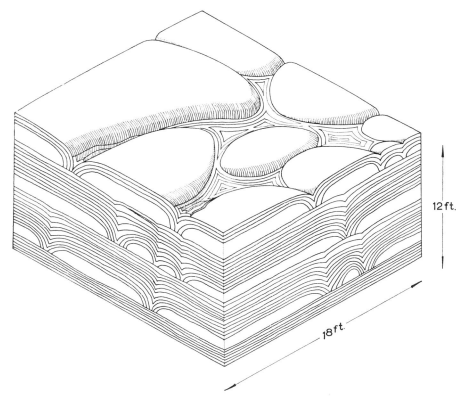

FIG. 22. *Isometric diagram showing idealized flat-topped dome structures in alga-laminated beds at the top of the Middle Magnesian Limestone reef*

This structure is seen in three dimensions on the shore platforms, where the domes are roughly oval in plan (Fig. 22). The longer axes of the domes tend to be aligned north-west to south-east, i.e. normal to the reef-front at this locality. In diameter they range from less than 1 yd to over 20 yd, those less than 4 yd in diameter having rounded tops. Domes are low and poorly defined in the upper 20 ft of the alga-laminated sequence which include intercalations of oolitic and non-organic granular dolomite at the northern and southern boundaries of the reef outcrop. Some of the alga-laminated beds are autobrecciated, and can be traced for several hundred yards by this characteristic. D.B.S.

Hart. Dolomites of reef facies are seen in two old quarries [448 340 ; 449 340] near Whelly Hill House, 1½ miles W.S.W. of Hart, and were cut in bores at the Naisberry Waterworks of the Hartlepools Water Company. The quarries contain 30 to 40 ft of thinly bedded alga-laminated dolomite interbedded with granular unlaminated, possibly inorganic, dolomite. Upper strata are fairly evenly bedded, but have broad undulations of the type seen in the uppermost 18 ft of the reef sequence at Black Halls Rocks. Some deeper, more pronounced downfolds on the north face of the east quarry [4500 3406] demarcate rounded domes up to 3 ft high, but because of widespread brecciation and alteration, the characteristic fine lamination is only locally preserved. There is less brecciation in the west quarry where finely crenulate alga-lamination simulating ripple marks is seen. Fine examples of small algal domes, rising through otherwise level-bedded dolomite to heights of up to 2 ft, occur in complex groups at a number of horizons near the base of the exposed sequence, and are especially prominent on a 12-ft face (Pl. XIA) on the east side of the quarry [4486 3406]. D.B.S., G.D.G.

In the water boreholes at Naisberry finely laminated granular dolomite of algal origin, similar to that exposed near Whelly Hill House, occurs between depths of about 80 ft and 148 ft. The laminae are broadly undulate and in some parts finely crenulate, as at Black Halls Rocks and in Hesleden Dene. Below this, bedded granular shelly dolomite, thought to be of lagoonal facies, continues to about 160 ft, and is underlain by thickly bedded cream and buff porous ?reef-dolomite, with a fauna dominated by fragmentary stunted gastropods and lamellibranchs, and including: ?*Nodosaria kingii* Paalzow, *Nodosaria?* [smooth form] ; polyzoan fragments ; *Straparollus permianus, Naticopsis?, Omphalotrochus helicinus, O.* cf. *thomsonianus* ; *Bakevellia antiqua, B.* cf. *ceratophaga, ?Parallelodon striatus, Permophorus sp., Schizodus sp.* ; ostracods. This bed is similar to the bedded reef exposed at Seven Chambers (p. 139). Between about 190 ft and 245 ft there is massive reef dolomite with a hard, finely crystalline matrix and a varied and abundant fauna: *Acanthocladia anceps, Batostomella crassa, Fenestrellina retiformis, Thamniscus dubius* ; *Dielasma elongatum,* small valve of Strophalosiid ; *Dentalium* aff. *speyeri* Geinitz ; *Naticopsis?, Omphalotrochus* cf. *helicinus, O. taylorianus, Pleurotomaria linkiana* ; *Bakevellia antiqua, B. ceratophaga, Parallelodon striatus, Permophorus costatus, Pseudomonotis speluncaria, Schizodus schlotheimi* (Geinitz). On some weathered surfaces of this rock, a confused fine concentric stromatolitic lamination can be seen. Between 245 ft and 248 ft oolitic and pisolitic dolomite, rich in gastropod and lamellibranch shells, each with a stromatolitic coating, marks a transition to underlying beds of lagoonal facies.

West Hartlepool. Several boreholes at the Hartlepools Water Company's works [507 333] penetrated dolomites of reef facies. In two of these Trechmann (1932, p. 170 ; 1942, pp. 321-2) describes the strata immediately below the base of the Upper Magnesian Limestone as 'reef-like rock with many concentric coatings' and as 'well-bedded . . . and much contorted': it is likely that these beds correlate with bedded, predominantly algal, reef-dolomites like those at Hesleden Dene, Black Halls Rocks and Hart. Bedded rocks below 286 ft in the 1940 bore (*op. cit.* p. 322) may be sub-reef. D.B.S.

Basinal Facies

Under this heading, details are given of all Middle Magnesian Limestone deposits lying east of the reef-front, including transitional beds at the base of the division, bedded off-reef equivalents of the reef, and barren bedded dolomites (with interbedded evaporites) of post-reef age. The stratigraphical relationships of these sub-divisions are shown in Figs. 17 and 20.

Transitional beds and bedded off-reef deposits

Chourdon Point. Thickly bedded and massive, brown, bioclastic, dolomitic calcarenite crops out at the foot of the cliffs in a small bay [4416 4686] on the north side of Chourdon Point. The outcrop is about 25 yd long and 8 ft high, and has the form of a low dome. The dolomite is friable and highly porous except for an irregular 2½-ft mass near the centre of the outcrop, which may have been detached from the reef-front, some 600 yd W.S.W. of Chourdon Point. The friable dolomite contains a varied and abundant fauna (see also Trechmann, 1954, pp. 201–2) indicative of the middle period of reef-growth. Fossils identified include: *Fenestrellina retiformis* ; *Crurithyris clannyana* (King), *Orthothrix excavata, Stenoscisma schlotheimi* ; *Straparollus sp.* non *S. permianus* (King) ; *Bakevellia antiqua, Permophorus* cf. *costatus* and *Streblochondria pusilla.*

Horden Dene, Easington Colliery. Middle Magnesian Limestone beds are exposed in Horden Dene from a locality [4424 4354], 250 yd E.S.E. of Easington Colliery station to the coast. Highly recrystallized, brown dolomitic limestone, thickly and irregularly bedded and extensively brecciated, is poorly exposed for about 120 yd in the bed of the stream. No diagnostic fossils have been found, but the lithological affinities of this rock are with off-reef beds as seen in other exposures, including one in an underground tunnel at Easington Colliery (p. 136) and outcrops in the Sunderland (21) district farther north.

Dene Holme, Horden. Middle Magnesian Limestone crops out on the south bank of Castle Eden Burn [454 404] from 150 to 300 yd W.S.W. of Dene Holme, where about 20 ft of granular and oolitic dolomite dips eastwards on the flank of a small anticline and overlie about 20 ft of irregularly bedded and massive cream shelly bioclastic dolomite. The latter

contain many derived fragments of shelly reef-facies dolomite, carbonate sands, silts and oolite, as well as many polyzoa, brachiopods and lamellibranchs. The following fossils have been identified: algal filaments, some forming crusts about shells ; *Miocidaris keyserlingi* ; *Acanthocladia anceps, Batostomella crassa, Fenestrellina retiformis* ; *Dielasma elongatum, Orthothrix excavata, O.* cf. *lewisiana, Spiriferellina cristata multiplicata, Stenoscisma globulina* ; *?Loxonema fasciatum* King, *Omphalotrochus sp., Pleurotomaria sp.* ; *Bakevellia antiqua, B. ceratophaga, Palaeolima? permiana* (King), *Parallelodon striatus, Pseudomonotis speluncaria, ?Streblochondria pusilla* ; ostracods. A faunal list for this locality is also given by Trechmann (1913, pp. 215–6). The fauna is notable for the paucity of cryptostome polyzoa, elsewhere almost ubiquitous in rocks associated with the reef phase. The core of the anticline is a low dome of massive cream dolomite lithologically similar to reef-facies rock.

Boreholes around Hartlepool. Basinal equivalents of the reef phase were penetrated in a number of boreholes around Hartlepool and in N.C.B. Offshore Nos. 1 and 2 bores. Trechmann (1913, pp. 187–8) found brachiopods and lamellibranchs below a depth of about 390 ft in cores of an 1888 boring [5163 3455] at the Warren Cement Works. The rock in which they were found is described as dark grey calcitic dolomite associated with varying quantities of gypsum and anhydrite. One mile E.S.E. of Warren Cement Works, the Hartlepool Lighthouse Bore [5320 3386] penetrated about 75 ft of beds in which a typical shelly reef fauna (lists, p. 176) occurred at three levels, separated by barren fine-grained dolomites (Magraw and others 1963, pp. 175, 199). The highest fossiliferous bed, lying between depths of 636 and 667 ft, appears to be equivalent to the main reef-building phase ; it consists of grey, fine, bioclastic calcarenite, with scattered larger shell fragments, interbedded with cream or buff, finely crystalline or porcellanous dolomite, with scattered well preserved whole shells. The calcarenitic beds are similar to those in bore holes in Hesleden Dene (p. 142) and contain irregular streaks and fragments of dark grey siltstone and silty mudstone, and abundant foraminifera, crinoids, polyzoa, brachiopods, gastropods and lamellibranchs. Thin sections show that the matrix of the calcarenite consists largely of finely comminuted organic debris. The light-coloured fine-grained interbeds are largely colour-mottled in shades of buff. Both types of bed contain fairly abundant small-amplitude stylolites, and sporadic angular derived fragments of dolomite, mudstone and siltstone. Other beds of bioclastic dolomite containing derived rock fragments occur at depths of 710 and 744 ft, and are thought to be basinal equivalents of the early phases of reef-building, the lower horizon being taken as the base of the division. A succession like that of the 1888 Bore was cut in another bore [5003 3430] at Howbeck Pumping Station (Trechmann 1942, pp. 319–20). In both boreholes dips of 20° or over occur throughout the off-reef sequence.   D.B.S.

The off-reef beds thin northwards and eastwards, and are interpreted to be only 37 ft thick in Offshore No. 1 Bore, 4 miles N. of Hartlepool (Magraw and others, 1963, pp. 175, 201). From between 1071 ft and 1074 ft in this bore the following were collected: ' Adhaerentina' permiana, ' Ammodiscus' ?, ' Glomospira' gordialis, uniserial chambered foraminifera [common ; all Nodosaria or Lingulina form of chambers] ; Horridonia ?[juv.], Neochonetes? davidsoni, Orthothrix excavata. ?Strophalosia morrisiana ; Bakevellia cf. antiqua ; ostracods. Neochonetes? davidsoni is common, and though some have radial ribs, most are smooth except for concentric growth lines. There is great variation in the length of the hinge line, and many of the shells are as long as they are wide.   D.B.S., J.P.

Post-reef beds

Except at Black Halls Rocks and in Shippersea Bay, the outer edge of the reef lies behind the present shore, the coastal cliffs being composed of post-reef Middle Magnesian Limestone and Upper Magnesian Limestone, lying against the steep reef-face. A thin succession of post-reef Middle Magnesian Limestone also overlaps on to the reef in the Hesledon, Easington, Hesleden and Hartlepool areas.

Murton, Cold Hesledon and Hawthorn. The most westerly outcrops of post-reef Middle Magnesian Limestone are at the north end of South Hill, $\frac{2}{3}$ mile N.N.W. of Cold Hesledon, where a few feet of bedded, granular, partly oolitic dolomite are seen in several small quarries and natural exposures. Fragments of similar dolomite are scattered on the ridge which culminates northwards in Fox Cover Quarry (p. 133), and there are good exposures in railway cuttings east of Hesledon East House, around Kinley Hill, and in adjacent coastal sections. The railway cutting [435 473], 450 yd E. of Hesledon East House, is in highly faulted and brecciated bedded dolomite extending for about 300 yd in an asymmetric anticline. Where least altered the dolomite is predominantly thin-bedded and granular with traces of ripple-marks and cross-bedding. Where brecciated, the rock is harder and more calcareous. Northwards-dipping beds at the northern end of the cutting may be Upper Magnesian Limestone. Thin-bedded, granular dolomite is also exposed in a second cutting [436 474], 200 yd farther north-east. A little to the south small exposures and rock fragments in the soil indicate that the top of Kinley Hill is formed largely of thinly bedded oolitic dolomite. This is also exposed in several small quarries nearby, where ooliths are locally almost obliterated by recrystallization. The main exposures are : natural outcrop [4336 4558], 270 yd E.N.E. of Kinley Hill (farm), 10 ft ; shallow digging [4303 4670], 200 yd N.W. of Kinley Hill, 2 ft ; old quarry [4302 4636], 220 yd S.S.W. of Kinley Hill, 8 ft ; old quarry [4285 4609] 620 yd S.S.W. of Kinley Hill, 10 ft ; old quarry [4313 4609] 540 yd S. of Kinley Hill, 15 ft ; and old quarry [4327 4634] 300 yd S.S.E. of Kinley Hill, 8 ft. The rock exposed 200 yd N.W. of Kinley Hill contains a number of pisolitic beds. Thin sections (E 29422, 29451, 29489–90) of other rocks from this locality are described by Dr. Dearnley as oolitic calcitic dolomites of silt-grade, consisting of closely packed ooliths, with outer rims of fine-grained dolomite (0·01 to 0·02 mm) and cores of coarsely (0·25 mm) crystalline calcite and dolomite, set in a matrix of aggregated dolomite and calcite rhombs.

Oolitic dolomite also forms the uppermost 18 ft of beds exposed in a group of old quarries in a wood, 300 to 750 yd W.S.W. from Hawthorn Tower [4396 4613]. The oolitic rock is underlain by 15 to 18 ft of thin- and medium-bedded, granular or finely granular dolomite, with traces of cross-bedding at some horizons and with a 1-ft layer containing low domes up to 2 or 3 ft in diameter 6 ft above the base. These lower beds are the probable equivalent of interbedded algal and non-algal strata overlying the reef at Black Halls Rocks, Shippersea Bay and Hesleden Dene. This suggestion is supported by the section exposed nearby in Hawthorn Quarry [437 463], where the top of the reef-rock passes southwards at about the same level as the base of the section in the quarries in the wood. The post-reef beds in Hawthorn Quarry are about 50 ft thick ; they are mainly inaccessible, but appear, from fallen fragments, to consist largely of granular dolomites similar to the upper beds exposed in the wood. The eastern slope of the reef in Hawthorn Quarry is truncated by a reverse fault, which brings down grey oolitic post-reef dolomite. About 12 ft of these beds are exposed at the main entrance [4383 4631] to the quarry.

At the coast, post-reef dolomites are exposed intermittently from the northern margin of the district to 470 yd S. of Beacon Point, where they overlie alga-laminated beds. Southwards for about 1050 yd from the northern margin of the district the coastal cliffs are composed of undulating, evenly bedded dolomite of broadly constant lithology, with a total thickness of 120 ft ; traces of cross-bedding and some ooliths are seen in places. A prominent bed of soft, cream dolomite, 10 to 15 ft thick, can be traced southwards for several hundred yards from the northern margin and a well-defined lens of breccia lies near the base of the cliffs from 700 to 515 yd N.W. of Chourdon Point. About 125 yd farther south, bedded, saccharoidal sand-grade dolomite shows the first signs of brecciation which increases in intensity towards the south.

At Chourdon Point and immediately to the south, completely brecciated rock, hardened by dedolomitization and recementation, gives rise to bold cliffs in which fragments up to 8 ft across stand out in relief. Mechanical brecciation, due to a series of sub-parallel normal faults, has given rise to second-generation breccias at a number of places on the headland.

From Chourdon Point to Hawthorn Hive the breccias are interrupted only by a few patches of undisturbed, or slightly disturbed thin-bedded saccharoidal, sand-grade dolomite. Most of the breccias are formed of fragments of similar rock. The degree of brecciation varies from slight, where existing beds are broken up more or less *in situ* and fragments have moved only slightly relative to one another, to complete, where fragments of strata are incorporated from higher beds, some now removed by erosion. Some completely brecciated areas have ill-defined margins and occur near the middle of areas of lesser brecciation, but others form sharply defined ' gashes ' bounded by steep slickensided surfaces cutting sharply across undisturbed bedding. Both types can be seen in the cliffs around Hawthorn Hive, where several breccias contain fragments of Upper Magnesian Limestone and red and green marls. The latter were probably deposited some hundreds of feet above their present position as part of the Upper Permian Marls. Carbonate rock in the brecciated zones, both in the fragments and the matrix, is generally more calcareous and harder than the unbrecciated parent rock.

From Hawthorn Hive southward past Beacon Point brecciation at first decreases : from about 480 to 320 yd north of the point the cliffs are formed by about 45 ft of thinly bedded, soft, finely granular dolomite, only slightly brecciated and apparently barren. Still farther south, however, extensive brecciation reappears and continues until the outcrop is terminated by faulting, 470 yd S. of Beacon Point. The succession in this ground is altogether about 130 ft thick. Where least altered the dolomites are thinly bedded, cream, and locally bear traces of original lamination and structures resembling cross-bedding. Elsewhere much of the sedimentary structure is obliterated by recrystallization. No undoubted algal structures have been seen.

Easington Colliery to Horden. Small old quarries [4298 4503 ; 4322 4516] respectively 260 yd E. and 510 yd E. of Thorpe Lea West contain a few feet of massive, barren, grey, crystalline dolomitic limestone, the exact stratigraphical position of which is not clear but which are here assigned to post-reef beds on general lithological grounds. A little farther east 15 ft of hard thinly bedded, grey, fine-grained, saccharoidal dolomitic limestone are seen in an old quarry [4348 4552] 570 yd W.N.W. of Beacon House. Structures resembling small ripple-marks are preserved in those parts of the limestone which are not autobrecciated. Since this quarry is topographically on the same level as a group of quarries west-south-west of Hawthorn Tower (p. 147) the beds are presumed to lie on the upper surface of the bedded reef. An isolated exposure [4358 4446] in a colliery railway cutting, 370 yd W. of Low Grounds, Easington Colliery, contains 12 ft of brecciated grey-white massive crystalline dolomitic limestone of uncertain, but possibly post-reef, age. The other main inland exposure of post-reef beds comprises 18 ft of grey recrystallized cross-bedded oolitic dolomite at Couslaw Holes Quarry [433 436], 420 yd N.N.W. of Paradise, Easington Colliery. These beds lie slightly higher than the level of the reef-crest, 250 yd farther east (p. 139). Altered oolitic dolomitic limestone forms further small exposures [4271 4372 ; 4302 4377], respectively 940 yd and 750 yd N.W. of Paradise.

At the coast, post-reef beds crop out for about 1½ miles southwards from Shot Rock. Between Shot Rock (where they overlie interbedded algal and oolitic dolomite) and the Easington Fault, they are about 110 ft thick and are lithologically similar to the beds north of Chourdon

Point. There is fairly extensive autobrecciation at the northern end of the outcrop and immediately north of the Easington Fault. South of the Easington Fault, post-reef dolomites are almost continuously exposed in the cliff for about 1100 yd, and intermittently exposed between slipped masses of drift for a further 600 yd before passing beneath Upper Magnesian Limestone, 450 yd N. of the Horden Fault. Throughout this distance they are cream oolitic dolomites, 120 ft thick, in gently undulating partly collapse-brecciated beds. The oolite is largely recrystallized, and locally resembles the apparently non-oolitic beds at this horizon north of Shot Rock. Zones of complete collapse-brecciation occur at irregular intervals (Horden Point is an excellent example), and well-defined breccia-gashes with fragments of Concretionary Limestone and of red marl are seen 520 yd S. of Loom and 60 yd S. of Horden Point. No fossils have been seen in any of these post-reef cliff sections.

South of the Horden Fault the section is obscured by colliery spoil, but according to Trechmann (1925, pl. xiii) the beds are like those to the north. They reappear beyond the southern end of the colliery spoil at Whitesides Gill, 1000 yd S. of Horden Point, and continue southwards before eventually passing beneath Upper Magnesian Limestone beds about 700 yd N.W. of Dene Mouth. Their upper junction is obscured by recrystallization and their thickness is uncertain.

Blackhall Colliery to Hesleden. Late Middle Magnesian Limestone beds overlie the reef in Hesleden Dene and north and south of Black Halls Rocks, whilst beds thought to occupy an off-reef position are exposed in Castle Eden Burn and its tributary Hardwick Dene and have also been penetrated in a number of boreholes in the coastal and offshore areas.

In the dene south of Hesleden, post-reef dolomites succeed alga-laminated deposits of the reef-flat. At Jack Rock [447 376] finely granular dolomite is interbedded with thin dolomites of algal origin in a 2½-ft bed at the base, and an 8 ft algal bed is seen 22 ft higher. The intervening strata are predominantly of finely granular dolomite with locally preserved traces of small-scale cross-bedding, and with alga-laminated horizons at some exposures farther downstream. The following section, measured on the north bank [4545 3712], 470 yd W. of the church at Monk Hesleden, is representative:

	ft	in
Dolomite, massive, finely alga-laminated	2	0
Dolomite, fine-grained, largely autobrecciated, probably originally algal	5	0
Dolomite, thinly bedded, finely granular, even-textured, probably non-organic ...	3	3
Dolomite, thinly bedded, conspicuously alga-laminated ...	1	9
Dolomite, flaggy (bedding undulate), finely granular, probably partly algal ...	1	2
Dolomite, massive and thickly bedded, granular, partly cross-bedded	3	6
Dolomite, thinly bedded, granular, partly finely cross-bedded	5	4
Dolomite, thinly bedded, with closely crenulate algal laminae near base	2	3
Dolomite, medium-bedded, granular, probably non-organic	4	0

The interbedded sequence exemplified above is probably altogether about 40 ft thick ; it can be traced downstream through a series of discrete exposures to a position approximately south of Monk Hesleden Church, where it is succeeded by about 30 ft of thinly bedded, finely granular dolomites with small scale cross-bedding throughout. Some 12 ft of these beds are exposed [4580 3710] on the north side of the stream, 250 yd E. of the church, where they are overlain by 18 ft of thicker bedded dolomites which are inaccessible but are thought to be oolitic. Further exposures of these beds can be seen for another 300 yd downstream. Farther downstream beds of the Upper Magnesian Limestone form a broad shallow syncline, but the junction with the Middle division is nowhere exposed. Post-reef beds reappear some 750 yd farther east where a number of small

outcrops occur on both sides of the valley for a distance of about 180 yd upstream from the railway embankment. The following section [4581 3692] was measured 340 yd W.N.W. of Middlethorpe:

	ft
Dolomite, thinly bedded, soft, oolitic, highly weathered ...	8
Dolomite, thickly bedded, cream, soft, finely granular	6
Dolomite, thickly bedded, cream, oolitic and pisolitic, very porous	6

The texture of the lowest bed is remarkable. It is predominantly composed of coarse ooliths, pseudo-ooliths and pisoliths, set in a fine-grained oolitic matrix. Scattered profusely through the rock are large (up to 15 mm) irregular, commonly grotesque pisoliths, elongated approximately parallel to the bedding. Many are hollow and now consist of a thin envelope only, whilst others contain aggregates of ooliths and oolitic debris. Some of the smaller pisoliths, too, are compound, and many of the ooliths have markedly eccentric shapes. It is inferred from nearby surface outcrops and the Hesleden Dene boreholes (Appendix I, p. 142) that these beds occupy a position over, and slightly to the east of, the reef-crest.

Farther east, in Hardwick Dene, post-reef dolomites form the lower part of the sequence exposed near Hardwick Hall [450 392] and for about 1100 yd to the north. The lowest rocks form the cores of anticlines 600 and 750 yd N.N.E. of Hardwick Hall, and are of thin-bedded finely granular cream dolomites, over 75 ft thick. By analogy with other sections in the area they are the correlative, in part at least, of the uppermost, bedded part of the reef sequence, but because of advanced recrystallization most of their original structure has been obliterated and their exact stratigraphical position cannot now be established. They are overlain successively by about 25 ft of flaggy, finely granular, cream dolomite, and about 35 ft of cross-bedded cream oolitic dolomite. At the base of the latter a pisolitic bed, well-exposed 300 yd N.N.E. of Hardwick Hall, bears a striking resemblance to the lowest member of the Middlethorpe section above, and almost certainly lies at the same horizon. Uppermost beds of the Middle Magnesian Limestone are again exposed [4516 4001] at the northern end of the Hardwick Dene as follows:

	ft
Dolomite, cream, massive and irregularly bedded, soft, finely granular, weathering to cream mealy powder; possibly oolitic	c. 20
Dolomite, cream, vaguely bedded, soft, finely granular, with traces of small-scale cross-bedding; thin-bedded at base	c. 20

Few of the original structures in these beds have survived advanced alteration and their correlation with the upper members at the southern end of Hardwick Dene is tentative.

In a small old quarry [4623 3874], 50 yd W. of Mickle Hill, there are 15 ft of cream cross-bedded oolitic dolomite; and at a natural exposure [4660 3814], 300 yd N. of Tweddle Black Halls, about 6 ft of irregularly bedded autobrecciated dolomitic limestone appear to be altered oolite. These are the only exposures in the ground between the coast and Hesleden Dene.

Along the coast at Black Halls Rocks, beds of reef facies crop out for 1500 yd in the core of an anticline (pp. 142-4). Post-reef strata comfortably overlap the reef-crest on the southern limb, and are well exposed southwards for about 80 yd from a locality [4783 3827], 780 yd E. of Black Halls. The lowest visible post-reef beds are composed of massive cream oolitic dolomite. The oolites are succeeded by 20 to 30 ft of soft, granular, partly autobrecciated dolomite, and then by about 25 ft of soft, oolitic dolomite. The uppermost beds are also exposed for about 65 yd in the core of a subsidiary anticline, the axis of which meets the coast 930 yd E. 18° S. of Black Halls. Exposures on the northern limb of the major anticline are discontinuous, but indicate the presence of a basal transition zone in which alga-laminated dolomite is interbedded with cream, oolitic dolomite.

These beds and the overlying oolite are extensively auto-brecciated, and terminate northwards at the Blackhall Fault which brings down Concretionary Limestone.

D.B.S.

Hart, Hartlepool and offshore areas. Post-reef beds of Middle Magnesian Limestone age have been quarried at several places between Hart and West Hartlepool, but over most of this area they are buried beneath Upper Magnesian Limestone and are known only from boreholes. The largest exposures are in the west and south-west faces of an old quarry [476 345], 300 yd E. of Hart windmill, where about 40 ft of fine-, medium- and coarse-grained, saccharoidal, dolomitic limestone and calcitic dolomite are seen. They form undulating beds which are predominantly thick or massive except in the uppermost 13 ft, where thin-bedding predominates. Most of the rock is hard, dense and re-crystallized, but secondary intergranular porosity is high in the lowest 15 ft of beds, and larger cavities occur throughout. The cavities are often grouped parallel to the bedding and are especially concentrated in a 12-ft bed in the middle of the section. There are local traces of lamination, and vague indications of low-angle cross-bedding suggest that at least parts of this rock were formerly oolitic. Similar hard compact massive rock, 7 ft thick, is exposed in a small quarry [4792 3407] at the north-western corner of Craggy Bank Wood, 870 yd E.S.E. of Hart windmill. Although unrecognizable in hand specimen, undoubted ooliths can be seen in thin section (E 29922) of this rock, which is a fine-grained mass of dolomite grains, grouped in subspherical aggregates (about 0·10 mm in diameter), with a calcite matrix. Low-angle cross-bedding was noted in this quarry and in a number of small natural outcrops of similar rock nearby. Similar dolomitic limestone and calcitic dolomite are 15 ft thick in an old quarry [4725 3445], 70 yd S. of Hart windmill, and 8 ft thick in an old digging [4743 3431], 280 yd S.E. of the windmill. A shallow digging [4705 3359] on the south-west side of Naisberry crossroads is in 2½ ft of well-bedded, compact, finely porous dolomite which has traces of low-angle cross-bedding and is probably equivalent to the beds exposed at Craggy Bank. No fossils were found in any of these quarries during the resurvey and this fact, coupled with the distinctive lithology, strongly suggests inclusion of these beds with the Middle division of the Magnesian Limestone rather than Upper, as claimed by Trechmann (1913, p. 212 et seq.).

D.B.S., G.D.G.

At Naisberry, two bores [4662 3362 ; 4660 3364] penetrated about 50 ft of granular, saccharoidal, calcitic dolomite, overlying reef-limestone. In both bores small flat cavities occurred in the highest part of these beds, and it is possible that these represent shell-casts indicating the presence of a few feet of Upper Magnesian Limestone. Yellow dolomite, 33 ft thick, was proved in a similar stratigraphical position in a bore [5075 3339] at Hartlepool Waterworks, where the base of the Upper Magnesian Limestone was clearly defined (Trechmann 1932, p. 170).

Most of the remaining borings into post-reef beds of the Middle Magnesian Limestone are located east of the reef-front and penetrate barren dolomites of basin facies. These rocks, closely resembling their equivalents cropping out on the coast north of Black Halls Rocks, are widely brecciated. In a bore [5003 3430], near Howbeck Hospital, Hartlepool (Trechmann 1942, p. 319) they are represented by about 63 ft of bedded yellow dolomite, brecciated in the upper part. This thickness is anomalously small for this area, and is probably due to the removal of interbedded soluble rocks. Brecciated oolitic dolomite is reported from a short core taken in post-reef beds at 432 ft in the North Sands Bore [5027 3534], and 118½ ft of rock described as ' yellow limestone, broken ' and ' yellow limestone, very much broken ' are recorded in the Hart Bore. According to Trechmann (1942, p. 313) most of the post-reef Middle Magnesian Limestone beds at the Thorpe Bulmer Bore are composed of brecciated dolomite, yellow above and grey below, but core recovery was poor and no thicknesses are available. Two beds 11 and 17 ft thick, and similarly described as ' broken ', form part of a post-reef

sequence of yellow and grey limestones at depths of 67½ and 103½ ft respectively below the base of the Upper Magnesian Limestone in an 1893 Bore [approximately 511 319] at Burn Road, West Hartlepool. Other bores at Palliser Works [5061 3525], at the outskirts of Hartlepool, and at Villiers Street [5082 3233], Smalley's Pulp Works [5130 3186], Durham Paper Mills [5114 3187] and Nelson Street [5154 3197], all in West Hartlepool, penetrated a short distance into these beds, and all record yellow limestone or dolomite. The Nelson Street Bore was examined by Trechmann (1942, p. 325), who records 11 ft of porous dolomite, partly oolitic, below the Upper Magnesian Limestone. The position of the West Hartlepool bores relative to the reef-front is not known, but the presence of dolomite rather than anhydrite suggests that some, at least, of the bores are above the reef.

In the easternmost land areas and in the offshore area, thick deposits of anhydrite (the Hartlepool Anhydrite) have been proved above beds thought to be basinal equivalents of the reef. Anhydrite was first found under Hartlepool by a bore [5163 3455] sunk in 1888 at Warren Cement Works (Marley 1892, p. 94 ; Trechmann 1913, p. 187), and has since been confirmed by three 1923 borings at the Cement Works (Trechmann 1925, p. 140), and by the Hartlepool Lighthouse Bore (Magraw and others 1963, pp. 176–8. 199). The general sequences are shown in Fig. 23, in an attempt to reconstruct the relationships of the Permian beds immediately north of Hartlepool Bay. In each bore a group of barren grey crystalline dolomites, containing abundant gypsum and anhydrite, occurs below the thick anhydrite, and in two cases the barren beds overlie fossiliferous dolomite. In Offshore No. 1 Bore and the Hartlepool Lighthouse Bore the dolomite immediately underlying the thick anhydrite has a nodular structure (Pl. XIIIA) which appears to be of algal origin when seen in thin section (E 29259). Although exact correlation is not yet possible, it is likely that, as in the Blackhall Colliery–Hesleden area, the barren beds are the equivalent of early post-reef beds, whilst the thick anhydrite passes laterally and with interdigitation into the later post-reef beds. The former presence of soluble interbeds in this latter group of strata was inferred by Trechmann (1913, p. 195) from the evidence of widespread collapse breccias and traces of sulphate. In the easternmost 1923 bore the anhydrite is overlain by 17½ ft of 'limestone', and a sequence of interlaminated gypsum and fine-grained dolomite occurs at this horizon in the Lighthouse Bore. Although no confirmatory fossils were found in either bore it is thought (Magraw and others 1963, p. 177) that the beds immediately overlying the anhydrite are of Upper Magnesian Limestone age.

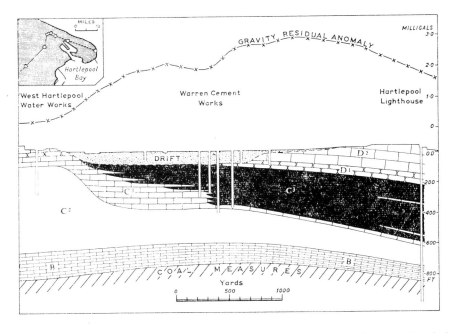

FIG. 23. *Section from West Hartlepool to Hartlepool, based on surface and borehole data, showing the inferred relationships of the Hartlepool Anhydrite. The upper line shows the approximate gravity residual-anomaly along the line of section. For key see Figs. 17 and 20*

In Offshore Nos. 1 and 2 bores respectively anhydrite 429 and 500 ft thick is associated with a sequence of dolomites like that in the Hartlepool Lighthouse Bore. Full details of these three bores are given in Magraw and others (1963), and abridged records are given in the Appendix (pp. 294, 313). The top of the anhydrite in both offshore bores is defined by the smooth sharp base of the Upper Magnesian Limestone.

The lithology of the anhydrite is broadly constant in all the bores and in the workings of the Warren Cement Works Mine (Trechmann 1932, p. 168; Sherlock and Hollingworth 1938, p. 37). It is a hard, bluish grey translucent rock, finely to coarsely crystalline, in which fine-grained grey or brown dolomite forms an extensive, delicate mesh of fretted stringers. Locally these may compose up to 50 per cent of the mass, but they generally range between 5 and 30 per cent. Dolomite also forms beds up to 14 ft thick spaced at wide intervals in all the bores, but is especially prominent towards the base of the anhydrite in Offshore No. 2 Bore. In some beds the dolomite is very finely and evenly laminated, the laminae in some cases showing conspicuous contortion (Pl. XIIIB). At the top and bottom of the Hartlepool Anhydrite the rock has been hydrated to gypsum up to 11 ft thick,

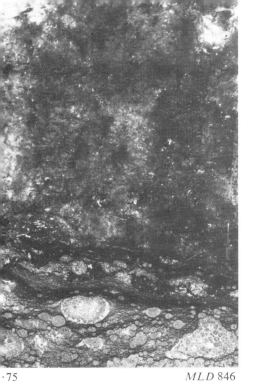

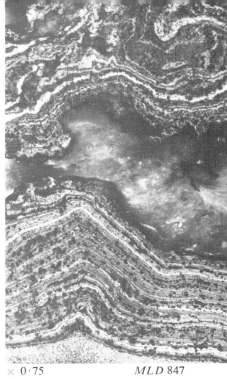

·75   *MLD* 846

$\times$ 0·75   *MLD* 847

A. GYPSUM WITH PATCHY ANHYDRITE ON
NODULAR DOLOMITE

B. LAMINATED DOLOMITE CONTORTED
BY FLOW. HARTLEPOOL ANHYDRITE

Geology between Durham and West Hartlepool (*Mem. Geol. Surv.*)   PLATE XIII

. CONCRETIONARY LIMESTONE, PARTIALLY BRECCIATED BY SOLUTION OF UNDERLYING EVAPORITES.
BLACK HALLS ROCKS

and tabular selenite crystals, up to $\frac{1}{2}$ in long, project into the anhydrite at many sulphate-carbonate interfaces. In the Warren Cement Works Mine, Trechmann (1932, p. 168) observed that much of the anhydrite was hydrated to gypsum at the north end of the workings, where it tended to thin out ; and Hollingworth (*in* Sherlock and Hollingworth 1938, p. 37) reported that gypsum veins became more prominent towards the south end of the workings.

The macroscopic appearance of the rock in the mine and in boreholes indicates extensive replacement of dolomite by anhydrite, and this is confirmed by examination of thin sections. A typical section (E 29256), from 844 ft in Offshore No. 1 Bore, is described by Dr. Dearnley as a dolomite-anhydrite rock consisting largely of anhydrite, with irregular lenses of small dolomite rhombs which are locally partly replaced by anhydrite. Large gypsum plates enclose anhydrite crystals, but also occur as replacements of dolomite. A thin gypsum vein cuts one dolomite band. The order of formation of the minerals appears to be first dolomite, then anhydrite and finally gypsum. Fluxion structures were noted by Dr. J. Phemister in anhydrite specimens (E 17876, 20804) from the Warren Cement Works Mine, and are clearly visible in weathered cores of anhydrite from Offshore No. 2 Bore. D.B.S.

Upper Magnesian Limestone

The Upper Magnesian Limestone is preserved only in the eastern part of the district (Pl. V). North of Black Halls Rocks the division is restricted to coastal areas where it forms four isolated shallow synclines. South from Black Halls Rocks to the West Hartlepool Fault it crops out in a coastal belt 1 to 2 miles wide and over 4 miles long, but is exposed only at the north-western and south-eastern extremities. South of the West Hartlepool Fault these rocks are known only from boreholes.

This account follows the long-established practice of sub-dividing the Upper Magnesian Limestone of Durham into Concretionary Limestone below, and Hartlepool and Roker Dolomites above. This sub-division usefully takes account of two distinctive lithologies, but the junction between them has never been precisely defined. A rough criterion is the presumed alga *Tubulites permianus* [Pl. VI, fig. 2], which has been only doubtfully recorded in the upper subdivision, but is very common in the lower. The remaining limited fauna of the Upper Magnesian Limestone (King 1850, table 1) is common to both groups and is dominated by euryhaline gastropods and lamellibranchs. Apart from foraminifera and ostracods only seven species are known, and of these only the lamellibranchs *Liebea squamosa* [Pl. VI, fig. 1] and *Schizodus schlotheimi* [Pl. VI, fig. 3] are at all common. The recognition by Mr. J. Pattison of a small, possibly punctate shell fragment (*Dielasma?*) from the Hartlepool and Roker Dolomites at Hartlepool is noteworthy, as brachiopods have hitherto not been recorded in Durham above the Middle Magnesian Limestone.

Writers from Kirkby (1861, p. 316) onwards have observed that all the forms in the Upper Magnesian Limestone are stunted and have attributed this to an inhospitable environment. The absence of polyzoa and brachiopoda (apart from the possible example noted above), both of which are generally intolerant of hypersaline water, supports this view. Plant remains are locally common throughout the Upper Magnesian Limestone.

At the close of Middle Magnesian Limestone times it is probable that the Zechstein Sea in this district had become almost filled and the Upper Magnesian Limestone was laid down on a surface of low relief. The buried reef protruded slightly above this surface and continued to exercise some local control over sedimentation: for instance, there are signs of overlap within the Upper Magnesian Limestone at Horden, Nesbitt Dene and West Hartlepool. It seems unlikely, however, that the reef acted as a western limit of deposition in a regressive sea as suggested by Woolacott (1912, p. 260 ; 1919b, p. 490) and Trechmann (1914, p. 260 ; 1925, p. 141), for it is overlain by Concretionary Limestone at Blackhall Colliery and West Hartlepool (Trechmann 1913, p. 213 ; 1932, p. 170).

It is clear from Fig. 20 that the distribution of Upper Magnesian Limestone in the district is controlled by a combination of structure and the present erosional surface, accentuated as Woolacott (1919b, Fig. 1) showed, by differential movement between the reef and the younger surrounding strata. This movement has lowered the base of the Upper Magnesian Limestone relative to the eastern face of the reef and brought the two into lateral juxtaposition. This is well seen at Shippersea Bay, where the base of the Concretionary Limestone has been lowered relative to the reef-crest by about 150 ft, of which about 20 ft can be accounted for by normal faulting. Some of the remainder may have been caused by differential compaction, but it is likely that most of the subsidence was effected by the solution of sulphate layers in the Middle Magnesian Limestone (p. 122). Collapse-breccias like those formed by solution in the Middle Magnesian Limestone are common also in the lower part of the Concretionary Limestone.

Gypsum occurs interstitially in granular beds and as irregular replacement patches, isolated porphyroblastic crystals and veins throughout the Upper Magnesian Limestone. It reaches a maximum in Offshore No. 1 Bore, where up to 50 per cent of a laminated dolomite in the Concretionary Limestone has been selectively replaced along the bedding. Generally, however, the gypsum is present in isolated blebs up to 3 in across, forming 3 to 5 per cent of the rock. At surface exposures the gypsum has been dissolved out, leaving stellate cavities and in some instances contributing to the collapse of the surrounding beds.

The top of the Upper Magnesian Limestone is not exposed in the district, and the full thickness of the division is seen only in boreholes. Its maximum proved thickness is 416 ft in Offshore No. 2 Bore, but inland it is not known to exceed 200 ft and is usually less than 150 ft thick. At a boring in the southern part of West Hartlepool [5154 3197], a short distance north of the West Hartlepool Fault, the Upper Magnesian Limestone is about 190 ft thick, but only a short distance south of the fault the thickness is reduced to only about 80 ft. Since the main subdivisions are recognizable in the records of these bores, the sharp reduction is probably depositional, and may perhaps indicate penecontemporaneous movement along the fault. The reduced thickness south of the fault is maintained by the Upper Magnesian Limestone throughout the brinefield area around Middlesbrough and over wide areas of Yorkshire.   D.B.S.

Concretionary Limestone

The Concretionary Limestone Group is intermittently exposed in coastal areas in the northern part of the district, but south of Hart it is known only from boreholes. It is composed mainly of thinly bedded granular dolomites of silt- to fine sand-grade, but the rock is locally recrystallized and in some such places contains many concretions. When freshly broken these rocks always smell strongly of petroleum (Trechmann 1931, p. 251) and both concretions and matrix yield a quantity of oily and carbonaceous matter on solution in acid (Trechmann 1913, p. 198). Fine lamination is common in the lower beds, particularly at the base, where the Flexible Limestone, $1\frac{1}{2}$ to 12 ft thick, is represented by dense impure dolomite, cream or grey in surface outcrops, and grey, foetid, bituminous, and in places shaly at depth. Although this bed has yielded fish remains from Fulwell (in the Sunderland district) none has been found in the present district. Plant debris, however, is common ; and poorly preserved lamellibranchs are recorded from a bed doubtfully referred to the Flexible Limestone at one locality (p. 158).

In coastal exposures the Concretionary Limestone falls into a lower group of beds containing abundant concretionary structures, and an upper group in which such structures are generally absent. The lower beds, 40 to 45 ft thick, are so laterally variable in lithology that exact correlation of adjacent sections is often difficult. The variation is due to irregular distribution of concretions, which locally combine to form massive hard grey crystalline limestone quite distinct from the normal thin-bedded cream soft granular dolomites into which they pass within a few feet either horizontally or vertically (Fig. 24). The concretions lack the wide range of patterns found in the Sunderland area (Abbot 1903, pp. 51–2 ; Holtedahl 1921, pp. 195–206), and are generally subspherical, mutually interfering aggregates of radially crystalline calcite, $\frac{1}{4}$ to $1\frac{1}{2}$ in across. The calcite crystals are brown or grey due to the presence of impurities, and commonly have scalenohedral terminations. Some interspaces between concretions are filled by yellow granular dolomite, others are empty. At many exposures where concretionary structures are rare or only incipient, the rock is composed largely of brown or grey crystalline calcite in which *Tubulites permianus* and species of *Liebea* and *Schizodus* are preserved together with traces of small scale cross-bedding, ripple-marks and thin slumped beds.

The upper subdivision of the Concretionary Limestone consists of cream soft granular dolomite which is essentially similar to non-concretionary parts of the lower beds. The fauna, also, is similar, but fossils are more abundant and *T. permianus* occurs in great numbers on many bedding planes. Trechmann (1913, p. 206) originally regarded these beds as part of the Concretionary Limestone, but he later (1931, p. 251) allocated them to the lower part of the Hartlepool and Roker Dolomites, and still later (1954, p. 205) gave them a separate status. Because of faunal and lithological similarities, however, Trechmann's original classification is preferred in this account.

In the coastal sections the total exposed thickness of the Concretionary Limestone is about 105 ft. Inland, in Castle Eden Dene and Nesbitt Dene, the absence of concretions and of *T. permianus* from outcrops of Upper Magnesian Limestone suggests that here the Concretionary Limestone may be partly or wholly overlapped. This evidence cannot be regarded as conclusive, however, in view of the range of lateral variation seen in coastal exposures. There is a further suggestion of overlap in boreholes at West Hartlepool Waterworks (Trechmann 1932, p. 170 ; 1942, p. 3) where no more than 35 ft of bedded non-concretionary dolomite separates Flexible Limestone from undoubted Hartlepool and Roker Dolomites. Under the southern part of West Hartlepool and also in the Amerston Hall Bore, Embleton, the Concretionary Limestone is a hard grey impure foetid commonly bituminous dolomitic limestone, 19 to 65 ft thick. Small quantities of free mineral oil have been reported at this horizon from two of the bores under West Hartlepool.   D.B.S.

DETAILS

Easington Colliery. The northern of two outcrops near Easington Colliery is at the top of 80-ft coastal cliffs just inside the northern margin of the district, where 8 ft of hard, grey, thin-bedded, crystalline, dolomitic limestone overlies very soft highly weathered cream oolitic dolomite of the Middle Magnesian Limestone in a syncline 50 yd across. The southern outcrop also takes the form of a shallow syncline, occupying about 650 yd of cliff section from 60 yd N. of Loom [444 443] to the Easington Fault. For most of this distance exposures are relatively small and are separated by slipped masses of drift, by a thick mantle of weathered rock, or by colliery waste. It is not, therefore, possible to follow the base of the Concretionary Limestone with any accuracy, or to measure a composite section, though it can be estimated that the exposed thickness of the subdivision here is about 50 ft. At the northern end of this outcrop the lower part of the cliff is formed of Middle Magnesian Limestone, the top of which is marked by a discontinuous layer of small calcareous concretions of a type which, at Sunderland and Seaham Harbour, occurs immediately beneath bedded solution breccias. At Loom the bed immediately overlying the concretions is a 15-ft breccia composed of angular fragments of hard, grey crystalline limestone with poorly developed concretionary structures, in a matrix of soft, mealy dolomite. For practical purposes the base of the Upper Magnesian Limestone is here taken at the base of the breccia. The evidence, however, strongly suggests that this is an abnormal junction and that the removal of a soluble bed, probably anhydrite, from the top of the Middle Magnesian Limestone, has caused brecciation in the overlying Concretionary Limestone and brought it into juxtaposition with a bed formerly some distance below it. Brecciated beds are also seen at this level at the southern end of the outcrop, where they are 40 ft thick in places. Higher Concretionary Limestone beds exposed in a series of small isolated outcrops in the northern limb of the syncline consist of alternations of thin-bedded, cream granular dolomites and grey-brown crystalline dolomitic limestones. The latter in some cases contain poorly-defined botryoidal structures, and are commonly brecciated, wholly or in part, by the collapse of underlying beds. Poorly preserved lamellibranchs are present in slightly brecciated granular dolomite 83 yd N. of the Easington Fault, and in fragments of crystalline dolomitic limestone in the fault-breccia.

Horden to Black Halls Rocks. Immediately north of the Horden Fault Concretionary Limestone extends for about 210 yd along the coast from a locality [4465 4249] 540 yd S. of Horden Point to the mouth of Warren House Gill. Exposures are discontinuous and the contact with the underlying beds is nowhere seen. The beds are extensively brecciated, as between Loom and the Easington Fault, and their total thickness cannot be determined. Where unbrecciated they consist of thinly bedded, cream granular dolomite of silt- to fine sandgrade, with scattered traces of lowangle cross-bedding and with poorly-preserved moulds of *Schizodus* and *Liebea*. About 110 yd N. of Warren House Gill part of the rock is laminated, and tends to split into paper-thin sheets–often a characteristic of the Flexible Limestone in the Sunderland area. Neither crystalline beds nor concretionary structures are found in any of these exposures, but the fauna and lithology resemble non-concretionary parts of the Concretionary Limestone north of Sunderland. The southern part of the outcrop, from Warren House Gill to the Horden Fault, is obscured by colliery waste. Concretionary Limestone, apparently overlapped by Hartlepool and Roker Dolomites, is thought to occupy a synclinal area beneath Horden, but is nowhere exposed.

Farther south Concretionary Limestone is exposed for about 950 yd along the coast north-west of the Blackhall Fault. The section [4631 4005], 720 yd E.N.E. of Blackhall Colliery Station is:

	ft
Limestone, dolomitic, flaggy and partly laminated, cream coloured and grey-cream, hard, finely granular or crystalline, with abundant *Tubulites permianus* (some fasciculate), abundant plant remains[1], and scattered moulds of gastropods, *Liebea squamosa, Schizodus rotundatus, S. schlotheimi* ...	5 +
Dolomite, vaguely bedded, cream-white, very soft, granular and finely granular ...	12 +

Traced southwards to an unnamed valley [4639 3993] the Concretionary Limestone, intermittently exposed, consists of partially or wholly fragmented finely granular or crystalline dolomite and dolomitic limestone in a matrix of cream, mealy dolomite. Some fragments contain moulds of *Schizodus* and *Tubulites*; others contain plant remains. Incipient globular concretionary structures occur at intervals. Further examples of collapse-brecciation are seen between this valley and Blue House Gill, where the lowest member of the gently dipping sequence is an 8ft bed of granular dolomite, and the highest [4655 3968] comprises a group of thin partly crystalline beds containing poorly preserved lamellibranchs (chiefly *Schizodus*

and *Liebea*) and scattered fasciculate groups of a tubular fossil resembling *Tubulites*. The fauna and lithology of these beds indicate a position low in the Concretionary Limestone sequence; it is possible that the basal 8-ft dolomite is Middle Magnesian Limestone.

South of Blue House Gill as far as the Blackhall Fault higher beds are exposed. They comprise thin-bedded dolomitic limestones containing abundant concretionary structures. The following section [4682 3941] was measured 30 yd north of the position of the Blackhall Fault:

	ft
Limestone, dolomitic, grey, fine-grained, hard, with scattered small globular concretions and abundant *Tubulites permianus*	2 +
Limestone, grey, massive and medium-bedded, laminated, crystalline, with many small radial concretions and scattered small pockets of cream powdery dolomite ...	8
Limestone, dolomitic, cream, finely granular, with scattered lamellibranch shells and *T. permianus*; fresh faces are grey and finely crystalline ...	6 +

Inland, at Blackhall Colliery South Shaft, Trechmann (1913, p. 213) recorded 63 ft of Concretionary Limestone overlying the Middle Magnesian Limestone reef.

Crimdon to Hartlepool. The most complete section south of the Blackhall Fault extends for about 1090 yd in the cliffs midway between Black Halls Rocks and Crimdon Beck. The following composite section was measured in a syncline [477 382] at Limekiln Gill:

	ft
Dolomite, cream, vaguely thin-bedded, soft, finely granular	5 +
Limestone, dolomitic, grey-brown, flaggy, partly wedge-bedded, crystalline	2

---

[1] Dr. W. G. Chaloner remarks: " Some of the plant remains are *Algites virgatus* (Münster) Stoneley and show the rather characteristic feature of filaments diverging from basal ' knots '. One specimen bears similar weakly carbonaceous filaments associated with casts of *Tubulites*-like bodies. Howse (1890, p. 233) regarded *Tubulites permianus* and *Algites virgatus* as synonyms. Both taxa are present on one specimen and I think it conceivable that they are different preservation states of one organism."

Limestone, dolomitic, grey-brown, thin- and medium-bedded, crystalline, with poorly developed concretionary structures 5

Limestone, dolomitic, grey-brown, massive, hard, crystalline, with prominent concretionary structures 1½

Limestone, dolomitic, cream-grey, thin- and medium-bedded, crystalline, with poorly defined concretionary structures ; scattered moulds of *Liebea squamosa, Permophorus sp., Schizodus schlotheimi* and ostracods and abundant *T. permianus* where least recrystallized 4

Limestone, dolomitic, grey, massive, with small concretionary structures ... ... 2 to 3

Limestone, dolomitic, cream-grey, thin- and medium-bedded, crystalline, partly brecciated, with many incipient and some well-defined globular concretions ... 10

Limestone, dolomitic, grey-brown, massive, crystalline, with concretionary structures 2

Limestone, brown, thin-bedded, crystalline, with small concretionary structures ... ... 0 to 2

Dolomite, calcitic, cream, thin- and medium-bedded, finely granular, with traces of small scale cross-bedding on weathered surfaces and scattered incipient globular concretions near top ; a few fossils including *T. permianus, Omphalotrochus sp., L. squamosa* and *S. schlotheimi* are scattered throughout ... 6 to 8

Dolomite, cream, flaggy, laminated, soft, finely granular ... 3

Dolomite, cream, thin-bedded and flaggy, finely granular, partially brecciated, with scattered small lamellibranch

moulds and abundant *T. permianus,* including fasciculate groups ... ... 2 +

Unexposed ... ... ?2

MIDDLE MAGNESIAN LIMESTONE —

The two lowest units of the Upper Magnesian Limestone in this section are similar in thickness and lithology to the Flexible Limestone, but differ from the latter in containing a shelly fauna.

In the cliffs south of Limekiln Gill Middle Magnesian Limestone crops out in the core of an anticline. The Concretionary Limestone, dipping off the southern limb of this structure, is a repetition of the sequence measured at Limekiln Gill. The lowest 12 ft are non-concretionary and contain moulds of small lamellibranchs and of *T. permianus.* The overlying 31 ft of strata contain prominent concretionary structures which are particularly abundant in the more massive beds. These massive beds are usually about 1½ to 3 ft thick, but laterally they give way abruptly along step-like interfaces to thin-bedded rocks containing only scattered concretions (Fig. 24). Most beds contain internal moulds of lamellibranchs, often in abundance and sometimes uncompressed, while on some bedding planes there are large numbers of sub-parallel *T. permianus.* Partial collapse brecciation (Pl. XIIIc) is widespread and breccia gashes incorporating fragments of red mudstone occur at intervals.

In the cliffs extending between 285 and 330 yd S.E. of Limekiln Gill the upper part of the above sequence appears to be represented by cream soft bedded dolomite, containing fossils only in the uppermost 6 ft. Farther south-east this is overlain by about 30 ft of similar thin-bedded and flaggy cream soft dolomite exposed in a shallow syncline 120 yd across. Moulds of lamellibranchs (chiefly *S. schlotheimi*) and of *T. permianus* are abundant. In the southern limb of the syncline, beds of this lithology are about 50 ft thick, the lowest 20 ft being the lateral equivalent of the soft dolomite exposed 285 to 330 yd S.E. of Limekiln Gill and the massive concretionary limestones into which they grade. Exposures

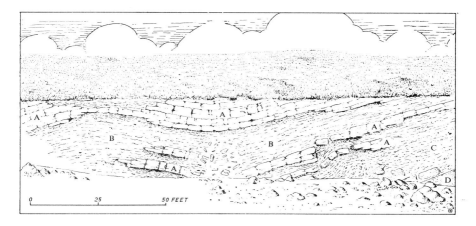

FIG. 24. *Field sketch of beds of the Concretionary Limestone Group, showing lateral variation and collapse structures caused by solution of underlying beds. A. Massive crystalline concretionary limestone. B. Thin-bedded crystalline limestone, locally developing concretionary structures. C. Thin-to thick-bedded granular dolomite, locally calcareous and with incipient concretionary structures. D. Massive granular shelly dolomite. This section of the coastal cliffs is about 1100 yd south of Black Halls Rocks*

are scarce between 450 and 820 yd S.E. of Limekiln Gill. At one locality [479 378] Middle Magnesian Limestone oolite forming the core of a small anticline is directly overlain by hard shelly thin-bedded dolomite locally forming the lowest member of the Concretionary Limestone. There is no brecciation either at the junction or in the conformable overlying beds. Brecciation is widespread, however, in most of the isolated exposures of finely granular (silt-grade) cream dolomite farther south, which contains small lamellibranchs and *T. permianus,* but in which concretionary structures are absent.

Concretionary Limestone, flexed in a gentle anticline, is continuously exposed from 820 yd to 1090 yd S.E. of Limekiln Gill. The northern limb is formed by about 21 ft of cream finely granular dolomite in flaggy and medium- to thick-bedded alternations containing moulds of lamellibranchs, and, in some beds, abundant sub-parallel *T. permianus.* Towards the crest of the flexure the dolomites become progressively harder and more calcareous, and midway along the exposure most beds pass laterally into hard crystalline concretionary limestone. Almost all the concretions are globular, radially

crystalline, and $\frac{1}{2}$ to $1\frac{1}{2}$ inches in diameter. Many are compound. Along the southern limb of the anticline only scattered concretionary structures are seen. The composite section at the southern end of the exposure is:

	ft
Dolomite, cream, finely granular; thin beds alternating with thick beds which predominate towards base; moulds of *Liebea squamosa, Schizodus schlotheimi* and *T. permianus* are common throughout and ostracods are also present	c. 45
Limestone, dolomitic, cream-grey, finely crystalline, thin-bedded, with scattered irregular lenses of harder concretionary limestone	5
Dolomite, cream, finely granular, with scattered globular calcareous concretions	4
Limestone, dolomitic, grey-brown, crystalline, flaggy, with incipient small concretions and large uncrushed moulds of *S. schlotheimi*	3

ft

Limestone, dolomitic, brown weathering to grey, crystalline, medium- and thick-bedded, with widespread incipient concretionary structures; flaggy at base ... ... 8+

Unexposed interval spanning 35 yd collapse-zone; vertical interval probably negligible —

Dolomite, cream, medium- and thick-bedded, finely granular 10+

Limestone, grey-brown, crystalline, in alternations 2 to 4 ft thick of massive and thin-bedded or flaggy units; small radially crystalline, sub-spherical concretions, $\frac{1}{2}$ to $1\frac{1}{2}$ in diameter, make up most of the rock; lamellibranchs preserved only in isolated pockets of non-concretionary dolomite; some small-scale slump structures ... ... 15+

Analysis (Trechmann 1914, p. 236) shows the uppermost member of this section to be slightly impure dolomite with only 2·81 per cent excess calcite over the dolomite ratio. Trechmann described the rock as a " compact aggregate of very minute dolomite-grains ". The lowest bed of the section passes northwards into non-concretionary dolomite. The base of the Concretionary Limestone is here about 15 to 20 ft below the lowest exposed bed so that the total thickness of the sub-division must locally be 105 to 110 ft.

Inland the Concretionary Limestone can now be seen only alongside Crimdon Beck, where it is exposed intermittently for 530 yd upstream and 60 yd downstream from the Blackhall Colliery–Hartlepool railway viaduct. All exposures are of cream soft bedded finely granular dolomite, containing moulds of L. squamosa and S. schlotheimi and large numbers of T. permianus. In a small outcrop in the stream bed, 440 yd W. of the viaduct, flaggy dolomite rests on cream oolitic dolomite, possibly at the top of the Middle Magnesian Limestone. In Nesbitt Dene, a mile or so to the west, the Concretionary Limestone appears to be locally overlapped by Hartlepool and Roker Dolomites. During 1960, a

few feet of flaggy grey crystalline limestone and dolomitic limestone containing poorly-preserved lamellibranch moulds (chiefly Liebea) and T. permianus were exposed in shallow excavations [4876 3576], 250 yd W.N.W. of Springwell House.

Boreholes have proved Concretionary Limestone at Hartlepool, West Hartlepool and offshore. In a bore [5003 3430] at Howbeck Hospital, 1 mile S.E. of Springwell House, a section of the strata encountered beneath 140 ft of drift and Hartlepools and Roker Dolomite is:

	ft
Hard platy yellow rock	10
Dolomite, hard, thinly bedded ...	10
Dolomite, very hard, grey, foetid	20
Dolomite, hard, dense, with shaly bands and traces of oil ...	17
MIDDLE MAGNESIAN LIMESTONE	—

Further details of this bore were given by Trechmann (1942, pp. 319–320) who assigned the hard platy rock at the top to the Hartlepool and Roker Dolomites.

In a bore at the Palliser Works [5061 3525], 1200 yd N.N.E. of the hospital, the Concretionary Limestone appears to be only about 10 ft thick. It is only 5 ft thick in a bore [5070 3343] at West Hartlepool Waterworks, where it consists of hard yellow platy dolomite, correlated by Trechmann (1932, p. 170) with the Flexible Limestone of Sunderland. From analysis of this dolomite Trechmann (1914, p. 238) calculated that there is 3·57 per cent of $CaCO_3$ in excess of that required to form pure dolomite: most of the 42·65 per cent insoluble residue is said to be silica. Farther south in Nelson Street No. 2 Bore [5154 3197], West Hartlepool, the Concretionary Limestone, including Flexible Limestone, is represented by $15\frac{1}{2}$ ft of hard grey dolomite at a depth of 241 ft (Trechmann 1942, p. 325). The same beds are 29 ft thick, 173 ft below the surface in a bore [5082 3233] at Villiers Street, West Hartlepool, where they are described by Fowler (1944, p. 201) as ' grey calcareous and micaceous dolomite interbanded with laminated dolomite, rich in bituminous streaks '. Quartz grains and muscovite flakes are seen in thin section to be widely distributed, and chert, fluorite, calcite and pyrite are also present. Bituminous material is concentrated chiefly in darker laminae, and is locally a major constituent of the rock. A specimen from 144 ft, rich in such laminae, was heated in a closed tube and yielded water and yellow oily material.

Because of its distinctive lithology the Concretionary Limestone can be recognized in drillers' records of the following old boreholes in the southern part of West Hartlepool:

Bore	N.G.R.	Thick-ness ft	Depth to base ft
Smalley's Pulp Works	... 5130 3186	37	225
Durham Paper Mills	... 5114 3187	75	235
Seaton Carew	5184 3001	19	606
Burn Road No. 1	511 319	45	223

The Flexible Limestone is represented by 2½ ft of 'blue shaly limestone' in Burn Road No. 1 Bore, and by 3 ft of 'dark blue shale with small feeder of rock oil and sulphur water' in the Seaton Carew Bore.

In Offshore Nos. 1 and 2 bores the Concretionary Limestone is respectively 66 and 50 ft thick, including 3¼ ft and 6 ft of grey thin-bedded dolomite correlated with the Flexible Limestone. The sequence in Offshore No. 1 Bore is:

	ft	in
HARTLEPOOL AND ROKER DOLOMITES, base at 525 ft 4 in ...	–	–
Interlaminated fine-grained dolomite and gypsum ...	9	11
Dolomite, very fine-grained		3
Dolomite, interlaminated and thinly interbedded with gypsum, which forms less than 30 per cent of the lowest 8 ft of the bed	9	6
Dolomite, very thin-bedded fine-grained, with 20 per cent anhydrite, mainly in irregular lenses parallel with bedding	28	6
Interbedded dolomite and anhydrite	3	6
Dolomite, thin- and medium-bedded, foetid, slightly bituminous	10	6
Dolomite, thin-bedded, bituminous (FLEXIBLE LIMESTONE)	3	9

Because no fossils were found in this bore the top of the Concretionary Limestone is taken at a marked lithological break. In the interlaminated and interbedded sequences, the sulphate appears to replace dolomite along selected bedding planes, particularly those corresponding in horizon to the beds containing concretionary structures in coastal exposures. In Offshore No. 2 Bore the 44 ft of Concretionary Limestone above the Flexible Limestone consist of grey, massive dolomite, with no concretionary structures and only small amounts of secondary gypsum. D.B.S.

HARTLEPOOL AND ROKER DOLOMITES

The Hartlepool and Roker Dolomites crop out in a syncline under the southern part of Horden, within an embayment of the Middle Magnesian Limestone reef at Thorpe Bulmer, and in the south-eastern part of the district. Inland, these beds are not known to exceed 150 ft in thickness, but they are 366 ft thick (with the topmost beds missing) in Offshore No. 2 Bore, 6 miles N. of Hartlepool. They are composed almost entirely of soft granular and oolitic cross-bedded ripple-marked dolomites which undergo no significant lateral variation in lithology.

Many of the ooliths in the oolitic beds are hollow ; others have cores of crystalline calcite (Trechmann 1914, p. 250) ; whilst fluorite forms the cores of pisoliths in a boring at West Hartlepool (Trechmann 1941, p. 321). According to Woolacott (1919, p. 490) the ooliths were formed inorganically by the deposition of dolomite around gypsum nuclei. He quotes, as evidence, the gypsum still remaining in the oolith cores in borings around Billingham, 6 miles S.S.E. of West Hartlepool. Dunham (*in* discussion of Stewart 1963, p. 41), however, claims that the dolomite crystals forming the shells of the ooliths cut the concentric banding and that the original material was probably calcite. In the offshore area of the Durham district, oolitic dolomite with a gypsum cement contains many empty ooliths. Multiple ooliths and pisoliths are common, especially between 150 and 200 ft above the base of the sub-division.

The limited fauna (p. 153) of the Hartlepool and Roker Dolomites is unevenly distributed, some bedding-planes being crowded with moulds of shells whilst intervening beds are nearly barren. In the offshore area, where the highest beds are preserved, the shelly fauna dies out about 200 ft above the presumed base of the sub-division. Plant remains, represented by obscure carbonaceous films, are present at all levels.

DETAILS

At the coast the Hartlepools and Roker Dolomites form a series of isolated outcrops between slipped masses of drift extending for about 450 yd on the north side of Dene Mouth [457 408]. They consist of thin-bedded and flaggy cream finely granular (silt- to fine sand-grade) dolomites showing little vertical or lateral variation in lithology. Some highly porous beds may be altered fine oolites. Much of the bedding is slightly irregular, with local traces of shallow-angle cross-bedding and ripple marks. Thin discontinuous beds of dolomite mudstone occur also in places, and moulds of small gastropods and lamellibranchs (especially *Liebea* and *Schizodus*) are common at some horizons. The greatest thickness at any single exposure is about 20 ft, and the aggregate thickness is probably about 40 ft. Since Concretionary Limestone is not seen at this locality, the stratigraphical position of these rocks is determinable only by their fauna and lithology; these suggest a position low in the Hartlepool and Roker Dolomites.

Inland the Hartlepool and Roker Dolomites are exposed on the south side of Castle Eden Burn for about 250 yd upstream from the Horden-Blackhall Colliery road. Here they comprise about 50 ft of unevenly thin- and medium-bedded granular and oolitic dolomite; wedge-bedding is seen at some levels, and ooliths, many of them irregular or compound, are found chiefly in the upper parts of the sequence. About 20 ft of similar

cream dolomite siltstone are exposed [4467 4039] on the opposite side of the Burn. Small-scale cross-bedding is widespread, but recognizable ooliths are rare. Since no diagnostic fossils were found among these rocks their correlation is based entirely on lithology and is thus tentative.

The following section is exposed [4516 4001] on the east side of Hardwick Dene, 900 yd N. of Hardwick Hall:

	ft
Dolomite, thin-bedded and flaggy, finely granular ...	12
Dolomite, cream, soft, medium-bedded, finely granular	9 to 10
Dolomite, cream, thin-bedded, finely cross-bedded, finely granular (possibly partly an altered oolite)	40

Much of the original texture of these beds has been obliterated by recrystallization. Trechmann (1913, p. 211) records *Schizodus schlotheimi*, ' *Liebea septifer* ' and small gastropods from the uppermost 50 ft, and refers these beds to the Hartlepool and Roker Dolomites. Discontinuous exposures of these beds are also seen in a syncline between 780 and 600 yd N. of Hardwick Hall.

Two miles farther south, in Nesbitt Dene [460 370] beds thought to be Hartlepool and Roker Dolomites are represented by 50 to 60 ft of soft finely granular oolitic dolomite. Their base is nowhere visible, but it is inferred that the Concretionary Limestone is almost overlapped against the partially buried foreslope of the Middle Magnesian Limestone reef. The following section was measured in a small quarry [4619 3798], 1230 yd N.N.E. of Thorpe Bulmer:

	ft
Dolomite, cream, thin- and medium-bedded, f i n e l y granular, partly oolitic ...	12
Dolomite, massive, finely granular, with close-set joints	10
Dolomite, massive to thick-bedded, with *Tubulites?* and scattered small lamellibranchs including *Liebea squamosa, Schizodus sp.* and *Permophorus sp.* (?CONCRETIONARY LIMESTONE)	6+

Some 300 yd downstream these beds yielded the following: *Tubulites permianus*; *L. squamosa* and *Schizodus schlotheimi*: Trechmann (1942, p. 313) obtained '*Liebea hausmanni*' and '*S. schlotheimi*' from soft dolomite near the top of this sequence in the Thorpe Bulmer Bore.   D.B.S.

At West Hartlepool the subdivision is now visible only in two old quarries [508 333], 800 yd N.N.W. of the station, and in a neighbouring railway cutting. The quarries contain a composite sequence of about 20 ft of cream granular and oolitic dolomite from which Trechmann (1913, p. 204) recorded '*Mytilus septifer*', '*Pleurophorus costatus*' and '*S. schlotheimi*.' Current-bedding and thin lenses of sub-porcellanous dolomite occur throughout. Boreholes have proved beds of similar lithology to a depth of 34 ft below the floor of the more northerly of the two quarries (Trechmann 1913, p. 204 ; 1932, p. 170) and the beds in the railway cutting are also similar.

The old town of Hartlepool stands on a promontory of Hartlepool and Roker Dolomites, thinly covered with drift except on the seaward periphery where a sequence, 60 to 80 ft thick, is exposed in the cliffs and on wide wave-cut platforms.

97544

The beds are composed chiefly of granular and oolitic dolomite forming a series of low domes and shallow basins, with a low sheet-dip slightly north of east. The main bedding units are constant in thickness but cross-bedding, mostly small-scale, is widespread. Trechmann (1913, p. 204) measured 85 ft of beds at this locality and records '*Natica cf. leibnitziana*', '*Turbo helicinus*', small indeterminate gastropods, '*Mytilus septifer*', '*Pleurophorus costatus*' and '*Schizodus schlotheimi*' from the lowest 35 ft. In addition *Tubulites?, Omphalotrochus sp.* and a fragment of *Dielasma?* (see p. 153) were collected during the resurvey from outcrops [5284 3340] on the beach, 320 yd S. of St. Hilda's Church, Hartlepool. An analysis (Trechmann 1914, p. 236) of an oolitic rock from Hartlepool Scars shows that it is an almost pure carbonate rock with 1·06 per cent $CaCo_3$ in excess of that required to satisfy pure dolomite. In the Hartlepools Lighthouse Bore, at the eastern end of the promontory, small specimens of *Liebea* and *Schizodus* were abundant in buff oolitic dolomite at 157½ ft, proving that the Hartlepool and Roker Dolomites are here at least 200 ft thick.

G.D.G., D.B.S.

The Hartlepool and Roker Dolomites extend southwards beneath West Hartlepool and out to sea, where they have been proved in Offshore Nos. 1 and 2 bores. Under West Hartlepool they have been proved in the following bores: Howbeck Hospital No. 3 Bore [5003 3430], 107 ft (Trechmann 1942, p. 319) ; Hartlepool Corporation No. 9 Bore [5078 3249], 6 in ; Villiers Street Bore [5082 3233], 118 ft ; Smalley's Pulp Works No. 1 [5130 3186], 125 ft ; Nelson Street No. 1 [5154 3197], 175 ft (Trechmann 1942, p. 324) ; Seaton Carew Bore ?65 ft ; Casebourne Cement Works Bore 65½ ft. Trechmann recorded small specimens of '*Liebea*' and '*Pleurophorus*' from the lower part of the Hartlepool and Roker Dolomites in the Nelson Street Bore, and *Chondrites logaviensis* Geinitz, *Liebea squamosa* and *Schizodus schlotheimi* were collected from these beds in Offshore No. 1 Bore.

The maximum proved thickness of the subdivision is 366 ft in Offshore No. 2

M

Bore (Magraw and others 1963, pp. 202–3), where it comprises:

		ft
Dolomite, soft, granular	...	161
Dolomite, mainly oolitic	...	60

			ft
Dolomite, in granular and oolitic alternations	...	...	50
Dolomite, granular, silt-grade	...		100

D.B.S.

UPPER PERMIAN MARLS

The Upper Permian Marls are the deposits of the final silting up of the Zechstein Sea in Durham, and include portions of two evaporite cycles. The generalized sequence is:

Red silty mudstone, interbedded with red sandstone at top and with abundant gypsum towards base	190 to 390
UPPER ANHYDRITE	6 to 20
Red silty mudstone, locally gypsum-rich	15 to 30
BILLINGHAM MAIN ANHYDRITE	25 to 62

These beds have been preserved only in the south-east of the district, around Embleton, where they crop out over an area of more than three square miles. Surface exposures are here obscured by drift and all the available information is from boreholes. These show that the subdivision thins westwards, and that most of the thinning is in the upper mudstones which are reduced from about 370 ft under the southern part of West Hartlepool to about 210 ft at Claxton. The top of the Upper Permian Marls is taken somewhat arbitrarily at the level where the intercalations of Bunter facies become predominant. This boundary is almost certainly diachronous; the scanty evidence from boreholes suggest that in addition to a gradual upwards increase in the proportion of sandstone, there is also a westerly increase, and this is taken to indicate that, as in Nottinghamshire (Sherlock 1926), there is lateral passage from mudstone to sandstone facies. Sulphates are absent at outcrop, presumably due to solution.   D.B.S., G.D.G.

DETAILS

At outcrop the Upper Permian Marls have been cored only in the Amerston Hall Bore, Embleton, where about 50 ft of dull brick-red sandy siltstone was proved. In uncored bores at Tinkler's Gill [4157 3054] and Middle Stotfold [4501 2987] these beds are respectively 75 ft and 206½ ft thick. They have also been proved in a number of shallow bores in adjacent parts of the Stockton (33) district. The Upper Anhydrite and the Billingham Main Anhydrite are not represented in the Amerston Hall Bore, and have not been recorded by the drillers of the uncored bores. Farther east, beneath Bunter Sandstone, the full thickness of the Upper Permian Marls is penetrated by three boreholes, of which the section in the Seaton Carew Bore (Bird 1888) may be taken as representative:

	ft	in
BUNTER SANDSTONE ...	93	0
Red sandy marl	47	0
Red and grey sandstone ...	10	0
Red marl	15	0
Red marl, with beds of grey and red sandstone	8	0
Red marl, with blue joints	35	0
Red marl, with beds of grey sandstone	24	0

	ft	in
Red marl, with beds of grey marl	33	0
Red marl, with blue joints	24	0
Red marl, with blue joints and veins of gypsum ...	171	0
Red marl and veins of gypsum	7	5
Anhydrite (UPPER ANHYDRITE)	3	0
Blue marl and veins of gypsum	3	0
Anhydrite	1	0
Red marl and veins of gypsum	10	0
Dark marl and gypsum, mixed	2	7
Anhydrite, with black joints (BILLINGHAM MAIN ANHYDRITE)	25	0
UPPER MAGNESIAN LIMESTONE	—	—

A similar section is recorded in the Casebourne's Cement Works Bore, West Hartlepool (Bird 1888), where the red mudstones are 390 ft thick, the Upper Anhydrite 18½ ft, and the Billingham Main Anhydrite 62 ft thick overlying 15 ft 8 in of anhydrite 'mixed with limestone'. Since the maximum thickness of the Billingham Main Anhydrite at the type locality, 6 miles farther south, is 33 ft (Wood 1950) the Casebourne figure is exceptional, and may consist in part of anhydrite which has replaced Upper Magnesian Limestone. In another bore [4473 5276], 1 mile S. of Brierton, the lithology and thicknesses of the Billingham Main Anhydrite and overlying marls and anhydrite are generally similar to those proved in the Seaton Carew Bore and in many boreholes (Marley 1892) around Billingham. In addition to these deep boreholes, the higher parts of the Upper Permian Marls were reached by a number of shallower holes, mainly in the Brierton and West Hartlepool areas.

North of the West Hartlepool Fault, the former presence of Upper Permian Marl is indicated by fragments of red and green silty marls which occur commonly in collapse-breccias within the Middle and Upper Magnesian Limestone in coastal areas north of Black Halls Rocks, and have been recorded as far north as Fulwell in the Sunderland (21) district (King 1850, p. xvi). Farther east, beds tentatively interpreted as Upper Permian Marl have been reported (Clarke and others 1961, p. 207) from the undersea area off Tynemouth and these probably extend southwards and crop out on the sea bed in the eastern part of the present district. D.B.S., G.D.G.

BUNTER SANDSTONE

Red sandstones of Bunter facies crop out, mainly beneath drift, over an area of about eight square miles between the West Hartlepool Fault and the southern margin of the district. Surface exposures in this area are found only on the foreshore at Seaton Carew and in the intertidal zone at Long Scar, ¾ mile to the north-north-east. Farther west, borings show that these beds grade by alternation into the underlying Upper Permian Marls. The maximum recorded thickness of Bunter Sandstone in the district is 185 ft, in the Casebourne's Cement Works Bore, West Hartlepool. At Billingham, 6 miles farther south, the sandstone is about 700 ft thick, and structural evidence suggests that a comparable thickness is present immediately south of the Seaton Carew Fault.

The Bunter of the district is composed mainly of soft fine-grained thick-bedded, dull red sandstones, with subsidiary grey sandstones and red mudstones and siltstones. Current-bedding, ripple-marks, contemporaneous (desiccation) breccias and ?sun cracks indicate shallow-water semi-continental deposition. Preferred orientations of fore-set current-bedding laminae indicate a south-westerly provenance which is in accord with the westerly increase in the proportion of sand noted in the transitional beds at the base of the Bunter, and with the suggestion of a diachronous base falling westwards (p. 164). The siltstone and mudstone intercalations at outcrop contain wisps and irregular lenses and patches of pale grey-green colour: and at some localities there is extensive yellow and grey colouration which may be due to the former cover of peat—possibly an extension of the thick bed still preserved in a number of small patches at West Hartlepool.

G.D.G., D.B.S.

DETAILS

About 15 ft of thick-bedded and massive dull red soft fine-grained sandstone are exposed at low spring tides on Long Scar [520 310], where minor dome and basin structures are superimposed on a gentle anticline. The sandstone grains are predominantly subangular, and range from 0·125 to 0·250 mm except in thin coarser bands where they range up to 0·50 mm. Beds are inconstant in thickness, ranging from 6 in to 3 ft and are commonly current-bedded. The Long Scar rocks appear to lie about the middle of the Bunter Sandstone.

Little Scar [527 305] is the name given to a group of small outcrops on the foreshore at Seaton Carew. These beds lie near the base of the Bunter Sandstone and contain, in addition to red sandstone similar to that at Long Scar, beds of siltstone and mudstone. The siltstones and fine-grained sandstones are predominantly flaggy and thin-bedded, and are red, yellow and grey in colour; they exhibit abundant small-scale current-bedding and rare ripple-marks. The mudstones are darker coloured than the sandstone, contain thin grey-green bands,

lenses and patches, and locally show signs of contortion and contemporaneous brecciation. Breccias of flakes and rolled fragments of mudstone in a fine sandstone matrix occur in several places. Current-bedding in the sandstones shows a marked preferred orientation indicating current-flow from the south and west, but ripple-marks are symmetrical and of the type normally ascribed to wave oscillation. In addition to forming individual beds, red mudstone also occurs infilling minor surface hollows in sandstones, and in some places contains desiccation cracks filled with sand.

Inland, Bunter Sandstone is proved in the following bores, all situated around the southern margin of West Hartlepool: Owton Manor [4931 2956], 234 ft; C.W.S. Lard Works [5118 3158], 201 ft; Casebourne's Cement Works, 185 ft; and Seaton Carew, 93 ft. Due to the difficulty of recognizing the base of the Bunter facies, figures given are approximate. The Cement Works Bore, situated 1 mile N.W. of Little Scar, shows that red mudstone ('marl') is common in the lower parts of the Bunter sequence.

G.D.G.

KEUPER MARL

On structural grounds, it is inferred that more than 700 ft of strata are preserved above the base of the Bunter near the coast immediately south of the Seaton Carew Fault. By comparison with the succession at Billingham, where the Bunter is only 700 ft thick, the uppermost beds are likely to be red Keuper Marl, though they are neither exposed nor proved by boring in the district.

D.B.S.

EPIGENETIC MINERALS

Some of the many records of epigenetic minerals from the Permian rocks of Durham are summarized by Fowler (1943, 1957) who found that they tend to be concentrated in two parts of the succession. The lower comprises the Marl Slate and the immediately overlying beds, where the most common minerals are calcite, galena, chalcopyrite, pyrite, sphalerite and baryte ; the higher lies near the top of the Lower Magnesian Limestone and is characterized by calcite, baryte and fluorite. Minerals also occur, however, throughout the remainder of the sequence : dolomite, baryte and pyrite are the most common, apart from gypsum and anhydrite, which are ubiquitous cavity-fillings in areas protected from solution.

Pyrite commonly replaces rock fragments, particularly mudstone, in the Basal Permian Breccias ; and Phemister (*in* Fowler 1943, p. 43) records baryte and fluorite locally replacing country rock near the top of the Lower Magnesian Limestone. Most mineralization, however, is found in and along cavities, bedding-planes, joints and fissures ; and this association is emphasized in the Ferryhill–Garmondsway area by a concentration of metallic ore-minerals close to the Butterknowle Fault.

A list of minerals recorded before and during resurvey is given, together with selected localities, in Table 3.

The origin and age of the epigenetic mineralization is in dispute. Fowler (1943, p. 50 ; 1957, p. 265) considered it to be contemporaneous with mineralization seen in Coal Measures and followed Trotter (1944, p. 227) in assuming a Tertiary age. Dunham (1952, p. 4) agreed with this, though he believed the probable age of the workable deposits of fluorite in the north Pennines to be late Carboniferous or early Permian. Westoll (1943), however, claimed that in the Marl Slate some, at least, of the blende, galena, chalcopyrite and malachite is syngenetic ; and that, as in the Kupferschiefer, these minerals show signs of diagenetic mobilization within the deposit and through overlying strata.

With the possible exception of baryte none of the epigenetic minerals in the Permian is present in commercial quantities, but these minerals are locally important in some dolomite quarries in the Lower Magnesian Limestone where they represent concentrations of impurities.　　　　D.B.S.

TABLE 3

DISTRIBUTION OF EPIGENETIC MINERALS IN THE PERMIAN

Minerals	Basal Permian Sands & Breccias	Marl Slate	Lower Magnesian Limestone	Middle Magnesian Limestone	Upper Magnesian Limestone	Localities
Copper ...			X			Raisby Hill Quarry
Sphalerite ...		C	X	X	X	Many quarries along scarp; Hesleden Dene No. 3 Bore
Galena ...		C	C	X		Many quarries along scarp; most boreholes; Black Halls Rocks
Chalcopyrite...			X	X		Several quarries along scarp, including Raisby Hill Quarry
Pyrite ...	C	C	C	X	X	Many quarries along scarp and most boreholes
Quartz ...				X		Hesleden Dene
Pyrolusite ...				X		Whelly Hill Quarry, Hart
Magnetite ...			X			Mainsforth Low Main Series, No. 9 Bore
Goethite ...			X			Chilton Quarry; many boreholes
' Limonite ' ...			X			Most surface exposures and boreholes
Fluorite ...			X	X	X	In most boreholes and in several quarries along scarp
Malachite ...		X	X	X		Several quarries along scarp, especially Raisby Hill
Azurite ...				X		Railway cutting south of Raisby Hill Quarry
Calcite ...	X	X	C	C	C	Most surface exposures and boreholes
Dolomite (with ankerite) ...			X	X		Many surface exposures and boreholes
Kaolinite ...			X	X		Fishburn Nos. 1 to 4 bores; Hesleden Dene No. 3 Bore
Dickite ...			X	X		Fishburn Nos. 1 to 4 bores; Hesleden Dene No. 3 Bore[1]
Collophane (Francolite)				X		Hesleden Dene No. 3 Bore[1]
Anhydrite ...			X	X	X	Offshore Nos. 1 and 2 bores; Seaton Carew Bore
Gypsum ...			X	X	X	Raisby Hill Quarry; Offshore Nos. 1 and 2 bores
Celestine ...			X			Several quarries along scarp
Baryte ...		X	C	X	X	Several quarries along scarp, especially at Chilton and Thrislington; many boreholes

X = recorded   C = common

[1] Called Hesleden Dene No. 2 Bore in descriptions and analyses given by Dunham and others (1948) and Guppy and Sabine (1956, pp. 65, 72–3)

PERMIAN FOSSILS

In addition to the fossils listed in the stratigraphical account the following have been identified from underground exposures and boreholes.

TUNNELS AT EASINGTON COLLIERY

See pp. 136–7 for stratigraphical details

1. Horizontal tunnel, 70 yd W.S.W. of West Pit
Ht. above O.D. about 8 ft. National Grid Ref. 4367 4412

Middle Magnesian Limestone, early reef-talus at foot of reef-slope.

Algal filaments; ' *Glomospira* ' *pusilla*; crinoid columnals; *Acanthocladia anceps, Batostomella crassa, Fenestrellina retiformis, Protoretepora ehrenbergi* (Geinitz), *Thamniscus dubius*; *Cleiothyridina pectinifera, Crurithyris clannyana, Dielasma elongatum, Horridonia horrida* [abundant] [Pl. VI, fig. 6], *Orthothrix excavata, Pterospirifer alatus, Spiriferellina cristata multiplicata, Stenoscisma schlotheimi* [abundant], *Streptorhynchus pelargonatus.* Strophalosiid [ventral valve with vermiform spines, asymmetrical pointed umbo: *?Craspedalosia* or *Dasyalosia*]; *Straparollus permianus, Loxonema fasciatum, L.* cf. *retusum* Dietz, *L. sp.* [with at least 12 whorls], *?Macrochilina symmetrica* (King), *Omphalotrochus sp.* [including a fine-ribbed form intermediate between *O. thomsonianus* and *O. taylorianus* in number of whorls and spiral angle], *Pleurotomaria linkiana, P. verneuili* Geinitz, *Strobeus geinitzianus*; *Bakevellia ceratophaga, Cardiomorpha modioliformis* (King), *Palaeolima? permiana, Parallelodon striatus, Permophorus costatus, Pseudomonotis speluncaria, Schizodus truncatus, Solemya biarmica, Streblochondria pusilla*; ostracods.

Polyzoa and Strophalosiid brachiopods are rare.

2. Horizontal tunnel, 55 yd W.S.W. of West Pit
Ht. above O.D. about 7½ ft. National Grid Ref. 4368 4413

Middle Magnesian Limestone, reef-talus at foot of reef-slope; slightly younger than at locality 1.

Algal debris; *Fenestrellina retiformis*; *Crurithyris clannyana, Dielasma elongatum, Orthothrix excavata, O.* cf. *lewisiana, Stenoscisma schlotheimi*; indet. gastropod; *Bakevellia antiqua, Parallelodon striatus, Pseudomonotis speluncaria.*

In contrast to that from locality 1, this fauna is dominated by cryptostome polyzoa and Strophalosiids.

3. Inclined adit, 400 yd W. of West Pit and 180 yd from adit mouth
Ht. above O.D. about 8 ft. National Grid Ref. 4341 4410

Middle Magnesian Limestone, coquina at base of reef.

' *Glomospira* ' *pusilla*; crinoid stems; *Batostomella crassa*; *Dielasma elongatum, Horridonia horrida, Stenoscisma schlotheimi*, Spiriferid fragments; *Straparollus?, Macrochilina symmetrica, Naticopsis sp., Omphalotrochus spp.* [including *O.* cf. *taylorianus*], *Strobeus geinitzianus*; *Parallelodon striatus, Permophorus costatus, Pseudomonotis speluncaria, Schizodus sp.*; *Peripetoceras freieslebeni.*

This fauna is akin to that from locality 1 in the relative abundance of *Stenoscisma* and gastropods and the fact that Strophalosiids and, more remarkably, cryptostome polyzoa are unrecorded.

4. Inclined adit, as 3 but 150 yd from mouth
Ht. above O.D. about 115 ft. National Grid Ref. 4344 4411

Middle Magnesian Limestone, massive reef-dolomite.

Acanthocladia anceps, Fenestrellina retiformis ; Dielasma elongatum, Stenoscisma schlotheimi [abundant], *Spiriferellina cristata multiplicata ; Naticopis leibnitziana ; Bakevellia?, ?Parallelodon striatus, Pseudomonotis speluncaria, Streblochondria pusilla.*

5. Inclined adit, 125 yd from mouth
Ht. above O.D. about 135 ft. National Grid Ref. 4346 4412

Middle Magnesian Limestone, massive reef-dolomite.

' *Adhaerentina* ' *permiana,* ' *Glomospira* ' *sp.* ; ?' *Spongia* ' *schubarthi* Geinitz ; *Batostomella crassa, Fenestrellina retiformis, Synocladia virgulacea ; Dielasma elongatum, Stenoscisma schlotheimi ; Bakevellia antiqua, Pseudomonotis speluncaria ;* ostracods.

6. Inclined adit, 100 yd from mouth
Ht. above O.D. about 165 ft. National Grid Ref. 4348 4413

Middle Magnesian Limestone, massive reef-dolomite.

' *Glomospira* ' *milioloides* ; ?' *Spongia* ' *schubarthi ; Synocladia virgulacea ; Orthothrix sp.,* coarsely spinose Strophalosiid, *Stenoscisma sp. ; Pseudomonotis speluncaria.*

7. Inclined adit, 50 yd from mouth
Ht. above O.D. about 215 ft. National Grid Ref. 4353 4416

Middle Magnesian Limestone, late-stage reef-dolomite.

Batostomella crassa, Fenestrellina retiformis, ?Synocladia virgulacea ; Dielasma elongatum, Orthothrix excavata, O. cf. *lewisiana ; Bakevellia?* [juv.] ; ostracods.

There are at least three distinct types of fossil assemblages in these Easington Colliery collections. In the first type (localities 1 and 3) *Stenoscisma* is abundant and *Horridonia,* gastropods and lamellibranchs are fairly common. The lamellibranchs include such genera as *Cardiomorpha, Solemya* and *Schizodus.* Strophalosiids and cryptostome polyzoa are rare or absent. The second type of assemblage (localities 2 and 7) is dominated by Strophalosiids and includes abundant cryptostome polyzoa. *Horridonia* and gastropods are rare or absent. Lamellibranchs are fairly common but are almost limited to *Bakevellia, Pseudomonotis* and *Parallelodon.* The third type (localities 5 and 6) is a sparser fauna dominated by *Synocladia virgulacea* which is rare or absent in the other two assemblages.

BOREHOLE AT MILL HILL RESERVOIR, EASINGTON

See pp. 113, 125, 137–8 for stratigraphical details
Ht. above O.D. 509 ft. National Grid Ref. 4122 4248

(a) Dolomite, reef-top sub-facies.

Depth 53 ft to 54 ft : Filamentous algal debris ; ' *Glomospira* ' *sp.* ; ?*Naticopis alterodensis* Dietz, *N. leibnitziana,* ?*Omphaloptycha pupoidea* Dietz, *Strobeus?,* small gastropod with an angular lower edge to the last whorl, indet. turreted gastropod ; *Bakevellia antiqua.*

Depth 56 ft 6 in to 63 ft 6 in: Filamentous algae, algal nodules [about 10·0 mm diameter] ; *Naticopsis?*, *Omphalotrochus sp.* ; *Bakevellia* cf. *antiqua* ; ostracods.

The shelly material occurs in pockets between sheets of algal filaments.

(b) Massive reef dolomite.

Depth 69 ft 9 in: Algal filaments ; *Omphalotrochus spp.* ; *Bakevellia* cf. *antiqua, B. ceratophaga, Parallelodon striatus, ?Permophorus costatus.*

Depth 70 ft 6 in: Algal filaments ; ?coiled foraminifer ; indet. turreted gastropod ; ostracod.

(c) Soft bioclastic dolomite of reef facies.

Depth 72 ft 6 in to 75 ft: Algal filaments [abundant] ; ' *Adhaerentina* ' *permiana* [abundant] ; *Fenestrellina retiformis,* indet. cryptostome polyzoa ; *?Craspedalosia lamellosa* (Geinitz), *Dielasma elongatum, Orthothrix excavata* [abundant] ; *Omphalotrochus spp.* including *O.* cf. *helicinus* and *O.* cf. *taylorianus,* smooth gastropods including *Naticopsis sp.* ; *B. antiqua, B. ceratophaga, Cardiomorpha modioliformis, Pseudomonotis speluncaria* ; ostracods.

This fauna is dominated by ' *Adhaerentina* ' and Strophalosiids. Gastropods and lamellibranchs are rare. The algal filaments are not in sheets as found higher in the borehole, but are (i) wound irregularly about the abundant shells and (ii) form series of nodules in single planes, the surfaces of which are covered with a brown crust. The nodules are up to 40·0 mm diameter.

Depth 76 ft to 82 ft : *Loxonema fasciatum, Omphalotrochus sp.,* indet. turreted gastropods ; *Bakevellia sp.* [juv.], *Permophorus?* [large lamellibranchs with extended posterior and concentric ornament but no radial costae. They are more tumid than normal *Permophorus costatus*].

(d) Soft granular dolomite, transitional between reef and lagoonal facies.

Depth 82 ft 6 in to 83 ft 6 in: *Spirorbis sp.* ; *Strobeus?* ; lamellibranch fragment ; rod-like structures, straight or curved with no regular internal structure seen [the dimensions are: diameter 8·0 to 13·0 mm and length 60·0 to 70·0 mm. They are possibly remains of worm burrows].

(e) Lagoonal dolomite adjacent to reef.

Depth 86 ft 6 in to 90 ft ; Algal filaments ; *Loxonema fasciatum, ?Naticopis leibnitziana, Omphalotrochus spp.* including *O. helicinus, Strobeus geinitzianus,* indet. turreted gastropod ; *Bakevellia antiqua, ?Permophorus costatus, Pseudomonotis speluncaria.*

(f) Soft granular lagoonal dolomite.

Depth 90 ft to 120 ft: ' *Ammodiscus* ' *roessleri* ; *Loxonema fasciatum, Omphalotrochus sp., Straparollus permianus, ?Strobeus geinitzianus* ; *Bakevellia sp.* ; ostracods.

Depth 164 ft to 171 ft 3 in: Coiled foraminifera including ' *Glomospira* ' *milioloides* ; *Horridonia horrida* ; *Astartella?* ; ostracods.

(g) Fine-grained saccharoidal lagoonal dolomite.

Depth 204 ft: ' *Ammodiscus* '?, ' *Glomospira* ' *pusilla* ; ?Productid spines ; *Bakevellia antiqua, Schizodus schlotheimi.*

(h) Fine-grained saccharoidal dolomite ; transitional beds at base of Middle Magnesian Limestone.

Depth 243 ft to 263 ft : Coiled foraminifera including ' *Adhaerentina* ' *permiana,* ' *Ammodiscus* ' *roessleri* [abundant], ' *Glomospira* ' *milioloides,* ' *G.*' *pusilla,* ?uniserial, chambered foraminifer ; *Astartella vallisneriana* [abundant], *Permophorus sp.* ; ostracods.

(i) Lower Magnesian Limestone.

Depth 280 ft: Coiled foraminifera ; *Cleiothyridina pectinifera.*

Depth 304 ft: *Schizodus* cf. *schlotheimi ;* ostracods.

Depth 313 ft: Coiled foraminifera ; coarsely-ribbed lamellibranch.

Depth 373 ft: ' *Glomospira* ' cf. *pusilla.*

HESLEDEN DENE NO. 1 BORE

Ht. above O.D. about 85 ft.   National Grid Ref. 4672 3697

(a) Reef-dolomite.

Depth 176 ft to 180 ft: Algal debris ; ' *Adhaerentina* ' *permiana,* ' *Glomospira* ' *gordialis,* ' *G.*' *pusilla* ; cryptostome polyzoa including *Thamniscus dubius* ; Strophalosiid [juv.] ; *Naticopsis leibnitziana,* indet. turreted gastropod ; *Astartella sp., Bakevellia ceratophaga* ; nautiloid fragments ; ostracods.

Depth 187 ft to 190 ft : *Batostomella crassa* ; *Orthothrix excavata* ; ?*Loxonema fasciatum, Omphalotrochus sp.* ; *Bakevellia sp., Permophorus sp.* [juv.] ; ostracods.

Depth 228 ft to 230 ft : ' *Glomospira* ' *pusilla* ; *Batostomella crassa* ; *Dielasma?* ; *Bakevellia antiqua, Parallelodon?*

HESLEDEN DENE NO. 2 BORE

Ht. above O.D. 92·6 ft.   National Grid Ref. 4654 3698

(a) Middle Magnesian Limestone.   Bedded off-reef calcarenite near foot of reef fore-slope.

Depth 246 ft to 250 ft: ' *Adhaerentina* ' *permiana* [abundant], ' *Glomospira* ' *gordialis,* ' *G.*' cf. *pusilla* ; ?*Lingula credneri Orthothrix sp.* and other Strophalosiids ; *Bakevellia* cf. *antiqua* ; ?ostracods ; fish spine.

Depth 251 ft to 257 ft: ' *A.*' *permiana,* ?*Geinitzina jonesi* (Brady), ' *Glomospira* ' *pusilla* ; ?sponge [very finely patterned mesh on a plane surface] ; *Horridonia horrida, Strophalosia morrisiana* ; *Dentalium?,* *Straparollus permianus,* ?*Naticopsis minima,* turreted gastropods including *Strobeus geinitzianus* ; *Bakevellia ceratophaga, Liebea sp., Permophorus costatus, Schizodus schlotheimi, S. truncatus, S. sp.* [with the rounded beak of *S. schlotheimi,* but the umbo is at extreme anterior end of the shell], *Solemya biarmica,* ?*Wilkingia elegans* (King) ; ostracods.

Depth about 270 ft: ' *A.*' *permiana,* ' *Ammodiscus* ' *roessleri,* ' *Glomospira* ' *sp.,* ?uniserial, chambered foraminifer ; ?trepostome polyzoa ; ostracods.

HESLEDEN DENE NO. 3 BORE

Ht. above O.D. 88·4 ft.   National Grid Ref. 4660 3703

(a) Middle Magnesian Limestone.   Bedded off-reef calcarenite near foot of reef fore-slope.

Depth 193 ft to 194 ft: '*Adhaerentina*' *permiana*, '*Glomospira*' *gordialis*, '*G.*' *pusilla*, uniserial, chambered foraminifera including *?Spandelinoides geinitzi* (Reuss); crinoid columnals; *Fenestrellina retiformis, Protoretepora ehrenbergi*; *Horridonia horrida, Orthothrix sp., Stenoscisma sp.*; *Omphalotrochus sp., Strobeus geinitzianus, ?S. leighi* (Brown); ostracods.

Depth 207 ft 6 in: '*G.*' *gordialis*, '*G.*' *pusilla*, uniserial, chambered foraminifera; *Fenestrellina sp.*; *Dielasma sp.*, Strophalosiids including *Orthothrix sp., Stenoscisma schlotheimi*, ?small Spiriferid; *Bakevellia* cf. *ceratophaga, Parallelodon striatus*; ostracods; fish remains.

Depth 209 ft to 212 ft: '*Adhaerentina*' *permiana*, '*Glomospira*' *gordialis*, '*G.*' *pusilla*, uniserial, chambered foraminifera; *Batostomella crassa*, cryptostome polyzoa; *Horridonia?, Orthothrix excavata* [juv.]; *Strobeus geinitzianus* [abundant at 211 ft 6 in to 212 ft], indet. non-turreted gastropods; *Astartella?, Bakevellia antiqua, B. ceratophaga, Permophorus costatus, Schizodus sp.*; ostracods; *Janassa bituminosa* Schlotheim (Plate VI, fig. 11).

(b) Middle Magnesian Limestone. Hard crystalline oolitic dolomite.

Depth 250 ft: '*Ammodiscus*' *roessleri*, '*Glomospira*' *pusilla*, uniserial, chambered foraminifera; *Horridonia horrida*; ostracods.

FISHBURN NO. 3 BORE
Ht. above O.D. 298 ft. National Grid Ref. 3718 3084

(a) Middle Magnesian Limestone, lagoonal dolomite.

Depth 124 ft to 125 ft: ?Productid spines and *Omphalotrochus?*

Depth 146 ft to 172 ft: '*Glomospira*' *sp.*; *Spirorbis?*; crinoid stem; ?Productid spines; gastropods including *Omphalotrochus?*; *Astartella?, Bakevellia antiqua, B. ceratophaga, Schizodus sp.*; ostracods.

Depth 197 ft to 216 ft 6 in: '*Adhaerentina*' *permiana*, '*Ammodiscus*'?, '*Glomospira*' *milioloides*, '*G.*' *pusilla*; *?Fenestrellina retiformis*; *Horridonia horrida, H. horrida* [juv.] or small Strophalosiid; *?B. antiqua, B.* cf. *ceratophaga, Permophorus costatus, Schizodus sp., Solemya?*; ostracods.

Depth 220 ft to 227 ft: *Tubulites permianus* (King) [see p. 131]; *Astartella?, B. antiqua, B. ceratophaga, Permophorus sp., Solemya?, Schizodus schlotheimi, S. truncatus* [juv.]; ostracods.

(b) Lower Magnesian Limestone.

Depth 285 ft: *Strophalosia morrisiana.*

Depth 305 ft 6 in: '*Ammodiscus*' *roessleri*, '*Glomospira*' *pusilla*; *B. antiqua*; ostracods.

Depth 320 ft: Rootlets; '*Adhaerentina*' *permiana*, '*Ammodiscus*' *roessleri*, '*Glomospira*' *pusilla*; ostracods.

Depth 353 ft: *Strophalosia morrisiana.*

Depth 386 ft: *Lingula credneri*; ?ostracod.

FISHBURN NO. 4 BORE
Ht. above O.D. 310 ft. National Grid Ref. 3858 3120

(a) Middle Magnesian Limestone, lagoonal dolomite.

Depth 121 ft to 143 ft: *Tubulites?*; '*Adhaerentina*' *permiana*, '*Ammodiscus*' *roessleri*, '*Glomospira*' cf. *milioloides*; *Horridonia horrida*; indet. turreted gastropod; *Bakevellia antiqua, Permophorus sp.* [juv.], *Schizodus schlotheimi, Solemya?* ostracods.

(b) Lower Magnesian Limestone.

Depth 300 ft to 307 ft: Rootlets.

Depth 311 ft 6 in to 321 ft 6 in : Indet. plant remains ; ' *Glomospira* ' *pusilla* ; *Strophalosia morrisiana*.

Depth 330 ft to 359 ft: Plant fragments ; ' *Adhaerentina* ' *permiana*, ' *Ammodiscus* ' *roessleri*, uniserial, chambered foraminifera including *?Geinitzina jonesi*, ' *Glomospira* ' *pusilla* ; *Acanthocladia anceps, Batostomella crassa, Fenestrellina retiformis, ?Hippothoa* aff. *voigtiana* (King) [see note 1], Polyzoan *?Gen. et sp. nov.* [see p. 181] ; *Crurithyris?, ?Dielasma elongatum, Horridonia horrida, Neochonetes?* [see note 2], *Pterospirifer* cf. *alatus* (Schlotheim), *Strophalosia morrisiana*, Strophalosiid [juv.] [see note 3] ; *Bakevellia antiqua, Streblochondria pusilla* ; ostracods.

NOTE 1. The *?Hippothoa* differs from King's figured specimen (1850, pl. iii, fig. 13) in lacking a neck at the proximal end of each bead and in the regular increase in size of beads from the distal to the proximal end of the organism.

NOTE 2. Several ventral valves are preserved. They are smooth except for concentric growth lines and spines in a row along the posterior margin and sparsely scattered on the rest of the valve. The cardinal extremities are upturned but the greater part of the valve is moderately convex. The dimensions of this valve are about 10·0 mm wide and 7·0 mm long. The dorsal valve was not clearly seen.

NOTE 3. A flat, sub-triangular, ventral valve with numerous coarse spines which are neither vermiform nor adpressed.

FISHBURN ' D ' BORE
Ht. above O.D. 318 ft.   National Grid Ref. 3567 3106

(a) Middle Magnesian Limestone, lagoonal dolomite.

Depth 91 ft to 100 ft: *Tubulites?* ; indet. gastropod ; *Bakevellia antiqua* [stunted], *B.* cf. *ceratophaga* [stunted], *Permophorus sp.* [juv.] ; ostracods.

Depth 117 ft: *Schizodus* cf. *truncatus*.

Depth 120 ft: ' *Glomospira* ' *pusilla, ?*uniserial, chambered foraminifer ; cryptostome polyzoa ; *Bakevellia sp.* ; ostracods.

Depth 134 ft 6 in: ' *Adhaerentina* ' *permiana*, ' *Ammodiscus* '?, ' *Glomospira* ' *milioloides* ; *?Strophalosia morrisiana* [juv.] ; *B. antiqua, B.* cf. *ceratophaga* ; ostracods.

Depth 140 ft to 152 ft: ' *Adhaerentina* ' *permiana*, ' *Glomospira* ' *milioloides*, ' *G.*' *pusilla* ; *Bakevellia sp., Schizodus?* [juv.] ; ostracods.

(b) Lower Magnesian Limestone.

Depth 212 ft to 216 ft: ' *Ammodiscus* ' *roessleri*, ' *Glomospira* ' *milioloides*, ' *G.*' *pusilla*, uniserial, chambered foraminifer ; *Bakevellia* cf. *ceratophaga* ; ostracods.

Depth 239 ft to 257 ft: Plant remains including *Chondrites logaviensis* ; ' *Adhaerentina permiana*, ' *Ammodiscus* '?, ' *Glomospira* ' *milioloides*, ' *G.*' *pusilla, ?Spandelinoides geinitzi* ; *Acanthocladia anceps, Batostomella?* [at 257 ft] ; *Strophalosia morrisiana* ; *Bakevellia antiqua* ; ostracods.

Depth 292 ft to 318 ft: Plant remains ; ' *Adhaerentina*' *permiana*, ' *Glomospira*' *milioloides,* ' *G.*' *pusilla* ; *Batostomella columnaris* (Schlotheim) ; *Strophalosia morrisiana* ; *Bakevellia* cf. *ceratophaga.*

Depth 326 ft: Plant remains [abundant] ; ' *Adhaerentina*' *permiana,* ' *Ammodiscus*'?, ' *Glomospira*' *pusilla* ; *Batostomella crassa,* cryptostome polyzoa ; *Crurithyris?* [small valve].

Depth 342 ft: *Lingula credneri, Neochonetes? sp. nov.* [see note].

There are several specimens of *Neochonetes?*. The ventral valve is wide and flat with a long hinge-line almost equal to the width of the shell. The surface has concentric growth lines and sparse and irregularly spaced coarse spines plus a single row of ?4 coarse spines on each side of the umbo and parallel to the hinge-line. The beak does not overhang the hinge-line. There is a distinct area and delthyrium with a convex pseudodeltidium. The ventral valve is about 17·0 mm wide and 14·0 mm long. The dorsal valve is flat to slightly concave with coarse concentric growth lines becoming almost lamellose. No spines were seen on the dorsal valve.

Assignment of these brachiopods to *Neochonetes* is doubtful because of the lack of any radial ornament and the presence of spine-bases on the ventral valve in addition to those adjacent to the posterior margin. Somewhat similar but smaller shells were found in the Lower Magnesian Limestone of Fishburn No. 4 Bore from between 330 ft and 359 ft.

TEN-O'-CLOCK BARN NO. 1 BORE
Ht. above O.D. about 310 ft. National Grid Ref. 3992 2980

(a) Middle Magnesian Limestone, lagoonal beds.

Depth 422 ft: *Tubulites?* ; ' *Ammodiscus*' *roessleri,* ' *Glomospira*' *sp.* ; ?Productid spine ; *Astartella?, Bakevellia antiqua* ; ostracods.

Depth 458 ft 9 in to 462 ft: ' *Adhaerentina*' *permiana,* ' *Ammodiscus*' *roessleri,* ' *Glomospira*' cf. *milioloides,* ' *G.*' *pusilla* ; *Astartella vallisneriana, Bakevellia sp. Liebea squamosa, Palaeolima?, Schizodus schlotheimi* ; ostracods.

Depth 465 ft 6 in to 470 ft 3 in: ' *Adhaerentina*' *permiana,* ' *Ammodiscus*' *roessleri,* ' *G.*' *pusilla* ; cryptostome polyzoa including *Thamniscus dubius* ; *Horridonia horrida* ; indet. lamellibranch ; ostracods.

WHIN HOUSES BORE
Ht. above O.D. 307 ft. National Grid Ref. 3997 3058

(a) Middle Magnesian Limestone, lagoonal dolomites.

Depth 224 ft to 226 ft 8 in: *Tubulites?,* strap-like plant remains ; ' *Glomospira*' *milioloides* ; *Bakevellia sp.*

(b) Middle Magnesian Limestone, transitional beds.

Depth 254 ft 6 in to 260 ft 9 in: ' *Ammodiscus*' *roessleri,* ' *Glomospira*' *pusilla* ; *Neochonetes? davidsoni, Strophalosia morrisiana* ; ostracods.

(c) Lower Magnesian Limestone.

Depth 275 ft to 287 ft 6 in: ' *A.*' *roessleri,* ' *Glomospira*' *milioloides* and ?uniserial, chambered foraminifera.

Depth 365 ft to 366 ft: ' G.' pusilla, uniserial, chambered foraminifera ; cryptostome polyzoa ; S. morrisiana ; ostracod.

Depth 403 ft to 405 ft: Plant remains ; ' Adhaerentina' permiana, ' G.' pusilla ; Batostomella crassa, cryptostome polyzoa ; indet. brachiopod ; ostracods.

Depth 414 ft: Strophalosia morrisiana.

Depth 418 ft 6 in to 428 ft 6 in: ' A.' roessleri, ' G.' pusilla ; Polyzoan, ?Gen. et sp. nov. [see note on p. 181], ?Penniretepora waltheri (Korn) [one stem with numerous short side branches. The finely ribbed reverse side only is visible] ; ?Howseia latirostrata (Howse).

The last named has a convex ventral valve, 10·0 mm wide and 9·0 mm long with a long hinge-line, an umbo which is tumid but does not overhang the hinge-line and spines including a row round the anterior margin.

Depth 444 ft 6 in to 459 ft 6 in: ' G.' pusilla ; crinoid columnals ; Acanthocladia anceps ; Crurithyris clannyana, Dielasma sp., Productoid? [juv.] ; Bakevellia cf. antiqua ; ostracods.

HARTLEPOOL LIGHTHOUSE BORE
Ht. above O.D. 25·5 ft.   National Grid Ref. 5319 3387

(a) Upper Magnesian Limestone, Hartlepool and Roker Dolomite.

Depth 147 ft to 150 ft: Liebea squamosa, Permophorus sp., Schizodus schlotheimi ; ostracods. The lamellibranchs are stunted to varying degrees.

(b) Middle Magnesian Limestone. Bedded off-reef calcarenite.

Depth 635 ft 9 in to 637 ft: ' Adhaerentina' permiana, ' Ammodiscus' roessleri, ' Glomospira' gordialis, ' G.' pusilla ; crinoid columnals ; Acanthocladia anceps, Batostomella crassa ; Horridonia horrida, small valve of indet. Productid ; ostracods.

Depth 647 ft 3 in to 653 ft: ?' Adhaerentina' permiana, ' Glomospira' milio-loides, ' G' pusilla, uniserial, chambered foraminifer, ?coiled foraminifer [very small: c.0·1 mm long] ; crinoid columnals ; cryptostome polyzoa ; Horridonia?.

Depth 655 ft 6 in to 658 ft 10 in: ' Ammodiscus' roessleri, ' Glomospira' gordialis, ' G.' pusilla, uniserial, chambered foraminifera including ?Nodo-saria kingii, ?Spandelinoides geinitzi ; crinoid columnals ; Calophyllum profundum (Germar) ; Acanthocladia or Thamniscus sp., Batostomella crassa, Fenestrellina retiformis, Protoretepora ehrenbergi, ?Synocladia virgulacea ; Crurithyris clannyana, Horridonia?, Spiriferid fragment, Stenoscisma?, Streptorhynchus pelargonatus ; Strobeus geinitzianus ; Bakevellia? [juv.].

Depth 659 ft to 664 ft 3 in: Coiled foraminifera, uniserial, chambered fora-minifera ; Acanthocladia anceps, Fenestrellina retiformis, Thamniscus dubius ; Crurithyris?, Orthothrix exavata, Spiriferellina cristata multiplicata, Stenoscisma sp., small indet. Strophalosiids ; ostracods.

Depth 664 ft 4 in to 667 ft 6 in: Crinoid columnals ; ?echinoid spine ; Acanthocladia or Thamniscus, Batostomella crassa ; Horridonia horrida, Strophalosia morrisiana ; indet. gastropod ; ostracods ; fish debris.

(c) Middle Magnesian Limestone, transitional beds.

Depth 707 ft 6 in to 717 ft 6 in: ' Glomospira' gordialis, ' G' pusilla, uniserial, chambered foraminifer ; ?cryptostome polyzoa and ostracods.

Depth 744 ft: 'Glomospira' pusilla; Batostomella crassa; cryptostome polyzoan fragments including Acanthocladia?; Productoid brachiopods, indet. Spiriferid.

(d) Lower Magnesian Limestone, near base.

Depth 841 ft: Coiled foraminifera including 'G.' pusilla; crinoid columnals; Batostomella?.

J.P., D.B.S.

TABLE 4

The following table summarizing the distribution of fossils through the Permian sequence is based only on the Survey Collection from the district. It gives, for instance, no indication of the extent of the Marl Slate fish fauna which is much better represented in the Collection of the Hancock Museum. Additional faunal records from the district are to be found in King (1850) and Trechmann (1925).

The lists given for the lagoonal and basinal facies of the Middle Magnesian Limestone include many species collected from close to the reef and some may be derived from it. Indigenous lagoonal and basinal faunas are very limited in the number of species and individuals (see pp. 120 and 122).

The numbers against certain names refer to the palaeontological notes following the table. The presence of a species is indicated in the table by c = common; x = recorded; ? = doubtfully recorded.

Explanation of abbreviations:

MS Marl Slate, LML Lower Magnesian Limestone, MML Middle Magnesian Limestone, L and T Lagoonal beds and transitional beds at base of Middle Magnesian Limestone, B Basinal beds, UML Upper Magnesian Limestone, CL Concretionary Limestone, H & RD Hartlepool and Roker Dolomite.

			MML			UML	
	MS	LML	L & T	REEF	B	CL	H & RD
Plantae							
Algites virgatus (Münster) ...						x	
Chondrites logaviensis Geinitz ...		x					x
Tubulites permianus (King)[1] ...			x			c	?
Plant remains [undetermined] ...		x	x	c	x	x	x
Foraminifera							
'Adhaerentina' permiana Paalzow[2]		c	c	x	c		
'Ammodiscus' roessleri (Schmid)		x	c		x		
'Glomospira' gordialis (Jones and Parker)			x	x	c		
'G.' milioloides (Jones, Parker and Kirkby)		x	c	x	x		
'G.' pusilla (Geinitz)		x	c	x	c		
Uniserial, chambered foraminifera		x	x	x	x		
including ?Geinitzina jonesi (Brady)		x			x		
?Nodosaria kingii Paalzow (non Jones) ...				x	x		
?Spandelinoides geinitzi (Reuss)		x			x		

			MML			UML	
	MS	LML	L & T	REEF	B	CL	H & RD
Porifera							
Eudea tuberculata (King) ...				x			
' Spongia ' schubarthi Geinitz ...				?			
Anthozoa (Rugosa)							
Calophyllum profundum (Germar)					x		
Crinoidea							
Crinoid columnals		x	x	x	x		
Echinoidea							
Miocidaris keyserlingi (Geinitz) ...				x	x		
Annelida							
Spirorbis sp.			x				
Polyzoa							
Acanthocladia anceps (Schlotheim)		x	?	c	x		
Batostomella columnaris (Schlotheim)		x					
B. crassa (Lonsdale)		x	x	c	x		
Fenestrellina retiformis (Schlotheim)		x	?	c	x		
Hippothoa aff. voigtiana (King) ...		x					
Penniretepora waltheri (Korn) ...		?					
Protoretepora ehrenbergi (Geinitz)				x	x		
Synocladia virgulacea (Phillips) ...				c	?		
Thamniscus dubius (Schlotheim) ...			x	c	x		
Polyzoan ?Gen. et sp. nov.[3] ...		x		?			
Brachiopoda							
Cleiothyridina pectinifera (J. de C. Sowerby)[4]		x		x			
Craspedalosia lamellosa (Geinitz)[5]				?			
Crurithyris clannyana (King) ...				x	x		
Dielasma elongatum (Schlotheim)...		x		c	x		?
D. elongatum sufflatum (Schlotheim)				x			
Horridonia horrida (J. Sowerby) ...		x	x	c	x		
Howseia latirostrata (Howse) ...		?					
Lingula credneri Geinitz	x	x			?		
Neochonetes? davidsoni (Schauroth)		?	x		x		
Neochonetes? sp. nov.		x					
Orthothrix excavata Geinitz[5] ...				c	x		
O. cf. lewisiana (de Koninck) ...				x	x		
Pterospirifer alatus (Schlotheim) ...		x		x			
Spiriferellina cristata multiplicata (King)				x	x		
Stenoscisma globulina (Phillips) ...				x	x		
S. schlotheimi (von Buch)				c	x		

	MS	LML	MML			UML	
			L & T	REEF	B	CL	H & RD
Streptorhynchus pelargonatus (Schlotheim)			x	x	x		
Strophalosia morrisiana King[6] ...		c	x	x	x		
Spiriferids, indet.			x		x		
Gastropoda							
Cyclobathmus? permianus (King)[7]			x	?			
Loxonema fasciatum King			x	x	?		
L. cf. *retusum* Dietz				x			
Macrochilina symmetrica (King) ...				x			
Naticopsis alterodensis (Dietz) ...				?			
N. leibnitiziana (King)			?	x			x
N. minima (Brown)			x	x	?		
Omphaloptycha pupoidea Dietz ...				?			
Omphalotrochus helicinus (Schlotheim)			x	x			
O. taylorianus (King)				x			
O. thomsonianus (King)				x			
Omphalotrochus sp.					x	x	x
Pleurotomaria linkiana King ...				x			
P. verneuili Geinitz				x			
Pleurotomaria sp.			?		x		
Straparollus permianus (King)[8] ...			x	x	x		
S. sp. non *S. permianus*					x		
Strobeus geinitzianus (King) ...			?	x	x		
S. leighi (Brown)					?		
Turreted gastropods, indet. ...						x	
Gastropods, indet.						x	x
Scaphopoda							
Dentalium aff. *speyeri* Geinitz ...				x			
Dentalium sp.					?		
Lamellibranchia							
Astartella vallisneriana (King)[4] ...			c				
Astartella sp.				x	?		
Bakevellia antiqua (Münster)[9] ...		x	c	c	x		
B. ceratophaga (Schlotheim) ...		?	x	c	x		
Cardiomorpha modioliformis King			x	x			
Liebea squamosa (J. de C. Sowerby)[10]			x	x	?	c	c
Palaeolima? permiana (King) ...			?	x	x		
Parallelodon striatus (Schlotheim)	?			c	x		
Permophorus costatus (Brown) ...			x	x	x		
Permophorus sp.						x	x
Pseudomonotis speluncaria (Schlotheim)			x	c	x		
Schizodus obscurus (J. Sowerby)[11]			x				

| | MS | LML | MML | | | UML | |
			L & T	REEF	B	CL	H & RD
S. rotundatus (Brown)		x				x	
S. schlotheimi (Geinitz)		?	x	x	x	c	x
S. truncatus King			x	x	x		
Solemya biarmica de Verneuil ...				x	x		
Solemya sp.			?				
Streblochondria pusilla (Schlotheim)		x		x	x		
Wilkingia elegans (King)					?		
Nautiloidea							
Peripetoceras freieslebeni (Geinitz)				x			
Ostracoda							
Ostracods	?	x	c	c	c	x	x
Pisces							
Acrolepis sedgwickii Agassiz ...	x						
Janassa bituminosa (Schlotheim) ...					x		
Fish remains [undetermined] ...	x				x		

1. The type of *Tubulites* [*Filograna*] *permianus* comes from " the south end of Black Hall Rocks " (King 1850, p. 56); it was originally thought to be a calcareous worm tube, although Howse (1890) included it with a query in his synonymy of *Algites virgatus*. Trechmann (1942) also thought it to be an alga because the tubes appeared to branch; Stoneley (1958) considered it identical with the Upper Zechstein form *Tubulites articulatus* Bein, a species erected by Bein in ignorance of the earlier name *F. permiana* and which he also considered algal.

Stoneley failed to detect any unmistakable branching, nor has any been seen among the specimens examined in the Geological Survey collections from this district. However, the close association of *Tubulites* with carbonaceous plant remains (p. 157) in Upper Magnesian Limestone, 720 yd E.N.E. of Blackhall Colliery station, provides further evidence of its probable algal origin. Specimens of *Tubulites* collected from the lagoonal beds of the Middle Magnesian Limestone in most cases have been found as isolated, randomly-orientated internal moulds, but a specimen from 227 ft in Fishburn No. 3 Bore bears a great number of internal moulds with a sub-parallel arrangement, the form of preservation common in the Upper Magnesian Limestone.

2. The coiled foraminifera assigned to ' *Adhaerentina* ', ' *Glomospira* ' and ' *Ammodiscus* ' have very little agglutinated material and should, therefore, probably be placed among the imperforate, calcareous Miliolacea. Of Palaeozoic Milioloidea, such as these, Wolanska (1959) stated ' it seems reasonable to suppose that the Palaeozoic Milioloidea constitute a distinctly separate group of relatively primitive forms and justify the establishment of a family to fit them.'

The above generic names, used normally for members of the Ammodiscidae, a family of foraminifera with agglutinated walls, would thus be inapplicable here. Although some names, such as *Agathammina* and *Calcitornella,* are already available for some of these forms, the erection of new species and, possibly, new genera, would be needed for the systematic nomenclature of all of them. Therefore, '*Adhaerentina*', '*Glomospira*' and '*Ammodiscus*' have been used throughout (e.g. '*G.*' *pusilla* has been preferred at this time to *Agathammina pusilla*) as the use of both Ammodiscid and Milioloid names for tests of similar wall composition would be confusing.

3. This cryptostome polyzoan [Pl. VI, fig. 10] is recorded in Lower Magnesian Limestone from Fishburn No. 4 Bore and Raisby Hill Quarry and questionably from Whin Houses Bore. It is also questionably recorded in Middle Magnesian Limestone reef near Horden.

The main stem has two, rarely three, rows of zooecia on the obverse side. The reverse is longitudinally striated, although faintly in some examples. Branching is rare and dichotomous, but the organism is mostly found as unbranched stems. It has several of the characteristics of *Penniretepora waltheri,* but differs from that species in the nature and rarity of its branching.

4. The determinations of the brachiopods and lamellibranchs have been largely based on the unpublished revision of these groups within the Magnesian Limestone by Dr. A. Logan (1962).

5. Muir-Wood (in Muir-Wood and Cooper 1960) has revived Geinitz's genus *Orthothrix* for those Zechstein Strophalosiids with fine, straight spines on both valves. *Strophalosia* has been retained for coarsely-spinose shells such as *S. morrisiana. O. excavata* Geinitz is the commonest species of *Orthothrix* found in the Magnesian Limestone of Durham. It is very variable in form. Dorsal valves are flat to concave and faintly lamellose or non-lamellose. Ventral valves are triangular to circular, with or without a median sulcus. The angle between the interarea of the ventral valve and the dorsal valve varies considerably, but in no case was the beak seen to overhang the dorsal valve.

O. lewisiana (de Koninck) is a name reserved by A. Logan for circular *Orthothrix* with beaks overhanging the hinge line. Those shells identified here as *O.* cf. *lewisiana* are circular and finely spinose but incomplete and it is not possible to ascertain the disposition of the beak. The shells questionably assigned to *O.* cf. *lewisiana* may well belong, therefore, to a variety of *O. excavata.* This may also be true of most of those questionably assigned to *Craspedalosia lamellosa* (Geinitz) as no vermiform spines were seen except on a fragment of one of them.

6. A distinctive feature of many of the specimens of *Strophalosia morrisiana* from the Lower Magnesian Limestone is the presence of a depression in the umbonal region of the ventral valve exterior. This is unlike the simple flattening of the beak (known as a cicatrix) which is caused by attachment of the shell to some anchorage. It is a regularly concave depression symmetrically positioned about the centre line of the valve. In juvenile shells it may form the greater part of the ventral valve, leaving only a rim of normal convex surface adjacent to the anterior and lateral margins. This depression is accompanied, although it is less often observed, by a corresponding elevation in the umbonal region of the dorsal valve. A good example of a dorsal valve with such a blister-like elevation was collected from Lower Magnesian Limestone at 134 ft 6 inches in Fishburn ' D ' Bore.

7. Many of the gastropods occur in the upper part of the Middle Magnesian Limestone reef and in the Upper Magnesian Limestone where the specimens are more or less stunted. Consequently specific identification is difficult and this is especially so as the type-specimens of most of the Magnesian Limestone species are from the lower parts of the reef. In addition to stunting it is possible that

species with external shell ornament in the earlier reef limestones lost their ornament in the less favourable conditions prevailing when the higher beds were deposited (Trechmann 1925). King (1850) was doubtful about most of the generic names he used and no complete revision of the Magnesian Limestone gastropods has since been undertaken. In Germany, Dietz worked on the Zechstein gastropods and erected several new species (1909), but the lack of synonymies in his paper makes comparison with King's species difficult and his descriptions and illustrations are not sufficiently detailed to allow extensive use of them for the identification of Durham Magnesian Limestone forms.

8. *Straparollus* is preferred to *Euomphalus* for the smooth-walled, low-spired shells called *E. permianus* by King (1850) and Branson (1948) and *S. permianus* by Dietz (1909). *Euomphalus* is now reserved for shells with an angulated outer-upper edge on each whorl.

9. The five species of *Bakevellia* described by King are reduced to two— *B. antiqua* Münster) and *B. ceratophaga* (Schlotheim)—by A. Logan (1962), and these two names have been used here exclusively. Both are variable species and there are many intermediate forms. It is possible that, in fact, there is only one highly variable species of *Bakevellia* in the Magnesian Limestone.

10. A. Logan's merging of *Liebea squamosa* and *L. septifer* has been followed, but there seem to be some grounds for keeping the species separate. Except for the one occurrence of *L. squamosa* at the mouth of Castle Eden Dene in the Upper Magnesian Limestone, all the *Liebea* examined from the Upper Magnesian Limestone are as *L. septifer* (King), and from the Middle Magnesian Limestone as *L. squamosa* (J. de C. Sowerby).

11. King's four species of *Schizodus* have been retained but it seems likely that *S. rotundatus* (Brown) with an almost central umbo and *S. truncatus* King with a pointed umbo are merely variants of *S. schlotheimi* (Geinitz). If any division of *S. schlotheimi* were to be adopted, it would preferably be between the rounded shells common in the Upper Magnesian Limestone and including *S. rotundatus* and the more angular forms (including *S. truncatus*) found in the Middle Magnesian Limestone. *S. obscurus* (J. Sowerby), fairly common in Yorkshire, has been recorded only once during examination of the Survey collections from this district—in lagoonal beds of the Middle Magnesian Limestone from Tinkler's Gill Bore.   J.P.

REFERENCES

ABBOT, T. G. 1903. The Cellular Magnesian Limestone of Durham. *Quart. J. Geol. Soc.,* **59**, 51–2.
ANDERSON, W., and DUNHAM, K. C. 1953. Reddened beds in the Coal Measures beneath the Permian of Durham and south Northumberland. *Proc. Yorks. Geol. Soc.,* **29**, 21–32.
BAGNOLD, R. A. 1941. *The physics of blown sands and desert dunes.* London.
BIRD, C. 1881. *A short sketch of the geology of Yorkshire.* London and Bradford.
BIRD, W. J. 1888. The South Durham Salt Bed and associated strata. *Trans. Manch. Geol. Soc.,* **19**, 564–84.
BRANSON, C. C. 1948. Bibliographic index of Permian invertebrates. *Mem. Geol. Soc. Amer.,* **26**.
BROWELL, E. J. J., and KIRKBY, J. W. 1866. On the chemical composition of various beds of the Magnesian Limestone and associated Permian rocks of Durham. *Trans. Nat. Hist. Soc. Northumberland, Durham* and *Newcastle-upon-Tyne,* **1**, 204–30.

BULLERWELL, W. 1961. In *Sum. Prog. Geol. Surv.* for 1960, 41, 60–1.

BURTON, R. C. 1911. Beds of Yellow Sands and Marl in the Magnesian Limestone of Durham. *Geol. Mag.,* **8,** pp. 299–306.

CALVERT, R. 1884. *Notes on the geology and natural history of the County of Durham.* Bishop Auckland.

CLARKE, A. M., CHAMBERS, R. E., ALLONBY, R. H., and MAGRAW, D. 1961. A marine geophysical survey of the undersea coalfields of Northumberland, Cumberland and Durham. *Trans. Inst. Min. Eng.,* **121,** 197–215.

D'ANS, J. 1947. Über die Bildung and Umbildung der Kalisalzlagerstätten. *Naturw.,* **34,** 295–301.

DAVIES, W., and REES, W. J. 1944. British resources of steel moulding sands, Part 5. The Permian Yellow Sands of Durham and Yorkshire. *J. Iron and Steel Inst.,* No. II for 1943, 104P–111P.

DEANS, T. 1950. The Kupferschiefer and associated lead-zinc mineralization in the Permian of Silesia, Germany and England. *Rept. XVIII Int. Geol. Congr.* (London), pt. 7, 340–52.

DIETZ, E. 1909. Ein Beitrag zur Kenntis der deutschen Zechsteinschnecken. *Jahrb. d. k. preuss. geol. Landes.,* **30,** 1, 444–94.

DUNHAM, K. C. 1948. A contribution to the petrology of the Permian evaporite deposits of north-eastern England. *Proc. Yorks. Geol. Soc.,* **27,** 217–27.

—— 1952. Fluorspar. *Mem. Geol. Surv. Min. Resources,* **4.**

—— 1960. Syngenetic and diagenetic mineralization in Yorkshire. *Proc. Yorks. Geol. Soc.,* **32,** 229–84.

—— 1961. Black shale, oil and sulphide ore. *Adv. Sci.,* **18,** 284–99.

——, CLARINGBULL, G. F. and BANNISTER, F. A. 1948. Dickite in the Magnesian Limestone of Durham. *Mineral. Mag.,* **28,** 338–42.

EDWARDS, W., WRAY, D. A., and MITCHELL, G. H. 1940. Geology of the country around Wakefield. *Mem. Geol. Surv.*

FOWLER, A. 1943. On fluorite and other minerals in Lower Permian rocks of south Durham. *Geol. Mag.,* **80,** 41–51.

—— 1944. A deep bore in the Cleveland Hills. *Geol. Mag.,* **81,** 193–206.

—— 1945. Evidence for a new major fault in north-east England. *Geol. Mag.,* **82,** 245–50.

—— 1957. Minerals in the Permian and Trias of north-east England. *Proc. Geol. Assoc.,* **67,** 251–65.

GUPPY, EILEEN M. and SABINE, P. A. 1956. Chemical analyses of igneous rocks, metamorphic rocks and minerals 1931–1954. *Mem. Geol. Surv.*

GUY, E. 1911. On the formation of cavities in the Magnesian Limestone. *Trans. Leeds Geol. Soc.,* pt. 16, 10–15.

GREEN, A. H. 1872. On the method of formation of the Permian beds of south Yorkshire. *Geol. Mag.,* **9,** 99–101.

HANCOCK, A., and HOWSE, R. 1870. On a new labyrinthodont amphibian from the Magnesian Limestone of Middridge, Durham. *Quart. J. Geol. Soc.,* **26,** 556–64.

HICKLING, G. 1931. The Summer Field Meeting (1931) in Northumberland and Durham. *Proc. Geol. Assoc.,* **42,** 378–85.

—— and HOLMES, A. 1931. The brecciation of the Permian Rocks. *Proc. Geol. Assoc.,* **42,** 252–5.

HIRST, D. M., and DUNHAM, K. C. 1963. Chemistry and petrography of the Marl Slate of S.E. Durham, England. *Econ. Geol.,* **58,** 912–40.

HODGE, M. B. 1932. The Permian Yellow Sands of north-east England. *Proc. Univ. Durham Phil. Soc.,* **8,** 410–58.

HOLLINGWORTH, S. E. 1942. Correlation of gypsum-anhydrite deposits and the associated strata in the north of England. *Proc. Geol. Assoc.,* **53,** 141–51.

HOLTEDAHL, O. 1921. On the occurrence of structures like Walcott's Algonkian algae in the Permian of England. *Amer. J. Sci.* (5), **1,** 195–206.

HOWSE, R. 1848. A catalogue of the fossils of the Permian system of the counties of Northumberland and Durham. *Trans. Tyneside Naturalists Field Club*, **1**, 218–64.

—— 1857. Notes on the Permian system of the counties of Northumberland and Durham. *Ann. Mag. Nat. Hist.* (2), **19**, 33–52, 304–12 and 463–73.

—— 1890. Catalogue of the local fossils in the Museum of Natural History. *Trans, Nat. Hist. Soc. Northumberland, Durham and Newcastle-upon-Tyne*, **11**, 227–288.

JOHNSTON, J. H. 1942. Permian lime-secreting algae from the Guadalupe Mountains, New Mexico. *Bull. Geol. Soc. Amer.*, **53**, 195–226.

KING, W. 1850. A Monograph of the Permian fossils of England. *Palaeont. Soc.*

KIRKBY, J. W. 1860. On the occurrence of *Lingula Credneri* Geinitz in the Coal Measures of Durham ; and on the Claim of the Permian Rocks to be entitled a System. *Quart. J. Geol. Soc.*, **16**, 412–21.

LEBOUR, G. A. 1902. The Marl Slate and Yellow Sands of Northumberland and Durham. *Trans. Inst. Min. Eng.*, **24**, 370–91.

LOGAN, A. 1962. A revision of the Palaeontology of the Permian Limestones of County Durham. *Unpublished Ph.D. thesis, King's College, Univ. of Durham.*

LOGAN, B. W. 1961. Cryptozoon and associate stromatolites from the Recent, Shark Bay, Western Australia. *J. Geol.*, **69**, 517–33.

LOTZE, F. 1958. Der englische Zechstein in seine Beziehung zum deutschen. *Geol. Jahrb.*, **73**, 135–40.

LOVE, L. G. 1962. Biogenic primary sulphide of the Permian Kupferschiefer and Marl Slate. *Econ. Geol.*, **57**, 350–66.

MAGRAW, D., CLARKE, A. M., and SMITH, D. B. 1963. The stratigraphy and structure of part of the south-east Durham coalfield. *Proc. Yorks. Geol. Soc.*, **34**, 153–208.

MARLEY, J. 1892. On the Cleveland and south Durham salt industry. *Trans. N. England Inst. Min. Eng.*, **39**, 91–125.

MUIR-WOOD, HELEN M., and COOPER, G. A. 1960. Morphology, Classification and Life Habits of the Productoidea (Brachiopoda). *Mem. Geol. Soc. Amer.*, **81**.

NAIRN, A. E. M., and THORNLEY, N. 1961. The application of geophysics to palaeoclimatology. In *Descriptive Palaeoclimatology*, Edit. A. E. M. Nairn, New York.

NALIVKIN, D. V. 1937. Scientific results of the Permian Conference. In Problems of Soviet Geology, 7. Translated by T. Storichenko with comments and foreword by J. S. Williams. *Bull. Amer. Assoc. Petrol. Geol.*, **22**, 771–6.

NEWELL, D., RIGBY, J. K., FISCHER, A. G., WHITEMAN, A. J., HICKOX, J. E., and BRADLEY, J. S. 1953. *The Permian Reef Complex of the Guadalupe Mountains Region, Texas and New Mexico.* San Francisco.

OELSNER, O. 1959. Bemerkungen zur Herkunft der Metalle im Kupferschiefer. *Freiburger Forsch.*, **58**, 106–113.

QUIRING, H. 1954. Permklima und Sonnentemperatur. *Neues. Jahrb. Geol. u. Pal.*, **7**.

RAMSAY, A. C. 1871. On the red rocks of England of older date than the Trias. *Quart. J. Geol. Soc.*, **27**, 241–56.

RUNCORN, S. K. 1956. Palaeomagnetic survey in Arizona and Utah: Preliminary results. *Bull. Geol. Soc. Amer.*, **67**, 301–316.

SCHOUTEN, C. 1946. The role of sulphur bacteria in the formation of the so-called sedimentary copper ores and pyrite ore bodies. *Econ. Geol.*, **41**, 517–38.

SEDGWICK, A. 1829. On the geological relations and internal structure of the Magnesian Limestone, and the lower portions of the New Red Sandstone Series in their range through Nottinghamshire, Derbyshire, Yorkshire_ and Durham, to the southern extremity of Northumberland. *Trans. Geol. Soc.* (2), **3**, 37–124.

SHERLOCK, R. L. 1926. A correlation of the British Permo-Triassic Rocks. Part I. *Proc. Geol. Assoc.*, **37**, 1–72.

—— and HOLLINGWORTH, S. E. 1938. Gypsum and anhydrite & celestine and strontianite. 3rd edit. *Mem. Geol. Surv. Min. Resources*, **3**.

SHOTTON, F. W. 1956. Some aspects of the New Red desert in Britain. *L'pool Manch. Geol. J.*, **1**, 450–65.

SMITH, D. B. 1957. *Proc. Geol. Soc.*, No. 1544, 13.

—— 1958. Observations on the Magnesian Limestone reefs of north-eastern Durham. *Bull. Geol. Surv. Gt. Britain*, No. 15, 71–84.

SORBY, H. C. 1856. On the Magnesian Limestone having been formed by the alterations of an ordinary calcareous deposit. *Rept. Brit. Assoc.* (2), 77.

STEWART, F. H. 1954. Permian evaporites and associated rocks in Texas and New Mexico compared with those of northern England. *Proc. Yorks. Geol. Soc.*, **29**, 185–235.

—— 1963. The Permian Lower Evaporites of Forden in Yorkshire. *Proc. Yorks. Geol. Soc.*, **34**, 1–44.

STONELEY, HILDA M. M. 1958. The Upper Permian flora of England. *Bull. Brit. Mus. (Nat. Hist.)* Geology, **3**, 295–337.

THOMAS, H. H., HALLIMOND, A. F., and RADLEY, E. G. 1920. Refractory Materials: Ganister and silica-rock—sand for open-hearth steel furnaces—dolomite. Petrography and chemistry. *Mem. Geol. Surv. Min. Resources*, **16**.

TRECHMANN, C. T. 1913. On a mass of anhydrite in the Magnesian Limestone at Hartlepool, and on the Permian of south-eastern Durham. *Quart. J. Geol. Soc.*, **69**, 184–218.

—— 1914. On the lithology and composition of Durham Magnesian Limestones. *Quart. J. Geol. Soc.*, **70**, 232–65.

—— 1921. Some remarkably preserved brachiopods from the Lower Magnesian Limestone of Durham. *Geol. Mag.*, **58**, 538–43.

—— 1925. The Permian Formation in Durham. *Proc. Geol. Assoc.*, **36**, 135–45.

—— 1931. The Permian, *in* Contributions to the Geology of Northumberland and Durham. *Proc. Geol. Assoc.*, **42**, 246–52.

—— 1932. The Permian shell-limestone reef beneath Hartlepool. *Geol. Mag.*, **69**, 166–75.

—— 1942. Borings in the Permian and Coal Measures around Hartlepool. *Proc. Yorks. Geol. Soc.*, **24**, 313–27.

—— 1944. On some new Permian fossils from the Magnesian Limestone near Sunderland. *Quart. J. Geol. Soc.*, **100**, 333–54.

—— 1954. Thrusting and other movements in the Durham Permian. *Geol. Mag.* **91**, 193–208.

WESTOLL, T. S. 1941. The Permian fishes *Dorypterus* and *Lekanichthys*. *Proc. Zoo. Soc.*, **111**, 39–58.

—— 1943. Mineralization of Permian Rocks of South Durham. *Geol. Mag.*, **80**, 119–20.

WILSON, E. 1881. The Permian Formation in the north-east of England. *Midland Naturalist*, **4**, 97–101, 121–24, 187–91, 201–208.

WOLANSKA, H. 1959. *Agathammina pusilla* (Geinitz) z Dolnego Cechsztynu Sudetow i Gor Swietokrzyskich. [*A. pusilla* (Geinitz) from the Lower Zechstein in the Sudeten and Holy Cross Mountains, with English summary]. *Acta. palaeont. pol.*, **4**, 27–59.

Wood, F. W. 1950. Recent information concerning the evaporites and the pre-Permian floor of south-east Durham. *Quart. J. Geol. Soc.*, **105**, 327–46.

Woolacott, D. 1912. The stratigraphy and tectonics of the Permian of Durham (northern area). *Proc. Univ. Durham Phil. Soc.*, **4**, 241–331.

—— 1919a. Borings at Cotefield Close and Sheraton, Durham. *Geol. Mag.*, **6**, 163–70.

—— 1919b. The Magnesian Limestone of Durham. *Geol. Mag.*, **6**, 452–65, 485–98.

Chapter IV

INTRUSIVE IGNEOUS ROCKS

General Account

Igneous intrusions in the Durham–West Hartlepool district take the form mainly of dykes and minor associated sills. They consist of quartz-dolerites and tholeiites allied in type to the Whin Sill (Holmes and Harwood 1928). Unweathered rocks are dark grey and fine-grained, but at outcrop they develop a light to rusty brown limonitic skin. In thin section (p. 190), they are seen to consist essentially of plagioclase feldspar, pyroxene and iron ores with accessory orthoclase, quartz and pyrite, and secondary carbonates. Alteration is most pronounced at fine-grained margins, where primary minerals are replaced by carbonates and clay minerals giving rise to 'white trap'.

The principal dykes of the district are the Hett, Brandon and Ludworth dykes (Pl. XIV), and it is with the last-named that sills are mainly associated. Shales and mudstones adjacent to the intrusions are indurated and lose fissility, while some sandstones take up volatile material released from the coals. As noted by Edwards and Tomlinson (1958, p. 54), coal is altered for only a short distance on either side of the Hett Dyke, but more extensively adjacent to the Ludworth Dyke. In the Easington and Horden area, for instance, the latter is flanked by belts, $\frac{3}{4}$ mile wide, of low volatile coals and associated effects can be detected for more than $1\frac{1}{2}$ miles to the north and south of it. To the east and west of this area, the zone of altered coal is much narrower.

Since the intrusions cut Middle Coal Measures, but are overlain by Upper Permian strata, they are clearly late Carboniferous or Lower Permian in age. The relationship of the intrusions to the structural features of the district is considered in Chapter V, and it will suffice here to note that the dykes have an east-north-easterly trend corresponding with that of some of the faults. According to Dunham (1948, p. 75), the dykes were intruded in tension fissures after the formation of the master joints and marginal faults and before the formation of the conjugate vein-fissures.　　　　E.A.F., D.B.S.

Details

Near the eastern margin of the Wolsingham (26) Sheet, the Hett Dyke is 12 ft wide trending N. 74° E. in Harvey Coal workings. Followed eastwards into the Durham district, it is offset 50 yd to the north by a north–south fault, beyond which it can be traced towards Hett by a line of old quarries clearly marked by a partially-filled trench. At Hett the dyke was once worked from six shafts, and Teall (1884, p. 228) recorded its width at this locality to be 10 ft with a steep inclination towards the north. The position of the dyke in the Brockwell Seam, however, 700 to 750 ft below, is virtually vertically beneath the surface position. The dyke was also quarried [2875 3705] on the south side of the railway, but it is now exposed only in the bed of Tursdale Beck [2885 3707], 500 yd W. of Tursdale House, where it is at least 9 ft wide. For a distance of about 1¾ miles east of the beck, the surface position of the dyke is obscured by drift, but its continuity has been proved in the Busty, Harvey and Durham Low Main workings. The trend underground is about N. 73° E. near Tursdale Beck, and about N. 70° E. near Bowburn.

The dyke is next seen at outcrop 500 yd N.N.W. of Heugh Hall Row. Thereafter, it trends N. 66° E. in workings in the Busty Coal, 400 yd E. of Hill Top Farm, in the Bottom Hutton Coal, 1050 yd S.S.W. of Old Cassop, and in the Five-Quarter and Main coals, 500 yd S. and S.E. of Old Cassop. Middle Coal Measures crop out in Cassop Vale above the position of the dyke in the Bottom Hutton Coal and a small feature on the southern side of the Vale may mark the outcrop of the dyke. Above the underground positions of the dyke in the Busty, Main and Five-Quarter coals, however, the outcrop is formed by Permian rocks which shown no sign of having been intruded, though the line of the dyke can be followed at a locality [3365 3895] 550 yd S. of Old Cassop and elsewhere by subsidence fissures marking the limit of coal extraction beneath. This finally supplies the evidence which Holmes and Harwood (1928, p. 524) were unable to find in support of the generally held belief that the dyke continues in Coal Measures beneath the Permian cover.

In the eastern area, the course of the dyke coincides with the plane of a small fault under Quarrington Hill near Hill Top Farm, and continues for about 1½ miles to the east-north-east on the same line. The dyke is apparently absent east-north-east of a locality [3466 3942] 600 yd N.N.W. of Dene House Farm, but the fault continues in the Bottom Hutton Seam and thence through Thornley Colliery workings.

The Brandon Dyke was formerly thought to die out before reaching the western margin of the district: in the Oakenshaw and Brandon collieries in the Wolsingham (26) district, it is about 6 ft wide, thickening eastwards to 9 ft, according to Holmes and Harwood (1928, p. 525). Approaching the western margin of the Durham–Hartlepools district, it has been proved as the northernmost of two thin dykes cut in the Durham Low Main Coal workings. The southern dyke is 12 ft wide at outcrop in the bed of the River Wear and is thought to continue eastwards into the district to an old quarry [2722 3954] 700 yd W.S.W. of High Houghall: it is probably an offshoot of the Brandon Dyke, though they are nowhere seen to join. The northern, main dyke was formerly exploited from two old quarries [2749 3965], now overgrown, in Saltwell Gill, 450 yd W. of High Houghall.

Farther east in the Hutton Seam workings, the Brandon Dyke has been proved [2799 3974], 150 yd N.E. of High Houghall, where it is 41 ft wide with 45 ft of cindered coal on the northern side and 210 ft of 'bad' coal beyond that. The mining plans suggest that strata on one side of the dyke are displaced with respect to those on the other side. The Brandon Dyke was proved in the Shincliffe Hutton workings between localities [2915 3995], 100 yd N.E. of West Grange, and [3044 4012] 450 yd W. of Whitwell House. Near to Shincliffe Colliery shafts, the dyke is 4 ft wide and has been proved in the Busty and Durham Low Main workings where a small fault on the southern side of the dyke throws down 7 ft to the north. The dyke in this eastern part of its course trends about N. 83½° E.

An unnamed dyke lies 250 to 300 yd S. of the eastern part of the Brandon Dyke betwen a locality [2917 3968], 200 yd S.S.E. of West Grange, and another [3061 3988], 450 yd N.E. of Low Grange. About midway between these localities, it was proved to be 9 ft wide in the Hutton Seam. Its extent to the west is not known, but continuous workings in the Hutton Coal show that it does not extend farther east. In the Durham Low Main and Busty workings, this dyke trends N. 82° E.

The western limit of the Ludworth Dyke is not known, but the presence of sills in certain bores in the area south of High Grange suggests that it may extend at least ½ mile and perhaps as much as 1 mile to the west-south-west of Shincliffe station (see below). The dyke itself has been proved in the Shincliffe Durham Low Main Coal workings [2973 3920], and also in the Bowburn Harvey and Busty workings, being 33 ft wide in the Busty according to Holmes and Harwood (1928, p. 524).

Farther east the dyke forms the southern boundary of the Sherburn Hill Colliery workings and it is traversed by two drifts [3417 4034], 1300 yd W.S.W. of Ox Close. The principal drift, inclined southwards at 1 in 6 from the Main Coal, intersects the Durham Low Main Coal and underlying measures on the north side of the dyke, but again meets the Durham Low Main Seam on the southern side. The presence of a fault is inferred, though its throw and position in relation to the dyke are not clear. The width of the dyke in the drift is 90 ft, with a 6-in stringer 30 yd to the north. This is contrary to an account by Holmes and Harwood (1928, p. 524), who state that the dyke is here in two portions 10 ft apart, and 18 and 27 ft wide respectively. This report was presumably derived from a contemporary colliery horizontal section, which erroneously represented the dyke as a single intrusion bifurcating upwards so as to show the seams above the Busty and Harvey traversed by two dykes, progressively farther apart at higher levels.

The northern side of the dyke has been proved in a heading from the Five-Quarter Seam [3475 4054], 650 yd W.N.W. of Ox Close. In this area the trend is about N. 73° E.   E.A.F.

The Ludworth Dyke was intersected in two parallel drivages in the Harvey Seam workings 900 yd S.S.W. of Ludworth Colliery shafts. It is 34 ft wide in one [3593 4085] and 32 ft wide in the other [3597 4087], and at both localities there is evidence of slight vertical displacement either along or near to the plane of intrusion. Another drivage [4065 4235], 1500 yd S.W. of Easington church, connects the Hutton workings of Shotton and Haswell collieries and proves the dyke to be 63 ft wide trending N. 72° E. along the plane of a fault, having a downthrow to the south of about 30 ft. The zone of cindered coal on the south side of the dyke is 86 ft wide. Farther east a drivage [4368 4356], 430 yd N.E. of Paradise, connecting the Durham Low Main workings between Easington and Horden collieries, proved the width of the dyke to be 49 ft along a trend of N. 73° E. and the throw of the fault to be about 50 ft. Still farther east two parallel drivages [444 446], in the Hutton Seam workings 3300 yd E. of Easington Colliery shafts, proved a width of 57 ft along the plane of a 40-ft fault. The trend is N. 69° E., and a minor off-shoot dyke is recorded 15 ft south of the main intrusion. The easternmost record of the dyke is in two parallel drivages [4756 4480; 4765 4483] in the Durham Low Main Coal, 4200 yd E. of Easington Colliery shafts, where it trends N. 70° E. and is 17 and 25¾ ft wide respectively. The throw of the fault coincident with the dyke is still down to the south, but the amount is unknown. All the provings of the Ludworth Dyke suggest that it is essentially vertical throughout its length.   D.B.S.

The intrusive sills of the district are all recorded from the south side of the Brandon and Ludworth dykes. In the district covered by the Wolsingham (26) Sheet, the 20-ft Brancepeth Sill (Holmes and Harwood 1928, p. 525) lies 52 ft above the Busty Seam and is probably a southern offshoot from the Brandon Dyke. A bore [2754 3875] about 650 yd S. of Butterby proved 17 ft 7 in of 'blue whin' at about 91 ft below the Bottom Hutton Coal. A bore [2888 3877] at High Butterby proved two sills, 1 ft 11 in and 7 ft 1 in thick, separated by 1¼ ft of seatearth-mudstone between 46 and 56 ft below the Bottom Hutton; and another bore [2809 3777], 500 yd W. of High Croxdale, proved a 6¼-ft sill at 75 ft below the same coal. At Cassop Moor, cindered coal in the Busty seam extends for at least 825 yd from the Ludworth Dyke, a typical section at the limit of workings [3195 3937], 900 yd N.E. of Cassop Grange, being: cindered coal

19 in, on band 3 ft 1 in, 'whin' 22 in, cindered coal 24 in. It is inferred that the 'whin' is a sill which continues in the roof of the Bottom Busty Coal over a wide area. In the Cassop Grange Bore [3166 3879], 300 yd E.N.E. of Cassop Grange, and 1000 yd from the Ludworth Dyke, the Durham Low Main Coal is cindered and the roof is baked. In a bore [3048 3923], 460 yd N.W. of Whitwell South Farm, a dolerite sill, 21¾ ft thick, lies 62¼ ft below the Bottom Hutton Coal. In another bore [3205 3998],

575 yd E.S.E. of Whitwell East House and only 50 yd from the dyke, a 1-in sill of 'white trap' and a 15-in sill of dolerite were proved 34 ft and 13½ ft respectively above the floor of the Main Coal ; and other dolerite sills, 5 ft 10 in, 1 ft and 6 ft 1 in thick were met at 52 ft 1 in, 65 ft 3 in and 85 ft 8 in respectively below the Main Coal. A further dolerite was entered 87 ft 2 in below the Main Coal and this continued to the bottom of the bore, 8 ft 5 in below.

E.A.F.

PETROGRAPHY

Earlier petrographical accounts of some of the dykes and associated sills have been given by Bell (1875) and Teall (1884). Teall recognized their similarity to the Whin Sill and distinguished them from the Cleveland Dyke because of their greater basicity, higher specific gravity, greater degree of crystallinity and paucity of phenocrystic feldspar. Holmes and Harwood (1928, pp. 526–8) gave modal and chemical analyses of the Hett, Butterby and Ludworth dykes and concluded they were comagmatic with the Whin Sill.

Fresh specimens [E 32391–3] obtained from the centre of the Hett Dyke, from the centre of the southern of the two dykes proved on the western margin of the district, and from the lower of two sills in Butterby No. 2 Bore (p. 189) are very similar dark grey, evenly medium-grained rocks with sparse phenocrystic feldspars up to 3 mm long. Plagioclase laths in the groundmass average 0·3 by 0·05 mm and show some alignment in two principal directions, although in places their orientation is at random. Their maximum extinction angle in symmetrical section is near 36° and the refractive index (β) is $1·560 \pm 0·003$, indicating a labradorite composition near $Ab_{48}An_{52}$. Slightly coarser plagioclase crystals are zoned with outer zones near labradorite in composition. Accessory orthoclase is very sparse. Pyroxenes are abundant mainly as anhedral granules averaging 0·09 mm across, and scattered singly or in groups amongst the feldspar laths. Pale green augite is predominant, with refractive index $\beta = 1·694$, 2V $(+) = 58 \pm 2°$.

Though mainly anhedral, slightly coarser, twinned subhedral to euhedral crystals also occur. Enstenite is less abundant forming elongated prisms up to 0·4 by 0·1 mm, and commonly rimmed with granular augite. The orthopyroxene is weakly pleochroic with Z=pinkish brown, X= pale green, 2V (−) large ; the negative sign indicating hypersthene. An intersertal pale brown mesostasis consists of a probably felsitic base with disseminated colourless to reddish brown microlites averaging 0·036 by 0·001 mm, specks of opaque iron oxide and globulites. Chlorite and carbonates also occur in the mesostasis. Titaniferous magnetite is prominent through the rock as subhedral to euhedral, or skeletal grains, averaging 0·08 mm, and is associated with finely granular pyrite. Quartz is a minor accessory mineral and part at least seems to be primary although, perhaps, of late formation. Small (0·5 mm diameter), rare spheroidal to irregular structures, possibly subvesicles, are infilled with quartz and carbonates, mainly dolomite (refractive index $\omega = 1·685$).

Labradorite laths commonly penetrate augite granules as noted by Teall (1884, p. 229), indicating an earlier crystallization of the feldspar. This relationship is not always clear, however, and in places penetration is mutual and crystallization of the two components may have overlapped. The overgrowth by augite of hypersthene indicates the normal discontinuous reaction process. Magnetite tends to penetrate along cleavages in augite thus giving rise to skeletal crystals and indicating a later crystallization.

In general, the above description agrees with the earlier work, although Holmes and Harwood (1928, p. 502) noted glomeroporphyritic aggregates of plagioclase, with highly calcic cores approaching An_{90} and with selvedges near An_{50} to An_{30}. A chemical analysis of the Hett Dyke from Tudhoe, a little to the west of the district, given by Holmes and Harwood (1928, pp. 527–8) shows: SiO_2 50·31 ; Al_2O_3 13·50 ; Fe_2O_3 3·86 ; FeO 8·29 ; MgO 4·98 ; CaO 9·63 ; Na_2O 2·39 ; K_2O 0·91 ; $H_2O + 110°C$ 0·99 ; H_2O—0·45 ; TiO_2 2·16 ; P_2O_5 0·27 ; MnO 0·12 ; CO_2 2·24 ; S 0·11 ; BaO 0·04 ; SrO 0·02 ; Cl tr ; Li_2O tr ; total less O for S (0·04), 100·28.

A specimen (E 27642) from the dyke exposed in the main drift in workings south of Sherburn Hill Colliery, was examined by Dr. K. C. Dunham who identified it as an altered dolerite. It consists predominantly of carbonate (mainly dolomite, with refractive index $\omega = 1·680$, with lesser calcite, $\omega = 1·658$) which almost entirely replaces the original rock. Clay minerals replace the feldspars of which only the characteristic lath-like outlines remain. Untwinned secondary feldspar is common. Accessory minerals include skeletal ilmenite and leucoxene in relatively large grains.

Of the Ludworth Dyke, one specimen (E 23744) from a fault zone [444 446] at the Hutton Seam level, Easington Colliery, is a white, highly altered basalt or chilled dolerite (cf. ' white trap '). Plagioclase has been pseudomorphed by kaolinite, and pyroxene by carbonate, the groundmass consisting of brown glass. Veinlets of opaque black glass are riddled with vesicles lined with dolomite. Altered dolerite (E 29426) from a vertical dyke in a roadway in the Durham Low Main Seam, 960 ft below O.D., 1800 ft S. of South Shaft, Easington Colliery, consists, according to Dr. R. Dearnley, of pseudo-morphed feldspar, pyroxene and leucoxenized ilmenite. The pseudomorphs after feldspar laths show ophitic texture, and comprise alkali feldspar, clay minerals, sericite, carbonates and quartz. Veins containing dolomite with a zeolite rimmed by pyrite, cut the dolerite. In an altered dolerite (E 27642) from the main drift in the Low Main Seam workings [2973 3920], S. of Sherburn Hill Colliery, feldspar has been replaced by clay minerals and most of the rock consists of dolomite and calcite. Untwinned secondary feldspar occurs and there is accessory leucoxenized skeletal ilmenite.

In general, the fresh specimens examined show very close similarity to each other and differ significantly from the Whin Sill only in the development of intersertal and intergranular patches of mesostasis which indicate a tholeiitic trend. This may reflect a more rapid cooling history, since the intrusions described here are relatively thin. However, they are clearly consanguineous with the Whin Sill. Modal analyses (in volume per cent) of the three fresh specimens described are compared in Table 5 below with quoted values (Holmes and Harwood 1928, p. 527) for the Hett Group. The analyses, by point count method, are of approximately 500 grain counts and the analytical error is about 1 per cent.

Comparison of the feldspar values shows close agreement and the difference between the pyroxene abundances is probably due to submicroscopic pyroxene inclusions in the mesostasis recorded. Comparison of the present modal analyses, however, with normative values given by the above authors (ibid., p. 527) shows close agreement in total pyroxene abundances. Some of the appreciable (5·56 per cent) normative orthoclase together with some of the quartz (9·77 per cent) may well occur in the cryptocrystalline mesostasis.   R.K.H.

TABLE 5

	Specific gravity	Quartz	Plagioclase (F)	Augite (P)	Hypersthene (P)	Titaniferous magnetite	Carbonate	Chlorite	Orthoclase	Mesostasis	$\dfrac{F}{P+F}$
E 32391 ...	3·06	3	51	20	1	8	4	tr	tr	13	71
E 32392 ...	2·03	3	49	24	1	11	5	tr	tr	8	66
E 32393 ...	2·98	3	48	26	1	9	3	1	tr	10	64
Mean ...	3·02	3	49	23	1	9	4	½	tr	10	67
				Pyroxene							
589 ...	2·96		49·7	30·3							62
631 ...	—		49·8	31·8							61
617 ...	—		48·5	31·1							61
Mean ...	—		49·3	31·1							61

E 32391: Hett Dyke [2884 3707], 750 yd E.N.E. of Hett church
E 32392: Sill associated with the Ludworth Dyke [2888 3877], Butterby No. 2 Bore at 443 ft 5 in
E 32393: Offshoot from the Brandon Dyke [2683 3942], River Wear, 550 yd S.E. of Farewell Hall
    589: Hett Dyke[1]
    631: Ludworth Dyke[1]
    617: Butterby Dyke[1]

[1] Holmes and Harwood 1928, p. 527.

REFERENCES

BELL, I. L. 1875. On some supposed changes basaltic veins have suffered during their passage through, and contact with, stratified rocks, and on the manner in which these rocks have been affected by the heated basalt. *Proc. Roy. Soc.*, **163**, 543–53.

DUNHAM, K. C. 1948. Geology of the Northern Pennine Orefield. Vol. I. Tyne to Stainmore. *Mem. Geol. Surv.*

EDWARDS, A. H. and TOMLINSON, T. S. 1958. A survey of low volatile coals in north-east and south-east Durham. *Trans. Inst. Min. Eng.*, **117**, 49–78.

HOLMES, A., and HARWOOD, H. F. 1928. The age and composition of the Whin Sill and the related dykes of the north of England. *Mineral. Mag.*, **21**, 493–542.

TEALL, J. J. H. 1884. Petrological notes on some north of England dykes. *Quart. J. Geol. Soc.*, **40**, 209–47.

Contours of Harvey Coal, broken where inferred; figures in hundreds of feet, those above O.D. preceded by + and those below O.D. preceded by -

Faults in Harvey Coal

Outcrop of Harvey Coal on sub-Permian surface

Intrusive igneous dykes

○ -1799

○ -1607

○ -934

N o r t h

S e a

Chapter V

STRUCTURE

The structural history of the district falls into three phases, namely pre-Upper Carboniferous, Carboniferous–Permian and post-Permian. The only evidence relating to the first phase is provided by gravity surveys undertaken by Bott and Masson-Smith (1957) and Bott (1961) who postulate an abrupt thickening of Lower Carboniferous and, possibly, of Devonian sediments towards the south-south-east across a penecontemporaneous depositional hinge-belt. This belt bisects the district along an east-north-easterly line extending from Ferryhill through Horden, swinging from there first to the east and then to the south-east as it is traced beyond the coast-line : it is presumed to be continuous with the southern margin of the Alston Block as defined by Trotter and Hollingworth (1928) and Dunham (1948) in the ground farther west. This control over sedimentation may have continued into Namurian times, but the Coal Measures show no signs of its influence.

The second phase is generally accepted to be either late Carboniferous or early Permian in age. It produced most of the faulting and folding in the Carboniferous rocks, and was accompanied by the emplacement of minor intrusions of quartz-dolerite and tholeiite (Chapter IV). One of the principal fractures—the Butterknowle Fault—was initiated at this time along a line which coincides approximately with the earlier hinge-belt. These earth movements were followed by a period of erosion during which the whole of the local Upper Coal Measures and varying amounts of the Middle Coal Measures and earlier rocks were removed. The Permo-Triassic sediments were then deposited unconformably on the eroded surface of the remaining Carboniferous rocks.

The subsequent flexuring, faulting and eastward tilting of these later sediments is attributed to the third phase of earth movements, which is probably Tertiary in age. During this episode, there was renewed movement along some of the earlier fractures as evidenced by the greater displacement in the Coal Measures than in the Permian along many of the faults : the Butterknowle Fault is an example.

The Butterknowle Fault separates two areas of contrasted structural aspect. To the north folding is gentle, but to the south folds with closures of several hundreds of feet are aligned with their axes parallel to the Butterknowle Fault (Pl. XIV).

STRUCTURE OF THE CARBONIFEROUS ROCKS

North of the Butterknowle Fault. The strike throughout most of the northern area is between 25° and 35° west of north, but along its southern margin it swings almost due east–west, parallel to the Butterknowle Fault. In the western part of the area are several broad folds with north-north-westerly axes and with closures of up to 200 ft though they are commonly less than 100 ft ; there are also a number of minor folds with axes lying either east–west or east–north–east. In the eastern part of the area, where the easterly dip increases to between 100 and 300 ft per mile, the amplitude of the minor folding is generally less than 40 ft, and these structures do not show on Pl. XIV.

Three main trends of faulting are recognized: north-easterly, north-north-westerly, and easterly.

The north-easterly faults trend between 45° and 55° east of north and are almost straight. Throws appear to remain constant at between 5 and 30 ft along most of their length and few exceed 50 ft. The north-north-westerly faults trend between 25° and 40° west of north. Individual faults are generally of shorter linear extent than those of the north-easterly group, though the few which are of comparable length have larger displacements. Thus, the Sherburn Fault has a maximum throw of over 200 ft, and the Durham Main Fault and others in the same area over 100 ft. Throws of more than 100 ft are also known in the faults bounding the Castle Eden Disturbance, the Wingate Trough, and along an undersea fault 3 miles east of Hawthorn Hive. The easterly faults, which trend between 80° and 100° east of north, include many minor faults arranged in line or *en échelon*. In the eastern part of the area, they also include larger structures such as the Easington, Horden and Blackhall faults, which have local displacements of over 100 ft.

Pairs of faults form troughs in each of the three groups, but those in the north-easterly group are small and shallow. The largest and deepest troughs are in the north-north-westerly group: they include the Wingate Trough, which is symmetrical and is 180 ft deep in the Harvey Seam, and the monoclinal Castle Eden Disturbance which has a western limb up to 200 ft higher than the eastern. Troughs are most numerous in the easterly group, but most of them are shallow.

A few small faults do not fit into any of the three major groups. The most notable are those associated with the dolerite dykes, which trend between 66° and 80° east of north. Apart from trend, these faults are very similar to those in the north-easterly group.

Butterknowle Fault and area to the south. The Butterknowle Fault can be traced in mine workings for eight miles between the western margin of the district and Hurworth Reservoir. On the southern, downthrow, side the measures dip towards the plane of the fault generally at angles up to 45°. Clarke (1962, p. 213) cited maximum dips of about 1 in 1·6 in parts of the Dean and Chapter Colliery take and 1 in 2·1 in parts of the Thrislington Colliery take, but he notes in addition that the dips can be irregular in magnitude close to the fault plane and that dips of 80° to the north occur adjacent to the fault near Thrislington Shafts. However, he observed that there is generally a slight diminution of dip within a few yards of the fault plane. In an examination of the workings from Dean and Chapter Colliery during the resurvey, no sign of compressional fracture was seen and all the minor faults associated with, and parallel to, the main structure are normal and highly inclined.

Where intersected in a cross-measures drift [2822 3373], about 50 ft W. of the main road east of Skibbereen, the Butterknowle Fault hades at 45° and has a throw of about 700 ft. Farther east, about ¾ mile N.E. of Trimdon, the throw of the fault in the Coal Measures diminishes to about 300 ft, and the steeply dipping measures on the southern side of the fault form the northern limb of the Trimdon Anticline. The axis of this anticline in the Harvey Seam lies between 1000 and 1200 yd south of the fault and the axial plane seems to dip south, parallel to the plane of the fault. This parallelism suggests that the fault and anticline are linked in origin. The anticline gives way southwards to a syncline, on the southern limb of which the Millstone Grit Series abuts against Permian rocks at the margin of the coalfield. The trend of the synclinal axis is roughly parallel to that of the anticline, but this parallelism does not extend to the axial plane which, in the syncline, has no clearly defined inclination. Moreover, in the anticline it is the northern limb which dips most steeply and in the syncline it is the southern limb: thus the two folds are of opposite asymmetry. Culminations and depressions lie along the axes of both folds (Pl. XIV).

The most numerous faults in the worked coalfield south of the Butterknowle Fault are those trending approximately north-north-west; they have throws of up to about 80 ft. Some of these faults lie along and adjacent to the Ferryhill Gap and two of them form a trough 60 ft deep, from which small south-easterly faults diverge. In the Garmondsway area the northern limb of the Trimdon Anticline is broken by short faults of north-north-westerly trend. The northern limb is also traversed by two easterly

faults: one—the Garmondsway Fault—has a southerly downthrow of 130 ft; the other fault, which has a northerly throw of 100 ft, is near the anticlinal crest. The ground to the south is traversed by two groups of generally minor faults: one group trends north-north-westerly; the trend of the other group varies from west-south-west to east-south-east and is parallel or sub-parallel to the Butterknowle Fault. One of the latter group, the Butterwick Fault, has a throw to the south of over 100 ft: it is thought to continue eastwards, first diminishing in throw around Ten-o'-Clock Barn and then becoming identified as the West Hartlepool Fault, which in the Embleton area is inferred from bores to have a throw of at least 500 ft in the Carboniferous. In this area the fault appears to be coincident with the axis of the main syncline.

STRUCTURE OF THE PERMIAN ROCKS

North of the Butterknowle Fault. Contours drawn on the base of the Permian rocks are shown on Pl. XV. Since the Permian rocks thicken downdip to the east, the dip of the higher beds is slightly less than the lower, but the average is about 110 to 150 ft per mile. North of the Easington Fault, the average dip of the Coal Measures exceeds that of the unconformity so that successively higher beds crop against the base of the Permian. South of the fault, the unconformity dips slightly more steeply than the Coal Measures and this, together with the difference in strike between Carboniferous and Permian strata, results in a closure of the coalfield towards the east.

Almost all faults of more than about 10 ft displacement in the Coal Measures are reflected in the overlying Permian rocks cropping out along the coast; either as clean, relatively simple breaks having a hade of about 20°, or as nearly vertical shatter belts up to 20 or 30 ft wide. In both cases the displacement in the Permian strata generally appears to be considerably less than in the underlying coal workings. Thus, the Easington Fault at the coast throws about 190 ft in the

Coal Measures, but only 30 to 50 ft in the Permian; similarly, the southern branch of the Blackhall Fault throws about 30 ft in the Coal Measures as compared with 3 ft in the Permian, though the relative displacements of its northern branch are 40 ft and 35 ft.

One of the few sizeable faults proved in the Coal Measures, but having no apparent extension into the Permian, is associated with the Ludworth Dyke. It has a throw of 40 to 50 ft in colliery workings, but no displacement can be traced in the incompletely exposed overlying Permian beds.

Along the sides of the escarpment spurs, the dip at the base of the Permian is low, but it increases near the heads of several of the re-entrants, and this is particularly apparent at the eastern ends of the Pittington Hill, Sherburn Hill and Quarrington Hill spurs.

In addition to the major structures shown in Pl. XV, most exposures of Permian beds contain low-amplitude rolls and flexures of non-tectonic origin. In many places, these appear to have been caused by local variations in sedimentation such as those caused by deposition

97544

O

over high points on the underlying Carboniferous or over dunes in the Basal Sands. Elsewhere, however, particularly east of the Middle Magnesian Limestone reef, such structures are thought to be due mainly to differential compaction and to differential subsidence caused by the solution of sulphates interbedded with carbonates. This process is also thought to account for a complex brecciated belt containing many minor strike faults which runs parallel with the reef-front and which is seen in many coastal exposures.

Butterknowle Fault and area to the south. The throw of the Butterknowle Fault appears to be much less in Permian strata than in Carboniferous: comparative figures near Skibbereen, for instance, are 350 ft (by extra polation) and 700 ft respectively.

Near High Hill House [2727 3292], 550 yd N.W. of the windmill, a fault with less than 25 ft throw in the Permian can be seen. The fault plane dips to the north, and on its northern side, the Permian is dragged down against Coal Measures, and dips at 50° to the north. However, within a yard or two the dip is reduced to between 8° and 20°, though in the adjacent quarries fissure-bounded blocks of strata dip in various directions. Although a throw of 35 ft is recorded in the Brockwell Seam workings underground, the same fault only throws a few feet in the Harvey Seam below this locality.

At Ferryhill, the anticlinal axis in the Permian is roughly parallel to the axis in the Coal Measures, but lies generally about 150 to 200 yd to the north of that in the Brockwell Seam. The folding in the Permian is commonly disturbed by the anomalous attitude of fissured blocks as seen, for example, in a quarry [2857 3278] at the northern end of the Ferryhill road cutting. A short distance to the south, where the axis crosses the cutting, three minor anticlines can be seen in a horizontal distance of 200 yd. East of the Ferryhill Gap, the structure of the anticlinal area is proved by bores to be more complicated. Contours on the base of the Permian suggest that instead of a single anticline, as in the Coal Measures, the Permian rocks contain several minor folds, the axes of which are commonly short (Pl. XV). However, it is not clear to what extent the present detailed form of the unconformity is affected by pre-Upper Permian erosion (see below).

The broad linear correspondence between the anticlines in the Carboniferous and the Permian rocks is not apparent in the synclinal structures. Basal Permian contours show that a 'low' opens to the south-east ½ mile N.W. of Rushyford, a similar 'low' falls southwards from Mainsforth, and a third falls southeastwards ½ mile S. of Fishburn Colliery. These 'lows' appear, at first sight, to represent the positions of valleys on the pre-Upper Permian peneplain, but this cannot easily be reconciled with other factors. For instance, the two western 'lows' lie over the Coal Measures synclinal axis, and the Fishburn 'low' falls southeastwards in close parallel to the same axis. Moreover, Basal Permian breccias resting in and on the margins of the hollows show no apparent preferential distribution along the bottoms of the 'lows', and commonly contain a significant proportion of Carboniferous Limestone debris which must have been derived from the south, i.e. against the present slope of the unconformity. These factors suggest that the depressions are at least partly tectonic in origin.

Beyond Hurworth Reservoir, the eastern extension of the Butterknowle Fault and the structures to the south of it have been inferred from the evidence of boreholes and a gravity survey. In some of the bores which do not reach the base of the Permian, its position can be estimated from the stratigraphy. The most doubtful area is a belt between Sheraton Grange [434 337] and West Pasture [467 289] where bores are widely spaced and the gravity information is open to several interpretations. At the western end of this belt, the West Hartlepool Fault at Embleton is considered to have a throw of about 500 ft in the Coal Measures and about 300 ft in the Permian. Between Dalton Piercy and West Hartlepool the throw can be estimated only in the Permian, where it seems to be about 600 ft. But there is virtually no indication of this fault on the gravity traverses, and its position is inferred mainly from the bore data. The Seaton Carew Fault is located mainly by the gravity survey and by two bores south of Low Stotfold which indicate displacement of at least 80 ft and by other bores which suggest that it is much greater–probably of the same order as that of the West Hartlepool Fault–and the Permo-Triassic outcrops are calculated on this basis. A fault linking these main faults is based almost entirely on gravity information. There is no information on the structure immediately south of the Seaton Carew Fault, but in the neighbouring parts of the Stockton (33) District to the south, the Permo-Triassic beds dip steadily eastwards and Keuper Marl is present in the coastal area 2 to 3 miles south of Seaton Sands. It is assumed that the strike swings westwards near the Seaton Carew Fault and the outcrop of the Keuper Marl has been calculated on this assumption.

AGE OF THE STRUCTURES

Most of the faults were initiated in late Carboniferous or early Permian times and the greater part of their displacement probably dates from this period. The folds of the western part of the district are probably also of this age, as may be the Mainsforth Syncline which has a pronounced north-north-westerly trend even after contours are corrected to the sub-Permian surface. However, many of the folds have an east-west trend when the inclination of the sub-Permian surface is subtracted.

The dykes are generally believed to have been intruded during a phase of this orogeny when compression was relaxed. It is here suggested that this phase may have been followed by one of compression in a different direction and that this resulted in the folding in the southern part of the district and in the formation of the Butterknowle Fault along earlier lines of weakness.

At a later period, probably in Tertiary times, the Permian was tilted and there was renewed movement along the Butterknowle Fault with accommodation structures in the adjacent folds. At least one of the cross-faults in the Ferryhill Gap throws both Coal Measures and Permian rocks an equal amount of 60 ft and it thus seems possible that some of the faulting in the district was initiated by these earth-movements.

E.A.F., D.B.S.

REFERENCES

BOTT, M. H. P. 1961. A gravity survey off the coast of north-east England. *Proc. Yorks. Geol. Soc., 33,* 1–20.
—— and MASSON-SMITH, D. 1957. The geological interpretation of a gravity survey of the Alston Block and the Durham Coalfield. *Quart. J. Geol. Soc.,* **113,** 93–117.
CLARKE, A. M. 1962. Some structural, hydrological and safety aspects of recent developments in south-east Durham. *Mining Engineer,* **27,** 209–31.
DUNHAM, K. C. 1948. Geology of the northern Pennine orefield. Vol. I. Tyne to Stainmore. *Mem. Geol. Surv.*
TROTTER, F. M., and HOLLINGWORTH, S. E. 1928. The Alston Block. *Geol. Mag.,* **65,** 433–48.

Chapter VI

PLEISTOCENE AND RECENT

INTRODUCTION

Most of the district is covered by drift which is generally thin in the central part of the area, but which exceptionally reaches a thickness of about 300 ft along buried valleys. The oldest deposits, which are preserved in fissures along the coast, consist of breccias and clays, partly pre-glacial and partly early glacial in age. They are followed by more extensive deposits formed during the Pleistocene glacial periods. These consist mainly of boulder clay, laminated clays, sands and gravels. Certain other stony clays may also have been deposited during these cold periods, but their mode of origin is uncertain. Laminated clays, terrace of warp deposits and associated raised storm beach and river terraces found in the south-eastern part of the district are probably late-glacial or early post-glacial. Widespread alluvium and some other river terrace deposits are thought to be mainly Recent in age.

Some of the earliest studies of the glacial geology and buried valleys in Durham were by Howse (1864) and Wood and Boyd (1864). Bore and shaft records published during the latter part of the nineteenth century form the basis of work by Woolacott (1905, 1906, 1909) on the buried valleys. General accounts of the glacial deposits have been given by Woolacott (1921), Raistrick (1931, 1934) and Hickling and Robertson (1949), while descriptions of the drift of the coastal area of east Durham have appeared in a series of papers published by Trechmann between 1912 and 1952. The district forms part of a much wider area dealt with by Carruthers (1939, 1953) in his accounts and interpretations of glacial drifts.   E.A.F., D.B.S.

BURIED VALLEYS

The buried valleys of the district (Fig. 25) form part of three major systems ; the River Wear system in the west, an eastern coastal system, and a River Tees system in the south. The watershed between the first two lies a mile or two east of the Permian escarpment, though some of the tributaries of the Wear system do not breach the escarpment.

The buried valley of the River Wear begins near Witton Park in the Wolsingham (26) district, 2 to 3 miles upstream from Bishop Auckland. Between Bishop Auckland and Finchale Priory, 3 miles N.N.E. of Durham, its average gradient is approximately equal to that of the modern valley and with certain exceptions its course is also similar. The longitudinal profiles of the buried valley and its tributaries show certain peculiarities, however. For instance, those of the River Deerness and Old Durham Beck seem to be hanging tributaries to the main Wear buried valley, which they join near Houghall and near Old Durham respectively. There is also a sudden fall in

198

gradient from -80 to -140 ft O.D. at the bottom of the main valley between Finchale Priory and Chester le Street, in the Sunderland (21) district. Similar irregularities in the Tyne Valley west of Newcastle upon Tyne have been explained (Hickling and Robertson 1949, p. 28) as due to 'glacial super-deepening' by the erosion of the valley bottom by the Tyne Glacier because the ice-flow was impeded by high ground at Gateshead and Low Fell. However, it is difficult to accept this explanation in Central Durham since there is no apparent reason for such deepening in this area. It is suggested instead that a more likely agent of erosion was sub-glacial melt-water and this accords with the presence of much gravel and sand believed to be sub-glacial in origin (p. 244) at or near the changes in profile ; some small buried valleys seem to be glacial drainage channels which have been wholly or partially filled (p. 237). The Ferryhill Gap also follows a buried valley for at least part of its course.   E.A.F.

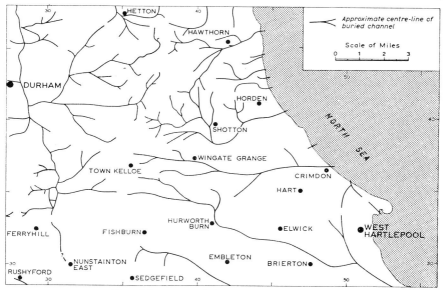

FIG. 25. *Map of the major buried valleys*

The buried valleys in the eastern area are less well proved by borings and shaft sinkings than those to the west, but their courses are commonly indicated by the distribution of solid outcrops and their intersection with the coastline is well exposed in cliffs. They rise close to the present watershed at about 400 to 500 ft O.D. and fall eastward at 50 to 100 ft per mile. Several of the smaller valleys seen in coastal sections lie well above present sea-level and are completely filled, but some of the deeper buried valleys, such as those at Hawthorn and Castle Eden, are only partly drift-filled and late-glacial or post-glacial valleys have been excavated along their courses. The bottoms of the deeper buried valleys at Hawthorn, Easington Colliery, Castle Eden, Limekiln Gill (Blackhall Colliery) and Crimdon cross the coast considerably below present beach level and probably continue for some distance out to sea. In cliff sections their sides are seen to slope at angles ranging from $8°$ to over $70°$, but are generally $10°$ to $20°$–considerably shallower than those of the present day valleys.   D.B.S.

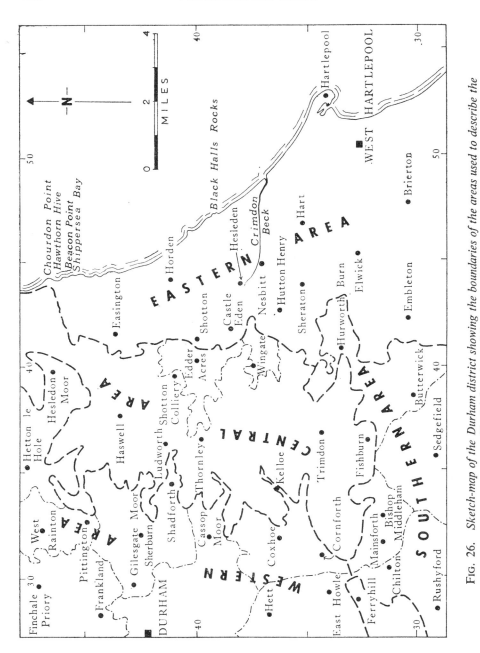

FIG. 26. *Sketch-map of the Durham district showing the boundaries of the areas used to describe the Glacial deposits, and important place names*

An important member of the River Tees system of buried valleys trends east-south-east through Hurworth Burn and Brierton. Apart from this, only the upper parts of a few buried tributary valleys of the Tees system lie within the district. One of these is proved by bores at Rushyford, a second continues southwards from the Ferryhill Gap through the ground near Nunstainton

East, and a third rises near Fishburn and falls to the east or south-east through the ground near Ten-o'-Clock Barn and Middle Swainston. In Ten-o'-Clock Barn No. 1 Bore, the drift is 300 ft thick with rockhead at about O.D. Farther east, a buried valley west of West Hartlepool was thought by Radge (1939, p. 184) to represent a former course of the River Tees. This view is not supported by more recent information which suggests a col within the valley near High Tunstall.

PLEISTOCENE DEPOSITS

For consideration of the Pleistocene deposits the district is divided into four areas, namely western, central, southern and eastern (Fig. 26). The western area corresponds with the River Wear Basin west and north of the Permian escarpment ; the central area corresponds with the Permian plateau and is mainly over 400 ft above O.D. ; the southern area lies in the northern part of the Tees Basin, south of the Ferryhill–Fishburn ridge ; and the eastern area comprises a strip following the coast and extending inland for a distance of 2 to 8 miles.

The character and relationships of the glacial deposits differ from one area to another and this has led to certain difficulties of interpretation and correlation. In the western area, for instance, there is no evidence of more than one glaciation, whereas in the eastern area there appears to be clear indication of three such episodes. In the following account, therefore, factual descriptions of the Pleistocene, area by area, are followed by sections in which conclusions are drawn concerning the sequence of glacial events and correlation and chronology. A further section attempts to reconstruct the main features of deglaciation in the Wear Valley around Durham.

D.B.S., E.A.F.

EASTERN AREA: GENERAL ACCOUNT

The Pleistocene succession in the eastern area is as follows:

	ft
'Prismatic Clay'	0 to 6
Tees Laminated Clays	0 to 30
Morainic Drift and Upper Gravels	0 to 50
Upper Boulder Clay	0 to 40
Middle Sands :	
Upper Division	0 to 20
Lower Division	0 to 80
Lower Boulder Clay	0 to 51
Lower Gravels	0 to 20
'Loess'	0 to 12
Scandinavian Drift	0 to 14
Fissure Deposits	

The inter-relationships of these deposits are shown diagrammatically in Fig. 27.

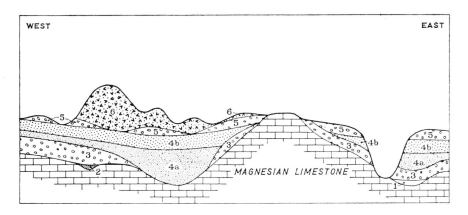

FIG. 27. *Diagrammatic section showing the relationships of drifts exposed in the eastern area. 1. Scandinavian drift. 2. Lower Gravels. 3. Lower Boulder Clay. 4a. Middle Sands (Lower Division). 4b. Middle Sands (Upper Division). 5. Upper Boulder Clay. 6. Morainic Drift and Upper Gravels*

The fissure deposits are preserved in numerous vertical or steeply inclined breccia-filled clefts in the Magnesian Limestone in the coastal area from Crimdon Beck northwards. Almost all extend vertically for the height of the solid rock in the cliffs (up to 80 ft) and Trechmann (1915, 1920) noted that some could be traced across parts of the foreshore now covered with re-sorted colliery waste. Most of the fissures are from 3 to 25 ft wide and are linear, but some are approximately circular in plan. Both varieties are found among beds partially or wholly brecciated by solution-collapse. Linear fissures also lie along fault-planes and are common in the brecciated zone (p. 122) parallel with and immediately east of the reef-front. Walls of the fissures commonly bear near-vertical slickensides and many of the breccia fragments are striated and polished.

The breccias are composed mainly of fragments of Magnesian Limestone, the fragments nearest the sides of the fissures generally being of wall rock, while those towards the centre are derived from higher beds, many of which (including Upper Permian Marls) have since been removed by erosion. Some fissures also contain clays, with wood and other plant material, mammalian bones and freshwater shells of early Pleistocene age (Trechmann, op. cit. ; Lesne 1920 ; Reid 1920). One fissure [4776 3813], 90 yd S. of Limekiln Gill, contains a number of erratic stones (Trechmann 1915, p. 57) and another [4478 4217], 180 yd S. of Warren House Gill, has yielded stones of Scandinavian origin (Trechmann 1920, p. 181).

The oldest glacial deposit of the district—the Scandinavian Drift (Trechmann 1915, p. 53)—occupies the floor of a buried valley at the mouth of Warren House Gill (p. 210). It consists of tough gritty grey clay containing abundant pebbles derived from Norway, but devoid of erratics from Scotland or Northern England. Shell fragments are abundant and include (Trechmann 1915, p. 65) many Arctic forms now extinct in British seas. The deposit has not been recorded anywhere else in the district and although a boulder of larvikite formerly seen in a temporary exposure in Castle Eden Dene

(Trechmann 1915, p. 54 ; Woolacott 1921, p. 64) suggests that pockets of Scandinavian Drift may be preserved beneath younger drift in other buried valleys, it is generally supposed that the Scandinavian ice did not extend much farther west than the present coastline. The shelly Scandinavian clay at Warren House Gill is overlain by a silty deposit, which Trechmann (1920, p. 174) claimed to be ' loess ', redistributed in the upper part by water.

The Lower Gravels occupy isolated depressions in the eroded surface of the Permian rocks in the eastern area. At the coast and in the adjacent deep denes most of these depressions are only a few yards across and the deposits are generally less than 5 ft thick ; they may represent infillings of closed hollows. By contrast a discontinuous sheet of gravel up to 8 ft thick extends above rockhead for several hundred yards in Castle Eden Dene. Moreover, up to 20 ft of interbedded sands, gravels and clays filling buried valleys, may belong to this subdivision. The Lower Gravels are characterized by a high proportion (commonly 50 to 60 per cent) of pebbles derived from the local Magnesian Limestone ; the remaining pebbles comprise a varied suite of Carboniferous limestones and sandstones, red, green and purple lavas, granites, gneisses, schists, flints, quartz, quarzites and dolerites.

The Lower Boulder Clay is a stiff, dark grey or grey-brown gritty clay which weathers brown or buff. As in other areas it contains many pebbles, cobbles and boulders of local Magnesian Limestone, Carboniferous sediments, Ordovician lavas from the Lake District, granites and Silurian greywackes from the Scottish Southern Uplands and blocks of dolerite of Whin Sill type. It also contains some Cheviot porphyries as well as boulders of Shap Granite, which are restricted to the southern part of the area.

Where complete sections through the Lower Boulder Clay are known only from bore and shaft sinkings, the deposit is undivided, but in the coastal cliffs between Chourdon Point and Hart Station four subdivisions are widely recognizable even where the total thickness of Lower Boulder Clay is as little as 7 ft. They are:

					ft
Dark grey-brown boulder clay	...	...	...	...	4 to 12
Red silty clay and clayey or sandy silt ...	...	...	...	0 to 3
Dark grey-brown boulder clay	...	...	...	...	2 to 5
Dark grey boulder clay	...	...	...	...	1 to 8

The junctions are not everywhere sharp and in some places the subdivisions are not easily discerned at close quarters. Seen from a distance, however, slightly different weathering tints and ' stepped ' profiles caused by differential resistance to erosion are locally apparent. The lowest subdivision appears to be the most resistant to erosion and is also the most variable in thickness and the most persistent. Scattered rounded and sub-rounded stones are mainly less than 2 in across, and are more numerous in the lower two subdivisions than in the highest. The two grey-brown clays which cannot be separately distinguished where the median red bed is absent, are somewhat more plastic than the lowest clay, but are otherwise similar in texture and stone content. The red bed is remarkably persistent throughout the cliffs, though seldom more than 2 to 6 in thick: it is generally stoneless but appears locally (as in the cliffs 1500 yd S.E. of Black Halls Rocks) to pass into a sandy or gravelly layer. Small chalky shell fragments are common in all three stony clays but are most numerous in the lowest, from which Trechmann (1952, p. 172) recorded *Cyprina islandica* (Linné), *Tellina balthica* (Linné) and *Saxicava sp.* at an exposure in Castle Eden Dene.

Lower Boulder Clay is absent from a number of places around Hawthorn, Easington Colliery, Peterlee, Horden and Hart, where drift is thin and higher members of the sequence rest directly on rockhead. Throughout the eastern area the Lower Boulder Clay commonly grades by interdigitation to the overlying Middle Sands. Elsewhere, as from 550 to 600 yd N. of Horden Point for example, the top of the boulder clay is sharp and apparently conformable to the overlying sands, and in yet other places (e.g. 600 yd N.N.E. of Horden station) the overlying sands rest on a smooth (?eroded) surface which truncates vague colour-banding in the upper part of the clay. More pronounced erosion is indicated by deep channelling in the top of the Lower Boulder Clay in the floor of buried valleys at Hawthorn Hive [440 460], ' Waterfall ' [445 435], Easington Colliery and Dene Mouth [457 407], where Lower Boulder Clay has been largely or wholly removed and the hollow so formed filled with later gravels.

The Middle Sands consist of two divisions. The lower is composed of fine bright red sand interbedded in many places with subordinate beds of red silt and clay, with beds of red and grey stony clay, and with localized beds of coarser sand and gravel. The deposit is unevenly distributed and appears to lie mainly in buried valleys and wide depressions in the surface of the Lower Boulder Clay. Between such areas it is thin or absent. In the fine red sand bedding is generally even and horizontal, and cross-bedding and ripple-marks are comparatively rare. Stones are mainly restricted to scattered thin lenses of fine gravel and are absent towards the top of the subdivision where the sand contains little or no mica. Interbedded red clays and silts form extensive thin even layers and contain neither stones nor clearly defined sedimentary structures. They are present at all levels in the deposit. In contrast, the beds of stony clay resembling boulder clay occur mainly in the zone of interdigitation at the base of the Middle Sands. Coarse gravel is also concentrated in the lowest part of the deposit where it commonly fills the bottoms of buried valleys and in some places rests directly on rockhead. In the few places where it can be seen the top of the lower division of the Middle Sands makes sharp contact with the overlying deposits and is locally channelled.

The upper division is composed mainly of gravel, although sand with a few stones is common in the lower part ; it also contains laminated clay in the area south of West Hartlepool. It is more extensive than the lower division, being commonly 5 to 20 ft thick and absent only on a few rockhead ' highs ', such as the west–east ridge south of Hart. The gravel is widely calcreted, especially in the uppermost 2 or 3 ft, and in some exposures sand of this division is strongly cemented by calcium carbonate. At Shippersea Bay (p. 208) calcreted gravel containing abundant marine shells appears to have become incorporated by re-working into this division. This is the Easington Raised Beach of Woolacott (1920, 1922) and Trechmann (1931).

Where the lower division of the Middle Sands is absent (pp. 207–9) the sands and gravels of the upper division rest with a sharp contact on the eroded surface of Lower Boulder Clay, but no re-working of the latter is evident. At Warren House Gill, where the lower division is relatively thick

(p. 211), the lowest member of the upper division is a thin bed of stony clay closely resembling boulder clay, and at some exposures such stony clay is interbedded with sand and gravel in this division. In other places (p. 213) such an interbedded sequence appears to represent the whole of the Middle Sands and it is difficult to separate the two divisions. In places the top of the Middle Sands is sharp and apparently not eroded, but elsewhere it grades by alteration into the overlying Upper Boulder Clay. Where a junction is sharp (p. 213), bedding in the uppermost gravel appears to be undisturbed, although numerous pebbles are shattered and fragments slightly displaced relative to each other. Where the upper junction is gradational (p. 213) the interbedded sequence is generally 2 to 3 ft thick, though exceptionally as much as 10 ft. In such sequences the top of the Middle Sands is taken at the base of the lowest thick bed of boulder clay. Stones in the upper division of the Middle Sands comprise a suite similar to that of lower parts of the Upper Boulder Clay, and contain a proportion of purple lavas probably derived from the Cheviot Hills and of fragments of Magnesian Limestone. Small chalky fragments of indeterminate shells are present in most exposures and the deposit has also yielded mollusca and vertebrate bones (p. 216).

Along the coast the Upper Boulder Clay reaches a maximum thickness of 40 ft in well-exposed cliff sections. It is 20 to 30 ft thick over large areas inland but is absent, probably due to late- or post-Glacial marine erosion, in a belt passing north-north-eastwards through West Hartlepool. Where unweathered, it is a dark brown to purplish brown stiff gritty clay texturally similar to the Lower Boulder Clay, but where it is weathered it is reddish brown and is thus readily distinguished from the brown lower deposit. The colour difference is also reflected in the derived soils and this, together with a scarp-like feature which commonly occurs at the margin of the Upper Boulder Clay, serves to delineate the two deposits at outcrop. The upper clay differs from the lower also in having a higher proportion of fragments of Upper Magnesian Limestone and a lower proportion of Lower Magnesian Limestone. Purple porphyritic erratics similar to the Lower Old Red Sandstone lavas of the Cheviot Hills are present in both clays, but Trechmann (1915, pl. viii) and Raistrick (1931, fig. 33) showed that they are restricted to a coastal area which corresponds approximately with the distribution of the Upper Boulder Clay. The base of the Upper Boulder Clay is sharp at some places and gradational by interdigitation at others (p. 213). Where it is gradational the lowest part of the deposit commonly contains streaks, lenses and thin beds of sand and stoneless clay (p. 211), but these are generally absent where the base is sharp. Higher parts of the deposit are less variable and lenses of sand and stoneless or laminated clay are rare. At some exposures (p. 213) the lower part of the deposit is darker brown than the upper part, but this colour difference appears to be local. The characteristic reddish weathering generally extends to only a few feet below the present surface. The top of the Upper Boulder Clay is defined only where it is overlain by Morainic Drift, mainly near the western margin of the eastern area. The few exposures show the clay there to be deeply channelled and locally eroded away completely. It would seem, however, that the deposit is persistent beneath most of the Morainic Drift.

The Morainic Drift is a heterogeneous deposit which occurs chiefly in a triangular tract between Easington, Elwick and Hart. This tract falls naturally into a western belt of north-north-westerly ridges (Fig. 28) comprising the Easington–Elwick Moraine (Francis and others 1963, p. 106) and an irregular eastern area in which the Morainic Drift is hummocky and is interpreted as kame and kettle moraine.

Several patches of gravel lie on Upper Boulder Clay in the coastal area north of Crimdon Beck. They are of uncertain origin, but because of their stratigraphical position they are here grouped tentatively with the Morainic Drift, and designated Upper Gravels.

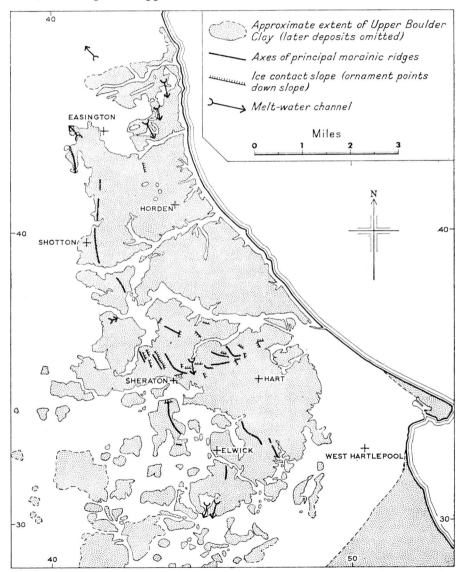

FIG. 28. *Sketch-map showing the distribution of Upper Boulder Clay in the eastern area*

The laminated clays, which lie at the surface in much of the lower Tees valley, extend northwards into the present district in a triangular area around Claxton. They are not exposed and are not shown separately on the one-inch map. Where seen in ground covered by parts of the neighbouring Stockton (33) Sheet, the clays range from chocolate brown to red and grey in colour, and are generally silty. Sand is concentrated into scattered ripple lenses and load or slump pouches along certain bedding planes. Small pale grey calcareous concretions, commonly grotesque, occur widely at the base of the soil developed on these clays.

The so-called 'Prismatic Clay', the highest division of the Pleistocene sequence in the eastern area, is a heterogeneous deposit which generally consists mainly of clay or silt with scattered erratic stones, but in many places has much of the character of, and grades down into, the underlying deposit. It derives its name from its characteristic columnar jointing which closely resembles that produced by desiccation of Recent stream alluvium. The deposit is widespread in the eastern area and most common on the Upper Boulder Clay. Where overlying gravel the 'Prismatic Clay' is very sandy and fills pipes extending down into the gravel.

EASTERN AREA: DETAILS

The eastern area differs from other parts of the district in that the inter-relationships (Fig. 27) of the glacial deposits are well seen in the deep denes and coastal cliffs; sections are commonly continuous for some hundreds of yards and show much lateral variation. A typical section towards the northernmost end of exposures in the coastal cliffs is seen [4428 4646] 70 yd S. of Chourdon Point, as follows:

	ft
Superficial clay and UPPER BOULDER CLAY: reddish brown stiff clay with scattered small stones	c. 4
MIDDLE SANDS (UPPER DIVISION): brown sand and gravel, calcreted at base	c. 15
LOWER BOULDER CLAY: roughly bedded, dark grey and dark grey-brown stiff clay, with scattered stones	c. 10

This sequence can be traced southwestwards to the head of Hawthorn Hive, where a deep buried valley, filled with Middle Sands, cuts through the Lower Boulder Clay and into the underlying dolomite. The section [4410 4508] here is:

	ft
'PRISMATIC CLAY': brown clay with scattered stones and columnar joints	2

	ft
UPPER GRAVELS: gravel, ill-sorted, coarse at top, finer below, partly calcreted; no shells seen	c. 10
UPPER BOULDER CLAY: dark brown stony clay	c. 10
MIDDLE SANDS: gravel, ill-sorted, very coarse (stones up to 2 ft diameter), calcreted at top; base not seen	50

South of the buried valley the Lower Boulder Clay, the Middle Sands and the Upper Boulder Clay are all slightly thicker than at Chourdon Point, and are continuously exposed to Beacon Point and beyond. At a locality [4435 4547], 20 yd N. of Beacon Point, the section consists of:

	ft
UPPER BOULDER CLAY: red-brown stony clay, poorly exposed ...	c. 10
MIDDLE SANDS (UPPER DIVISION): brown sand and gravel, poorly exposed in uppermost 15 ft ...	c. 22
LOWER BOULDER CLAY:	
Dark brown stony clay ...	12
Brown-red silty sand	2 ft to 3
Dark brown stony clay ...	3
Grey stony clay, weathering pale grey-brown	4

Southwards from Beacon Point the Lower Boulder Clay thins progressively against rockhead around the head of Shippersea Bay [4425 4528], where 13 ft of sand and gravel regarded by Woolacott (1920, 1922) and Trechmann (1931) as a raised beach deposit (the 'Easington Raised Beach') rests on a bevelled platform of Magnesian Limestone at about 90 ft A.O.D. The lowest 2 to 3 ft of the deposit consist of poorly cemented coarse sand and fine gravel ; higher parts consist mainly of coarser gravel which is firmly cemented in the uppermost 8 ft. Woolacott (1922, p. 66) stated that the gravel was overlain 'by a few inches of soft sand containing shell fragments', but this section is not now visible. Pebbles in the gravels are well rounded, and many in the lower 6 to 8 ft are bored by *Saxicava* and other marine organisms. Woolacott (1922, p. 65) also recorded such borings in the surface of the underlying limestone, though none was seen during the resurvey. The gravels contain a high proportion of local limestone together with a suite of rocks of Scottish, Cheviot and Lake District origin similar to that in the Lower Boulder Clay. Trechmann (1952, p. 171) also recorded two Scandinavian pebbles.

Shells and shell fragments (listed by Woolacott 1920, p. 65) are abundant in the lower beds of the deposit, but are less common in the higher, calcreted beds and towards the north end of the exposure. Woolacott claims that these shells indicate a climate similar to that of the present day. This accords with the view expressed by Mr. D. F. W. Baden-Powell (in litt.) that the fauna indicates an interglacial rather than an interstadial climate, and that although the *Purpura* shells are very worn, none has the characteristics of ancient forms such as *P. carinata* Brown. Radiocarbon age determination, carried out by Dr. Meyer Rubin (U.S. Geological Survey, Washington D.C.) shows that shells (sample No. W-1426) from the beach are more than 38,000 years old. The C^{14}/C^{12} and O^{16}/O^{18} ratios of carbonate from one of the shells was determined by mass spectrometry by Dr. Irving Friedman in the Denver laboratory of the U.S. Geological Survey, who found them

to be typical of a marine organism. His results show that this shell from the sample dated has not been appreciably contaminated nor its C^{14} content altered by contact with groundwater.

The lateral extent of the beach is in doubt because critical sections are inaccessible. Northwards, the gravel thins to 3 to 4 ft against rising rockhead, but it is not clear whether it passes into or is older or younger than the calcreted gravel of the Upper Division of the Middle Sands, which rests on rockhead for about 80 yd north of the main exposure and can then be traced on to Lower Boulder Clay. Southwards the gravel can be traced for about 60 yd on rockhead before passing on to Lower Boulder Clay. Beyond this locality, however, no shells have been found and the gravel which overlies the boulder clay appears to be continuous with the upper part of the Middle Sands. This gravel is overlain by about 25 ft of red-brown Upper Boulder Clay which appears to be *in situ* and to be continuous with stony clay which overlies the beach gravel at the main exposure. Clay from this has worked down for about 2 ft into the top of the gravel, the bedding of which is undisturbed, although many of the pebbles are broken and their parts slightly displaced relative to each other. The field evidence thus suggests that the raised beach deposit is post-Lower Boulder Clay and pre-Upper Boulder Clay, and that it now forms part of the Upper Division of the Middle Sands.

The Lower Boulder Clay reappears on the south side of Shippersea Bay but farther south, from 830 to 310 yd N. of Shot Rock, the Upper Division of the Middle Sands again oversteps it and rests directly on Magnesian Limestone. Rockhead is highly irregular from 230 to 60 yd N. of Shot Rock, and has a profile like the present marine platform at the base of the cliff. A typical section [4432 4480] 80 yd N. of Shot Rock, is:

ft

'PRISMATIC CLAY' and UPPER
    BOULDER CLAY: red-brown
    gritty stony clay, columnar
    joints at top ...   .   c. 6

ft

MIDDLE SANDS (UPPER DIVISION):
Gravel, medium to coarse,
calcreted ... ... c. 2
Not exposed, presumed non-
calcreted gravel ... ... c. 6
Brown sand and fine gravel,
thinning northwards 3 ft to 15
Gravel, medium-grained at top,
coarse to very coarse at
base, banked against an
overhanging cliff of Mag-
nesian Limestone about 20 ft
high ; inaccessible 2 ft to 20

MAGNESIAN LIMESTONE ... ... —

The gravel at the base of this section is
composed predominantly of local Magnesian Limestone cobbles and boulders
and is thought to be a beach deposit like
that at Shippersea Bay. After re-
appearing 210 yd S. of Shot Rock, Lower
Boulder Clay maintains a thickness of
2 to 10 ft for about 1100 yd before thinning out under a buried gravel-filled valley
at 'Waterfall' (Horden Dene). For most
of this distance the Lower Boulder Clay
is immediately overlain by the Lower
Division of the Middle Sands. A representative cliff section [4439 4431], 40 yd S.
of Loom, is:

ft

'PRISMATIC CLAY' and UPPER
BOULDER CLAY: red-brown
(dark brown where fresh)
gritty stony clay ... ... c. 10

MIDDLE SANDS (UPPER DIVI-
SION):
Gravel ... ... c. 4
Brown clay 4
Gravel 2

MIDDLE SANDS (LOWER DIVI-
SION):
Interbedded fine red sand and
silt ... ... c. 20
Interbedded yellow and red
sand 0½
Interbedded brown and red
stoneless clay 0½
Brown sand 0¼

LOWER BOULDER CLAY: dark
brown gritty stony clay ... 2

MAGNESIAN LIMESTONE ... ... —

One of the more significant inland sections in the northern part of the eastern
area is seen in low cliffs [4264 4584] on
the north side of North Dene, Hawthorn:

ft

'PRISMATIC CLAY': brown stony
clay with close irregular
joints ... ...up to 2
?UPPER GRAVELS: gravel 6 in to 1
?UPPER BOULDER CLAY: light
brown stony clay ; base dips
at about 20° to west 1 ft to 4
MIDDLE SANDS (?UPPER DIVISION):
fine gravel, current-bedded,
with lens of brown clay 3 ft
above base ; contains some
large boulders ; undulating
(?erosional) base ... 2 ft to 6
LOWER BOULDER CLAY: dark
brown gritty stony clay,
weathering grey 7

A little farther south patches of Upper
Gravels overlie Upper Boulder Clay at
Spring Bank [435 428] and Paradise
[434 432]. The Upper Gravels are also
recorded in a borehole [4310 4230], 150
yd S.W. of Horden Hall, which penetrated
soil 1 ft, sandy clay 5 ft, gravel 2 ft, Upper
Boulder Clay 21 ft, Middle Sands 3 ft,
Lower Boulder Clay 21 ft.

A buried valley seen in the cliff at
'Waterfall' is about 75 yd across and
12 to 20 ft deep. It cuts through the
Lower Boulder Clay into the underlying
limestone and is filled with sands and
gravels of the Lower Division of the
Middle Sands, here about 25 ft thick.
Southwards to the mouth of Warren
House Gill the Lower Boulder Clay ranges
in thickness from 2 to 15 ft, the Middle
Sands from 10 to 70 ft and the Upper
Boulder Clay from 5 to 20 ft. The fol-
lowing section was measured in the
coastal cliffs [4450 4341] 150 yd S. of
'Waterfall':

ft

Soil 1

'PRISMATIC CLAY':
Brown stony clay with vertical
columnar joints ... 4 ft to 6
Brown bedded clay, lenticular
bed ... ... up to 1

ft

UPPER BOULDER CLAY:
Brown stony clay, weathering
purple   6
Brown sand   ...   ...   up to 1
Brown sandy stony clay, with
a thin median lens of brown
sand and with streaks of red
clay near base   ...   ...   c. 10

MIDDLE SANDS (UPPER DIVI-
SION): sand, evenly bedded,
with subordinate gravel and
some thin clay beds   ...   c. 12

MIDDLE SANDS (LOWER DIVI-
SION):
Bedded red silt and fine sand...   c.   8
Fine-grained brown sand   ...   3

LOWER BOULDER CLAY: dark grey-
brown gritty stony clay   ...   c. 14

MAGNESIAN LIMESTONE ...   ...   —

North of Horden Point sands of the Lower Division of the Middle Sands are mainly brown, but they become increasingly redder southwards and are almost entirely red from 350 yd S. of the Point. Between 50 and 200 yd S. of the Point a buried valley is cut through the Lower Boulder Clay into the Magnesian Limestone and is filled with cross-bedded gravels of the Lower Division. Trechmann (1915, p. 68) recorded a few small fragments of shells and hornblende- and biotite-schist from these gravels. The gravels pass upwards into evenly bedded red and brown sand and silt, which are overlain successively by flat-lying gravel of the Upper Division and by about 20 ft of Upper Boulder Clay. A section representative of the sequence between the buried valley and Warren House Gill is exposed [4465 4250], 540 yd S. of Horden Point, as follows:

ft

BLOWN SAND: brown sand
with Recent gastro-
pods   6

'PRISMATIC CLAY': brown
clay with small stones
including red sandstones
and ironstone nodules ;
close cuboidal joints   ...   4

ft

?UPPER GRAVEL: fine gravel,
no shells seen   2

UPPER BOULDER CLAY: dark
brown gritty stony clay
with many small pebbles
of local rocks and thin
sand beds   ...   ...   c. 10

?MIDDLE SANDS (UPPER DIVI-
SION): red-brown clay
with a rough sub-vertical
close joint pattern and a
layer of slickensided
leathery clay at base ...   2

MIDDLE SANDS (LOWER DIVI-
SION): evenly bedded
red sands, red silts,
and fine yellow-brown
gravels ; cross-bedding
foreset laminae dip pre-
dominantly to north ;
base sharp and slightly
irregular (?erosional) ...   35

LOWER BOULDER CLAY: very
dark grey-brown gritty
stony clay with scattered
small pebbles ...   ...   18 to 20

LOWER GRAVELS: irregularly
interbedded yellow sand
and brown clay   ...   2½

MAGNESIAN LIMESTONE   ...   —

At Warren House Gill the Lower Boulder Clay is underlain by a buried valley in the lowest part of which Trechmann (1915, p. 53) recorded a deposit which he termed Scandinavian Drift. His section is now largely obscured by colliery waste but about 4 ft of tough plastic gritty grey clay containing small pebbles and numerous fragments of thick-shelled arctic lamellibranchs and gastropods can still be seen [4470 4234] at beach level on the north side of the present valley. Trechmann recorded (op cit., p. 61) that the width of the outcrop was slightly less than a quarter of a mile and he observed a maximum thickness of '12 or 14 ft in the cliff section'. He analysed (op cit., pp. 64-5) 500 pebbles and found 7·9 per cent to be of larvikite and 1·3 per cent of nordmarkite, and that a large proportion of the rocks are characteristic of the Oslo area of Southern Norway. A number of large boulders of larvikite, rhomb-porphyry and nordmarkite are scattered on the foreshore opposite Warren House Gill, but the clay now exposed yields only pebbles ¼ to 2 inches in diameter. Shells in the clay are described by Trechmann as 'thoroughly Arctic' in character, and as constituting a fauna closely comparable with that of the Bridlington Crag of the Yorkshire coast. Trechmann (1920, pp. 173-8) recorded that the boulder clay is overlain by a lenticular brown silty deposit banked against the southern slope of the buried valley. The lower 6 ft of this deposit is unbedded, and was interpreted by Trechmann as loess. It is overlain by a similar but layered deposit containing thin beds of sand and fine gravel and interpreted as loess redistributed by water.

The sequence overlying the Scandinavian Drift is well exposed in a gully [4464 4224] on the south side of Warren House Gill:

	ft
'PRISMATIC CLAY': brown clay with columnar joints	3
UPPER BOULDER CLAY:	
Brown gritty stony clay, weathering bluish purple, with lenses of red silty sand near base ; the lowest 10 in are dark purple	c. 30
Fine brown sand ...	0½
Brown gritty stony clay with pockets of gravel	up to 0½
MIDDLE SANDS (UPPER DIVISION):	
Dark red fine sand, in impersistent lenses ...	up to 0½
Fine to medium gravel ...	3 in to 0¾
Coarse gravel, with thin lenses of red sand and clay	3½
Fine red-brown sand ...	0¾
Medium gravel with pockets of fine brown sand	up to 2
Dark purple to red slightly stony clay, ½ to 1 in, overlying grey-brown stony clay, with an undulate (?erosional) base	up to 0½

97544

MIDDLE SANDS (LOWER DIVISION):	ft
Sand, dark brown to red, very fine, silty, argillaceous and compact in upper 6 to 9 in : no stones	4½
Not exposed (a slip plane here cuts out about 30 ft of bedded red sand seen in nearby sections)	4
Sand, predominantly red	2
LOWER BOULDER CLAY: dark grey-brown gritty stony clay with thin beds of red stoneless clay and lighter brown stony clay in top 6 in	10

Colliery waste obscures much of the drift section between Warren House Gill and Whitesides Gill, but the sequence to the south of the latter is similar to that in the gully (above) and both divisions of the Middle Sands are continuously represented. However, between about 900 and 420 yd. north of Dene Holme the Lower Division thins progressively southwards and for at least 50 yd the gravel of the Upper Division rests unconformably on a few inches of irregularly bedded clays and sands, and these in turn upon an irregular (?eroded) surface of Lower Boulder Clay. A section [4553 4086], 390 yd N. of Dene Holme, is:

	ft
MIDDLE SANDS (UPPER DIVISION): hard calcreted gravel ; uneven base ...	15 to 18
?MIDDLE SANDS (LOWER DIVISION):	0½
Sand, light yellow-brown, silty, with some small stones and thin beds of silt 2 in	
Clay, dark red, silty, no stones 1 in	
Clay, dark brown, silty, with scattered small stones 2 in	
Clay, dark red, silty, no stones 1 in	

P

ft

LOWER BOULDER CLAY:
Clay, light brown, gritty,
with scattered stones;
very uneven slightly gra-
dational base 0¾
Clay, dark brown, gritty,
with scattered stones   10

In the above section it is possible that
the top of the Lower Boulder Clay
should be taken 3 in higher, at the top
of the 2-in bed of dark brown stony
clay which, like the 9-in bed below, has
a leached or weathered appearance
reminiscent of the lower part of a clay
soil profile. The lower part of the Lower
Boulder Clay is exposed 15 yd N. of
the above section, where the lowest few
inches of the deposit are weakly
laminated; the base is smooth and rests
on a varied deposit up to 2 ft thick
which occupies a rockhead hollow and
consists mainly of gravel with wisps and
lenses of orange and red sand and dark
brown stony clay. Quartz, Carboniferous Limestone, coal and dolerite are
amongst the pebbles in the gravel.

Farther south the unconformity rises
again and in the cliffs [4573 4057], 200 yd
E.N.E. of Dene Holme, separates 4 ft of
partly calcreted fine gravel from 8 ft or
more of interbedded fine red sand and
sandy silt containing thin beds of yellow-
brown gravel and one thin bed of grey
clay. The base of the Middle Sands is
not seen here, but is exposed [4590 4039]
in cliffs 380 yd E.S.E. of Dene Holme:

ft

MIDDLE SANDS (LOWER DIVISION)
Alternations of dark brown stiff
semi-plastic stony clay and
brown sand 1½
Dark brown stiff semi-plastic
clay, with scattered stones and
thin beds and lenses of
purple-red incoherent silty
clay, especially near base ... 2¼
Evenly interbedded fine red
sand, sandy silt and clay with
thin beds of dark grey stoneless clay, especially near the
base 1¼
LOWER BOULDER CLAY: dark brown
stiff stony gritty clay, nearly
black in the lower part ...   17
MAGNESIAN LIMESTONE ... ...   –

In the cliffs between this locality and
Black Halls Rocks indifferent exposures
of drift indicate a sequence generally
similar to that farther north. The Lower
Division of the Middle Sands, however,
has not been traced south of a locality
about 400 yd S.E. of Dene Mouth.

A short distance farther south Trechmann (1915, 1920) described a number
of fissures (now obscured by colliery
waste) in the Magnesian Limestone on the
foreshore near Blackhall Colliery. In
describing the contents of two of the
fissures [approx. 463 400] he distinguished
(1920, p. 184) an earlier clay containing
much partly pyritized plant material from
a later clay containing fresh-water shells,
large masses of peaty material and trunks
of oak, elder and pine. In an exhaustive
treatment of plant material from the
earlier clay, Reid (1920, p. 106; in
Trechmann 1920, p. 184) distinguished 114
species of plants, from which she inferred
derivation from a woodland habitat on
limestone some distance from the sea. She
considered that the flora is post-Reuverian,
pre-Tiglian in age, and thus earliest
Pleistocene according to present classification. Reid stated (1920, pp. 108, 111)
that the flora of the later clay indicates
that it is of Cromerian or slightly later
age and that it was deposited in an essentially aquatic habitat in a climate generally similar to that of today. Nine species
of insects were identified by Lesne (1920)
in material from the fissures supplied by
Mrs. Reid. Lesne stated (pp. 487–8) that
three of these species are now extinct in
western Europe, four are closely related
to forms living today, and that two are
common insects which could have been
introduced into the sample during or after
collection. He suggested that the insects
he identified as *Trechus* and *Argutor* may
be the direct ancestors of forms now living in parts of western Europe and that
both these genera 'sont des formes
hygrophiles qui recherchent les stations
fraiches ou froides.'

Farther down the coast, in cliffs
[4732 3858], 670 yd E.S.E. of Blackhall
Rocks Station, the following section can
be seen:

ft

'PRISMATIC CLAY': brown stony
clay with colunmnar joints   1½

ft

UPPER BOULDER CLAY:

Stiff bluish grey (weathering to red-brown) stony clay, with thin beds and streaks of red clay ... ... c. 15

Dark brown stiff stony clay, with prominent vertical joints ; the almost flat base is apparently not erosional c. 8

MIDDLE SANDS:

Gravel, mostly fine, partially calcreted, with a few chalky shell fragments ; upper layers are undisturbed, although a number of stones are broken and the fragments slightly rotated ... ... c. 5

Brown sand with scattered pebbles and with thin lenses of coal granules ... ... c. 17

LOWER BOULDER CLAY: dark grey-brown stony clay, with thin median bed of red silty clay ; top obscured ... ... c. 20

MAGNESIAN LIMESTONE ... ... —

Stones in the ' Prismatic Clay ' include shelly Carboniferous limestones, sandstones and shales with a few stones, including quartz, of non-local origin. The uppermost layer and overlying soil contain Neolithic chipped flints. The lower part of the sequence is repeated in cliffs [4763 3825], 1150 yd E.S.E. of Blackhall Rocks Station, where the multi-layered nature of the Lower Boulder Clay is well seen. The section is inaccessible and thicknesses are estimated:

ft

MIDDLE SANDS: brown sand ; top not seen 5

LOWER BOULDER CLAY:

Red-brown stony clay 5
Dark grey-brown silty clay ... 1
Red-brown stony clay 3
Dark grey-brown stony clay ... 1

MAGNESIAN LIMESTONE ... ... —

About 100 yd to the south-south-east, at the mouth of Lime Kiln Gill [4769 3818], a depression in rockhead, 85 yd wide at beach level, contains over 15 ft of cross-bedded partly calcreted

shelly gravel (Lower Gravels) overlain by Lower Boulder Clay. The gravel thins sharply against the steep southern side of the depression, but thins out gradually against the shallower northern slope up which it extends as a 1- to 3-ft bed to about 30 ft above beach-level. Trechmann (1915, p. 68) noted that pebbles of Lake District rocks are common in the gravel which he later (1920, p. 191) interpreted as being the deposit of a subglacial stream.

On the southern side of the depression, in cliffs [4771 3814], 1300 yd S.E. of Blackhall Rocks Station, the following section is seen:

ft

UPPER BOULDER CLAY:

Dark brown (weathering redbrown) stony clay ... c. 25

Dark brown leathery clay ... 1

Brown sand with thin lenses of dark brown and red clay... 0½

Dark brown clay with scattered small pebbles and thin gravel lenses 5

MIDDLE SANDS:

Brown current-bedded sand with coal granules 0¼

Dark brown stony clay, of uneven thickness ... ... c. 1

Brown clayey gravel 0¼

Dark brown stony clay, of uneven thickness ... ... c. 1

Brown sand with scattered small pebbles 1

Dark brown and red gravelly clay 0¼

Very fine brown gravel ... 0⅔

LOWER BOULDER CLAY: dark red clay with scattered small stones 0½

The lowest member of this section overlies about 20 ft of poorly exposed dark brown stony clay, with red clay beds near the top. These rest, in turn, on the Lower Gravels filling the buried valley at Lime Kiln Gill. The contact between the Lower Boulder Clay and the gravels is not exposed at this locality, but on the northern side of the depression it is clear and sharp and is marked by locally-preserved traces of orange-coloured silt or clay.

A short distance inland, several patches of Upper Gravels have been mapped [462 392, 471 388 and 472 380], 700 yd W.N.W., 350 yd E.S.E. and 1000 yd S.S.E. of Blackhall Rocks Station. Several of these contain small exposures of clean coarse ill-sorted gravel in which no shells have been found.

To the north-west of Blackhall the Lower Division of the Middle Sands has been exposed in many temporary excavations in and around the new town of Peterlee where the deposit is bright red and contains thin beds and lenses of red and brown laminated clay. Red sand of this division is also exposed at many places in the lower reaches of the deeper valleys, particularly Castle Eden and Crimdon denes. A borehole [4323 3952], 730 yd N.W. of Dene Leazes, Castle Eden, proved:

	ft
UPPER BOULDER CLAY:	
Firm red sandy stony clay ...	7½
Red sandy silt	2½
Stiff brown sandy stony clay ...	15½
MIDDLE SANDS:	
Sand and gravel	2½
Red sandy silt	12
Sand and gravel	7
Stiff grey sandy stony clay ...	23½
Sand and gravel	2
Red silt	2½
LOWER BOULDER CLAY:	
Dark brown boulder clay ...	7½
Brown boulder clay and red sand	1½
Brown boulder clay	14
MAGNESIAN LIMESTONE	—

Middle Sands are exposed farther south in the north bank of Heads Hope Dene [4219 3723], 540 yd W. of Eden Vale, Castle Eden, where, beneath 2 ft of soil and weathered red-brown pebbly clay, they consist of the following: irregularly bedded brown clay 6 in, red sandy silt 5 in, on brown laminated clay thinly interbedded with yellow sandy silt 1 ft 2½ in. The lowest member of the sequence overlies dark brown Lower Boulder Clay exposed in a separate section nearby. Middle Sands are also seen on the north bank of Crimdon Beck

[4498 3613], 1060 yd W.N.W. of Thorpe Bulmer, near Hart, where they are represented by 8 ft of clayey and sandy gravel poorly calcreted near the base. Here, four subdivisions can be recognized in the Lower Boulder Clay: dark brown stony clay with thin layers of brown-red sand near the base 4 in, on dark grey-brown stiff stony clay 1 ft, dark grey-brown (locally dark reddish) stiff stony clay with uneven laminae of brown sand 1 ft 8 in, dark reddish brown stony clay with poorly defined bedding visible on weathered surfaces 1 ft. This section is in a slipped block and the beds dip eastwards at 10 to 30°. Middle Sands also crop out farther downstream in Crimdon Beck and its tributary, Thorpe Bulmer Beck, where they occupy a deep buried valley. Farther to the south-east, between Hart Station and Springwell House, 2½ ft of interbedded sand and stony clay resting conformably on thin Lower Boulder Clay were seen in temporary exposures. In one of these [4869 3575], 240 yd N.N.W. of Springwell House, the following section was measured:

	ft
Soil: red-brown sandy loam, few stones	1
Red-brown gritty stony clay; pebbles, small and well-rounded, rare at top; indefinite base	2
?Beach deposit: fine clayey gravel; no shells seen ...	1
MIDDLE SANDS:	1⅔
Brown sandy stony clay 1 ft	
Brown-red clayey sand with coal granules 8 in	
LOWER BOULDER CLAY:	1⅓
Brown gritty stony clay 2½ in	
Coarse red sand with coal granules 1 in	
Brown grittty stony clay 2 in	
Coarse red sand with coal granules 3 in	
Brown gritty stony clay, with many fragments of local Magnesian Limestone 6 in	
MAGNESIAN LIMESTONE	—

Sections in adjacent excavations show that the Lower Boulder Clay is 2 to 6 ft thick.

In Fifty Rigs Plantation Bore [4337 3492], 780 yd W.S.W. of Sheraton, deposits, thought to be those of a buried valley or glacial lake, underlie 18 ft of stony clay as follows:

	ft
Muddy sand	13
Clay and stones	2
Clay and sand beds	2
Clay and stones	19⅝
Bound (i.e. calcreted) sand and gravel	5
Clay and stones	9¾
Bound sand and gravel	3
Clay and stones	12½
Bound sand	1½
Clay and stone	22½
Fine clay	8
Soft sand	4
Clay and stones	45
MAGNESIAN LIMESTONE	—

The borehole lies a short distance to the west of the Easington–Elwick Moraine, the most northerly exposure of which is in a small quarry [4421 3257] near Shotton Hall, where 15 ft of gravel are seen. The topography between Easington and Hutton Henry is morainic, but there are no exposures; it is inferred from the nature of the soils, however, that the deposit consists mainly of gravelly clay north of Shotton and of sand and gravel between Shotton and Hutton Henry. Exposures of morainic drift between Hutton Henry and Sheraton consist of sand and gravel. The most southerly is an old quarry [4470 3538] at the south end of Sedgewick Hill, Sheraton, where 6 ft of brown and red-brown gravel are now exposed and where Trechmann (1915, p. 74) recorded worn fragments of *Cyprina islandica* (Linné), *Turritella sp.* and *Mactra sp.*, and noted 'Cheviot porphyrites' among the erratics.

Between Sedgewick Hill and Sheraton a deep cutting has proved that a prominent hill mapped as morainic drift is in fact formed of a mantle of Upper Boulder Clay resting discordantly on a thick lens of laminated clay and sand. A generalized section is:

	ft
UPPER BOULDER CLAY: red-brown stiff gritty clay with scattered stones ...	up to 25
MIDDLE SANDS: dark brown laminated clay, passing southwards into brown sand	5 to 40
LOWER BOULDER CLAY: dark brown stiff stony clay ...	up to 6

Bedding in the laminated clay at the northern end of the cutting is horizontal, and is truncated by the overlying boulder clay. This also truncates a number of faults in the laminated clay, presumed to have been caused by the pressures exerted by overriding Upper Boulder Clay ice. One of these, a reversed fault, demonstrates that such pressure acted from a north-easterly direction. The lateral passage from clay to sand is accomplished in a distance of about 40 yd. The sand dips to the north at about 5° to 10°.

South of Sheraton the moraine appears to be formed mainly by gravelly clay or clayey gravel, capping the prominent Farden, Whangdon and Beacon hills; it is exposed only in a small quarry [4421 3257] on the east side of Beacon Hill where 20 ft of gravel with a clay matrix are visible. It is possible, however, that some of these hills are of similar origin to that between Sedgewick Hill and Sheraton.

East of the Easington–Elwick Moraine morainic drift gives rise to a hummocky (?constructional) topography best seen in the area between Castle Eden and Hart. It rests on a rolling surface of Upper Boulder Clay and consists of coarse ill-sorted gravel, of which 8 ft are exposed [4473 3531] 1100 yd E. of Sedgewick Hill. The gravel, which has been worked at several localities near Thorpe Bulmer [450 350], gives rise to very stony soils. Several small enclosed hollows, some peat-filled, occur within this hummocky area, and others are also found around Castle Eden and Shotton where the deposit consists mainly of fine red sand. Between Hart and Elwick the morainic drift is less hummocky than north of Hart and appears to be mainly clayey. South of Elwick, however, the main constituent is gravel which forms a number of low mounds [450 310] around Dove Cote. Where exposed, 120 yd E. of Elwick church, it is a heterogeneous deposit consisting mainly of ill-sorted sand and gravel with streaks and lenses of reddish brown stony clay inclined at angles up to 40°. Pebbles here are similar in type to those in the Lower and Upper boulder clays, although farther north Trechmann (1952, p. 172) recorded 'much Cheviot andesite' in a small conical hill south of Hesleden. Small chalky shell fragments are common in less clayey parts of the morainic drift.

Steep primary slopes, here interpreted as ice-contact faces, are found mainly around Hutton Henry, Nesbitt and Sheraton (Fig. 28), where they range up to 40 ft in height. In the area of the Easington–Elwick Moraine they are on the eastern side of the ridges, excellent examples being on the flank of Sedgewick Hill, Sheraton, and on a number of gravel mounds north and south-east of Fleet Shot Farm. In the hummocky area between Sheraton and Nesbitt, where drainage is partly internal, the ice-contact slopes are more variable in direction and are generally less than 20 ft high. Meltwater channels are uncommon here, but about ¼ mile east of Sheraton an excellent example trends north–south for about 300 yds. The walls of this channel and adjacent ice-contact slopes consist of unconsolidated sand and gravel lying at or near the angle of rest; they appear to have undergone little modification by subsequent erosion.

West of the moraine, exposures of Lower Boulder Clay are uncommon, but the following layered sequence at the top of the Lower Boulder Clay is seen beneath 5 ft of alluvial sand and gravel in the east bank of Bedlam Gill [4191 3137], 1700 yd N. of Embleton:

	in
Brown stony clay...	10
Red silty clay	0½
Dark brown stony clay... ...	3
Red silty clay	0½
Brown sand	0½
Dark brown stony clay	3
Red stony clay	2
Dark brown stony clay	15
Red stony clay	3
Dark brown stony clay	8

To the south and east Middle Sands, the Lower Division of which is bright red and about 15 ft thick, overlie Lower Boulder Clay in deep valleys [420 290] immediately east and north of Embleton church. Farther east the sand is mainly brown and it contains thin beds of red and brown laminated or stony clay where exposed in valleys [450 290] near Lower Stotfold and along Dalton Beck. A typical section is found in the west bank of Dalton Beck [4696 3030], 850 yd N.W. of Brierton:

	ft
UPPER BOULDER CLAY (probably reworked); red-brown gritty stony clay, with columnar joints at top	5
MIDDLE SANDS:	
Red-brown sand with thin layers of coal granules ...	2
Purplish red-brown stoneless clay	0½
Red-brown sand with thin layers of coal granules ...	0¾
LOWER BOULDER CLAY:	
Purplish red-brown stony clay	1½
Black stony clay	0½
Red-brown stony clay ...	1
Medium-size gravel with sand matrix and pebbles coated with black films	1⅓
Black stony clay, with thin purple band, brown at top...	1
Purple-brown stony clay with thin bands of brown stony clay 1 ft 8 in and 2 ft 4 in below top, and two thin sand beds near base	13

D.B.S.

About 1000 yd upstream, at Dalton Piercy Crag [4653 3142], the upper subdivision of the Middle Sands is represented by calcreted gravel from which mollusca and vertebrate bones have been obtained. The mollusca include *Helix nemoralis* Linné and *Oxychilus?*; the vertebrates are identified by Mr. C. J. Wood as *Bufo vulgaris* Laurent, *Rana temporaria* Linné, and *Arvicola sp.* The gravel at this outcrop contains a number of deep natural cavities and it is possible that the bones were introduced into these after the deposit had formed. Farther east an old sand pit [479 301], 400 yd E. of Brierton, contains 23 ft of sand from which Trechmann (1939, p. 99) obtained a humerus of *Coelodonta antiquitatis* (Blumenbach). This deposit is believed to belong to the Upper Division of the Middle Sands. About ½ mile to north-east these sands are overlain by Upper Boulder Clay in a pit [484 306], 1000 yd N.E. of Brierton. The section is as follows:

	ft
UPPER GRAVELS or MORAINIC DRIFT:	
Red-brown stony clay, with pockets and lenses of sand near base	15
Coarse gravel, brown to red-brown, with sand and sandy clay matrix ; some pebbles cracked *in situ* ; this deposit passes westward into fine gravel and sand with inter-bedded red plastic stoneless clay ... ...0 to 25	
UPPER BOULDER CLAY: blue-grey stiff stony clay, with markedly irregular (erosional) top and flat base ; thin lenses and stringers of fine sand in basal khaki-brown 12 in ... ... up to 10	
MIDDLE SANDS: fine to coarse brown sand, weakly-cemented at top, with fine gravel below ; evenly bedded throughout ; bedding undisturbed at top ; small-scale cross-bedding and imbricate structure in gravels	15
	G.D.G.

Middle Sands are well exposed in the south-west bank of Claxton Beck [4629 2875], 1980 yd S.W. of Brierton, where beneath 1 ft 8 in of superficial red-brown gravelly clay they consist of the following:

	in
Brown level-bedded sand ...	72
Red clay with closely spaced cuboidal joints	4
Brown clay with closely-spaced cuboidal joints	1
Red clay with closely-spaced cuboidal joints	2
Brown laminated clay	4
Brown sand with ¼-in bed of coal granules	6
Red semi-plastic stoneless clay ...	1½
Brown sand with streaks of coal granules	3
Alternating thin beds of brown sand and red and brown semi-plastic stoneless clay	15
Brown sand	4
Red clay with thin sand partings	3
Red stony clay, base not seen ...	1
Unexposed	24
Brown sand with streaks of coal granules	144

A section on the east bank of the stream, 70 yd E.S.E. of this locality, is in sand which is the equivalent of the lowest member of the above sequence and which overlies dark grey Lower Boulder Clay. Red-brown Upper Boulder Clay is present *in situ* about 20 ft above the top of the measured section.   D.B.S.

The terrace of warp (p. 246) on which most of West Hartlepool has been built is cut mainly in solid rock and Lower Boulder Clay, the Upper Boulder Clay having been removed by marine erosion. It is broken 700 yd S.S.W. of the railway station by a gravel mound [509 321] which is believed to be post-Upper Boulder Clay in age. At Hartlepool Docks [520 340] boulder clay exposed in a temporary excavation contained stones which Trechmann (1915, p. 77, footnote 3) believed to be indicative of a westerly or north-westerly provenance. Since the field relations of the clay suggest that it is older than the Upper Boulder Clay capping the nearby promontory of Hartlepool, it is probably Lower Boulder Clay, though Trechmann's description of it is not fully consistent with this interpretation. Alternatively, though less likely, it may represent an isolated and possibly unique patch of an earlier drift.

Farther south a group of closely spaced boreholes have proved thick sequences of drift. In most of these the Middle Sands are only 2 to 20 ft thick and are represented by laminated clay which grades downwards by alternation to Lower Boulder Clay. Most of the bores lie just beyond the southern margin of the district, but one [5100 2845], 1410 yd S.S.W. of Seaton Carew station, penetrated the following succession :

ft

UPPER BOULDER CLAY:

Soil, 6 in, over brown and grey mottled clay	...	...	...	$7\frac{1}{2}$
Red and grey mottled clay	...			$0\frac{1}{2}$
Red, brown and grey stiff boulder clay with sandy pockets	...	...	...	$25\frac{1}{2}$

MIDDLE SANDS:

Laminated clay	...	...	$0\frac{1}{2}$
Laminated clay on sand	...		$3\frac{1}{2}$

ft

LOWER BOULDER CLAY:

Red laminated clay ...	...	1
Laminated boulder clay	...	2
Stiff red sandy clay ...	...	$3\frac{1}{2}$
Boulder clay ...	...	...	$21\frac{1}{2}$

?BURIED VALLEY DEPOSITS:

Red sand ...	...	...	$2\frac{1}{2}$
Silty sand ...	...	...	4
Brown sandy clay ...	...	$1\frac{1}{2}$
Fine red sand ...	...	...	4
Stiff boulder clay ...	...	$6\frac{1}{2}$
Red sand and gravel ...	...	$3\frac{3}{4}$

TRIASSIC ('BUNTER'): sandstone, weathered at top ... ... —

G.D.G., D.B.S.

CENTRAL AREA: GENERAL ACCOUNT

In the central area the earliest and most extensive drift deposit is Lower Boulder Clay which ranges in thickness up to 120 ft, but is generally much less. It is overlain in places by small isolated patches of sand and gravel, and in other places (especially in the ground bordering the eastern area) by laminated clays which are locally as much as 160 ft thick.

The Lower Boulder Clay is similar to and continuous with the deposit of the same name in the western area, though it includes stones derived locally from the Magnesian Limestone in addition to those named (p. 225) in that area. The sands and gravels are closely associated with meltwater channels. Deposits of sand in the eastern part of the area are extensions of the Middle Sands farther east (p. 204), but the sand and gravels elsewhere in the area cannot be correlated certainly with similar deposits in other areas.

The laminated clays fill or partially fill broad depressions in the surface of the Lower Boulder Clay at Edder Acres and Wingate (Fig. 29). The clays are dark brown and contain numerous films and laminae of fine brown sand which, near Edder Acres, increase in thickness and number as they are traced eastwards. Pebbles are rare. The outcrop of the laminated clays is bounded to the east by the Upper Boulder Clay and Morainic Deposits of the eastern area (p. 205). In most places the nature of the contact is uncertain, but there are some indications that the laminated clays are locally overlain by both Upper Boulder Clay and Morainic Deposits.   D.B.S.

CENTRAL AREA: DETAILS

Haswell to Fishburn. The characteristic drift deposit of the plateau is brown boulder clay, which exceeds 25 ft only in buried valleys and where banked against slopes. The lower parts of the deposit, exposed mainly near the crest of the scarp, are lighter brown and more gritty than higher parts, and a slightly higher proportion of the pebbles are derived locally from the Magnesian Limestone. The deposit gives rise to brown stony clay-loam soils which, in most

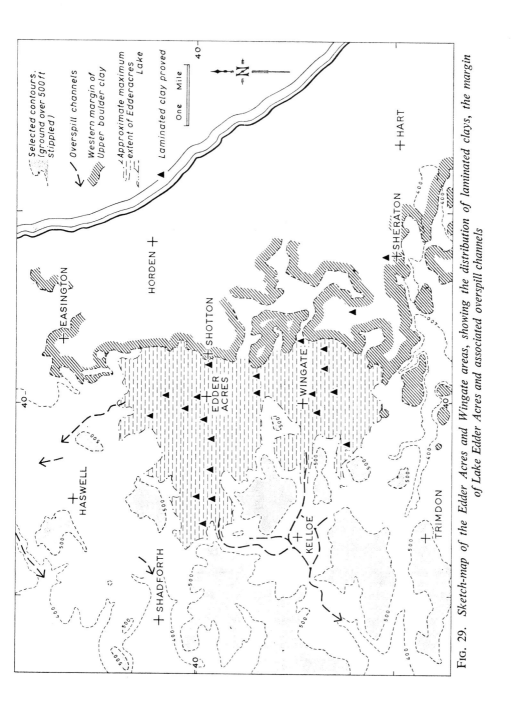

Fig. 29. *Sketch-map of the Edder Acres and Wingate areas, showing the distribution of laminated clays, the margin of Lake Edderacres and associated overspill channels*

places, are readily distinguishable from reddish soils formed on the Upper Boulder Clay of the eastern area (p. 205).

The boulder clay rests on an undulating rock surface and fills a number of buried valleys. Most of these have been located by bores and shafts and an additional V-shaped valley up to 12 ft deep has been traced for about 150 yd in ground cleared for quarrying operations [330 360], 750 yd N. of Coxhoe East House, Coxhoe. The maximum recorded thickness of clay is 120 ft in Ludworth Colliery Shaft, and 69 ft were penetrated in Deaf Hill Shaft. In surface exposures the clay rarely exceeds 15 ft and the following may be taken as representative: Crime Rigg [342 418], 1 to 5 ft; east lip of Tuthill Quarry [390 420], Haswell, 3 to 12 ft; Shotton Quarry [399 418], 4 to 8 ft; Wingate Quarries [370 370], up to 8 ft; old quarry [368 360] at West Moor Lime Works, Trimdon Grange, up to 12 ft; West Cornforth Quarry [319 346], 3 to 10 ft. D.B.S., E.A.F.

Where drift is less than 25 ft thick, it is generally composed solely of boulder clay, but thicker sections, known almost entirely from bores and shafts, commonly contain one or more beds of sand and gravel. In Shotton Colliery Engine Pit [3974 4124], 2½ ft of gravel lie at the base of the drift, and in Hetton Colliery Blossom Pit [3596 4702] 21 ft of sand and gravel rest on rockhead. These examples are exceptional, however, for most beds of sand and gravel are underlain by 10 to 25 ft of boulder clay.

Outcrops of sand and gravel in the western part of the central area are restricted mainly to the ground north of Haswell where they form low rolling mounds flanking a meltwater channel [370 440 to 370 450] at South Hetton. The best section available at the time of writing is in excavations [377 447], 350 yd N.N.W. of Fallowfield, South Hetton, where foreset beds in up to 20 ft of very poorly sorted, irregularly bedded gravel dip mainly westwards. Workings in up to 30 ft of rather more sandy gravel, 250 yd N.W. of Fallowfield, are no longer visible. About 8 ft of poorly sorted gravel are seen in an old quarry [3759 4508] at the south-western end of

Cockhill Plantation, South Hetton, and 1¼ miles farther east, 3 ft of clayey gravel rest on Magnesian Limestone in a small quarry [4035 4511] at the top of Coop Hill. The only other good sections are in disused sand pits [401 421] at Thorpe Moor Farm, ¾ mile N.N.E. of Shotton Colliery; here up to 12 ft of well-sorted fine and medium gravel appear to pass below dark brown boulder clay at the margins of the pits.

South of Haswell, sand and gravel are present mainly in small widely separated patches, generally resting on boulder clay, but also forming narrow outcrops immediately west of the edge of the Upper Boulder Clay of the eastern area. The two largest patches are between 200 and 800 yd N.W. of Kelloe church, and near Kelloe Law [365 375], where gravel occupies several acres of arable land dominated by a prominent gravel hill known as The Banks.

Sand and gravel at the eastern margin of the central area are in most cases an extension of the Middle Sands farther east (p. 204). Principal exposures are in an old pit [3979 3476], 400 yd W.N.W. of White Hurworth, where 11 ft of sand and fine gravel are seen; and in silage pits [4044 3459] at the east side of Stob Hill, 650 yd W.N.W. of Hurworth Burn station, where 2 ft of brown stony clay and soil overlie 1½ ft of brown clayey sand, 1 ft of white (?leached) sand and 2 ft of brown clayey gravel.

Edder Acres. A triangular tract of thick drift occupies ground below the 450-ft contour between Wheatley Hill, Shotton Colliery and Wellfield. The drift comprises Lower Boulder Clay which appears to be continuous with that of the remainder of the central zone, overlain by thick deposits of lacustrine laminated clay, sand and gravel. The greatest known thickness is in a shaft [4073 3855] at Wellfield, where the section comprises 'blue clay' and 'leafy clay' [laminated clay] 98½ ft, loamy sand 54½ ft, on 26 ft of boulder clay, which rests on rockhead. In general the laminated clay is found only below 435 ft O.D., though the sands and gravels locally extend to slightly above this level.

Because of the cover of later deposits, the boulder clay is seen only in deep excavations such as those at Shotton brickworks [398 403], in the bottom of deeply entrenched valleys such as Calf-pasture and Edder Acres denes, and in shafts and boreholes. Records of bore-holes and inferences based on field ex-posures suggest that this clay ranges from about 10 to 30 ft in thickness.

It is overlain in almost the whole of the Edder Acres area by laminated clay, best exposed in workings at Shotton brickworks. At the west end [3975 4016] of the workings, 1200 yd S.S.E. of Shotton Colliery station, the section is:

	ft
Clay, dark brown (weathering to purplish grey in top 6 to 8 ft), with many small fragments of coal and shells and scattered pebbles ; the proportion of pebbles decreases progressively downwards in the lowest 3 ft	24¼
Clay, dark brown, plastic, stone-less, almost homogeneous ...	2½
Clay, dark brown, plastic, lami-nated, in beds 2 to 3 in thick	2
Clay, brown and dark brown, plastic, laminated (15 to 30 laminae per inch), with thin lenses and beds of brown fine sand containing abundant mica and coal granules ; the sand beds commonly contain ripples, and small flat calcareous con-cretions occur in some clay laminae	22
Dark grey-brown boulder clay	—

There is local discordance at the base of the higher stony clay and in places the laminae of the underlying bed are overturned in an east-south-easterly direction. Pebbles in the higher clay are similar to, though less abundant than, those in the boulder clay at the base of the section. Mr. P. Beaumont (in litt.) suggests that the deposit may be a mud flow. The higher clay is apparently not present at the eastern end [3999 4023] of the workings where 42 ft of laminated clays contain much more sand than at the western end and include a 14½-ft bed

in which sand and clay are evenly inter-bedded. Sand laminae in this bed con-tain current-ripples and foreset-bedding, indicative of movement towards the west, i.e. in the direction of decreasing sand content.

Because of undulations in the surface of the Lower Boulder Clay the laminated clays range widely in thickness, reaching a recorded maximum of 98¼ ft in Well-field Shaft (p. 220). They are indifferently exposed in the valley between Swan Castle and Edder Acres and have been proved in boreholes near Edder Acres Farm. The deposit was formerly worked for brick-making 500 to 800 yd N.E. and 700 to 900 yd S.S.E. of the present brick-works. Further west, filled-in brick clay workings [368 400 and 382 398], north of Thornley and Wheatley Hill, occupy a similar stratigraphical position with respect to the boulder clay and lie at about the same topographical level. South of Edder Acres the laminated clays were formerly worked for brick- and tile-making around Wellfield [408 384]. At Shotton, temporary excavations [411 397] along a new road revealed 8 ft of lamin-ated clay which appeared to pass east-wards beneath morainic drift and Upper Boulder Clay, and a similar relationship is inferred from poor exposures in the flanks of Castle Eden Dene and its tribu-taries a short distance south and west of Shotton.

Sand and gravel, most of which lies topographically higher than and appears to overlie the laminated clay, forms low mounds 900 yd N.N.E., 850 yd N.W., 750 yd E. and 700 yd S.E. of Edder Acres Farm.

Wingate. Drift around Wingate occu-pies a buried valley, and is considerably thicker than in most of the central zone. The greatest recorded thickness is 191 ft 8 in in Hutton Henry Marley Pit [4141 3692] ; it is 130 ft in the Per-severance Pit, 330 yd farther south and only 62 ft in Wingate Colliery Lord Pit [3967 3718] towards the northern edge of the tract.

The lowest deposit is the dark grey-brown Lower Boulder Clay. The record of the Lord Pit makes no distinction between boulder clay and the overlying deposit, but 28 ft 8 in of clay—presumed to be boulder clay—is reported overlying rockhead in the Marley Pit. It is now exposed only in a railway cutting [4120 3715], 930 yd N.E. of Wingate station, where it is overlain by 3 ft of sand ; it was formerly seen in a disused brick-pit [403 371], 600 yd N.N.W. of Wingate station (Trechmann 1952, p. 176).

The deposit at the surface over most of the Wingate tract is laminated clay, which was formerly worked extensively, but is now exposed only in temporary sections. It is generally found only below about 435 ft O.D. and may be continuous with the laminated clay around Edder Acres. At brick pits 600 yd N.N.W. of Wingate station Trechmann (loc. cit.) recorded 15 ft of laminated clay lying on 1 ft of sand, and collected numerous fresh-water lamellibranchs from a 2 ft bed towards the base. Laminated clay from this pit is illustrated by Carruthers (1953, fig. 2b ; pl. i). During the resurvey 11½ ft of red-brown sandy laminated clay was noted in a small temporary exposure [4094 3684], 500 yd N.E. of Wingate station. Similar clay was seen in a number of shallow excavations around the ruins of Hutton Henry Colliery, and in the banks of Heads Hope Dene [4219 3722], 480 yd S. of Castle Eden station, where it was 3½ ft

thick. Laminated clay is not separately distinguished in the records of the three colliery shafts but 84 ft of clay immediately below the surface at the Marley Pit is presumed to be laminated from the evidence of the nearby excavations.

In the eastern part of the Wingate tract, sand and gravel up to 25 ft thick form several acres of moundy ground between Wellfield House and the railway line, and are exposed in a number of old workings around Beech House [4084 3719]. Gravel also forms moundy ground 800 to 1200 yd E. of Wellfield House, and occupies a smaller area 500 yd E.N.E. of Wingate Station. Most of the sand and gravel appears to lie above the top of the laminated clay, but there is some overlap of levels and lower parts of the gravel may pass locally into upper parts of the laminated clay. As in the Edder Acres tract, the laminated clays are poorly exposed in the flanks of valleys (Heads Hope Dene and its tributaries) entering the eastern area, but here too, they seem, from their general field relationships, to be at least partly overlain by Upper Boulder Clay and Morainic Drift. The localities at which laminated clays have been proved are shown on Fig. 29, which also shows the presumed maximum extent of the lake in which they were deposited and the position of the channels which drained it.

D.B.S.

SOUTHERN AREA : GENERAL ACCOUNT

A tripartite subdivision of the drift can be recognized in the southern area as follows : Upper Stony Clay 0 to about 30 ft, Middle Sands and Gravels 0 to about 50 ft, on Lower Boulder Clay 0 to about 54 ft. The Lower Boulder Clay closely resembles that of the western and central areas, but contains boulders of Shap Granite as well as other stones of local and distant origin. The sands and gravels form isolated mounds overlying the Lower Boulder Clay and also occur intermittently between the two clay deposits. The Upper Stony Clay is of uncertain origin and is only poorly exposed over a limited area.

E.A.F., G.D.G., D.B.S.

SOUTHERN AREA : DETAILS

Chilton to Rushyford. In the tract southwest of Kay's Hill, drift is thin and consists mainly of Lower Boulder Clay. It is exposed only in the sides of a swallow hole [2952 2912], 225 yd E.S.E. of Standalone, where it appears to be rather more than 14 ft thick. A buried valley trends south-eastwards under the alluvium of the Rushyford Beck at Rushyford, where the surface seems to be composed of very stony boulder clay, and where an old clay pit was worked [2807 2887], 200 yd W.N.W. of the inn. A small isolated patch of sand and gravel caps a low hill about 1150 yd W.N.W. of the inn.

Between Kay's Hill and Nunstainton East the gently undulating ground seems to be entirely composed of Lower Boulder Clay which is 54 ft thick in a bore [3104 2932], 650 yd N.W. of Nunstainton East.

Two patches of sand and gravel rest on the boulder clay and cap low hills about 350 yd S.W. and 450 yd S.S.W. of Nunstainton East. Sand is exposed to a thickness of 25 ft in an old sand pit [3122 2941], 550 yd N.W. of the same farm and crops out round the western margin of the hill to the north-north-west. At the eastern end of the pit brown Upper Stony Clay rests on the westwards-sloping surface of the sand and forms the upper part of the hill. The sand can be traced round the southern flank of the hill where it has been worked in old pits about 100 yd N.W., 75 yd W. and 125 yd S.E. of Nunstainton East. The outcrop of sand can be followed into the western side of a valley about 250 yd E. of the farm, but on the eastern side the sand seems to wedge out and Upper Stony Clay is presumed to rest on Lower Boulder Clay. On the northern flank of the hill, the sand also seems to wedge out. Upper Stony Clay overlies sand in a railway cutting about 680 yd N.E. of Nunstainton East, and also forms the surface of two hills on either side of the railway. North-eastwards from there, over rough ground known locally as Linger and Die, it rests directly on Lower Boulder Clay. Sand and gravel form the surface of the south-eastern part of the hill around Stony Hall and apparently rest on Lower Boulder Clay in a railway cutting 400 yd farther north. Their relationship to the Upper Stony Clay, however, is obscure. A bore [3261 2952], about 60 yd S.S.E. of Stony Hall, penetrated a thick sequence of sand and clay overlying presumed Lower Boulder Clay as follows:

					ft
Soil	...	...	...	...	1½
Sand	...	...	...	...	4
Leafy clay		...	...	...	15
Sand	...	...	...	...	3
Leafy clay ...		...	...	...	1½
Loamy sand		...	...	...	10
Brown clay		...	...	...	7
Sand	...	...	...	...	3
Brown clay		...	...	...	3
Loamy sand		...	...	...	6
Leafy clay ...		...	...	...	3
Brown sand		...	...	...	12
Leafy clay		...	...	...	2
Loamy sand		...	...	...	9
Brown stony clay			...	...	25
Loamy sand		...	...	...	15
Brown stony clay			...	...	18
'Marl' and 'limestone'				...	—

At the northern end of the railway cutting [3172 3002], about 1025 yd N.N.E. of Nunstainton East, the Upper Stony Clay and Lower Boulder Clay are separated by sands and gravels, which are up to about 40 ft thick to the west, where they form a rather hummocky ridge. extending to Great Chilton through Chilton Grange. Between this ridge, which rises to over 400 ft O.D. at Great Chilton, and that extending south from the south-eastern part of Ferryhill there is a glacial drainage channel with its floor at about 340 ft O.D. An isolated patch of sand and gravel forms two low mounds, 650 and 850 yd N.W. of Chilton Hall. Exposures of sand and gravel in this hummocky ground are now rare and the sand pits west of Ferryhill station are obscured. The only good section is at Chilton Quarry where most of the 40-ft face consists of sand with sandy clay bands, gravel being largely confined to the upper layers. These deposits rest on Magnesian Limestone, but farther west, north of West Close, the sand and gravel seem to rest on boulder clay.

A small patch of gravelly sand about 325 yd long lies about 200 yd S. of Ferryhill church. Along its northern margin it rests on Magnesian Limestone, but to the south and downslope it appears to rest on boulder clay.   E.A.F., G.D.G.

Mainsforth to Bishop Middleham. In this tract drift is generally 3 to 15 ft thick, increasing to 50 ft in a number of rockhead hollows and to more than 100 ft in a buried valley along the south-western margin. It consists of patchy Lower Boulder Clay overlain by more widespread sand and gravel. The boulder clay crops out mainly in the lower ground around Bishop Middleham where, in an old pit [3370 3087] 500 yd S.E. of Island Farm, it is 12 to 15 ft thick and contains lenses of sand. Farther north, 26 ft 4 in of dark grey boulder clay are recorded in a bore [3361 3242] 170 yd N.E. of Farnless, and the deposit reaches a maximum known thickness of $37\frac{1}{2}$ ft in a bore [3319 3132] 330 yd N.N.W. of Island Farm.

Around Bishop Middleham the Lower Boulder Clay is patchily overlain by up to 15 ft of sand and gravel. This deposit extends westwards to Ferryhill in an almost unbroken, though poorly-exposed, sheet. Bores in this ground show that the sand and gravel are generally less than 15 ft thick, except along the margin of the buried valley lying to the south-west, and that in most places they rest directly on rockhead. North and west of Mainsforth they give rise to a hummocky topography, but to the south-east the topography is more subdued and summits appear to be related to a dissected level at about 290 ft above Ordnance Datum. G.D.G.

Sedgefield. Boulder clay is locally thin or absent west and north of Sedgefield, and is probably less than 20 ft thick in much of a $\frac{1}{2}$-mile belt flanking the River Skerne, where it is dark brown and apparently continuous with clay which forms most of the lower-lying ground south and east of Sedgefield. Although formerly worked in pits [3400 2950 ; 336 290], 1 mile W. of Sedgefield, this clay is not now exposed, but it gives rise to characteristic brown stony clay soils. It has been proved in a number of boreholes which show the clay to be locally 120 ft thick and generally to include one or more thin beds of sand.

The boulder clay is overlain by up to 30 ft of sand and gravel, which form a single large patch of undulating ground

west and north of Sedgefield and a number of isolated smaller moundy patches farther east. Over limited areas the base of the deposit appears to be more or less level, so that the height of the mounds can be taken as an indication of the thickness of sand and gravel. Closed hollows, some floored by clay, have been noted between some of the mounds. The sand and gravel is best seen in disused sand pits [3651 2997 ; 3643 3038] near Ryal Farm, and in excavations in and around Sedgefield. From the scanty evidence of field exposures, it appears that the deposit is increasingly finer and less gravelly towards the north of the Sedgefield area, and current-bedding, where seen, indicates transport from the south. G.D.G., E.A.F.

Butterwick to Hurworth Burn. The area has an undulating surface of low overall relief, and has a uniform brown stony clay-loam soil. The drift is poorly exposed, however, and details of it are known almost entirely from boreholes. These show that it ranges from about 100 ft to over 300 ft in thickness, consisting of stony clay overlying 40 to 50 ft of interbedded red, yellow and brown clays (some stony), silts, sands and gravels. In a cored bore [3988 3111] at Whin Houses, 1 mile N.E. of Butterwick, the succession is :

	ft
Clay, slightly reddish light brown, with grey weathering along cracks in upper part ; sandy, with scattered pebbles of grit, sandstone, coal, ironstone, etc.	$19\frac{2}{3}$
Clay, chocolate-brown, apparently thin-bedded, with light buff silty partings	$15\frac{1}{4}$
Boulders and sand ; rock fragments include dolerite, Magnesian Limestone and sandstone	5
Clay, brown, finely laminated, with sandy partings and thin sand beds ; laminae partly contorted	10
Boulders and sand, as above ...	8
Clay, brownish grey, with abundant subangular fragments of sandstone and Magnesian Limestone	15

	ft
Boulders and sand, as above ...	7
Clay, brown plastic, probably with sand partings	25
Clay, dark brown, with abundant subangular erratics ...	9
Clay, reddish brown, thin-bedded, with reddish buff silty partings and 2- to 3-in beds of fine brown sand towards base	51
Sand (not recovered)	2¼
Clay, dark grey-brown, roughly horizontally bedded, with abundant small subangular erratics	
Clay, dark grey-brown, thin-bedded, with light buff silty partings and contorted beds of fine sand	12¾
'Limestone' (not cored) ...	25
'Clay, with heavy sandy partings' (not cored)	65
MAGNESIAN LIMESTONE	–

From this record, rockhead is assumed, uncertainly, to lie at the top of the 25 ft 'Limestone' at a depth of 180 ft. A similar uncertainty as to the position of rockhead arises in several other bores in the area, because of the tendency of the Middle Magnesian Limestone to break down during drilling to sand-grade fragments, which are contaminated with clay and fragments of stones from higher parts of the hole before being brought to the surface. Despite these uncertainties, it is clear that drift is abnormally thick in a number of deep buried valleys, including the Skerne Valley west of Hurworth Burn, and that the sub-drift surface is of much higher relief than the present surface. The bedded character of the drift in the buried valleys suggests a complicated Quaternary history, with a number of stony clays separated by melt-water deposits. Laminated and bedded clays such as those proved in the Whin Houses Farm Bore may be of lacustrine origin and broadly comparable with those around Edder Acres and Wingate.

D.B.S., G.D.G.

WESTERN AREA: GENERAL ACCOUNT

In the western area the following threefold subdivision of the Glacial deposits is recognized: Upper Stony Clays 0 to 50 ft, Middle Sands, Gravels and Clays 0 to 260 ft, Lower Boulder Clay 0 to 120 ft.

The Lower Boulder Clay is a stiff, dark grey or grey-brown clay which weathers brown or buff and has a gritty texture. It contains many pebbles, cobbles and boulders derived from Carboniferous sediments as well as far-travelled erratics such as Ordovician lavas from the Lake District, granites and Silurian greywackes from the Scottish Southern Uplands and blocks of dolerite of Whin Sill type. The deposit seems to be continuous with, and is lithologically similar to, the Lower Boulder Clay of the areas to the east, contrary to the view of Raistrick (1934, fig. 3) who correlated it with the Upper Boulder Clay of the eastern area.

The Middle Sands, Gravels and Clays form bedded sequences which show pronounced lateral variation and transition from one deposit to another. The sands are sharp, well sorted and commonly contain a high proportion of clay: those formerly used for moulding sand near Durham, for instance, contain up to about 9 per cent (Boswell 1918, pp. 142–4). Some of the sands are cross-bedded in units up to about 2 ft thick; others are flat-bedded. The sands also form layers, lenses and wisps within gravels and laminated clays. The gravels, which contain a similar range of erratics to the Lower Boulder Clay, occur as lenses within the sands or in beds up to 60 or 80 ft thick. The sands and gravels are interdigitated with and locally pass laterally into laminated clays (Pl. XVI) which have been worked extensively for brick making. Most of these clays are silty and contain partings of silt or fine sand in which sedimentary structures such as low-amplitude ripple-marks and flow marks are well preserved. Lamination is very fine in places, and in samples obtained near East Moor Leazes is less than 1/500 of an inch. Contorted beds, many of them less than an inch thick, can be traced for considerable distances, particularly at this locality. Thin bands resembling boulder clay occur in places and xenoliths of boulder clay up to a few millimetres across are also common. The clays worked near East Moor Leazes contain boulders, including Carboniferous Limestone and dolerite, in addition to abundant smaller stones. Many of the latter are of limestone and these lead to a high rate of wastage since they calcine during the brick-firing process. In the upper part of the sequence near Croxdale Wood House and High Butterby, trails of organisms have been found on some of the bedding planes.

In some places the Middle Sands, Gravels and Clays are associated with terrace features. For instance, a short distance south of Durham, where hummocky ground rises to 347 ft. above O.D., and at Framwellgate Moor, just beyond the western boundary of the district, sand forms flattish terrace features up to about 350 ft. Just within the western margin of the district terrace-like features successively fall in altitude until north of Durham they are only just over 300 ft O.D. Parts of these features, especially where the surface is composed of sand and gravel, are hummocky, and resemble kame-terraces. On the eastern slope of one such feature near Newton Grange and Red House, there are a number of closed ill-drained hollows which appear to be kettle-holes. Terrace features also occur on interbedded sequences of sands and laminated clays as depicted in Pl. XVI, Section 2 and supported by mapping and bore information in the valley east of Bishop's Grange. Here laminated clays can be traced between bands of sand along the valley side (Pl. XVI, Section 1) from the area largely made up of laminated clays. The lower parts of the buried valley east of Bishop's Grange are filled predominantly by gravel and sand, on which are small benches with surfaces standing at about 140 to 155 ft above O.D. The sands, gravels and clays to the south and south-east of the ridge which now extends across the western area from East Rainton to Gilesgate Moor are built up to approximately 300 ft O.D.

The Upper Stony Clays succeed the Middle Sands, Gravels and Clays. Generally, they are thin and consist of light brown sandy and gleyed clays with bleached and weathered pebbles, cobbles and boulders together forming a suite of hard sandstone, dolerite and volcanic rocks which resembles the stone content of the Lower Boulder Clay and Middle Sands of the area. The clays are thickest between Shincliffe Colliery, Croxdale Hall and Heugh Hall (Pl. XVI, Sections 3 and 4), reaching a maximum of 50 ft in one bore. Nearby, in other bores, they are about 20 ft thick. Elsewhere, they are commonly less than 10 ft thick. Where thick, the deposit looks like boulder clay, particularly in the lower part, but where thin its affinities are less apparent.

From the East Rainton–Gilesgate Moor ridge southwards, the Upper Stony Clays are absent in the valleys trenched into the bedded sequences of the Middle Sands, Gravels and Clays, and they are also absent on the flat or gently undulating surfaces lying a little below or above 300 ft around Sherburn. They can be mapped continuously on most of the upper surface of the East Rainton–Gilesgate Moor ridge and over the north-western flank near Carrville. On the western side of the area, similar clays cap the ridge extending west and south-west of Durham City, but to the north-west they are absent in much of the ground near Aykley Heads. At Framwellgate Moor, however, they rest on the surface of one of the marginal terraces, extending eastwards down the eastern flank of the feature south of Low Newton.

WESTERN AREA: DETAILS

Frankland to Finchale Priory. Along the eastern margin of the area, between Wilson's Row and Moor End (Carrville), the drift is generally less than 30 ft thick and consists essentially of Lower Boulder Clay resting on rockhead. It is best exposed in the gorge and tributary gills of the River Wear between Frankland Farm and Finchale Wood west of Finchale Priory; other exposures in the gorge north-east of Ford Cottage are less accessible. The clay is up to 25 ft thick resting on bedrock in Winch Gill [3072 4572], 600 yd S.W. of Wood Side, and in adjacent gills and riverside bluffs. Similar stony clay with many boulders is exposed in the upper part of the gill [3065 4472], 200 yd S.S.W. of Low Grange, and in the southern tributary of the Bow Burn [2933 4762], 600 yd N.W. of Finchale Priory. It is 6 to 18 ft thick in several bores and pits in the vicinity of Grange and Low Grange.

The sequence becomes thicker and more complex farther west, where several easterly tributaries to the buried valley of the River Wear cross the course of the modern stream and break the continuity of rock outcrop along the banks. Two of these, extending from about ¼ mile N.W. of Wilson's Row and from near Rainton Adventure Pit, converge and cross the present river about 300 yd S.E. of Finchale Banks, continuing from there to meet the main buried valley north-east of Red House. Farther south, another buried tributary valley, proved by borings near Low Grange, seems to link eastwards between Ramside and The Rift with an important north-north-easterly buried valley east of Broomside (Fig. 25).

Within these buried valleys the Lower Boulder Clay passes under bedded clays and sands, but although it is commonly recorded as a single stony clay in bore records, precise definition at outcrop is not everywhere clear. This is exemplified by the following section [2870 4689], 1050 yd W.S.W. of Finchale Priory:

	ft
Brown sandy stony clay, gravelly in places; thin gravelly layer at base ...	15
Grey laminated clay with silty bands	1½ ft to 2
Grey boulder clay with rounded stones	6
Grey clay with sandy bands	6
Sandy boulder clay, with large boulders in upper part	11
Sandstone (above Top Brass Thill Coal)	—

Pl. XVI, Section 2 illustrates the relationships of the drifts between the northern end of Raintonpark Wood and the western margin of the district, 700 yd W.S.W. of Newton Grange. West of the River Wear, between East Moor Leazes and Low Newton, several bores on, or close to, the line of this section prove relatively thin Lower Boulder Clay overlain by a much more variable thickness of clays with some stony and sandy bands. These bedded clays belong to the Middle Sands, Gravels and Clays. The upper part of this sequence has been extensively worked for brick-clay at Brasside, Frankland, Newton Grange, and most recently at Finchale Brick Works. A bore in the north-eastern corner of Brasside Brick Pit [291 451], 950 yd S.E. of Low Newton, proved:

	ft
Soft clay	8
Soft clay with a little sand ...	8
Soft clay and stones	8½
Loamy sand	4½
Soft clay with a little sand and stones	8
Strong clay mixed with coal ...	10
Soft laminated clay	6
Sharp sand mixed with loam and clay	6
Soft clay	20
Brown stony clay with a little sand and coal	13
Soft clay	17
Stony clay and sandstone ...	11½
Sandstone	—

During the resurvey the sequence in the area of the clay pits was best exposed in Finchale Brick Works, where about 14 ft of laminated silty clays have been extracted. They consist of fine silt and clay disposed in thin sheets separated by partings of coarse silt and fine sand. In hand specimen layers appear to be about ¼ to ½ in thick, but under low magnification a finer lamination, ranging down to 1/500 in or so, can be seen within the layers (Pl. XVII). Bands of small stones and grit are present in the lowest 5 ft. Boulders are common towards the bottom of the section and include Carboniferous Limestone and dolerite. Above the silty clays and under 6 to 9 in of soil lie 2½ to 3 ft of brown and grey loamy and locally very sandy clay containing scattered subangular fragments of pale sandstone, and a few weathered boulders of dolerite and volcanic rocks. The lower part of this sandy clay is coarsely laminated, and appears at first sight to represent the weathered top of the underlying laminated silty clays. Careful examination, however, suggests that there is a discontinuity between the coarsely laminated sandy clay and the finely laminated underlying beds. This discontinuity is more easily detected in the nearby Newton Grange Pit, where the top of the silty clay is disturbed beneath flat-lying sandy beds. The origin of this upper sandy clay is not clear (p. 241). A bore

[2920 4595], 650 yd W.N.W. of Union Hall, proved loamy soil 8 in, on yellow-grey clay 3 ft 4 in, blue-brown clay 2 ft 6 in, and blue clay 27 ft 3 in. The section at Frankland Pit, 825 yd S.S.W. of Union Hall, is brown sandy soil 1 ft 3 in, on sand 4 ft 3 in, loamy clay 27 ft 4 in, strong stony clay 13 ft 6 in, on 'brown freestone ramble'.

West of the brickfields, a bore [2845 4583], 250 yd N.E. of Low Newton, proved 171½ ft of 'brown clay' with thin bands of 'loamy sand' and with 'stony' beds at the centre and base. Another bore [2848 4611], 525 yd N.N.E. of Low Newton, also penetrated loamy sand, though loamy, leafy and stony clays still comprise most of the sequence. Farther north, however, in the deep valleys falling to the River Wear, sand beds are commonly interbedded with the clays, while to the west clays are subordinate to sands. In the Newton Hall Bore [2797 4564], 250 yd S.E. of Newton Grange, for instance, there are 221 ft of bedded, dominantly sandy deposits underlain by 12 ft of stony clay with 'whin tumblers', representing the Lower Boulder Clay.

The ground under Low Newton, which lies along the main buried valley of the River Wear, hides a transition between the clay country to the east and the sand country to the west (Pl. XVI, Section 2). At Framwellgate Colliery, just beyond the western margin of the district and 1100 yd S.W. of Newton Grange, the full tripartite sequence can be distinguished:

	ft
UPPER STONY CLAYS—'sandy brown clay'	17½
MIDDLE SANDS, GRAVELS AND CLAYS—beds of leafy and loamy clay with partings and layers of sand	99½
LOWER BOULDER CLAY—strong black clay	26

In the ground east-north-east of Bishop's Grange (Pl. XVI, Section 1), clays are subordinate to sands and gravels. At the eastern end of this section, about 800 yd N.W. of Finchale Priory, the surface is formed by clay which rests on sand. To the south-west, where

Cocken Hall once stood, there is a terrace of sand at about 150 ft above O.D., and a bore [2870 4729], 975 yd W.N.W. of Finchale Priory, proved sand lying on Lower Boulder Clay. To the west this sand appears to pass laterally into a clay which forms the upper part of the valley side. The Lower Boulder Clay crops out lower in the valley side where it is locally as little as 3 ft thick. The flood-plain of the River Wear at Low Cocken is formed of sand and gravel, as is much of the underlying drift, though a bed of clay is proved in two bores [2802 4727 ; 2799 4739], 140 yd S.S.W. and 80 yd W. respectively of Ford Cottage. In the area north-north-west of Ford Cottage, in Harbourhouse Park, features, springs and zones of slipping suggest that the steep upper slopes to the west of the river consist largely of sand, underlain by clays. An underlying bed of gravelly sand seems to wedge out northwards, away from the centre-line of the main buried valley which here underlies the present river about 300 yd S.S.W. of Ford Cottage. The clay underlying these gravelly sands probably passes southwards under the river alluvium and thus corresponds to the bed proved in the bores near Ford Cottage. This implies a drop of about 50 ft in about 150 yd which is of the same order as that depicted in Pl. XVI, Section 1. The upper clay band is not known to extend farther south than about 250 yd W. of Ford Cottage, but it may continue just below the 100-ft contour and about 250 yd E. of Bishop's Grange, where it crops out under landslip and has been proved in three bores to the west of the river. The rear wall of the landslip is composed of sand forming a bluff corresponding in height and stratigraphical position to the upper steep slope in Harbourhouse Park. At a higher level in the Bishop's Grange area there is a flattish bench possibly representing the outcrop of a still higher bed of clay, also recorded in the nearby bores.

Towards the southern end of the area, on the eastern bank of the brook about 800 yd S.W. of Frankland Farm, the beds below the sands were uncovered during prospecting for brick clay. They comprise a layered sequence of boulder

clays and sands dipping north-eastwards into the slope. The layers, which are extremely variable in thickness, are as follows: soil to 1 ft, on sandy stony clay 1 to 5 ft, fine brown sand 6 ft, heavy dark blue boulder clay 8 ft, sand 6 ft, boulder clay 4 ft, sand 4 ft, argillaceous silt 8 ft, boulder clay with layers of sand to about 15 ft. To the south, between 100 and 500 yd N. of Crook Hall, a thick series of laminated clays has been worked for brick-making, but landslips have obscured the sections.

West Rainton to Pittington. The drift in the area north-west of West Rainton, Middle Rainton and East Rainton consists mainly of a single thin, poorly exposed boulder clay amounting to 13 ft 8 in at Rainton Adventure Pit, 20 ft at Resolution Pit and only 5 ft 11 in at Meadows West Pit. Exceptionally, however, the drift includes sand and gravel, as in the railway cutting about $\frac{1}{4}$ mile N. of the Adventure Pit, where sand underlies 6 ft of sandy stony clay. Bores [3133 4676 ; 3092 4704], 400 yd S.S.W. and 650 yd W. of the Adventure Pit, proved respectively sandy boulder clay $11\frac{1}{2}$ ft, on sand and gravel $13\frac{1}{2}$ and $16\frac{1}{4}$ ft, and very stony boulder clay $2\frac{3}{4}$ and $18\frac{1}{2}$ ft. The sand and gravel of these bores is probably equivalent to that proved in the railway cutting and if these deposits belong to the middle division, the underlying stony boulder clay would appear to be Lower Boulder Clay. However, it is not clear whether the single boulder clay represents the lower division alone or whether it also includes clay from the middle or upper division.

At Moor House, sandy stony clay thought to be Lower Boulder Clay, 6 ft thick, overlies sandstone. The drift thickens to about 30 ft farther south in two old bores 'in the Close House Grounds'.

In most other parts of the area the drift is thin and is composed mainly of clay again thought to be Lower Boulder Clay except along the ridge near Pitfield. This ridge, the top of which extends a little over 300 ft O.D., trends north-eastwards and is hummocky near Field House. Farther north-eastwards it rises to over 325 ft O.D. near Middle Rainton and East

Rainton where the drift appears to consist of thin boulder clay resting on rockhead. On the south-east flank of the ridge there are a few rounded hillocks and flattish terraces lying a short distance above the 300-ft contour. Small hills of sand lie 450 and 650 yd N.N.E. of Pittington Station.

On the continuation of this ridge to the south-west of Pitfield, near The Rift, Belmont Furnace Pit proved clay with sandy gravel 27 ft, on sand 7 ft, clay 5 ft, sand and loamy clay 14 ft, clay 8 ft, strong pebbly clay 22 ft. This sequence is thought to represent Upper Stony Clay 27 ft on Middle Sands, Gravels and Clays 34 ft, Lower Boulder Clay 22 ft. At Lady Seaham Pit, 2 ft of soil and sand rest on 60 ft of clay thought to represent Lower Boulder Clay.

Hetton le Hole to Hesledon Moor. The Hetton Burn is thought to follow the line of a buried valley (Fig. 25), which continues north-westwards beneath the Rainton Burn. Low hills of gravelly sand rising to just over 300 ft O.D. lie along the floor of the valley of the Rainton Burn. North of Low Moorsley, at a locality [3422 4714], 175 yd S.S.E. of Stobley Moor House, an old clay-pit was worked for bricks and tiles. The pit is now overgrown apart from 2 ft of very sandy and silty clay which may have been derived from laminated clay. North of the road crossing the burn near Stobley Moor House, the drift seems to consist entirely of Lower Boulder Clay. East of Low Moorsley flattish terrace-features at 310 to 315 ft O.D. are covered with sandy and gravelly soils or very sandy clay with more sandy patches, and a rail cutting through one of the terraces reveals a sequence of interbedded sand, clay and clayey gravel bands. These deposits are taken to be the Middle Sands, Gravels and Clays. The terraces are dissected by dry valleys. Near Coal Bank the edge of one of the terraces marks the southward limit of the sand, but farther east, sand and gravel extend up the slope and seem to vee into the valleys. Much of the ground farther up these valleys west of Elemore Colliery consists of clay, presumably Lower Boulder Clay, seen resting on rockhead in an old quarry at Coal

Bank and proved to a thickness of 50 ft at Elemore Colliery George Pit. However, at the now obscured Pemberton's Quarry [356 465], 1100 yd E. of Coal Bank, rockhead is directly overlain by sand and gravel. In the valley of the Hetton Burn the sand overlies clay and to the east forms a terrace at about 270 ft O.D. An excavation in this terrace [3506 4809], 1125 yd N.N.W. of Hetton station and just beyond the northern margin of the district, proved 3 ft of brown argillaceous sand on 10 ft of bluish grey Lower Boulder Clay.   E.A.F.

Along the margin of the area east of Hetton le Hole drift is generally thin and is composed mainly of dark brown boulder clay. This is exposed at only a few places, of which the following are representative: old brick pits [369 465], 400 yd S.E. of Little Eppleton, 8 ft; railway cutting [3952 4594], 150 yd N.W. of Hesledon Moor Cottages, 8 ft. It is 16 and 10 ft thick respectively at South Hetton and Hawthorn collieries.

Away from the margin and in buried valleys the drift is thick–amounting to 64 ft at Elemore Colliery Isabella Pit and 68 ft at Hawthorn Colliery Shaft–and in many places is made up of an interbedded sequence of clays, stony clays, sands and gravels. The sands and gravels form low mounds and give rise to gravelly clay-loam soils. They are exposed only at the following localities: railway cutting [3567 4650] at Hetton le Hole, 15 ft; railway cutting [3752 4726], 400 yd S. of Carr House, Eppleton, 20 ft (fine, with much sand); and railway cutting [3952 4594], 150 yd N.W. of Hesledon Moor Cottages, 6 ft, on boulder clay. The railway cutting contains the only section now visible in the large spread of sand and gravel at Hetton le Hole, but in the Blossom Pit [3597 4702] 9 ft of sand rest on 4 ft of gravel on rockhead, and the deposit is proved to have a thickness of up to 12 ft in a number of shallow bores in a field 300 to 450 yd S.S.E. of the Blossom Pit. Farther east, 3 ft of sand lie on boulder clay in South Hetton Colliery Shaft, and in Hawthorn Shaft the sequence is clay with some sand 12 ft, on sand and gravel $5\frac{1}{2}$ ft, clay $1\frac{1}{2}$ ft, boulder clay 3 ft, sand and clay 16 ft, boulder clay 5¼ ft, sand and gravel 11¾ ft, boulder clay (as noted above) 10 ft, on rockhead. The sand and gravel below the uppermost clay was also proved in a number of foundation boreholes drilled in a field 200 to 400 yd E. of Hawthorn Shaft, where it is generally 1 to 4 ft thick, but reaches 19 ft in one borehole [3906 4596] ; it is correlated with a bed of sand and gravel exposed in the railway cutting 600 yd S.S.W. of Hesledon Moor East. Temporary exposures in rising ground 150 to 200 yd W. and N. of the shaft revealed 20 to 30 ft of brown very stony clay with pockets and lenses of sandy clay and clayey sand.

The spread of sand and gravel at Hetton le Hole extends over the northern margin of the district and appears to be genetically related to Curlew Hope, a meltwater channel at Eppleton Colliery, 1 mile N. of Little Eppleton. Where exposed immediately north of the district boundary the gravel rests on a channelled surface of boulder clay, and fore-set bedding dips mainly westwards (i.e. downslope). The deposit seems everywhere to be the uppermost member of the drift sequence. D.B.S.

Gilesgate Moor to Shadforth. The western margin of this area is flanked by a terrace-feature extending from Gilesgate Moor to Ramside, which between Broomside and Ramside is at 313 ft O.D. A small slip on the western side of the Carrville railway cutting [3110 4388], 1000 yd N.W. of Broomside House, shows soil 1 ft, on sandy stony clay 6 ft, on sand. The clay is one of the Upper Stony Clays. The boundary between the clay and sand seems to drop away at the northern end of the cutting so that the north-western slope of the terrace-feature is apparently faced by sandy clay. Towards The Rift, where the terrace joins a similar feature extending through Pitfield to East Rainton (p. 229), the clay includes many small pebbles. To the south-west around Gilesgate Moor, the ground is gently undulating, being generally a little over or under 300 ft above O.D., but rising to about 320 ft south-west of Moor End. The sandy clay soil hereabouts is probably derived from Upper Stony Clay. A clay pit was formerly worked [2975 4245] 600 yd N.N.E. of Bent House Farm.

Sand proved in Broomside railway cutting extends north-eastwards and south-westwards along the terrace-slope and was formerly worked [311 434] 800 yd W. of Broomside House. At the time of resurvey only the uppermost portion of the section was visible: soil 8 in, on sandy stony clay 1½ to 3 ft, sand and small gravel 0 to 1 ft, sand 8 ft. East of Gilesgate Moor, where sand is not evident, the soil on the slopes falling gently to Pittington Beck seems to be derived from sandy clay. The slopes below the 200-ft contour are commonly marked by small slips, and the material in these suggests that the ground consists of interbedded clays and thin bands of sand. The clays include both stony and laminated varieties, and are probably the lateral equivalents of the Broomside sands and form part of the Middle Sands, Gravels and Clays. An Upper Stony Clay appears to cover the top of the terrace-feature whether the underlying material is clay or sand.

On the eastern side of the Pittington Beck, north of the Sherburnhouse Beck, the Upper Stony Clay is absent, and the sands form extensive outcrops. The junction between the sand and the underlying Lower Boulder Clay can be traced up the valley of the Coalford Beck east of Broomside House and is last seen at a locality [3252 4309] 650 yd S.W. of Hallgarth church, where a bore proved 111 ft of clay with sand partings. Lower Boulder Clay with a veneer of sand and gravel can also be traced south-westwards from near Broomside House into the valley of the Pittington Beck. Between Broomside House and Sherburn sands and gravels form a terrace at 290 to 295 ft O.D. and are about 15 ft thick. A small gravel pit was formerly worked [3205 4252] 425 yd N.E. of Sherburn church. The disused Sherburn Colliery Lady Durham Pit is sited at the western margin of the terrace and proved 62 ft of clay with thin sand beds on 66 ft of clay. The north-western part of the terrace has been worked for gravel which is at least 8 ft thick and contains lenses of sand in disused pits north-west of Broomside House.

North-east of Sherburn the ground rises to just over 300 ft above O.D., and contains low hills with gravelly and sandy soil. Recent workings show 6 ft of gravel resting on sand. The gravel contains an ice-wedge pseudomorph at least 4 ft deep. A bore [3271 4282] 1225 yd N.E. of Sherburn church proved clay 3½ ft, on sand and gravel 15½ ft, clay 9 ft, sand 32 ft, clay 27 ft. A pit opened in an adjacent hill contains a similar bedded sequence, dipping gently to the north-west and consisting of a thin layer of sandy gravel, on 6 to 8 ft of sandy stony boulder clay, on sand. A bore [3287 4301] 1475 yd N.E. of Sherburn church, just beyond the margin of this spread of sand and gravel, proved 106 ft of clay with sand patches; another nearby bore [3297 4321] 400 yd S.S.E. of Hallgarth church, proved 3 ft of sand (probably alluvial) resting on 109 ft of boulder clay with sand. These deposits form the bulk of the material filling the buried valley of Coalford Beck. Over 30 ft of sand and gravel have been worked in a pit on the north side of the beck between 200 and 600 yd S.W. of Hallgarth church. Northwards for more than ½ mile, the sand and gravel forms a terrace with gently undulating surface having a maximum height of about 315 ft O.D. on the north-eastern margin. Sand and gravel also form the surface for about 1500 yd on the south-eastern side of the Coalford Beck. At its northern extremity, 650 yd S.W. of Hillside, Littletown Engine Pit [3385 4335] proved gravel 31 ft, on sand 8 ft, loamy clay 15 ft, strong blue clay (Lower Boulder Clay) 12 ft.

Sand and gravel extend north-eastwards from the Hallgarth area as far as the slope south of North Pittington Colliery, but the shaft there and at the Buddle Pit met only Lower Boulder Clay; the former proved 30 ft of gravelly clay, and the latter soil and clay 6 ft, on strong blue clay 18 ft.

Much of the re-entrant between Littletown and Hillside is probably filled with bedded clays including boulder clay. The only visible section [3433 4392] 900 yd W.N.W. of Elemore Grange comprises 1 ft of grey clay with small stones on 4 ft of brown sandy stony clay. The 104½-ft

drift section at Littletown Colliery comprises yellow clay 2 ft, on blue clay 13 ft, blue clay and stones 56½ ft, gravel 6 ft, fine blue clay 12 ft. The clays forming the upper part of this sequence have been worked for brick making; there is an overgrown pit [338 434] 200 yd N. of Littletown House, and clays at a similar level were also formerly worked [341 430] 450 yd S.E. of Littletown House. A northwards-draining channel, 25 to 30 ft deep, cuts the spur 200 yd W. of Hastings House.

Between the re-entrants the drift is thin and apparently uniform. Thus, a bore [3349 4334] 300 yd W.N.W. of Littletown House proved boulder clay and sand 8 ft, on 'sandstone' 11 ft, boulder clay and sand 14 ft; and another bore, 500 yd to the south-west, proved 9 ft of boulder clay on 54 ft of soft dark clay with boulders. Two further bores [3310 4266; 3290 4249] 975 yd and 1250 yd S.W. of Littletown House, recorded 68½ ft of boulder clay and 78 ft of sand and boulder clay respectively, while Sherburn Hill Colliery West Pit passed through 91 ft of clay.

In the south-western part of the area a bore [3057 4038] 400 yd N.W. of Whitwell House proved loamy clay 2 ft, on dark blue clay 7 ft, loamy clay 5 ft, dark blue clay with cobbles 23½ ft, dark blue clay 26 ft, laminated clay 15 ft, brown stony clay 1½ ft, laminated clay 1 ft, stony blue clay 6 ft, laminated clay 3 ft, boulder clay 10½ ft. About ¼ mile to the north-east, Whitwell 'A' Pit proved loamy clay 12 ft, on blue clay 10½ ft, sand 0 to 9 ft, strong blue clay 9 ft, laminated clay 42 ft, strong blue clay 31 ft. It is presumed that the lowest item in the bore is Lower Boulder Clay, but it is impossible to trace any of the overlying units laterally. It is likely that much of the ground to the north extending into the valley of the Pittington Beck is made up of these bedded clays, for small slips [3096 4115], 900 yd W.N.W. of Byers Garth, include sandy stony clay, laminated clay and undoubted boulder clay. Farther upstream, in the valley of the Chapman Beck about 750 yd S.W. of Byers Garth and above about 275 ft O.D., these clays are overlain by sands which also form a flat-topped outlier at just over 300 ft O.D. near Whitwell House. There are no sections to show whether or not clays are interbedded with the sands. On the north-eastern side of the beck, however, the sands are clearly overlain by clay which is in turn over-lain by sand at the top of a hill rising to 315 ft O.D., 350 yd S.S.W. of Byers Garth. The clays include thin bands of sand and gravel, and the sands include thin bands of clay. The lower band of sand can be traced eastwards to a locality [3220 4028] ½ mile S.S.E. of Byers Garth Farm, where it seems to wedge out. It can also be traced round the hill on which this farm stands, though in places the boundaries are indefinite.

To the north-east of Byers Garth the outcrop widens and at Sherburnhouse Colliery forms a flattish area at just over 300 ft O.D. The colliery shaft section shows sand and gravel 17½ ft, on soft brown clay 2½ ft, sand 24½ ft, stiff strong clay (Lower Boulder Clay) 4 ft. To the east of the colliery the sand and gravel seems to wedge out in the banks of the present valley, but there is probably a lateral passage into clay with sandy part-ings found at the same level in a bore [3261 4145] 550 yd E. of Sherburnhouse Colliery; this proved sandy clay and stones 5½ ft, on stony clay with sand partings 29 ft, stony clay 21½ ft. Another bore [3311 4146], 550 yd farther east, proved 20 ft of boulder clay with thin bands of laminated clay on 31 ft of boulder clay, at about the same level. The Lower Boulder Clay is recognizable in these and other nearby bores as a stony clay 15 to 30 ft thick, but the overlying sequence in the deeper part of the buried valley con-sists of a complex succession of bedded clays similar to that proved elsewhere. Thus a bore [3303 4100], 650 yd N.N.W. of Running Waters, proved 177½ ft of drift as follows: loamy clay 2 ft, on yellow and brown stony clay 3½ ft, brown stony clay 57 ft, 'limestone ramble' 3 ft, brown stony clay and sand 54 ft, stony clay 31 ft, laminated clay 3 ft stony clay 24 ft. Another bore [3318 4105], 150 yd to the east-north-east, recorded clay with beds of sand 65 ft, on boulder clay 117 ft. The detail of these bores is

probably less reliable than that of a bore [3372 4132] 450 yd S.S.W. of Crime Rigg, which proved brown sandy laminated clay 4 ft, on laminated clay with bands of sand 75½ ft, boulder clay 4½ ft. Bores near Shadforth village prove 60 to 120 ft of 'sand and boulder clay'. It is possible that a patch of Upper Stony Clay lies across the southern two-thirds of the valley north-west of Running Waters, but the evidence is very meagre.

Durham to Cassop Moor. Near the northern extremity of this area Kepier Florence Pit proved gravel 3 ft, on sand ½ ft, clay 60 ft. A temporary exposure [2853 4379], 70 yd to the south, revealed at the time of resurvey 2¼ ft of soil and sandy stony clay, on 2¼ ft of rather argillaceous sand. Another exposure [2848 4386], 60 yd W. of the pit, consists of gravelly sandy clay with a few layers or partings of brown clay; and in the river bank close by [2847 4387] there are 6 ft of slipped laminated clays. An old brick-clay pit [2835 4349], 300 yd N.E. of the remains of Kepier Hospital, formerly worked laminated clays. The slopes to the east of this show small slips which suggest that bedded clays may extend to higher levels. These are not differentiated in Kepier Pit, 850 yd S.E. of Kepier Hospital, which recorded clay 6 ft, on sand 12 ft, clay 77 ft. The top clay is interpreted as Upper Stony Clay but most of the lower clay is likely to be laminated, with only a little Lower Boulder Clay at base. Farther south, in a bore [2920 4201] 400 yd E.N.E. of Old Durham, the drift sequence is laminated clay 25½ ft, on sand 6 ft, stony clay 1½ ft. This is evidently below the limit of the Upper Stony Clay which seems to be largely confined to the top of the terrace feature extending from Gilesgate Moor westwards as far as Gilesgate church. The clay is proved to be at least 6 ft thick in the graveyard, and it overlies pebbly sand which is seen cropping out in the slope immediately south of the church. At the top of the south bank of Pelaw Wood Beck the sequence is brown sandy clay to 3 ft, on dark grey clay 2 ft, brown sandy clay 2 ft, on sand. At the base of the bank is sandy soil 2 ft, on gravel and sand 6 ft, fine gravel with sand 9 ft. Sand with bands of clay, typical of the sequence immediately below the Upper Stony Clay, are exposed [2867 4243 ; 2882 4233] 400 yd and 575 yd S.E. of the church. The gravelly sands and sandy gravels are more characteristic of the ground below 200 ft O.D. and are exposed at scattered points in Pelaw Wood, on the steep north bank of the River Wear between 250 yd S.S.W. and about 600 yd S.S.E. of the church. These deposits are seen to best advantage in Durham Gravel Pit between 250 yd S. and 500 yd S.E. of Old Durham: in the old, south-western, part of the pit, where the gravels lie in belts up to 300 yd long and trending S. 35° E., they are rudely arch-bedded in cross-section with intervening trough-bedded sands which interdigitate with the gravels along the margins of the belts. At the time of the resurvey the gravels along the south-western margin of the pit were exposed up to 20 ft thick. Later workings in the north-eastern part have proved as much as 60 ft of gravel, in which no arch-bedding is seen. The gravels, which contain thin lenses of sand and the usual suite of erratics, are coarse in many places and include blocks of sandstone up to 6 ft × 3 ft × 2 ft in the older part of the pit. The arch-bedding and linear deposition suggest that the gravels originated as subglacial eskers.

The gravels and sands continue under the alluvium of the River Wear and a bore [approx. 285 410] said to be near the waterworks, about 700 yd W.N.W. of Shincliffe church, proved sand 12 ft, on sand and gravel 10½ ft, sand 11½ ft, sandy clay 2½ ft, sand 2 ft, gravel 12 ft, sand and gravel 2½ ft. The deposits thin rapidly to the east, however, against the side of the buried valley of the Old Durham Beck, and at Old Durham Colliery 600 yd S.E. of Old Durham the succession is soil and clay 9 ft, on gravel and sand 4½ ft, sand 2½ ft, strong stony blue clay (Lower Boulder Clay) 16 ft. Sands can be traced eastwards in cuttings of the disused railway to ¼ mile S.E. of Bent House and in the soil for some 350 yd farther east, where they seem to pass laterally into the sequence of bedded clays which lie in the valley of the Pittington Beck (p. 232). Clays were formerly worked in Shincliffe Old Brickyard and between ¼ and ½ mile W.S.W. of

Sherburn Hospital. At the time of resurvey the workings were being filled by refuse and the only section seen [3023 4130] was 600 yd W.S.W. of Sherburn Hospital: light sandy soil 1 ft, on brown and grey sandy clay with sandstone boulders 2 ft, brown plastic clay, sandy in lower part 1½ ft, fine yellow sand with thin black partings 8 ft.

Much of the eastern part of Durham City is built upon interbedded sands and clays with thin beds of gravel. Old bores in property bordering on Claypath proved sections as follows: (i) soil 1½ ft, on sand and gravel 5½ ft, sandy clay 5 ft, sand and gravel 8½ ft, stony clay 5 ft, loamy clay and sand 4½ ft ; (ii) forced ground 11¼ ft, on sand and gravel 32¼ ft, sand and silt 10 ft, hard stony clay 8½ ft, sand and silt 8 ft, hard stony clay 1½ ft. Farther south-west the basement of Messrs. Doggarts Ltd., on the east side of the market place, is said to be in sand, and the centre of the market place is underlain by sandy gravel. Dominantly sandy drift seems to extend southwards on the Cathedral peninsula, where a bore in the Castle Yard proved sand under made ground. Excavations at Hatfield College [276 422], about 100 yd N.E. of the Cathedral, revealed the following section (examined by Mr. G. D. Mockler):

	ft
Grey clay with stones	6
Hard gravel	2¼
Fine yellow sand, thickness varying laterally from 4 in to 7 ft	2¼
Brown clayey sand with coal debris ; pebbles in basal 1 ft	4
Fine yellow sand with coal debris	1
Brown clay ... ... to	1
Unexposed	3¼
Sandstone	—

The drift thickens rapidly northwards under the northern side of the Castle into the buried valley of the River Browney which is aligned east–west and joins that of the River Wear under the eastern part of the city. The slope of the side of the buried valley can be seen by the riverside path, 70 yd S. of the western end of Elvet Bridge.

Sands with interbedded clays and gravels extend southwards from the disused Elvet station on the south-eastern

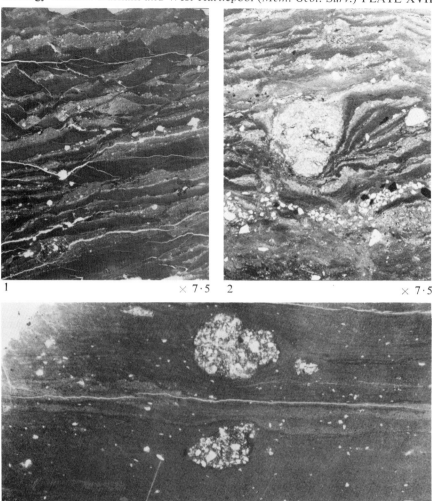

1   × 7·5   2   × 7·5

3   × 7·5

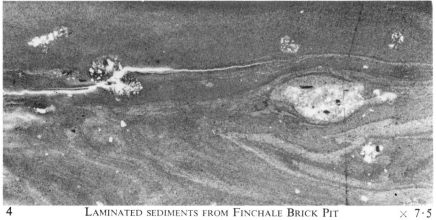

4   LAMINATED SEDIMENTS FROM FINCHALE BRICK PIT   × 7·5

(*For explanation see pp. xi-xii*)

PLATE XVIII

Geology between Durham and West Hartlepool (*Mem. Geol. Surv.*)

GLACIAL GRAVELS AT DURHAM GRAVEL PIT

side of the city as far as Houghall. From there, sandy gravels and gravelly sands continue at least as far as the river immediately north of Low Butterby and wedge out near Croxdale Wood House as proved by bores (Pl. XVI, Sections 3 and 4). Gravelly sand is exposed at a number of places in the steep banks west of the alluvium near the School of Agriculture. Near the top of the bank, about 300 yd W. of the northern end of the School of Agriculture, 2 ft of sand and fine gravel overlie 2 ft of dark brown clay. Clay bands were also proved in foundation bores and excavations for Grey College: they can be traced at the surface by topography of the banks and by scattered springs such as those occurring round the sandy hills rising to a maximum of 347 ft above O.D. to the north-east of Oswald House. Sand and gravel persist under the alluvium at Houghall and at Houghall Colliery they consist of sand 9 ft, on small gravel 6 ft, sand 16 ft. Sandy gravel also crops out in the slopes between the terraces around High Houghall. The section at a locality [2817 3963] 300 yd E. of the farm is terrace alluvium 3 ft, on sand 3 ft, gravel to 1 ft.

Sands also crop out on the eastern side of the River Wear from north-east of Shincliffe to as far south as High Butterby. Fine sand with coaly layers is exposed in old sand pits [2924 4081 ; 2949 4061], 150 yd N.E. and 400 yd E.S.E. of Shincliffe church, and also in the slopes east of Shincliffe Hall, where it is locally gravelly. Between Shincliffe Hall and High Butterby the sands are successively overlain first by laminated clays, then by one of the Upper Stony Clays. This upper clay is extensive between Shincliffe, Croxdale Wood House and Bowburn (Pl. XVI, Section 4), but is absent along the lower valley of the Skip Beck (Pl. XVI, Section 3), which joins the Tursdale Beck near Croxdale Hall. Under the Upper Stony Clay the sands pass laterally into a succession of interbedded clays and sands. The sands, without clay bands, are exposed in the slopes north of Shincliffe Bank, but at the roadside [2942 4013], 700 yd E. of Shincliffe Hall, laminated clays are present between them and the Upper Stony Clay. At Shincliffe Colliery, 300 yd farther east, the sinking recorded:

		ft
Yellow clay		1
Blue clay		1½
Loamy sand		3¼
Loamy clay and sand		5
Strong blue clay		14
Loamy clay and sand		17
Loamy sand		23½
Sand		22
Strong blue clay with sand partings		42½
Brown leafy clay		8½
Strong brown clay		2
Loam		20
Sand		2½
Strong brown stony clay (LOWER BOULDER CLAY)		9

A section in a bore 975 yd farther to the east-north-east is similar, but less sandy. Another bore [3003 3920], 180 yd S.S.E. of Shincliffe station, proved clay and cobbles 53½ ft, on loamy and brown clay 47 ft, sand and loam 33 ft, brown clay 56 ft, sand and loamy clay 10 ft, boulder clay 17 ft. This sequence thins eastwards (Pl. XVI, Section 4) and rockhead is overlain by heavy dark Lower Boulder Clay in the valley of Whitwell Beck, between 450 yd N.E. and 300 yd E.N.E. of Low Grange. The immediately overlying beds are not exposed but at a locality near the top of the west bank, 350 yd S.E. of Low Grange, there is a very sandy clay with sand lenses, thought to lie close below the base of the Upper Stony Clay. Interbedded clays and sands were proved by Whitwell 'B' and 'C' pits farther east. At the top of Whitwell 'B' Pit 31 ft of blue clay were penetrated, and this may represent Upper Stony Clay. In a bore [3220 3962] 1400 yd W. of Old Cassop, the sequence is Upper Stony Clay 27 ft, on laminated clay 2 ft, brown stony clay 2 ft, stony laminated silt ?10 ft, brown silt with clay partings ?4 ft, argillaceous sand ?4 ft, Lower Boulder Clay 1 ft ; another bore [3224 3920], 250 yd farther south, showed Upper Stony Clay 37 ft, on clay with laminated bands 3 ft, Lower Boulder Clay 6 ft. The drift in the two re-entrants father east is known only from a bore [3400 3965] 400 yd S.S.W. of Strawberry Hill, and from Cassop 'A' Pit, which proved 52 ft of clay and 48 ft of blue stony clay respectively. These deposits are probably boulder clay, as in the headward parts of other re-entrants.

The sands north of High Butterby extend below river level to join with similar deposits cropping out in the High Houghall area (p. 235 and Pl. XVI, Section 3). To the south they wedge out against the Lower Boulder Clay and are overlain under Butterby Wood by laminated clays which give rise to extensive landslips west and east of Croxdale Wood House. Up to 20 ft of these clays are exposed in a small cliff [2775 3903], about 600 yd N.N.W. of the house, where they are silty and contain abundant partings of sand with current structures. The southern margin of the sands and gravels is proved in a bore [2806 3831], 225 yd S.S.E. of the house, as follows: Upper Stony Clay 15½ ft, on laminated clay 22 ft, sand 20 ft, clay 32 ft, laminated clay 52½ ft, sand 32½ ft, gravel 12 ft, Lower Boulder Clay 9½ ft. The upper sand in this bore probably represents the sandy and silty laminated material which commonly lies in the upper part of the sequence. Another bore [2833 3823], 500 yd S.E. of Croxdale Wood House, proved Upper Stony Clay 50 ft, on laminated clay 118 ft, and Lower Boulder Clay 9 ft. These bores are respectively second and third from the left on P. XVI, Section 4. Rockhead is exposed [2733 3770] in the banks of the Skip Beck, 150 yd. S. of Croxdale Hall, and Lower Boulder Clay is thought to occupy the lower part of the valley upstream from this locality. The Lower Boulder Clay is 50 ft thick in a bore [2809 3778] 475 yd W. of High Croxdale, and the Upper Stony Clay is seen to rest on sand in a roadside silage pit [2821 3778] 375 yd W. of High Croxdale. The intervening beds presumably vee out in the valley of the Skip Beck downstream from a bore [2894 3818] 600 yd N.E. of High Croxdale, which proved Upper Stony Clay 13 ft, on sand 75 ft, laminated clay 36 ft, Lower Boulder Clay 11 ft. Much of the sand in this bore is thought to represent laminated silty clays with abundant sandy partings and bands of sand. To the south-east a bore [2929 3787], 800 yd E.N.E. of High Croxdale, proved a similar, but thinner,

sequence as follows: Upper Stony Clay 15 ft, on sand 12 ft, laminated clay 41 ft, Lower Boulder Clay 22 ft.

In the area between Shincliffe Colliery, Croxdale Hall and Bowburn, Upper Stony Clay forms a gently undulating plain. Bowburn Colliery tip stands on a flat which has a relief of less than 4 ft (between 305 and 309 ft O.D.) over a lozenge-shaped tract of nearly 100 acres. The flat-topped western and southern parts of the area rise to about 315 ft between High Butterby and Skip Beck, and to 325 ft between Skip Beck and Tursdale Beck. Between 200 yd S.W. and 650 yd S.S.W. of South Grange is a bench at 312 ft to 316 ft O.D. and this rises gently east-north-eastwards to another bench at 320 to 325 ft. Another flattish area lies to the west of Whitwell South Farm at about 335 to 340 ft O.D.

E.A.F.

Cornforth to Coxhoe. This area includes the buried tributary valleys which fall westwards to join a larger buried valley extending northwards along a sinuous course from the Ferryhill Gap to join the main buried valley of the River Wear near Shincliffe (Fig. 25). Two bores of uncertain location near Coxhoe station proved about 60 ft of stony clay which is presumed to extend beneath low hills of sand and gravel 200 yd N.N.W. and 300 yd N.E. of the station. At West Hetton Colliery the section is yellow clay 12 ft, on gravel 9 ft, and stony clay 36 ft. It is possible to interpret this section as representing Upper Stony Clay and Lower Boulder Clay with intervening Middle Sands and Gravels, but this tripartite sequence cannot be recognized from any of the other records in the area. Lower Boulder Clay is certainly present and locally may approach a thickness of 100 ft. In places, however, it seems to split into a number of layers separated by laminated clays, sand or gravel, and it is then difficult to define. At Clarence Hetton Pit, for instance, the sequence is Upper Stony Clay 18 ft, on sand 3 ft, blue clay 30 ft, sand 15 ft, blue clay 9 ft; and the Lower Boulder Clay may be equated either with the lowest three items or with the basal 9 ft of clay alone. In a bore [3133 3605], 750 yd N.N.W. of

East Pasture, the sequence is boulder clay ?32 ft, on laminated clay 4½ ft, boulder clay 1 ft, laminated clay 1½ ft, boulder clay 2 ft, laminated clay 6 ft, boulder clay 15 ft. Near the centre of the buried valley, a bore [3053 3614], 150 yd E.N.E. of Standalone, proved yellow sandy clay 4 ft, on laminated clay and sand 26 ft, Lower Boulder Clay 75 ft, argillaceous sand 21 ft. The yellow clay, which is probably an Upper Stony Clay, persists to the south-west, and at Tursdale Colliery it is 7 ft thick, overlying 25 ft of sand and gravel resting on blue clay 30½ ft, sand 3 ft, blue clay 20 ft. The last three items probably comprise Lower Boulder Clay with a parting.

A 15-ft bed of sand and gravel in a bore [3022 3571], 350 yd S.W. of Standalone, is overlain by 10½ ft of sandy clay and boulders, and is underlain by 12½ ft of sandy clay. In a bore 100 yd farther south the sequence—believed to represent the normal tripartite succession—is sandy clay 14 ft, on sand and gravel 20 ft, clay 45 ft. The sand and gravel crops out to the west, and was formerly worked in a pit [2998 3616], 600 yd E.N.E. of Broom Hill. South of Brandon House it seems to form part of a lenticular body elongated to the north-north-west. A disused pit [3042 3505], 150 yd N.W. of Brandon House, shows a passage from sand to gravel from east-north-east to south-west; the gravel is up to 30 ft thick 150 yd W. of the same house. Pits 500 to 600 yd N.E. of the house seem to have been worked in boulder clay for brick-making. The flat-topped sandy and gravelly hill upon which Brandon House stands, rises to just over 300 ft O.D. and is composed entirely of drift. To the west, the Coxhoe Beck separates it from a hill of similar shape again rising to just over 300 ft O.D. This hill is formed mainly by solid rock with an impersistent veneer of boulder clay and with a patch of gravelly sand at the northern end. It is entirely surrounded by alluvium, beneath which the drift is at least 30 ft thick. Another smaller hill to the south is similarly surrounded. The steep-sided valleys isolating these hills are probably partially-filled buried glacial drainage channels which are connected to the Ferryhill Gap to the south.

Much of West Cornforth stands on a gently sloping triangular feature composed of sand and gravel. The deposit, which may be linked under the alluvium of the Coxhoe Beck with the sand and gravel at Brandon House, also extends under the present Thrislington Colliery tip, and along the eastern flank of the Ferryhill Gap into the southern area. The sinking at West Cornforth Colliery (Thrislington Coke Oven Pit) recorded the following sequence at the western edge of the outcrop: brown soil 2½ ft on yellow clay 4½ ft, soft loamy clay 1 ft, sand and gravel 6½ ft, strong brown clay 17 ft, freestone gravel 17½ ft, strong brown clay 14 ft, loamy clay, with sandy partings 2½ ft, limestone gravel 4½ ft.

The area lying between Bowburn Beck, Tursdale Beck, Coxhoe Beck and Coxhoe undulates gently between 270 and 280 ft O.D. with a central alluvial patch lying somewhat lower. Upper Stony Clay is thought to form much of the surface, but the relationship of the surface clay to the clay north-west of Bowburn (p. 236) is not clear.

North of Bowburn Beck, Lower Boulder Clay is clearly indentifiable. It is 10 to 30 ft thick and is overlain by a thick sequence of bedded clays with partings and bands of sand. In a bore [3018 3792], 1050 yd N.N.W. of Crow Trees Farm, the succession is loamy clay 7 ft, on brown clay 16 ft, blue clay and sand 37 ft, stony clay 20 ft, blue clay 21 ft, stony clay 10 ft, laminated clay and sand 15 ft, sand 10 ft, stony clay 15 ft. Bowburn Colliery, 250 yd to the east, proved clay and sand 6 ft, on clay, silt and sand 6 ft, stony clay 25½ ft, laminated clay 9½ ft, sand 6 ft, strong clay 6½ ft, laminated clay 8 ft, loamy clay 10 ft, gravel and sand 8 ft, stony clay 11 ft. In the Old Quarrington area farther east, a single boulder clay seems to be present and this is presumed to be Lower Boulder Clay. Laminated clays have been worked in pits 200 yd N.E. and 550 yd S.S.W. of Coxhoe level crossing, and possibly also at Coxhoe Pottery, 400 yd N.W. of Coxhoe station.

Hett to East Howle. The glacial stratigraphy of this area seems to be relatively simple for most of the bores record only boulder clay or boulder clay and sand, up to about 75 ft thick. The ground rises to over 450 ft O.D. around and southwest of High Butcher Race, where it forms a flattish plateau.   Bores have proved a small buried valley falling south-south-eastwards under the main A1 road about 250 yd S. of High Butcher Race Farm. Excavations for petrol storage tanks at the filling station about 350 yd S. of the same farm were entirely in boulder clay. Drift is very thin in the ground between Mount Huley and Low Butcher Race, and also along the top of the ridge to Hett. The southern slopes of Tursdale Beck are also relatively free from drift.   To the west and east of the ridge, the drift thickens, especially in small buried valleys which trench the slopes.   West of Broom Hill sandstone crops out on both sides of a glacial drainage channel which falls to the south-east and is followed by a small stream which joins Tursdale Beck. A bore [2937 3590], 75 yd S.W. of Broom Hill, proved 13 ft of drift consisting largely of boulder clay.   A bore [2976 3564], on the alluvium 490 yd S.E. of Broom Hill,

proved 37 ft of boulder clay, and a comparable thickness was proved in a bore [2981 3601] 400 yd E.N.E. of the same farm.   In a further bore [2959 3646], 550 yd N.N.E. of the farm, the sequence is soil 3 ft, argillaceous sand 12 ft, laminated clay 15 ft, ?Lower Boulder Clay 27 ft. This lies near the western margin of the bedded clays of the Cornforth–Coxhoe area (p. 237).

Thin boulder clay is mapped over much of the northern slopes of Ferryhill below 500 ft O.D.   It is seen to be at least 12 ft thick 800 yd N.W. of Ferryhill Reservoir, where there may be a buried valley with its head near Dean and Chapter Colliery.   At the colliery site, excavations proved up to 25 ft of boulder clay without the base being reached.   At one locality, the boulder clay overlies yellow sand about 3 ft thick on white clay 3 ft, which has been interpreted (Brown 1905) as a lacustrine deposit.   A small glacial drainage channel is seen [280 336], 450 yd E.S.E. of Skibbereen. Its floor stands at about 360 ft O.D.   E.A.F.

INTERPRETATION OF THE GLACIAL SEQUENCE

The earliest Pleistocene deposits are those preserved in fissures on the coast between Hawthorn and Crimdon Beck.   Trechmann (1915, p. 60) suggested that the fissures lay open before the Scandinavian ice reached the area, and that they were filled with material carried to the area by the ice itself.   He considered that the rock fragments in the fissures were slickensided at this time by movement during compaction of the breccia under the weight of overlying ice.   Most fissures, however, contain only local Permian rock fragments, and these form the bulk of the breccia even where mixed with later material.   For this reason, and also because of their general association with collapse-brecciated country rock, it seems more likely that the fissures first formed during and after the solution of sulphates in the Middle Magnesian Limestone following post-Permian (?Tertiary) uplift and that most of the material from beds now eroded away fell in at about this time. Subsequent compaction of the resulting breccia would have provided space for the gradual accumulation of additional material towards the top of the fissure, and it is suggested that such compaction gave rise to surface depressions into which Pleistocene material was deposited ahead of or under the advancing Scandinavian ice.

The extent of the ice-sheet from which the Scandinavian Drift was deposited is uncertain, but in Co. Durham it seems likely that it moved only a short distance inland.

From the presence of the deposit which he interpreted as loess immediately overlying the Scandinavian Drift at Warren House Gill, Trechmann (1920, pp. 187–91) inferred that the withdrawal of the Scandinavian ice was followed by a prolonged period of sub-aerial denudation during which many of the deep buried valleys of the coastal area were eroded.

The Lower Gravels contain erratics similar to those in the overlying Lower Boulder Clay, but are unlike those found in the Scandinavian Drift. The gravels may be relics of a glacial episode intermediate in age between the Scandinavian Drift and the Lower Boulder Clay or even, locally, beach deposits derived from such an intermediate drift. A more acceptable interpretation, however, is that the gravels were deposited sub-glacially by Lower Boulder Clay ice as suggested by Trechmann (1920, p. 191).   D.B.S.

The Lower Boulder Clay is regarded here as the deposit of a major ice-sheet which covered the whole district. Striae in the central area (Trechmann 1915, p. 78 ; Woolacott 1921, p. 27) indicate that this ice moved towards the south or south by east, and this is consistent with the predominance of Carboniferous rocks and the absence of Magnesian Limestone in most of the Lower Boulder Clay in the western area, and with the scarcity of Upper Magnesian Limestone in the central area. In the southern part of the eastern area the ice may have been deflected eastward by Stainmore ice occupying the Tees Basin, for Trechmann (1915, p. 77, footnote 3) inferred a westerly or north-westerly derivation for stones in the Lower Boulder Clay at Hartlepool docks (p. 217). Supporting evidence for this suggestion comes from the distribution of erratics of Shap Granite, which are seen on the surface of the Lower Boulder Clay in the southern area and also in the eastern area as far north as Elwick (Trechmann 1915, p. 77). The subdivisions of the Lower Boulder Clay can be interpreted as evidence of recurrent movement of the ice-sheet, but they seem equally likely to have been formed by overriding of different streams of ice.   D.B.S., E.A.F.

In the eastern area, the upward passage by alternation from Lower Boulder Clay to the Lower Division of the Middle Sands suggested that a time interval between the two is unlikely and the sands are interpreted as sub-aerial outwash of waning Lower Boulder Clay ice. The cutting of channels into and locally through the boulder clay and the filling of these with conspicuously cross-bedded coarser gravels, may be taken to indicate that at first water was abundant and fast-flowing. The scarcity of pebbles and level bedding of the silty deposits towards the top of the division, however, suggest a later decrease in turbulence and perhaps a reduction in supply of sediment and flow of water. It is not known how far the present distribution of the Lower Division of the Middle Sands reflects their former extent, for part of the variation in their thickness is clearly due to subsequent erosion which in some places, as in the cliffs north of Dene Mouth, Horden, removed them completely before the Upper Division of the Middle Sands was deposited. The latter is interpreted as outwash from, or partly beneath, the advancing Upper Boulder Clay ice. The implication is, therefore, that there was a considerable interval between deposition of the two divisions of the Middle Sands. The Easington Raised Beach (pp. 204, 208) may have been formed during that interval.   D.B.S.

In the western area most, if not all, of the Middle Sands, Gravels and Clays was probably formed during the decay of the Lower Boulder Clay ice (pp. 243–5) and can thus be correlated with the Lower Division of the Middle Sands of the eastern area. In the west, however, no evidence has been found of an interval corresponding with that in the east.   E.A.F.

The Upper Division of the Middle Sands and the Upper Boulder Clay contain no pebbles derived from the central area, but they carry abundant fragments of Concretionary Limestone of a type found only near Sunderland, and this is taken to indicate that this ice advanced from the north. Similarly, the distribution of the Upper Boulder Clay and the lack of Concretionary Limestone pebbles in the central area suggest that the ice moved inland to only a short distance west of the present 400-ft contour before halting and depositing a north–south ridge of morainic material extending between Easington and Hutton Henry. The advancing ice is believed to have blocked existing streams draining eastwards and thus to have formed lakes in which the laminated clays of the central area were deposited (Fig. 29). The earliest of these clays occupy individual valleys, but field relationships suggest that they were overridden as the ice continued to advance and that the lakes coalesced and deepened before eventually overflowing westward through a series of impressive channels which unite before breaching the Permian escarpment near Town Kelloe. The lakes may at one time have extended as far east as Sheraton, for there a large mass of laminated clay and sand, locally overthrust towards the south, has been smoothed-off and apparently largely eroded away by the ice from which some 20 ft of Upper Boulder Clay were subsequently deposited. Alternatively the whole mass may be a raft forming part of a push-moraine.

South and West of Sheraton the Upper Boulder Clay and associated morainic drift are patchy, and there is no well-defined terminal moraine, but according to Flint (1957, p. 161) such ' attenuated ' drift margins are at least as extensive in North America and Europe as margins marked by end-moraine. The only linear morainic ridges in this part of the district lie north and south of Sheraton itself, and it is suggested that these mark the line of a temporary stand or slight readvance during melting. Subsequent melting is believed to have taken place *in situ* while the ice was stagnant, giving rise in the area between Castle Eden, Sheraton and Hart to kame and kettle moraine. There, steep slopes interpreted as ice-contact faces appear to be randomly oriented, but on the eastern slopes of the ridges at Sheraton such faces all slope east-north-eastwards.

During the Upper Boulder Clay glaciation of the eastern area, the evidence strongly suggests that the central area and parts of the southern area remained unglaciated, and, that except where covered by ice-dammed lakes, they were subjected to intensive sub-aerial weathering. Because of the ice to the east it may be supposed that the drainage was temporarily deflected to the west, and the east–west glacial drainage channels (Fig. 29) in the western part of the central area may be evidence of this. The gravels which are associated with the channels and which overlie Lower Boulder Clay may also belong to this period. As might be expected, the flora of the central area, as recorded by pollen in the lake clays, indicates cold conditions, but interpretation of the evidence is complicated by the presence of pine-pollen which may be derived.   D.B.S.

The sequence of events in the western area during the Upper Boulder Clay glaciation is uncertain. Because the Upper Stony Clays may represent a return to cold conditions, a case can be made for correlating these deposits with the Upper Boulder Clay. This is discussed further in a later section (p. 243) together with the further possibility that some of the laminated clays of the west may be formed in a manner analogous to, and contemporaneously with, those of the central area. There are grounds for supposing, however, that the deposition of the Upper Stony Clays took place as a sheet during the same glaciation that was responsible for the Lower Boulder Clay and the Middle Sands, Gravels and Clays. Because some at least of the material appears to be boulder clay, the agency may have been a local readvance of ice. Some support for this is found in local disturbance, including thrusting, in the laminated clays north-east of Low Newton and in a bore near Croxdale Wood House. The possibility remains, however, that the patches now surviving may have been formed in various ways: some may be ground moraine, others may be englacial tills and still others may be solifluxion deposits. They are not thought to include ablation moraine, because they generally contain a high proportion of clay and no angular debris.

Evidence taken to indicate that the deposition of some, at least, of the Upper Stony Clays followed soon after the formation of the Middle Sands, Gravels and Clays is found near Tursdale Colliery. There the eastern side of a ridge of gravel and sand flanking a patch of alluvium is partly covered by stony clay, whereas the western side is bare of clay and this suggests that it was either protected by the detached mass of stagnant ice invoked on p. 244 or trimmed by meltwater passing from the relict mass of ice to the north of Durham on its way to the Ferryhill Gap. Such meltwater would have followed a devious course round the margin of the suggested readvance along the present line of the Tursdale Beck, though in the opposite direction to the present flow.

A single frost-wedge pseudomorph has recently been found in the gravel cap north-east of Sherburn and this is the only undoubted example of periglacial activity in the western area. Its date and the date of the deep weathering of the Upper Stony Clays are unknown. E.A.F.

Part of the sequence of events following the withdrawal of the Upper Boulder Clay ice can be inferred from the relationship of late-Glacial and post-Glacial deposits to morphological features such as the 80-ft marine planation (p. 246). The laminated clay around Claxton which extends into the district from neighbouring Tees-side is up to 40 ft thick in the latter area, where it overlies unweathered and apparently only slightly eroded Upper Boulder Clay and passes upward by alternation into sand up to 30 ft thick (Smith 1965, p. 58). From the nature of the contact between the laminated clay and the Upper Boulder Clay it is inferred that the laminated clay around Claxton is of Late-Glacial age. Woolacott (1921, pp. 67–9) and Radge (1939, fig. 2) have suggested that the laminated clay of Lower Tees-side was deposited in a large lake held up by retreating Cheviot (Upper Boulder Clay) ice lying to the east, and this interpretation is accepted here. The clays may thus be contemporaneous with, or only slightly younger than, the Morainic Drift and Upper Gravel. It is not known whether the laminated clay is older or younger than certain weakly developed surfaces at about 140 and 95 ft O.D. (p. 246), but it is certainly older than deposits which rest on the 80-ft planation and which are dated by radiocarbon measurement (p. 247). D.B.S.

The origin of the 'prismatic clay' is controversial. In places it is rudely bedded and retains many of the features of the underlying deposit and here derivation *in situ*, perhaps by deep frost action, seems likely. Elsewhere, however, it contains stones of a type not found in the underlying deposit, and may there represent a concentrate from a bed now eroded away or be an allogenic solifluxion deposit.

GLACIAL CHRONOLOGY AND CORRELATION

Trechmann (1920) suggested that the Scandinavian Drift is post-Cromerian because although nowhere seen to overlie fissures containing undoubted early Pleistocene deposits it does rest on similar, but unfossiliferous fissure deposits. He further suggested (op. cit.) that its deposition was closely associated in time with the final infilling of the fissures and for this reason it may be assigned to the Lowestoft Stage. This is also the age tentatively assigned to it by West (1963, pp. 161–3), who follows Lamplugh (in discussion of Trechmann 1915, p. 80) in correlating the Scandinavian Drift with the lithologically similar shelly Basement Clay of Holderness and extends this correlation also to the Lowestoft Till and North Sea Drift of East Anglia. It follows that the overlying interglacial ' loess ' deposits are likely to be Hoxnian.

The Lower Boulder Clay, which overlies the ' loess ' and is therefore post-Hoxnian, is correlated with reasonable confidence throughout all four areas of the district by virtue of its lithology and physical continuity. Field mapping and borehole records show it to continue northwards into the Sunderland district and southwards over much of Tees-side. Still farther south it bears a strong resemblance to the Drab Boulder Clay of Holderness which Penny (1959, 1964) assigned to the Main Würm (late Weichselian). The evidence of the Easington Raised Beach, discussed below, conflicts with this by suggesting a Gipping (Saale) age for the Lower Boulder Clay. The Weichselian age of the Upper Boulder Clay of the eastern area, however, does not appear to be in question. This deposit can be correlated with the (upper) Red Boulder Clay of the Cleveland Hills (Radge 1939), with the uppermost boulder clay of the north-east Yorkshire coast, and farther south with the Hessle and Hunstanton tills which are generally accepted to be of Weichselian age (Penny 1959, 1964 ; West 1963). Moreover at Neasham in the Stockton (33) district its correlative is overlain with no apparent break by late-Glacial clays of late Zone I age (Blackburn 1952).

The Easington Raised Beach may be evaluated as follows. It appears to underlie, and would seem, therefore, to be older than, the Upper Boulder Clay. Its temperate fauna indicates interglacial rather than interstadial conditions and this, taken in conjunction with the established minimum age of 38,000 years (p. 208) implies that the beach must be pre-Weichselian, that is not younger than Ipswichian. The field relationships suggest strongly, though not conclusively, that the beach is younger than the Lower Boulder Clay, and this suggestion is supported by its suite of pebbles which could not have been derived from the Scandinavian Drift though they could be from the Lower Boulder Clay. A strong case can therefore be made for claiming that the Lower Boulder Clay, together with its associated sands and gravels, are products of a Gipping (Saale) glaciation. The alternative view—namely that both boulder clays are Weichselian—can only be upheld on the assumption that the Easington Raised Beach pebbles were derived from deposits of an earlier non-Scandinavian glaciation which, with the doubtful exception of the Lower Gravels, are nowhere known to be preserved in Durham.

The correlation of the deposits overlying the Lower Boulder Clay in the western area with those of the eastern, central and southern areas is closely bound up with this question of whether the Lower and Upper Boulder clays in the east represent separate glaciations of Gipping and Weichselian ages respectively or whether both are deposits of the later glaciation. If both are Weichselian then the break in the eastern area between the Lower and Upper divisions of the Middle Sands is of only minor significance and there is no difficulty in correlating the Upper Boulder Clay of that area with some, if not all, of the western Upper Stony Clays as deposits of readvancing ice. If, on the other hand, the Lower Boulder Clay is Gipping, two alternatives must be considered. One is based on the suggestion made above, that the Upper Stony Clays were deposited from a readvance of Lower Boulder Clay ice and if this is so the whole of the Middle Sands, Gravels and Clays and Upper Stony Clays of the western area are equivalent merely to the Lower Division of the Middle Sands of the eastern area. Moreover, there would be no deposits in the western area to represent the Upper Boulder Clay (Weichselian) glaciation in the east. The other alternative is that the western succession does contain a break like that of the east, though it is nowhere apparent, and in this event some of the laminated clays and interbedded sands of the western area may be later than the sands and gravels demonstrably deposited by decaying Lower Boulder Clay ice. They might thus be, like the laminated clays of Wingate and Edder Acres, deposits formed in a pro-glacial lake dammed by Upper Boulder Clay ice along the eastern seaboard. This alternative interpretation is reminiscent of the ' Lake Wear ' postulated by Raistrick (1934), though his concept of the Durham Gravels as delta deposits within that lake is no longer acceptable (p. 244).

D.B.S., E.A.F.

DEGLACIATION OF THE WEAR VALLEY

An early phase in the decay of the Lower Boulder Clay ice in the Wear Valley around, and to the south of, Durham City is probably represented by glacial material rising to between 360 and 375 ft O.D. and partially blocking tributary valleys at Littletown and Running Waters. The glacial margin at this time probably extended from Littletown, where there is a small marginal drainage channel, round the flanks of Sherburn Hill to Running Waters and thence to abut against the western tip of Quarrington Hill. It continued around the flanks of the embayment between Coxhoe, Cornforth and Hett, and passed beyond the boundary of the district south-west of Hett. The ice would thus have formed a lobe which occupied much of the ground below the present 400-ft contour and which would have owed its initial formation to the emergence of the high ground immediately to the east and south of the area through the downwasting ice. This high ground is now breached to the south by the Ferryhill Gap (p. 245).

97544   R

After this early phase, the lobe of ice may be supposed to have contracted rather rapidly until the margin lay along the line now marked by a ridge extending across the area from East Rainton to Gilesgate Moor. Although this ridge ends at Durham City at a gap now traversed by the River Wear, it has a western counterpart extending from south of the city northwards along the boundary of the district to Plawsworth. During this major phase of contraction, the marginal ice was probably thin, and separate masses of stagnant ice are thought to have become isolated, as for instance west of Tursdale Colliery and immediately south of Hallgarth, where tributary valleys are now floored by broad tracts of alluvium, followed downstream by narrow valleys. One of these alluvial patches is joined to the west by a glacial drainage channel, and is flanked to the east by a ridge of gravel and sand, whose form and bedding suggest that it originated as an esker beneath stagnating ice. Sediments were probably deposited around these stagnant masses as the margin of the ice lobe drew back, and meltwater began to fill up the ground between the margin and the high ground.

In the area lying to the east of the margin, the supply of sediment was periodic judging from the interbedded sequences of sands, clays, and subordinate gravels built up to approximately 300 ft O.D. In the area northeast of Sherburn the high proportion of Magnesian Limestone in the topmost layer of gravel, together with the sedimentary structures in the immediately underlying sands, suggests derivation from the east. The presence of a number of drainage channels (not those shown in Fig. 29) dissecting the central area between Hetton le Hole, Haswell and South Hetton accords with this conclusion.

To the west of the lobe, however, sedimentation was probably more continuous and gave rise to the thick sequence of sands with subordinate clays extending north and south on the western side of Durham City. At Framwellgate Moor they are built up to flattish or gently undulating terrace features at about 350 ft O.D. which fall successively in altitude eastwards to just over 300 ft O.D. north of Durham City. Parts of these features, especially where capped by sand and gravel, are hummocky, and resemble kame-terraces. The surface of the ridge to the east of the River Wear, between East Rainton and Gilesgate Moor, also stands at various levels between about 330 ft and 300 ft O.D. The falling levels of these features are interpreted as reflecting the downwasting of the ice as the lobe contracted. That the features are at least in part marginal is supported by the presence of a number of ill-drained hollows, which appear to be kettle-holes, on the eastern slope of the lowermost feature near Newton Grange and Red House.

It may be supposed that as the southern limit of the lobe drew back towards Old Durham, large quantities of meltwater were liberated and the marginal ice became stagnant. The arch-bedding of linear belts of gravel in the lower part of Durham Quarry suggests a sub-glacial origin with transport and deposition effected by meltwater streams flowing under the wasting ice. These streams were probably disgorged below the surface of the rising waters of a pro-glacial lake to the south, and the finer fractions of sediment were largely carried out and deposited on the floor of this lake. The sub-glacial gravels, where not terraced by later river activity, are overlain by sands in which slumping and contortion (Pl. XVIII) can be explained as the result of the melting of relict masses of ice enclosed within the sediments. The gravels, however, are not disturbed and this supports the conclusions that the ice was stagnant. The gravel sheets interbedded with sands in the upper part of Durham Quarry were probably deposited by streams flowing subaerially from the ice. Near Old Durham, the gravels are proved to over 150 ft above O.D. and they can be traced north-westwards to Pelaw Woods at about the same level. To the south, the gravels seem to be replaced first by sand and then by laminated clay (Pl. XVI, Sections 3 and 4) suggesting that as the supply of coarse sediment waned, the deposition of clay began successively closer to the source of supply until eventually it covered the site of Durham Quarry. It seems likely that by this time the pro-glacial lake was largely filled by sediments, and trails of organisms found on some of the bedding planes in the upper part of the sequence in bores near Croxdale Wood House and High Butterby suggest that sedimentation was interrupted from time to time.

The water level of the pro-glacial lake during this phase of deglaciation may have been controlled by the height of the Ferryhill Gap, which now stands at 265 ft O.D. The gap is interpreted as having been eroded by southward flowing meltwater. This follows Raistrick (1931, p. 288), who believed that the gap was cut by water escaping from a 'Lake Wear' which occupied the whole region between Weardale and the River Tyne, blocked by an ice-margin lying across north-east Durham. This view was contrary to Woolacott's (1921) suggestion that the gap had been formed by meltwater escaping northwards from a glacier occupying the Tees Valley. It is now clear that the gap has had a more complicated history than previously thought. East and north of Swan House, it is floored by thick drift thought to be largely Lower Boulder Clay because it is overlain to the south by sands and gravels which can be traced into deposits mapped elsewhere as Middle Sands. It continues southwards as a buried gap or valley at least as far as Mainsforth Colliery, which it passes closely to the east, while the present gap diverges from it south-east of Swan House. Boring and topographical evidence suggests that the buried gap has steep sides. It is possible that the gap was in existence before the Lower Boulder Clay glaciation. Alternatively it may have been formed contemporaneously, either by pro-glacial meltwater in front of the advancing ice or by a river flowing beneath that ice.

The distribution of post-Lower Boulder Clay deposits farther to the north of Durham City is in many respects similar to that of the equivalent deposits around and to the south of the city, particularly in respect of the apparent lateral passage from sands to laminated clays (Pl. XVI, Sections 1 and 2). If the deposits of these two areas are contemporaneous and were formed by comparable processes it would seem to follow that the contraction of the margins of the lobe of ice south of Durham must have been accompanied by contraction to the north, and that the lobe broke up to form one or more detached masses. Because these deposits extend into the ground farther to the north and west, further discussion of their interpretation is deferred until field survey is completed.   E.A.F.

Terrace of Warp and Other Coastal Planations

Ground interpreted as Terrace of Warp forms a continuous belt from Crimdon Beck to the southern margin of the district. The terrace falls gently seawards from the base of the backslope at about 80 ft O.D., and is composed of a number of ill-defined sub-terraces distinguished by slight changes of slope. Coastal recession has bitten deeply into the terrace north of Hartlepool, and it is only 500 to 600 yd wide at Hart station. Farther south, however, it is up to 2 miles wide and at its seaward end grades almost insensibly into recent alluvial flats at about 12 ft O.D. The terrace is the northward extension of a marine planation recognized by Agar (1954, pl. xvi) in the Tees Valley, where its landward margin is defined by a 'late-glacial' shoreline at 82 ft O.D. Agar's shoreline at 41 ft O.D. has not been traced within the district.

The terrace is essentially an erosional feature, cut into Magnesian Limestone at West Hartlepool and into drift elsewhere. Marine deposits occur only in small patches, rarely exceeding 3 ft in thickness and consisting mainly of brown to grey clay, commonly homogeneous and stoneless, or with small well-rounded stones. They appear to be re-worked material from the underlying drift, into which they grade insensibly. Locally, these deposits contain thin beds and lenses of silt, sand and gravel, as in temporary exposures [487 357] near Springwell House, Hart, where an extensive bed of sand and gravel rests on an eroded surface of boulder clay and Magnesian Limestone (p. 214). No marine shells have been found in this sand and gravel, but small incrustations of calcium carbonate, thought to have been produced by marine worms, are present in cracks and on the upper surface of the limestone. No incrustations have been found where the limestone is overlain by boulder clay. Gravel lying at the foot of the backslope of the terrace was exposed during 1942 in temporary excavations near Springwell House, and was interpreted by W. Anderson (*in* discussion on Trechmann 1947, p. iii) as a raised beach deposit.

In addition to ground depicted as Terrace of Warp, remnants of further surfaces which may be older marine planations lie at about 140 and 95 ft O.D. The former is an ill-defined flat best preserved at Black Halls where streams are graded to its western edge. The latter extends for about 750 yd N. of Black Halls Rocks and is also found above Chourdon Point. Farther south, the rear edge of this surface may be represented by a low rounded slope between Crimdon Beck and Middle Warren Farm, but this feature cannot be traced through West Hartlepool where the inland edge of the 80-ft planation coincides with a prominent ridge interpreted as a raised storm beach (see below). No marine deposits have been found on the 95-ft and 140-ft surfaces, although redistributed drift locally fills depressions in the surface of the underlying deposits.

Near the southern margin of the district a number of alluvial flats lie between 80 and 110 ft O.D., but without detailed levelling it is not possible to relate these to either the 80-ft or 95-ft surfaces.   D.B.S., G.D.G.

The even surface of the 80-ft planation and the degree to which minor streams around West Hartlepool are graded to its rear edge suggest that sea-level was constant for a considerable period. The coast subsequently receded to the east of its present position, as is indicated by the presence of the submerged forest peat (p. 248) which lies on the surface of the planation and extends to at least 35 ft below present mean sea-level. The dating of an antler from the peat (p. 249) shows that the minimum age of the period of relatively low sea-level, and hence of the late stage of the period of planation, is $8,100 \pm 180$ years to $8,700 \pm 180$ years B.P. Because the formation of the peat must post-date the cutting of the planation on which it lies and also because highest parts of the planation were presumably cut well before the lowest (the known range of at least 130 ft would represent a period of nearly 4,000 years at an annual rate of relative sea-level change of 1 cm), the coastline established at 80 ft O.D. must be considerably older than 8,700 years B.P. This argument independently supports the suggested Late-Glacial age of the laminated clays into which the planation cuts around Claxton.

The dating of the 80-ft marine planation at more than 8,700 years B.P. and perhaps early Post-Glacial throws some light on the amount of isostatic depression suffered by the area during the Upper Boulder Clay (i.e. late Weichselian) glaciation. According to Godwin and others (1958, p. 1518) the general sea-level at 8,700 years B.P. was about 80 ft below its present position, having risen steadily from more than 200 ft below its present position at about 14,000 years B.P. (a rise of about 90 cm per century). From the date quoted by Godwin and others, and assuming the 80-ft shoreline to have a minimum age of 9,000 years B.P. and a mean sea-level 15 to 20 ft below the shoreline, an isostatic depression of at least 150 ft is indicated. Since isostatic recovery is exponential, however, it is clear that this figure is almost certainly far too low, since much of the isostatic recovery would already have taken place before the planation was formed. It would, in any event, be considerably greater if the 140-ft (?Late Glacial) planation should prove to be marine, for this pre-dates the 80-ft planation and was therefore formed when the general sea-level was well over 100 ft below its present level. An isostatic depression of 300 to 350 ft would be caused by about 900 to 1,050 ft of ice, a reasonable figure bearing in mind that Cheviot erratics from this ice have been found up to 1,000 ft on the Cleveland Hills (Radge 1939, p. 185), the top of which was unglaciated at this time.   D.B.S.

RAISED STORM BEACH DEPOSITS

Sand, with some gravel, forms a low ridge, 300 to 400 yd wide and trending north–south for over $3\frac{1}{4}$ miles between Throston and Greatham, on the neighbouring part of the Stockton (33) Sheet. The ridge lies at the landward edge of the Terrace of Warp, and is interpreted as a raised storm beach deposit genetically related to that terrace. It is widest (about 650 yd) a short distance north of Rift House [495 310], where it has an asymmetrical profile with the steeper slope facing west, but elswhere the ridge is rounded and roughly symmetrical in profile.

The material of the beach, exposed in many small quarries and temporary excavations, is mainly yellow, fine- to medium-grained sand, but fine, or rarely medium, gravel is relatively common. The pebbles in the gravel are well-rounded, and 'pea-gravel', made up almost entirely of rounded pebbles of 7 to 10 mm diameter, is locally abundant. Bedding is generally horizontal or dips gently eastwards, and graded-bedding can be seen at several localities. Shells are rare, but fragments resembling *Cardium sp.* were found in temporary excavations near Owton Manor, and Trechmann (1915, p. 74) recorded shells in current-bedded gravel in a road cutting [4935 3316] at Throston near the northern end of the ridge. From there to Catcote, 1 mile to the south, sand forms the top of the ridge, gravel the western slope, and gravel, passing eastwards into sand, the eastern slope. Temporary excavations in 1955–56 near Owton Manor [493 294] proved coarse gravel at the western edge of the deposit ; from there it passes eastwards into sand containing lenses of fine gravel, then into pure sand and finally into grey sandy and stoneless clay. The beach here lies upon boulder clay, but at Throston it lies against the eastern flank of a mound of morainic drift. Some of the stoneless sands of the deposit may originally have been redistributed by wind. South of Owton Manor the ridge is low and its upper surface locally falls below 80 to 82 ft O.D., the upper limit of the Terrace of Warp. Here the sand and gravel may represent a bar rather than a beach deposit.   G.D.G.

SUBMERGED FOREST

Beds of peat, termed ' submarine forests' (Howse and Kirkby 1863, p. 4) or ' submerged forests' (Raistrick 1931, p. 290 ; Trechmann 1936, p. 161), crop out in a discontinuous belt up to a mile wide from North Sands, Hartlepool, to the foreshore oppsite Long Scar, West Hartlepool (Trechmann 1936, fig. 1). The peat is generally $1\frac{1}{2}$ to 2 ft thick, but it reaches 17 ft in the dock area at Hartlepool. The base of the deposit ranges from about 35 ft below O.D. to 12 ft above O.D.

In the northern part of its outcrop the peat is known only in boreholes and excavations, the most northerly record being in a trench [504 354], 2 miles N.W. of Hartlepool Coast Guard Station, where Trechmann (1936, pp. 164–5) described 2 ft of lignite-like peat lying on boulder clay beneath beach and dune deposits. Slightly farther south, the peat is $1\frac{3}{4}$ ft thick in a bore [5061 3525] at the Palliser Works (Trechmann 1941, p. 324), where it is overlain by 51 ft of beach and dune deposits. Numerous boreholes in the dock area at Hartlepool prove over 10 ft of peat overlain by or interbedded with recent marine or estuarine alluvium. In dock excavations, Cameron (1878, pp. 351–2) recorded peat up to 8 ft thick which contained isolated horns of ruminants and tree trunks up to 20 ft long and $3\frac{1}{2}$ ft diameter. South of the docks, the deposit passes under Hartlepool Bay, where dredgers occasionally bring up peat, tree stumps and trunks, and hazel nuts.

Peat of the submerged forest is now seen at the surface only on the foreshore opposite West Hartlepool (see map and photograph in Trechmann 1947), where it forms a number of isolated patches in the intertidal zone. Here it is generally less than 2 ft. thick and is a dark brown mass of vegetable matter containing roots, stumps and trunks of trees. The roots in many places penetrate into the underlying deposit, the top of which is commonly leached to grey or grey-blue. Trechmann (1947, p. 25) recorded lenses of ' grey calcareous marl ' containing *Helix nemoralis* near the base of the peat in some exposures, and (1947, p. 27 ; 1952, p. 177) blue sticky clay containing *Scrobicularia piperita* [*sic*] near the top of the peat ' towards Longscar Rock '. This clay grades laterally and vertically into the peat, indicating that the peat accumulated close to sea-level. The most southerly record of peat within this district is on the foreshore opposite Long Scar, where $1\frac{1}{2}$ to $2\frac{1}{2}$ ft of this deposit was excavated during 1956 at about 7 ft below O.D. Farther south, its former presence is indicated by widespread grey-blue clay similar to that which underlies the peat north of Long Scar and remnants of the peat itself are preserved on the foreshore from 330 to 520 yd south of the southern margin of the district.

In addition to shells and plant material, the peat at West Hartlepool has yielded bones of *Bos* and bones and antlers of several species of deer. (Trechmann 1947, pp. 23–32). Isotopic dating (Barker and Mackey 1961, pp. 41–42) of an antler collected by Trechmann gave ages ranging from $8,100 \pm 180$ to $8,700 \pm 180$ years B.P. The lower figures were obtained on material from the soft core of the antler, and are thought to indicate contamination by younger organic tissue. The isotopic age agrees well with Trechmann's estimate (1947, p. 29) of 8,000 to 9,000 years B.P. based on the cultural stage of Mesolithic chipped flint implements from the peat (see also Trechmann 1936, pp. 161–8). G.D.B., D.B.S.

Marine or Estuarine Alluvium

Marine or estuarine alluvium forms much of the low-lying areas between Hartlepool and West Hartlepool and south of Seaton Carew. In the vicinity of Hartlepool Dock it is widely overlain by made ground, but in foundation bores and temporary excavations it consists mainly of blue and grey silt and mud with fairly abundant marine shells. Generally it overlies or is interbedded with peat, and is locally over 20 ft thick. In a number of boreholes sand and gravel are present at the top or bottom of the deposit. North of the docks the alluvium is a grey-brown or yellow-brown stoneless clay, seen only in rare excavations.

South of Seaton Carew, marine alluvium underlies a veneer of blown sand in a flat area between dune ridges, and is the surface deposit at Seaton Snook. Drainage ditches at the latter locality have penetrated up to 18 in of pale yellow or yellow-brown stoneless clay, overlying 6 ft of brown stoneless clay. Recent marine shell fragments are scattered sparingly throughout. G.D.G.

Calcareous Tufa

Calcareous tufa has been noted and is still being formed at two localities in eastern Durham—one on the east side of a small valley [4104 3991] at Shotton, and one in a field [395 344], 1000 yd N.E. of Dropswell, Trimdon. At Shotton the tufa is cream-coloured and forms sheets and pockets over and around a mass of calcreted gravel. At the other locality it forms a semi-circular sheet about 70 yd by 35 yd in extent, downslope from a spring issuing from the base of a large deposit of gravel. The tufa here is cream-grey and is over 3 ft thick where seen in temporary exposures near the middle of the outcrop. D.B.S.

CAVE DEPOSITS

Caves at a number of places in the Magnesian Limestone have been filled or partially filled with rock rubble in a matrix of brown clay and silt. In the lower part of one such deposit in a cave at Bishop Middleham, Raistrick (1933) recorded abundant mollusca and bones of small reptiles, which he considered to be of Atlantic age. Higher parts of this deposit contain bones of man and other large mammals, associated with Iron Age implements.

D.B.S.

BLOWN SAND

Deposits of recent blown sand fringe the coast from Horden southwards. In places between Horden and the mouth of Crimdon Beck it forms a veneer on the south-facing slopes of valleys cutting the coast and of depressions in the face of the cliffs, but the largest patches are on rolling land behind the cliff tops immediately north of Warren House Gill and Crimdon Beck. Auger-holes and temporary excavations prove the sand to be only 2 to 3 ft thick in most of these patches, though it is 6 ft thick and contains gastropod shells on the side of a valley [455 408], 200 yd N. of Dene Mouth.

From Crimdon Beck to Hartlepool, blown sand forms a continuous dune ridge up to 50 ft high and 300 yd wide. It is 47 ft thick in a borehole [5061 3525] 2 miles N.W. of Hartlepool Coast Guard Station. At Middleton, Hartlepool, a small patch of somewhat older blown sand is now built over. Farther west, similar sand forms two small mounds [507 347 ; 513 345], resting on alluvium.

South of West Hartlepool two generations of blown sand are distinguished. The earlier forms a narrow belt extending southwards into the Stockton (33) district. North of Seaton Carew it has been levelled and built over, but to the south it forms a low ridge on the east side of Seaton Snook and is exposed in small sections on the golf links. At the time of the original geological survey this ridge was shown at the head of the contemporary beach, but accretion of beach material following the building of the North Gare Break-water has extended the land eastwards for 200 to 300 yd. The new land so created is largely covered with a prominent active dune ridge—the later generation of blown sand—which is up to 30 ft high and 150 yd wide. Alluvium between the two ridges is covered with a veneer of sand up to 2 ft thick.

D.B.S., G.D.G.

LACUSTRINE ALLUVIUM AND PEAT

The sites of former lakes are marked in several parts of the district by spreads of lacustrine alluvium. In the western, central and southern areas they form linked or discrete areas of flat, generally poorly drained ground (' carrs ') along stream valleys where they fill or partially fill natural hollows subsequently drained by normal stream development. In the eastern area and parts of the central area, many of the former lake deposits lie in the morainic tract and occupy closed hollows or hollows drained on only one side ; these are generally unrelated to the present drainage pattern.

Because of the high water table, sections in the lacustrine alluvium are generally shallow and in most places they consist of poorly bedded yellow or brown silt or clay. Sand is uncommon, and gravel rare. Temporary sections and boreholes show that the clay and silt, locally laminated, continue in depth, and that peat is in many places an important constituent, especially from about 2 to 8 ft below the surface. In the closed hollows of the eastern area peat is thought to be a major component of several of these deposits, and it may be thick in some of the former kettle holes.

The age of the lacustrine alluvium in the linked basins of the western, central and southern areas has not been established, though it may be partly or wholly Glacial. Farther east, however, the lake deposits are assumed to be late-Glacial and post-Glacial because they lie on Upper Boulder Clay. They are probably comparable with the similar lake deposits at Neasham (Blackburn 1952) which contain a nearly complete post-Upper Boulder Clay pollen sequence. It is likely, however, that the bulk of the higher peat is of late Boreal or Atlantic age, and in some basins peat is still in process of formation.   D.B.S.

STREAM DEVELOPMENT

In the catchment area of the River Wear the regional contours, as lines joining equal summit heights, are concave to the west, so that they trend south-westwards on the south and north-westwards beyond the northern margin of the district. These lines suggest the strike of the surface upon which the consequent drainage was instituted. To the west, Weardale represents the course of the upper part of the proto-Wear, which flowed east-south-eastwards across what is now the Shildon Gap into the Tees basin. The proto-Browney flowed slightly south of east across the centre of the district and part of its course is represented today by the valley of Shadforth Beck, though the modern stream flows in the opposite direction. It is generally accepted that a subsequent tributary of the River Tyne cut back southwards and successively captured the consequent streams, including the Browney, and finally the Wear itself. The wind-gap of the proto-Wear is at about 550 ft O.D. south of Grange Hill, near Westerton, and the present Wear follows the valley of the subsequent stream from Bishop Auckland to Chester-le-Street. As a result, such streams as the proto-Shadforth Beck became obsequents with reversed drainage.

The major lines of the pre-glacial drainage system in this area are suggested by the map of the buried valleys (Fig. 25). Although some of the valleys may have been modified by meltwater or other glacial erosion the similarity of the modern drainage pattern to the buried valley system suggests that the latter was formed largely by normal fluvial and erosional processes. Certain peculiarities in their longitudinal profiles, such as a sudden fall between Cocken and Chester-le-Street from about -80 ft to about -140 ft O.D., and the hanging of tributaries such as the buried valleys of the River Deerness and the Old Durham Beck to the main valley, seem to be too fresh and abrupt to be satisfactorily explained as surviving nick-points. They are also difficult to explain by glacial erosive super-deepening, and may be at least in part due to erosion by glacial meltwater (p. 199). Apart from these irregularities, the average fall of the buried valleys is slightly greater than that of the present Wear Valley. The tributary buried valley profiles are significantly steeper, since they head at a similar level to the present valley and fall to the main buried valley at about 100 ft lower than the present river.

Terraces are preserved along the River Wear around High Houghall, Shincliffe, and the south-eastern part of Durham City, and a single terrace has been mapped near Frankland. At High Houghall, four terraces are cut into sand and gravel at 195 to 185 ft, 170 to 165 ft, 155 ft, and 145 to 140 ft O.D. The lowest level is somewhat generalized and may include parts of two terraces, but topography is confused by meander migration. The alluvium of the river here lies at about 130 to 135 ft and is bordered on the river bank by well-marked levees. These terraces can be traced inter-mittently downstream as far as Durham, the highest falling about 10 ft and the lowest 15 to 20 ft. North of Durham, flights of terraces are absent, but a feature at about 165 ft at Frankland is assigned to the highest terrace. A narrow strip in the gorge south-east of East Moor Leazes stands at 115 to 120 ft, 15 to 20 ft above the alluvium and is equated with the lowest terrace. The dominant feature in this area, however, is a large very gently undulating terrace which lies on both sides of the river gorge between Low Newton and Leamside. Its rear edge stands at about 190 ft and towards the river it stands at 176 to 178 ft O.D. The surface material is a light sandy or loamy clay with a low pebble and boulder content consisting mainly of fairly well-rounded and weathered sandstones, dolerites and volcanic rocks. The material weathers like alluvium, but although its rear edge near Leamside is embayed in places, no traces of abandoned meanders are evident in spite of its wide extent. It is tentatively thought to have originated immediately after deglaciation, and to have had an important effect upon the development of the river. The correspondence in level between this feature and the highest terrace upstream suggests that both are remnants of the immediately post-glacial valley bottom before the river became incised into it. The relatively small incised meanders between Finchale and Frankland indicate that wide meanders did not have time to develop. Farther upstream, the river at this stage was able to widen its flood plain, but the meanders did not reach the size they later attained. The drift was preferen-tially eroded, but where the river began to cut into solid rock, its course became fixed. As a result, the flights of terraces survive either immediately downstream or upstream of these fixed portions. Between them, meander migration has generally destroyed the higher remnants, and within the gorges terrace formation was virtually non-existent. During the formation of the High Houghall terrace sequence, the meander continued to migrate south-eastwards leaving a flight of terraces in its wake, and the river is now excavating the slopes under High Butterby.

In the tributary valleys, terraces are rare, and some may be of glacial origin or date from the period of deglaciation. Most contain only narrow impersistent strips of alluvium, but this deposit is more extensive along the Tursdale Beck upstream from Tursdale House, continuing thence into some tributary valleys, but more particularly through the Ferryhill Gap. The lower part of Tursdale Beck seems to be a tributary which has cut back and rejuvenated a glacial drainage channel. The floor of the channel in solid rock, relatively resistant to erosion, acted as a threshold above which the valley was liable to flood so that alluvium was deposited on its floor.

The flow of meltwater during deglaciation seems to have been southwards along this valley and through the gap, so that drainage in this area has been reversed twice. During the development of the buried valley system, the drainage was to the north ; at the end of the glacial period, it was reversed south ; and with the rejuvenation of the Tursdale Beck drainage channel, it was reversed north, though the confluence with the river was displaced upstream from Shincliffe to Low Burnhall.   E.A.F.

Drainage in most central and eastern parts of the district falls eastwards, down the dip of underlying Permian rocks. Minor streams in the central area lie in broad, relatively mature valleys cut in Magnesian Limestone and Lower Boulder Clay, and converge eastwards to form four main streams— Hawthorn, Horden and Castle Eden burns and Crimdon Beck. These continue eastwards through immature deep gorges ('denes') cut mainly into drift deposits filling former valleys, but locally into Magnesian Limestone. All have narrow flood plains lying above nick-points at the head of the incised part of their courses (a few hundred yards west of the western edge of the Upper Boulder Clay) and a further flood plain lying above a second group of nick-points within their gorges 1 to 2 miles farther east. Somewhat wider flood plains are present in the lower reaches of Castle Eden Burn and Crimdon Beck, but they are not seen in the two northern valleys, possibly due to truncation of their lower courses by coastal recession.

Terrace deposits in several of the coastal denes are thought to indicate a complicated history, beginning with initial downcutting of the streams as the Upper Boulder Clay ice withdrew eastwards and allowed the former easterly drainage to become re-established. Streams would have been graded initially to the higher coastal planations or to the level of ice-marginal lakes, but further downcutting would have taken place during early post-Glacial time as relative sea-level was then at least 50 ft below O.D. The intermittent nature of the fall in relative sea level, indicated by the multiplicity of coastal planations and the minor breaks of slope in the surface of the 80-ft planation, probably accounts for all the terraces noted. Subsequent (?post-Boreal) relative rise of sea level resulted in infilling by alluvium of the deeper parts of the valleys and in the burial of the more easterly parts of the lowest terrace in Castle Eden and Crimdon denes.

An early stage in the downcutting of the Castle Eden Burn is evidenced by an abandoned course [444 402] at 160 to 185 ft O.D. on Upper Boulder Clay at Horden, and another such course [439 458] is found at 120 to 140 ft O.D. on the south side of Hawthorn Dene, near Hawthorn Tower. All four valleys contain terraces at 30 to 20 ft, 15 to 12 ft and 10 to 6 ft above present stream level. In different places each of these terraces is cut in rock, and the uppermost terrace (now preserved mainly on isolated spurs) is characteristically a rock terrace with a thin patchy cover of gravel. The middle and lower terraces are formed mainly of gravel, and together occupy much of the valley floor.

The profiles of the four streams and their relationship to the glacial sequence suggest that they became re-established along existing valleys following withdrawal of Lower Boulder Clay ice, but were later blocked by Upper Boulder Clay ice and its deposits. They then fed ice-marginal lakes which appear, at one stage, to have overflowed westwards (p. 219), but upon withdrawal of the ice the water was released eastwards to begin cutting the post-Upper Boulder Clay gorges. This phase, perhaps represented by the upper nick-points and abandoned high level courses, may have led to the establishment of profiles graded to the coastal planations at about 140 and 95 ft A.O.D. and later to the 80-ft sea-level represented by the Terrace of Warp. The lower nick-points and probably also the various stream terraces may be related to rejuvenation caused by the regression of the sea below the 80-ft level and to subsequent coastal recession. In the beds of two of the streams, however, the lower nick-points are at or near the junction of boulder clay and limestone and may owe something to local factors.

Some indication of the age of the terraces is found towards the seaward end of Castle Eden and Crimdon denes, where the lowest terrace appears to pass beneath the alluvium of the lower flood plains and is presumably graded to a sea-level appreciably lower than that of today. The period of relatively low sea-level is shown by the dating of the peat of the submerged forest to be early Atlantic or earlier, and the lowest and youngest of the terraces is therefore probably more than 9,000 years old. The alluvium which overlies it must date from or be younger than the period of relative rise of sea-level and is therefore appreciably less than 9,000 years old. D.B.S.

Between the four main valleys, and east and south-east of the Hart ridge, a number of minor streams flow eastwards in shallow parallel valleys cut mainly in Upper Boulder Clay. Between Hart station and Seaton Carew these streams are graded to the landward edge of the Terrace of Warp, and only the most vigorous have cut appreciably into this terrace in their lower courses. Farther north, however, coastal recession has removed the Terrace of Warp and rejuvenated the minor streams, the lower courses of which now occupy short steep gorges such as Warren House, Blackhills and Whitesides gills near Horden, and Blue House, Cross and Limekiln gills near Blackhall Colliery.

South of the ridge at Hart, surface drainage generally flows south-eastwards to the Tees. An exception is the River Skerne, the upper reaches of which lie north of the ridge and follow the surface position of the Butterknowle Fault. At Hurworth Burn, however, the stream turns sharply south-westwards, and continues in this general direction to the southern margin of the district. From the trend of its upper reaches the stream properly forms part of the drainage pattern north of the Hart ridge, and it seems likely that it was originally a tributary of Crimdon Beck, but was diverted westwards by ice or glacial deposits filling its lower course. From Holdforth downstream, it follows a devious course through old lake-basins, now largely filled with lacustrine and organic deposits, but still liable to flood. They are thought to date from the period of deglaciation and lie between the Ferryhill and Aycliffe gaps. Remaining streams in the area south of Hart are graded to alluvial flats at the landward edge of the 80-ft marine planation (p. 246) and resemble streams north of Hart in having wide and relatively mature valleys in their upper courses and narrow steep-sided valleys in their middle courses. These valleys widen again and contain a narrow flood plain a short distance before reaching the 80-ft planation. Only one terrace, at 8 to 12 ft above present stream level, is preserved in the valleys south of Hart and this grades downstream into or passes beneath alluvium. In their lower courses the streams meander through this alluvium and over the 80-ft surface at gradients of 1 in 250 or less. D.B.S., G.D.G.

References

AGAR, R. 1954. Glacial and Post-Glacial geology of Middlesbrough and the Tees Estuary. *Proc. Yorks. Geol. Soc.,* **29,** 237–53.

BARKER, H., and MACKEY, J. 1961. British Museum natural radiocarbon measurements III. Sample description C. Great Britain BM–80. West Hartlepool submerged forest. *Radiocarbon,* **3,** 41–2.

BLACKBURN, KATHLEEN B. 1952. The dating of a deposit containing an elk skeleton found at Neasham near Darlington, County Durham. *New Phytologist,* **51,** 364–81.

BOSWELL, P. G. H. 1918. *A memoir on British resources of refractory sands for furnace and foundry purposes. Part I.* London.

BROWN, J. C. 1905. On some lacustrine deposits in the drift near Ferryhill. *Trans. Nat. Hist. Soc. Northumberland and Durham,* **1,** 288–92.

CAMERON, A. G. 1878. Notes on some peat deposits at Kildale and West Hartlepool. *Geol. Mag.,* **5,** 351–2.

CARRUTHERS, R. G. 1939. On northern glacial drifts: some peculiarities and their significance. *Quart. J. Geol. Soc.,* **95,** 299–333.

—— 1953. *Glacial drifts and the undermelt theory.* Newcastle.

DWERRYHOUSE, A. R. 1902. The glaciation of Teesdale, Weardale and the Tyne Valley and their tributary valleys. *Quart. J. Geol. Soc.,* **58,** 572–607.

GODWIN, H., SUGGATE, R. P., and WILLIS, E. H. 1958. Radiocarbon dating of the eustatic rise in ocean-level. *Nature,* **181,** 1518–19.

FLINT, R. F. 1957. *Glacial and Pleistocene geology.* New York.

FRANCIS, E. A., PHILLIPS, L. S., and SMITH, D. B. 1963. Field meeting to central and south-east Durham, *in* Annual Report. *Proc. Yorks. Geol. Soc.,* **34,** 104–12.

HICKLING, H. G. A., and ROBERTSON, T. 1949. Scientific survey of north-eastern England: geology. *Brit. Assoc.*

HOWSE, R. 1864. On the glaciation of the counties of Durham and Northumberland. *Trans. N. England Inst. Min. Eng.,* **13** (1863–64), 169–85.

—— and KIRKBY, J. W. 1863. A synopsis of the geology of Durham and part of Northumberland. *Tyneside Naturalists Field Club,* 1–31.

LESNE, M. P. 1920. Quelques insectes du Pliocène Supérieur du Comté de Durham. *Bull. Mus. Natnl. Hist. Nat., Paris,* **26,** 388–94, 484–8.

PENNY, L. F. 1959. The last glaciation in East Yorkshire. *Trans. Leeds Geol. Assoc.,* **7,** 65–77.

—— 1964. A review of the last glaciation in Great Britain. *Proc. Yorks. Geol. Soc.,* **34,** 387–411.

RADGE, G. W. 1939. The glaciation of North Cleveland. *Proc. Yorks. Geol. Soc.,* **24,** 180–205.

RAISTRICK, A. 1931. Glaciation. *In* Carruthers, R. G. and others. The geology of Northumberland and Durham. *Proc. Geol. Assoc.,* **42,** 281–91.

—— 1933. Excavation of a cave at Bishop Middleham, Durham. *Arch. Aeliana,* **10,** 112–22.

—— 1934. The correlation of glacial retreat stages across the Pennines. *Proc. Yorks. Geol. Soc.,* **22,** 199–214.

REID, ELEANOR M. 1920. On two preglacial floras from Castle Eden (County Durham). *Quart. J. Geol. Soc.,* **76,** 104–44.

SMITH, D. B. 1965. In *Sum. Prog. Geol. Surv.* for 1964, 58.

TRECHMANN, C. T. 1915. The Scandinavian Drift of the Durham coast and the general glaciology of the South-East Durham. *Quart. J. Geol. Soc.,* **71,** 53–82.

—— 1920. On a deposit of interglacial loess and some transported preglacial freshwater clays on the Durham coast. *Quart. J. Geol. Soc.,* **75,** 173–203.

—— 1931. The Scandinavian Drift or Basement Clay on the Durham coast. *In* Carruthers, R. G. and others. The geology of Northumberland and Durham. *Proc. Geol. Assoc.,* **42,** 292–4.

—— 1936. Mesolithic flints from the submerged forest at West Hartlepool. *Proc. Prehist. Soc.,* 161–8.

—— 1939. A rhinoceros bone from Brierton, near West Hartlepool and a skeleton of an elk from Neasham, near Darlington. *Proc. Yorks. Geol. Soc.,* **24,** 99–102.

—— 1942. Borings in the Permian and Coal Measures around Hartlepool. *Proc. Yorks. Geol. Soc.,* **24,** 313–27.

—— 1947. The submerged forest beds of the Durham coast. *Proc. Yorks. Geol. Soc.,* **27,** 23–32.

—— 1952. On the Pleistocene of East Durham. *Proc. Yorks. Geol. Soc.,* **28,** 164–79.

WEST, R. G. 1963. Problems of the British Quaternary. *Proc. Geol. Assoc.,* **74,** 147–86.

WOOD, N., and BOYD, E. F. 1864. On the ' Wash ' or ' Drift ' through a portion of the coalfield of Durham. *Trans. N. England Inst. Min. Eng.,* **13,** 69–85.

WOOLACOTT, D. 1905. The superficial deposits and preglacial valleys of the Northumberland and Durham coalfield. *Quart. J. Geol. Soc.,* **61,** 64–96.

—— 1906. The preglacial ' Wash ' of the Northumberland and Durham coalfield. *Proc. Univ. Durham Phil. Soc.,* **2,** 205–12.

—— 1920. On an exposure of sands and gravels containing marine shells at Easington, Co. Durham. *Geol. Mag.,* **57,** 307–11.

—— 1921. The interglacial problem, and the Glacial and Post-Glacial sequence in Northumberland and Durham. *Geol. Mag.,* **58,** 21–32, 60–9.

—— 1922. On the 60-ft raised beach at Easington, Co. Durham. *Geol. Mag.,* **59,** 64–74.

Chapter VII

MINERAL PRODUCTS AND WATER SUPPLY

COAL: RANK AND QUALITY

The following information on the rank and quality of coals in the district has been supplied by Mr. A. H. Edwards, Coal Survey Officer, Northumberland and Durham Division, National Coal Board.

Durham coals exhibit a wide range of rank, quality and physical characteristics, ranging from soft, bright coking coals containing less than 30 per cent of volatile matter to hard, rather dull coals with greater than 36 per cent. The ranges of rank and volatile matter covering Durham coals, which include those heat-altered, are given below.

TABLE 6

Coal Rank Code			Description	Volatile Matter (d.m.m.f.)
200H	...	...	Heat-altered, non caking	9·1–19·5
303H	...	...		19·6–27·0*
302H	...	...	Heat-altered, weakly caking	19·6–32·0
301	...	...	Prime coking coals	24·0–32·0*
401	...	...	Very strongly caking carbonisation coals	32·1–36·0
402	...	...		Over 36·0
501	...	...	Strongly caking carbonisation coals and gas coals	32·1–36·0
502	...	...		Over 36·0
601	...	...	Medium caking, house and industrial coals ...	32·1–36·0
602	...	...		Over 36·0
701	...	...	Weakly caking, house and industrial coals ...	32·1–36·0
702	...	...		Over 36·0

* National Limits $\left.\begin{array}{l}303H \\ 301\end{array}\right\}$ 19·6—32·0

The variation of rank, from west to east, for any one seam is much greater, generally, than that between adjacent seams in any particular locality. The main reserves of 301 coals, confined mainly to the Harvey and seams below, are found in west Durham and those of 501 and 502 coals in the coastal collieries of east Durham. The lower rank 600 and small amounts of 700 coals are found principally in three of the upper seams of the sequence–the High Main, Five-Quarter and Main seams–and, mainly, in north-east Durham, though 600 type coals are found in the lower seams, which are near to the surface or the base of the Permian on the southern limits of the field. Many of the seams worked in the coalfield, and particularly the 301 coals of west Durham, contain less than 5 per cent of ash and 1 per cent of sulphur. These strongly caking, low ash and sulphur coals are becoming exhausted but there are large reserves of 401 and 501 coals, mainly found in central and east Durham, with ash and sulphur in amounts not exceeding 7·5 and 1·5 per cent. The following notes refer specifically to the district covered by this memoir.

Variation of Rank. A rank map of the Harvey Seam for the district is shown in Fig. 30 and the pattern shown, generally, is the same for all seams. The majority of the coals mined have Coal Rank Codes of 401, 402, 501 and 502. The main reserves are situated in the undersea areas to be developed by Easington, Horden and Blackhall collieries. It should be noted that two off-shore boreholes at Blackhall have shown that there is a large increase in caking power and a decrease in volatile matter in the sequence of seams from the High Main to the Brockwell, proved in these boreholes. The Bottom Brockwell Seam encountered in No. 1 Bore had a rank code of 301b, comparable with that of the prime coking coals of west Durham. The Ludworth Dyke (see Fig. 30), which follows a general east-north-easterly direction and is encountered in workings at Bowburn, Sherburn Hill and Thornley before continuing out to sea between Easington and Horden collieries, has a severe metamorphic effect (Edwards and Tomlinson 1958) on the coal on both its flanks. The effect is most marked at Easington and Horden where low volatile, heat-altered coals extend for about three-quarters of a mile on either side of the dyke in all seams being mined. The highly altered coals of rank code 200H to 302H (see Table 6 for characteristics) are found close to the dyke. These coals merge into 301 coals, which, however, are true to type only with regard to volatile matter as their caking properties have been partially destroyed. The effect of the Hett Dyke in the seams, where it has been encountered, extends for a distance of less than 500 ft on either side and the change of rank in the coal which results is of little practical importance.

Quality. These comments on quality are based on the ' seam less dirt ' in current workings but they can also be taken as a guide to the quality of the seams in the locality as a whole. The inherent ash of the seams is variable and values from less than 5 per cent to more than 10 per cent are encountered. Thirty per cent of the current output is won from seams containing less than 5 per cent of ash—the biggest producers of such coals being the Durham Low Main, Harvey and Busty seams. High ash coals account for only 8 per cent of the output. The amount of sulphur present in the nine seams worked is also variable and values cover the range from less than 1 per cent to 4 per cent, though only 8 per cent of the current output is from coals with over 2·5 per cent of sulphur and these are mainly confined to the High Main and Hutton seams. Almost 40 per cent of the present output, however, can be described as being low to moderately low in sulphur (coals containing less than 1·5 per cent). Two important minor constituents in coal-phosphorus

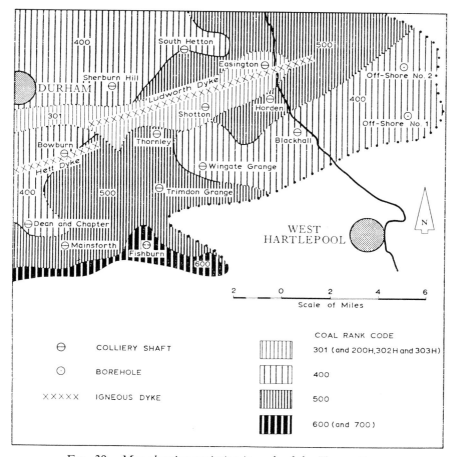

FIG. 30. *Map showing variation in rank of the Harvey Seam*

and chlorine–are mostly low, amounts of less than 0·01 and 0·2 per cent generally being present, but in some parts of the undersea area, particularly in the upper seams, chlorine may exceed the tolerated figure of 0·3 per cent.

Utilization. On account of its strong caking properties about 25 per cent of the present output is supplied to local coke ovens and to steelworks in the vicinity of Middlesbrough, for the manufacture of metallurgical coke. Some of the strongly caking 401 type is sent as far as the Midlands for blending with their less strongly caking coals. The coals also have a high gaseous thermal yield, and, consequently, about 30 per cent of the output is supplied to the gas industry, some of which is used locally, but most is shipped to the Thames and South Coast for use in southern gas works. A proportion of the larger coal is ideally suitable for household purposes and about 12 per cent. of the current output is supplied to domestic consumers. About 6 per cent is disposed of as smalls to power stations, part locally and the remainder to Thames-side where retort stokers are used. The balance of the output, about 27 per cent, some of which is cleaned, is sent to numerous plants for steam raising purposes, the steam smalls having proved to be suitable for the larger types of steam raising appliances and cyclone boilers. The cleaned coal is supplied, also, to brick works and sulphuric acid plants, and for use in small automatic stokers installed at industrial premises, schools and hospitals. In the future the gas and metallurgical coke markets may decline, but there should be no difficulty in increasing supplies from this district to the steam market which would include substantial tonnages for the electricity industry.   A.H.E.

COAL: MINING AND RESERVES

Coal mining still constitutes the major industry of the Durham–West Hartlepool district. In the exposed coalfield, most of the better quality seams were worked out from many small shafts and surface adits before or during the 19th century. Mining of the western part of the concealed coalfield waxed during the later part of that century and is now declining: in future, therefore, workings will inevitably be concentrated in the coastal and undersea areas. Thus, whilst about 8 million tons were produced in the district during 1964 from 20 deep mines and one surface drift, over half of this output came from the four large collieries on the coast and from the combined collieries of Elemore, Eppleton and Murton.

Apart from coal in inconvenient pockets, most of the inland reserves of thick coal are now located either between existing areas of abandoned or current workings, or in the takes of abandoned collieries such as Haswell, Castle Eden and South Wingate. Under the sea large areas of thick coal are known to lie ahead of existing workings in several seams.

One of the major factors limiting extraction of coal from the southern part of the concealed coalfield is the proximity of overlying water-bearing Permian strata. Water from this source percolates into some workings in such quantities as to exceed pump capacity and has, in the past, led to abandonment and flooding of collieries. In consequence, working within 150 ft of water-bearing strata is now prohibited by statute unless it can be shown that the hazard has been overcome. In a pilot test to determine the feasibility of de-watering the Permian strata, heavy pumping was carried out from a borehole at Rushyford (Armstrong and others 1959). From the results of this test it was calculated that sufficient water could be removed to facilitate the extraction of about 16 million tons of otherwise unworkable coal in the Mainsforth Syncline, but no full scale de-watering has yet been attempted. Projects of this kind, if economic, could add greatly to existing reserves in other parts of the district, especially in those western parts of the concealed coalfield where the dip of the Permian and Carboniferous rocks is almost equal. Farther east, in coastal and undersea areas, economic and practical limitations rule out de-watering.

COAL: SEA COAL

Coal derived from spoil dumped into the sea from coastal and some inland collieries and supplemented by coal derived from outcrops off the coast of Northumberland is sorted and redeposited at many places along the Durham coast. Much of the coal is inferior, having been rejected by colliery washing and grading plant, but it is combustible and forms the basis of a small but well-organized industry which annually recovers some thousands of tons for sale locally and to the continent.

OIL AND GAS

Small quantities of mineral oil were recorded in cavities at the base of the Upper Magnesian Limestone in the Seaton Carew Bore (Bird 1888, p. 570) and in a bore at Howbeck Hospital (Trechmann 1942, p. 319). Additionally, small quantities of an oily substance were distilled by Dr. J. Phemister (unpublished Geological Survey Report) from specimens at about this horizon in a bore at Villiers Street, West Hartlepool. In all of these bores the oil was found in a dark bituminous shaly dolomite or dolomitic shale, the local equivalent of the sapropelic Flexible Limestone. Neither oil nor gas has been recorded from this bed elsewhere in the district, but both it and the overlying Concretionary Limestone commonly smell strongly of oil when freshly broken, and Trechmann (1913, p. 198) records that the latter yields much oily and carbonaceous matter on solution in acid. It is from the Upper Magnesian Limestone that natural gas is derived near Whitby, about 25 miles south-west of the district (Falcon and Kent 1960, p. 30).

LIMESTONE AND DOLOMITE

Prior to about 1800 the rocks of the Magnesian Limestone were used mainly for building purposes, and most of the early settlements (p. 106) along the Permian escarpment were built of dolomitic limestone and dolomite worked in numerous small and a few large quarries. Most of these were opened in the evenly bedded Lower Magnesian Limestone which provided the most suitable building material. With the exception of reef-rock, which has been used on a small scale in buildings at Hawthorn, Easington, Peterlee and Hesleden, the dolomite of the Middle Magnesian Limestone is too soft and variable for building purposes, but the Upper Magnesian Limestone was formerly worked in large quarries at Hartlepool, and forms the substance of nearby churches, harbour works and private houses.

With the increasing use of lime for agricultural purposes in the early part of the 19th century, a number of quarries in the Lower Magnesian Limestone and some new ones, including the large Tuthill Quarry near Haswell, in the Middle Magnesian Limestone, supplied ground or burnt lime. Most of these are now abandoned and current needs are met mainly by a few large quarries between Thrislington and Sherburn Hill (a single firm at Sherburn Hill annually produces 130,000 tons of lime for agricultural use. Information by courtesy of Adam Lythgoe Ltd.). Some of these quarries also supply crushed limestone and dolomite for use as building aggregate, building lime and road metal.

The principal uses for the Magnesian Limestone today, however, are as a flux in steel-making around West Hartlepool and Middlesbrough, and in the manufacture of refractories. For use as a flux the less dolomitic varieties are preferred, and demand is met from a number of quarries between Mainsforth and Quarrington Hill and, farther east, from Hawthorn. The requirements of the refractories industry are best met by more dolomitic varieties, quarried and calcined mainly around Coxhoe. New quarries are now being developed south of Thrislington Hall. Reserves of both varieties are enormous, and are limited mainly by the overburden of Middle Magnesian Limestone to the east.

SAND AND GRAVEL

Sand is derived for building purposes from Basal Permian Sands, from Glacial sands and gravels, and from beach and dune deposits. The Basal Permian Sands are clean, well-graded and coarse, but are now worked in only two quarries at Sherburn Hill and in one north-west of Quarrington Hill. Glacial sands, generally poorly graded and mixed with gravel, are more widely distributed at the surface than the Basal Permian Sands, but are now worked at only six localities in the district. At most of these localities operations are on a small scale and in some the deposit is almost exhausted, but gravel is still being quarried intensively at Old Durham, near Sherburn, and at Fallowfield near South Hetton. There are good reserves of both Basal Permian Sands and Glacial sands and gravels, the former being limited mainly by wedging out and by increasing thicknesses of limestone overburden, and the latter, which are concentrated in the morainic area between Castle Eden, Sheraton and Hart, by unpredictably variable clay content. It is probable that good reserves of gravel also underlie the flood plain of the River Wear.

During 1963 output from all these sources amounted to 163,574 cubic yards of building sand, 39,472 cubic yards of concreting sand and 22,778 cubic yards of gravel.

Beach and dune sand form large reserves at the coast between Hart and Hart Warren, where they are now being worked, and to the south of Seaton Carew.

MOULDING SAND

According to Davies and Rees (1944) the Basal Permian Sands of Durham and Yorkshire are very variable, but at some places they are suitable for iron and non-ferrous moulding and are so used in Yorkshire. The sands in their natural state, however, are not sufficiently refractory to be used for steel moulding, although Davies and Rees suggest that some, including the deposit in Crime Rigg Quarry, could be improved sufficiently to be used for this purpose by the addition of bonding clay and by the removal of some silt.

BRICK CLAYS

Laminated clays used in the manufacture of bricks are now worked only at Shotton Colliery, though in the 18th and 19th centuries they were dug for the manufacture of bricks, tiles and pipes in many large pits extending over 60 acres between Durham and Finchale Priory, between Thornley and Shotton and between Wellfield and Station Town. Large reserves of laminated clay still remain in all three areas.

Bricks and tiles were also formerly manufactured from boulder clay south-east of Tursdale Colliery, between Nunstainton East and Sedgefield, and in Tilery Plantation, Pudding Poke Farm, Elwick. Bricks are also made from Coal Measures shales and mudstones, particularly those overlying the High Main Coal, which are worked for this purpose in a newly-opened quarry near the Adventure Pit, 1 mile N.W. of West Rainton.

ANHYDRITE

Anhydrite of late Middle Magnesian Limestone age was formerly worked at the Warren Cement Company's mine at Hartlepool. The mine is now abandoned and flooded, but large reserves of this mineral exist (p. 152) under Hartlepool where the deposit is up to 360 ft thick, and in parts of the offshore area where it reaches 500 ft.

Two other beds—the Upper Anhydrite and the Billingham Main Anhydrite—occur in the south-eastern part of the district, where the anhydrite is probably altered to gypsum and partially or wholly removed by solution at outcrop.

ROCK SALT AND BRINE

No deposits of rock salt have yet been proved within the district, but by comparison with the adjacent district to the south covered by the Stockton (33) Sheet, salt is probably present at the base of the Upper Permian Marls in the undersea area off Seaton Carew and may be present at depth beneath Seaton Sands. No salt was penetrated in the Seaton Carew Bore, but the record (Bird 1888, p. 571) shows that a feeder of brine was encountered in the Magnesian Limestone at a depth of 1,153 ft and the working of this brine was at one time contemplated. Brine springs have also been reported (Anderson 1945a, 1945b) in Carboniferous strata near Butterby, in the bed of the Wear, near the confluence of Saltwell Gill, and in offshore coal workings. The brines, though variable in composition, all contain relatively high concentrations of chlorides (1610 ppm in one case), those of sodium, calcium and magnesium being the commonest, although barium is present in many. Some undersea brines also contain appreciable quantities of sulphate, some of which is precipitated in pipes in the form of swallow-tail selenite crystals up to $1\frac{1}{2}$ in long.

MISCELLANEOUS

At Hartlepool, about 50 million gallons of sea water are treated each day for the extraction of magnesia. This is used in the production of furnace linings, of which the plant has an annual output of 220,000 tons. Minerals which are locally present in considerable quantities but are not economically workable include baryte in the Lower Magnesian Limestone (especially around Chilton and Thrislington) and sulphides of lead, zinc, copper and iron in the Marl Slate. Galena is also present in a fault breccia at Black Halls Rocks where it was once investigated in a trial adit.

WATER SUPPLY

Water supply in the district is distributed by three main public undertakings. The exposed coalfield and the south-western part of the concealed coalfield are supplied by the Durham County Water Board with soft water, mainly from surface sources farther west. Supplies elsewhere are drawn largely from wells and boreholes sunk into the Permian strata. Most of the northern part of the district is supplied by the Sunderland and South Shields Water Co., which has wells or boreholes at Seaton, North Dalton, Dalton (New Hesledon), Hawthorn, Mill Hill (Easington), Thorpe (Easington Colliery), Peterlee,, and New Winning (Wellfield) and also obtains water from underground bores drilled up into Permian strata from abandoned workings in Castle Eden Colliery. All the bores penetrate into or through the Basal Permian Sands, which are demonstrably a major source of water in at least one of them. In some of the bores however, the Sands may subsequently have been sealed off to reduce the risk of damage to the pumping machinery. The southern part of the district is supplied mainly by the Hartlepools Water Co. from bores in the Magnesian Limestone at Amerston Hall, Pudding Poke, Naisberry, Dalton Piercy, Howbeck and West Hartlepool. Yields from all these wells are consistently high, commonly more than 50,000 gallons per hour, generally with water-table depressions of only a few feet. Details of the quality of underground water from Permian strata are given by Anderson (1945a) who showed that the water is everywhere hard and in some coastal areas is polluted by sea-water. More recent figures (Wilkinson and Squire 1964) show that it ranges from hard to very hard: the water supplied to the Hartlepools for instance has a temporary hardness of 275 parts per million and a permanent hardness of 185. This is satisfactory for domestic use, but too hard for some local industries. The requirements of the latter are met from small surface reservoirs at Hurworth Burn, Crookfoot and Hart, supplemented with water from boreholes. Water is also drawn from the Magnesian Limestone by private installations owned by the National Coal Board, the Durham County Water Board, a number of commercial concerns and at least one domestic user.

Details of the output from all underground sources are given in a hydrometric survey of the Wear and Tees catchment areas (Anon 1961), in which the rate of replenishment of the aquifer is assessed. This assessment shows that abstraction in general exceeds replenishment in the northern part of the district and around West Hartlepool, but is substantially less elsewhere. The considerable quantities of water pumped from mines in the concealed coalfield and largely discharged at the surface, and the trial borehole at Rushyford which, on test, yielded up to 92,000 gallons per hour, testify to the potential of the southern area.

Other underground sources of water used in the district are the Bunter Sandstone south of the West Hartlepool Fault, and the superficial deposits. The former yields about 3 million gallons per day to industrial users on the southern outskirts of West Hartlepool, and is being tested at the time of writing for public supply purposes in a bore at Greatham, immediately south of the district boundary. Water in the superficial deposits has long been drawn from shallow wells and boreholes scattered throughout the district, but most of these latter sources fluctuate seasonally and have been generally subject to pollution. In consequence, only a few of these installations are still in use, and their life appears limited. Large volumes of water remain untapped in these superficial deposits, however, and there seems no reason why, with careful siting, this could not in some places be utilized more fully.

D.B.S., E.A.F., G.D.G.

References

ANDERSON, W. 1945a. Water supply from underground sources in north-east England. Part III. Supplement—General discussion for New Series One-Inch Sheets 21 (Sunderland) and 27 (Durham). *Geol. Surv. Wartime Pamphlet* No. **19.**

—— 1945b. On the chloride waters of Great Britain. *Geol. Mag.,* **82,** 267–74.

ANON. 1961. *Wear and Tees hydrological survey. Hydrometric areas 24 and 25.* Ministry of Housing and Local Government. Published H.M.S.O., London.

ARMSTRONG, G., KIDD, R. R. and BUCHAN, S. 1959. Dewatering scheme in the south Durham coalfield. *Trans. Inst. Min. Eng.,* **119,** 141–52.

BIRD, W. J. 1888. The south Durham salt bed and associated strata. *Trans. Manch. Geol. Soc.,* **19,** 564–84.

DAVIES, W. and REES, W. J. 1944. British resources of steel moulding sands. Part 5. The Permian Yellow Sands of Durham and Yorkshire. *J. Iron and Steel Inst.,* No. 2 for 1943, 104–11.

EDWARDS, A. H. and TOMLINSON, T. S. 1958. A survey of 16 low volatile coals in north-east and south-east Durham. *Trans. Inst. Min. Eng.,* **117,** 49–78.

FALCON, N. L. and KENT, P. E. 1960. Geological results of petroleum exploration in Britain 1945–1957. *Mem. Geol. Soc. London,* **2,** 1–56.

TRECHMANN, C. T. 1913. On a mass of anhydrite in the Magnesian Limestone at Hartlepool and on the Permian of south-eastern Durham. *Quart. J. Geol. Soc.,* **69,** 184–218.

—— 1942. Borings in the Permian and Coal Measures around Hartlepool. *Proc. Yorks. Geol. Soc.* **24,** 313–27.

WILKINSON, D. and SQUIRE, N. (Ed.) 1964. *The Water Engineer's Handbook.* London.

Appendix I

RECORDS OF SELECTED SHAFT AND BOREHOLE SECTIONS

Sections of 17 representative shafts and borings are given in full. Abridged logs are given of other selected sinkings. Many of the latter have already been published in detail in 'An account of the strata of Northumberland and Durham as proved by borings and sinkings' issued in 4 volumes by the North of England Institute of Mining and Mechanical Engineers between 1878 and 1910. These logs are referred to in the text below, by means of the abbreviation 'B. and S.' followed by the serial number of the entry. Other abbreviations used are A.O.D. and B.O.D. for levels above and below Ordnance Datum respectively.

In both down bores and up bores, depths or heights refer to the bases of seams.

Amerston Hall Trial Bore

> Surface 246 ft A.O.D. 6-in NZ 43 S.W. Site [4268 3028]
> 120 yd N.N.E. of Amerston Hall, Embleton. Drilled 1964 by
> Boldon Drilling Co. Ltd. for Hartlepools Water Co.

Drift to 231 ft, Upper Permian Marl to about 280 ft, Upper Magnesian Limestone to bottom at 380 ft.

Belmont Colliery, Furnace Pit

> Surface 290 ft A.O.D. 6-in NZ 34 N.W. Site [3166 4503] 270 yd
> E.N.E. of The Rift. Sunk 1840. For detailed section see B. and
> S. No. 101.

Drift to 85 ft, Durham Low Main 39 in at 222 ft, Hutton 56 in at bottom at 310 ft.

Bishop Middleham Colliery, East Pit

> Surface 298 ft A.O.D. 6-in NZ 33 S.W. Site [3368 3128] 640 yd
> W.S.W. of East House. Sunk 1846 by Wm. Coulson Ltd.

Drift to 30 ft, Magnesian Limestone to 198 ft, Marl Slate to 207 ft, ?Basal Permian Sandstone to 216 ft.

Bottom Hutton 36 in at 284 ft, Top Harvey 37 in at 423 ft, Bottom Harvey 33 in at 431 ft.

Boring from Bottom Harvey: Bottom Busty 30¾ in at 614 ft, Brockwell (banded) 21½ in at 697 ft.

Black Halls No. 1 Diamond Bore

Surface approximately 140 ft A.O.D. 6-in NZ 43 N.E. Site [4711 3842] 330 yd E.N.E. of Black Halls. Drilled 1894. For detailed section see B. and S. No. 2438.

Drift to 60 ft, Magnesian Limestone to 688 ft, Basal Permian Breccia to 690 ft. Main 44 in at 729 ft, Durham Low Main 59 in at 865 ft, Brockwell 27 in at 1309 ft, Ganister Clay, thin, at 1486 ft. Drilled to 1555 ft.

Blackhall Colliery South Shaft

Surface 173 ft A.O.D. 6-in NZ 43 N.E. Site [4604 3955] 1100 yd E.N.E. of Hardwick Hall. Sunk 1909–13.

Drift to 68 ft, Magnesian Limestone to 726 ft, Marl Slate to 731 ft, Basal Permian Sands and Breccias to 764 ft.

High Main (banded) 72 in at 797 ft, Metal 11 in at 804 ft, Five-Quarter 65 in at 897 ft, Main 58 in at 909 ft, Durham Low Main 46 in at 1062 ft, Bottom Hutton 39 in at 1150 ft. Sunk to 1189 ft.

Further details of the Permian beds in this shaft are given by Trechmann (1913, p. 213).

Blackhall Colliery No. O.B.2 Underground Bore

Down 338 ft from Five-Quarter workings at 764 ft B.O.D. 6-in NZ 44 S.E. Site [4706 4002] 4200 yd S.E. of Horden Point. Drilled 1919 by A. Kyle Ltd. for Horden Collieries Ltd.

Main 44 in at 65 ft, Hutton (banded) 96 in at 335 ft.

Blackhall Colliery Nos. 2 and 8 Underground Bores

Up 86 ft and down 380 ft from Durham Low Main at 634 ft B.O.D. 6-in NZ 43 N.E. Site [4696 3708] 380 yd N.N.W. of Middlethorpe. Drilled 1931 by A. Kyle Ltd. for Horden Collieries Ltd.

Up bore : no coals proved.

Down bore : Harvey (2 leaves) 36 in at 282 ft, Bottom Busty 24 in at 332 ft.

Blackhall Colliery Nos. 6 and 7 Underground Bores

Up 142 ft and down 514 ft from Durham Low Main workings at 766 ft B.O.D. 6-in NZ 43 N.E. Site [4817 3818] 1110 yd N.E. of Fillpoke. Drilled 1931 by A. Kyle Ltd. for Horden Collieries Ltd.

Up bore : no coals proved.

Down bore : Bottom Tilley 25 in at 350 ft, Top Busty 25 in at 362 ft.

Blackhall Colliery No. 9 Underground Bore

Down 314 ft from Hutton workings at 1163 ft B.O.D. 6-in NZ 44 S.E. Site [4741 4074] 3930 yd S.E. of Horden Point. Drilled 1932 by A. Kyle Ltd. for Horden Collieries Ltd.

Bottom Harvey 28 in at 155 ft, Bottom Tilley (banded) 34 in at 213 ft.

Blackhall Colliery Nos. 13 and 14 Underground Bores

Up 135 ft and down 277 ft from Hutton workings at 1093 ft
B.O.D. 6-in NZ 43 N.E. Site [4772 3981] 2540 yd N. of Fillpoke. Drilled 1932 by A. Kyle Ltd. for Horden Collieries Ltd.

Up bore: Durham Low Main (banded) 40 in at 104 ft.
Down bore: Harvey 50 in at 174 ft, Busty 27 in at 241 ft.

Blackhall Colliery No. 16 Underground Bore

Down 305 ft from Hutton workings at 1191 ft B.O.D. 6-in
NZ 44 S.E. Site [4831 4127] 4450 yd E.S.E. of Horden Point.
Drilled 1937 by A. Kyle Ltd. for Horden Collieries Ltd.

Harvey 23 in at 153 ft, Bottom Tilley (including 8-in band) 48½ in at 230 ft.

Blackhall Colliery No. 17 Underground Bore

Down 277 ft from Hutton workings at 1148 ft B.O.D. 6-in
NZ 43 N.E. Site [4844 3986] 2250 yd N.E. of Black Halls.
Drilled 1937 by A. Kyle Ltd. for Horden Collieries Ltd.

Harvey (2 leaves) 45 in at 166 ft, Bottom Tilley 30 in at 223 ft, Bottom Busty 36 in
at 245 ft.

Blackhall Colliery Nos. 18 and 19 Underground Bores

Up 116 ft and down 260 ft from Hutton workings at 1055 ft
B.O.D. 6-in NZ 43 N.E. Site [4798 3940] 1550 yd N.E. of Black
Halls. Drilled 1937 by A. Kyle Ltd. for Horden Collieries Ltd.

Up bore: Top Brass Thill 21 in at 106 ft.
Down bore: Harvey 41 in at 154 ft, Busty 24 in at 238 ft.

Blackhall Colliery No. 27 Underground Bore

Up 283 ft from Durham Low Main workings at 1079 ft B.O.D.
6-in NZ 44 S.E. Site [4917 4045] 5700 yd E.S.E. of Horden
Point. Drilled 1949–50 by N.C.B.

High Main 68 in at 219 ft, Metal (including 4-in band) 48 in at 213 ft, Five-Quarter
65 in at 163 ft, Main 16 in at 120 ft.

Blackhall Colliery No. 28 Underground Bore

Up 243 ft from Durham Low Main workings at 1013 ft B.O.D.
6-in NZ 43 N.E. Site [4923 3939] 2670 yd E.N.E. of Black Halls.
Drilled 1951 by N.C.B.

Base of Permian Basal Breccia at 242 ft, Metal and High Main 90 in at 216 ft,
Five-Quarter (including 6-in band) 57 in at 173 ft, Main (banded) 57 in at 87 ft.

Blackhall Colliery No. 30 Underground Bore

Up 379 ft from Durham Low Main workings at 1224 ft B.O.D.
6-in NZ 43 N.E. Site [4923 3939] 2670 yd E.N.E. of Black Halls.
Drilled 1952 by N.C.B.

Base of Permian possibly reached at top of hole, High Main (including 10-in band)
41 in at 266 ft, Metal (including 3-in band) 46 in at 212 ft, Five-Quarter 40 in at
172 ft, Main 41 in at 124 ft, Top Maudlin (including 9-in band) 35 in at 95 ft, Bottom
Maudlin 13 in at 42 ft.

Blackhall Colliery No. 34 Underground Bore

Up 390 ft from Durham Low Main workings at 1208 ft. B.O.D.
6-in NZ 44 S.E. Site [4954 4211] 5470 yd E. of Horden Point.
Drilled 1955 by N.C.B.

High Main (banded) 66 in at 259 ft, Metal (banded) 48 in at 215 ft, Main 60 in at 123 ft.

Blackhall Colliery No. 36 Underground Bore

Up and down from Durham Low Main. 6-in NZ 44 S.E. Site [4979 4104] 6090 yd E.S.E. of Horden Point. Drilled 1955 by N.C.B. Cores examined by D. B. Smith.

Up bore: measurements from roof of Durham Low Main at 1116 ft B.O.D.

	Thickness ft	in	Height ft	in		Thickness ft	in	Height ft	in
			269	8	Seatearth-mudstone ...		6½	211	6
Ironstone, red		5	269	3	Siltstone	6	0	205	6
Shale and siltstone, red at					Sandstone, argillaceous ...	1	6	204	0
top	9	2	260	1	Shale, silty	1	3	202	9
Ironstone	1	7	258	6	Sandstone	6	1	196	8
Sandstone	2	0	256	6	Siltstone		6	196	2
Siltstone	16	9	239	9	Shale	6	8	189	6
Sandstone	2	3	237	6	Siltstone, sandy	6	2	183	4
Siltstone and shale ...	3	4	234	2	Sandstone, argillaceous ...	3	0	180	4
Seatearth-mudstone ...		6	233	8	Shale, plants, '? mussels'	4	0	176	4
HIGH MAIN					FIVE-QUARTER				
Coal ... ?8 in					Coal, shaly 16 in	1	4	175	0
Seatearth-mudstone 6 in					Seatearth-mudstone ...		2	174	10
Coal ... ?30 in					Shale, carbonaceous ...		2	174	8
Seatearth-mudstone 10 in					Mudstone, plants... ...		10	173	10
Coal ... 5 in					Seatearth-mudstone ...		2	173	8
Coal, shaly 15 in					Shale and mudstone ...	1	5	172	3
Seatearth-mudstone 7 in					Sandstone and siltstone ...	5	7	166	8
Coal ... 1 in					Shale, carbonaceous ...		2	166	6
Seatearth-mudstone 2½ in					Coal, shaly 7 in		7	165	11
Coal ... ?1½ in	7	2	226	6	Seatearth-mudstone, possibly with small fault ...	9	11	156	0
Shale, partly carbonaceous	7	6	219	0	Shale and mudstone ...	7	8	148	4
Shale, plants	2	10	216	2	Shale, carbonaceous, with				
Shale, carbonaceous ...	1	0	215	2	thin coals	1	2	147	2
METAL					Seatearth-mudstone ...		1	147	1
Coal, banded 22 in					Shale, mudstone and siltstone	7	6	139	7
Coal, shaly, interbanded					Sandstone		7	139	0
with carbonaceous					Shale, plants		8	138	4
shale 15½ in	3	1½	212	0½	Sandstone		8	137	8
					Shale		5	137	3
					Sandstone	29	1	108	2
					Seatearth-mudstone ...		4	107	10
					Mudstone, plants... ...		3	107	7

		Thickness ft in	Height ft in			Thickness ft in	Height ft in
MAIN				Seatearth-mudstone ...		2 6	82 3
Coal ...	8 in			Mudstone and shale, plants		7	81 8
Coal, shaly	3 in	**11**	**106 8**	Siltstone, plants		8	81 0
Shale and siltstone ...		3	106 5	Sandstone, argillaceous ...		37 0	44 0
Shale, carbonaceous ...		6	105 11	Siltstone, plants		1 4	42 8
Sandstone, argillaceous ...		4	105 7	Shale, plants		2 11	39 9
Shale, carbonaceous ...		4	105 3	Seatearth-mudstone ...		2 7	37 2
Sandstone		4 3	101 0	Shale, carbonaceous ...		2	37 0
Shale conglomerate ...		5	100 7	Coal,			
Shale		3	100 4	pyritic	4 in	**4**	**36 8**
Sandstone		5 8	94 8	Seatearth-mudstone ...		1 6	35 2
Shale, carbonaceous, and				Shale, silty		7	34 7
sandstone		11	93 9	Seatearth-mudstone ...		3	34 4
Sandstone		7 1	86 8	Siltstone and mudstone ...		2 4	32 0
Shale, plants		4	86 4	Shale, silty; plants ...		5 0	27 0
Shale, carbonaceous ...		5	85 11	Siltstone, plants		11 0	16 0
Coal ...	6 in	**6**	**85 5**	Sandstone, argillaceous ...		2 0	14 0
Shale, carbonaceous ...		3	85 2	Excavation		14 0	0 0
Coal ...	5 in	**5**	**84 9**				

Down bore: measurements from floor of Durham Low Main at 1119 ft B.O.D.

	Thickness ft in	Depth ft in		Thickness ft in	Depth ft in
Excavation	3 0	3 0	HUTTON		
Sandstone	3 0	6 0	Coal ... 22 in	**1 10**	**151 4**
Shale, ' mussels ' ...	9 10	15 10	Shale, carbonaceous ...	4	151 8
Shale, many ' mussels ' ...	4 5	20 3	Mudstone and siltstone,		
Shale, silty; plants ...	1 3	21 6	roots	4 2	155 10
Seatearth-mudstone ...	3 0	24 6	Shale, plants, ' mussels '		
Sandstone	1 2	25 8	near base	13 2	169 0
Seatearth-mudstone ...	7 10	33 6	Coal and		
Siltstone, plants	3 2	36 8	carbonaceous		
Shale, plants	3 6	40 2	shale 8 in	**8**	**169 8**
Sandstone, coal films at			Shale and siltstone, roots	1 0	170 8
base	16 6	56 8	Sandstone...	2 6	173 2
Sandstone, roots at top ...	1 11	58 7	Shale, ' mussels ' at 179 ft	12 4	185 6
Shale	9	59 4	Shale, carbonaceous ...	7	186 1
Sandstone, some roots ...	4 5	63 9	Seatearth-mudstone ...	2	186 3
Siltstone, many plants ...	2 3	66 0	Shale, carbonaceous, coal		
Shale, silty	10	66 10	films	5	186 8
Sandstone, argillaceous ...	5 10	72 8	Seatearth-mudstone ...	2 3	188 11
Shale, silty	3 6	76 2	Shale, plants, and below		
Sandstone	4 11	81 1	205 ft ' mussels ' ...	33 3	222 2
Shale, sandy	1 6	82 7	Coal ... 8 in	**8**	**222 10**
Sandstone	4 10	87 5	Seatearth-mudstone ...	6	223 4
Shale, silty	1 2	88 7	Mudstone	6	223 10
Sandstone, broken above			Sandstone and shale ...	2 6	226 4
97 ft	21 5	110 0	Shale, ' mussels '	6	226 10
Shale, silty; plants ...	2	110 2	Sandstone, argillaceous ...	6	227 4
Sandstone, massive ...	39 4	149 6	Shale, sandstone beds ...	9 5	236 9
			Shale, many ' mussels ' ...	24 11	261 8

		Thick-ness		Depth				Thick-ness		Depth	
		ft	in	ft	in			ft	in	ft	in
HARVEY MARINE BAND						Mudstone and shale ...		1	3	280	9
Shale, black, many						Sandstone, argillaceous ...		1	0	281	9
Lingula		1	2	262	10	Sandstone...		3	0	284	9
Coal ... 6 in			6	**263**	**4**	Shale and sandstone ...			9	285	6
Seatearth-mudstone ...		3	4	266	8	Sandstone...		2	2	287	8
Mudstone and shale ...		3	5	270	1	Shale, silty			2	287	10
Coal ... 3 in			3	**270**	**4**	Sandstone, shale beds ...		4	8	292	6
Seatearth-mudstone ...		3	5	273	9	Shale, silty		1	2	293	8
Mudstone and siltstone ...		2	0	275	9	BOTTOM HARVEY					
Sandstone, argillaceous ...		1	3	277	0	**Coal** ... 15 in					
TOP HARVEY						Seatearth-					
Coal ... 21 in		1	9	**278**	**9**	mudstone 6 in					
Seatearth-mudstone ...			9	279	6	**Coal** ... 15 in		3	0	**296**	**8**

Blackhall Colliery No. 42 Underground Bore

Up 274 ft from Durham Low Main workings at 1163 ft B.O.D. 6-in NZ 44 S.E. Site [4854 4208] 4400 yd E.S.E. of Horden Point. Drilled 1962 by N.C.B.

High Main (including 10-in band) 79 in at 261 ft, Top Five-Quarter 11 in at 237 ft, Bottom Five-Quarter 34 in at 189 ft, Top Main 13 in at 161 ft, Bottom Main 34 in at 139 ft.

Blackhall Colliery No. 43 Underground Bore

Down 604 ft from level at 1230 ft B.O.D. 6-in NZ 54 S.W. Site [5079 4246] 6700 yd E. of Horden Point. Drilled 1962–3 by N.C.B.

Durham Low Main 46 in at 67 ft, Top Hutton 29 in at 180 ft, Bottom Hutton 29 in at 223 ft, Top Harvey 20 in at 352 ft, Bottom Harvey 40 in at 361 ft, Top Tilley 27 in at 401 ft, Three-Quarter (including 3-in band) 36 in at 575 ft.

Bowburn Colliery Downcast Shaft

Surface 282 ft A.O.D. 6-in NZ 33 N.W. Site [3041 3794] 1030 yd N.W. of Park Hill.

Drift to 150 ft, Durham Low Main 34 in at 230 ft, Top Hutton 24 in at 295 ft, Bottom Hutton 25 in at 325 ft, Harvey 24 in at 468 ft, Busty 54 in at 620 ft. Sunk to 659 ft.

Brasside Colliery Pumping Shaft

Surface 160 ft A.O.D. 6-in NZ 34 N.W. Site [3030 4596] 1220 yd W. of Moor House. Sunk 1867. For detailed section see B. and S. No. 289.

Drift to 12 ft, Maudlin 6 in at 25 ft, Durham Low Main 18 in at 123 ft, Top Brass Thill 20 in at 158 ft, Hutton 54 in at bottom at 199 ft.

Brasside Colliery, Woodside Pit

Surface 200 ft A.O.D. 6-in NZ 34 N.W. Site [3108 4605] 430 yd W.N.W. of Moor House.

Five-Quarter 46 in at 32 ft, Main 72 in at 72 ft, Durham Low Main 24 in at 204 ft, Hutton 54 in at bottom at 284 ft.

Broomside Bore

Surface 230 ft A.O.D. 6-in NZ 34 S.W. Site [3201 4356] 330 yd N.E. of Broomside House. Drilled 1961 by Boldon Drilling Co. for N.C.B.

Drift to 5 ft, Maudlin 3 in at 30 ft, Top Busty (banded) 5 in at 513 ft, Bottom Busty about 21 in at 530 ft. Drilled to 600 ft.

Broomside Colliery, Adelaide and Antrim Pits

Surface 245 ft A.O.D. 6-in NZ 34 S.W. Site [3167 4375] 500 yd N.N.W. of Broomside House. For detailed section see B. and S. No. 301.

Drift to 81 ft, Durham Low Main 35 in at 193 ft, Hutton 58 in at bottom at 279 ft.

Butterby Farm No. 4 Bore

Surface 306 ft A.O.D. 6-in NZ 23 N.E. Site [2929 3787] 800 yd E. of High Croxdale. Drilled 1958 by Cementation Co. Ltd. for N.C.B.

Drift to 132 ft, Durham Low Main 27 in at 274 ft, Top Hutton 22 in at 338 ft, Bottom Hutton (including 1½-in band) 33 in at 371 ft, Harvey 24 in at 513 ft. Drilled to 518 ft.

Casebourne Cement Works Bore

Surface approximately 25 ft A.O.D. 6-in NZ 53 S.W. Site [5140 3128] 1480 yd N. of Seaton Carew railway station. Drilled 1887–8 for Casebourne & Co. For detailed section see Bird 1888, p. 569.

Well (details of strata not known) 30 ft, Bunter Sandstone to 215 ft, Upper Permian Marl with thick anhydrite at base to ?730 ft, Upper Magnesian Limestone to bottom at 770 ft.

Cassop Colliery ' A ' Pit and Bores

Surface 458 ft A.O.D. 6-in NZ 33 N.W. Site [3406 3835] 1050 yd S.W. of Dene House Farm. Sunk 1836–64. For detailed sections see B. and S. 403, 407.

Drift to 54 ft, Metal 37 in at 339 ft, Five-Quarter 37 in at 343 ft, Main 52 in at 430 ft, Durham Low Main 30 in at 556 ft, Top Hutton 23 in at 659 ft, Bottom Hutton 38 in at 694 ft, Harvey 32 in at 831 ft.
Bore from Harvey: Top Tilley 37 in at 874 ft.

Cassop Grange No. 1 Bore

Surface 329 ft A.O.D. 6-in NZ 33 N.W. Site [3166 3879] 300 yd E. of Cassop Grange. Drilled 1955 by N.C.B.

Drift to 22 ft, Three-Quarter (banded) 44 in at 26 ft, Main (banded) 111 in at 87 ft, Durham Low Main (banded) 33 in at 213 ft, Top Hutton 20 in at 286 ft. Drilled to 289 ft.

Cassop Moor Pit

Surface 339 ft A.O.D. 6-in NZ 33 N.W. Site [3179 3925] 680 yd N.E. of Cassop Grange. Sunk 1840. For detailed section see B. and S. No. 2518.

Drift to 44 ft, Metal (banded) 82 in at 99 ft, Five-Quarter 38 in at 102 ft, Main 49 in at 180 ft, Durham Low Main 34 in at 307 ft, Top Hutton (banded) 48 in at at 387 ft, Bottom Hutton 20 in at bottom at 409 ft.

Castle Eden Colliery, Maria Pit

Surface 361 ft A.O.D. 6-in NZ 43 N.W. Site [4382 3806] 1400 yd E.S.E. of The Castle, Castle Eden. Sunk 1840. For detailed section see B. and S. No. 409. Outline section given by Anderson (1941, p. 15).

Drift to 51 ft, Magnesian Limestone to 632 ft, Marl Slate to 633 ft, Basal Permian Sands to 645 ft.

Five-Quarter 52 in at 713 ft, Main 59 in at 757 ft, Durham Low Main 54 in at 910 ft, Bottom Hutton 45 in at 1037 ft. Sunk to 1252 ft.

Castle Eden Colliery No. 1 Underground Bore

Down 335 ft from tunnel at 312 ft B.O.D. 6-in NZ 43 N.W. Site [4348 3653] 2580 yd S.S.E. of The Castle, Castle Eden. Drilled 1934 by A. Kyle Ltd. for Horden Collieries Ltd.

Bottom Hutton 30 in at 129 ft, Top Tilley (including 2-in band) 41 in at 294 ft.

Chilton Colliery

Surface 433 ft A.O.D. 6-in NZ 23 S.E. Site [2780 3068] 1070 yd W.S.W. of West Close. Sunk 1872. For detailed section see B. and S. No. 444.

Metal (banded) 47 in at 358 ft, Five-Quarter 53 in at 384 ft, Main (2 leaves) 52 in at 456 ft, Maudlin (banded) 61 in at 545 ft, Top Brass Thill 24 in at 675 ft, Top Hutton (banded) 30 in at 719 ft, Bottom Hutton 12 in at 749 ft, Harvey (2 leaves) 52 in at 933 ft.

Staple pit to Brockwell, sunk 1882. Busty (2 leaves) 41 in at 1069 ft, Brockwell 49 in at 1200 ft.

Chilton Colliery Surface Bore

Surface 387 ft A.O.D. 6-in NZ 23 S.E. Site [2891 3020] 870 yd W. of Chilton Hall. Drilled 1960 by N.C.B. Cores examined by A. M. Clarke.

	Thickness ft in	Depth ft in		Thickness ft in	Depth ft in
DRIFT			Dolomitic limestone, buff,		
Sandy boulder clay ...	19 6	19 6	semi-porcellanous,		
			flaggy, with sporadic		
MAGNESIAN			crystalline bands; many		
LIMESTONE					
Limestone (not cored) ...	115 6	135 0	small crystalline cavities	19 0	154 0

	Thickness ft in	Depth ft in
Dolomitic limestone, grey, porcellanous, flaggy, with dark grey bands and partings and sporadic massive layers; buff bands in top 1 ft; some black wavy bituminous partings...	6 0	160 0

MARL SLATE

	Thickness ft in	Depth ft in
Shale, grey, hard, dolomitic, finely laminated, 3-in massive limestone band at 163½ ft; low angle cross-bedding from 164 to 167 ft; fish scales ...	12 6	172 6

BASAL PERMIAN SANDSTONE AND BRECCIA

	Thickness ft in	Depth ft in
Sandstone, bluish grey, fine- to medium-grained, massive, with well-rounded frosted quartz grains set in finer light grey matrix; black ?pyrite on fracture planes; a few flattened dark grey shale or siltstone pellets at 174 ft	2 5	174 11

MIDDLE COAL MEASURES

	Thickness ft in	Depth ft in
Sandstone, pale grey, fine- to medium-grained, micaceous at top, carbonaceous partings below; reddened in bottom 6 in	6 1	181 0
Siltstone, grey, thick-bedded, with fine sandy partings; scattered rootlets ...	2 0	183 0
Mudstone, grey, with irregular red silty shale bands in top 1 ft, silty and micaceous below; a few 'mussels'; well-preserved plants 186 to 190 ft ...	7 0	190 0
Seatearth-mudstone, grey; reddish ironstone at base	2 3	192 3
Sandstone, purplish grey; discontinuous micaceous partings ...	2 9	195 0

	Thickness ft in	Depth ft in
Mudstone, grey and purple, with a few yellowish buff shaly patches and bands	5 0	200 0
Siltstone, grey and reddish grey; red ironstone nodules from 200 to 204 ft and at base ...	5 6	205 6
Seatearth-siltstone, red and grey ...	1 6	207 0
Sandstone, pale grey, fine-grained, with red wisps and wavy partings; some wavy cross-bedding and slump-bedding; carbonaceous bedding planes ...	9 0	216 0
Siltstone, grey, with sporadic reddish bands above 220 ft, generally thin-bedded; a few sandstone bands, argillaceous and massive below 225 ft	18 0	234 0
Mudstone, grey, shaly in top 6 in, dark grey below; clay ironstone bands ...	6 9	240 9

TOP MAIN

		Thickness ft in	Depth ft in
Coal ...	9 in		
Seatearth-mudstone	45 in		
Coal ...	39 in	7 9	248 6

	Thickness ft in	Depth ft in
Seatearth-mudstone, brownish grey, black at top; abundant sphaerosiderite especially in bottom 1 ft ...	5 11	254 5
Sandstone, light grey, fine-grained, grey siltstone partings; cross-bedded and wavy-bedded; some hard irony layers ...	46 7	301 0
Mudstone, silty; plant debris; layers with coalified plant debris 303 to 304½ ft ...	3 6	304 6
Seatearth-mudstone ...	1 6	306 0
Mudstone, grey shaly; ironstone bands throughout; abundant plant debris ...	17 7	323 7

	Thickness ft in	Depth ft in
MAUDLIN		
Coal ... 29 in		
Seatearth-mudstone 11 in		
Coal ... 10 in		
Mudstone, carbonaceous 4 in		
Coal ... 19 in	6 1	329 8
Seatearth-mudstone ...	2 10	332 6
Siltstone, thin bedded, with light grey sandstone partings ...	17 6	350 0
Sandstone, massive, fine-grained, micaceous; coalified fragmentary plants abundant at base	4 3	354 3
Coal ... 14 in	1 2	355 5
Sandstone, dark grey, medium-grained, micaceous; plant debris ...	3	355 8
Siltstone, grey, massive; some irony layers ...	12 4	368 0
Mudstone, grey, shaly, silty, with ironstone layers; pyritized 'mussel' fragments at 377 ft; pyritized 'mussels' in black carbonaceous shale in bottom 2 in ...	9 10	377 10
DURHAM LOW MAIN		
Coal ... 21 in	1 9	379 7
Seatearth-mudstone ...	3 5	383 0
Sandstone, grey, fine-grained, micaceous; scattered rootlets and plant debris ...	8 6	391 6
Sandstone, light grey, medium-grained, cross-bedded; becoming finer grained, carbonaceous and more micaceous downwards ...	41 8	433 2

	Thickness ft in	Depth ft in
TOP BRASS THILL		
Coal ... 16 in	1 4	434 6
Seatearth-siltstone, thin medium-grained sandstone bands ...	2 0	436 6
Sandstone and siltstone alternating ...	7 2	443 8
BOTTOM BRASS THILL (UPPER LEAF)		
Coal ... 12 in	1 0	444 8
Seatearth-mudstone ...	5 4	450 0
Mudstone, silty, with sandstone layers; plant debris	2 4	452 4
BOTTOM BRASS THILL (LOWER LEAF)		
Coal ... 2 in	2	452 6
Seatearth-sandstone, with siltstone bands ...	1 4	453 10
Sandstone, light grey, medium-grained, with micaceous partings and carbonaceous speckling	11 8	465 6
Sandstone, light grey, fine-grained, with irregular siltstone bands and partings ...	13 5	478 11
TOP HUTTON		
Coal ... 9 in		
Black shale 10 in		
Coal ... 6 in	2 1	481 0
Seatearth-mudstone ...	1 4	482 4
Siltstone, grey, interbedded at top with fine-grained sandstone; 1-ft sandstone band at 492 ft ...	10 2	492 6
Mudstone, grey; ironstone layers throughout; poorly preserved 'mussels' 502 to 503 ft; coalified plant debris at base ...	16 0	508 6
BOTTOM HUTTON		
Coal ... 26½ in	2 2½	510 8½
Shale, carbonaceous ...	3½	511 0
Seatearth-mudstone ...	1 6	512 6

Clarence Hetton Colliery Shaft

Surface 309 ft A.O.D. 6-in NZ 33 N.W. Site [3208 3624] 1140 yd W.N.W. of Coxhoe Hall. Sunk 1839. For detailed section see B. and S. No. 2540.

Drift to 75 ft, Main (banded) 67 in at 97 ft.

Cocken Colliery Upcast Shaft

Surface 184 ft A.O.D. 6-in NZ 24 N.E. Site [2978 4748] 380 yd
N.N.E. of Finchale Priory. Sunk 1869. For detailed section
see B. and S. No. 2550.

Drift to 48 ft, Maudlin 30 in at 113 ft, Durham Low Main 33 in at bottom at 186 ft.

Cocken Drift No. 1 Bore

Surface 172 ft A.O.D. 6-in NZ 24 N.E. Site [2939 4745] 410 yd
N.W. of Finchale Priory. Drilled 1951 by Cementation Co. Ltd.
for N.C.B.

Drift to 27 ft, Maudlin 30 in at 57 ft, Durham Low Main 32 in at 156 ft. Drilled to
162 ft.

Cocken No. 32 Bore

Surface 170 ft A.O.D. 6-in NZ 24 N.E. Site [2997 4741] 430 yd
N.E. of Finchale Priory. Drilled 1952 by John Thom Ltd. for
N.C.B.

Drift to 21 ft, Main (old workings) 108 in at 43 ft, Maudlin 36 in at 104 ft, Durham
Low Main (old workings) 42 in at 177 ft, Brass Thill (banded) 105 in at 222 ft.
Drilled to 226 ft.

Cold Knuckles No. 6 Bore

Surface 373 ft A.O.D. 6-in NZ 34 S.W. Site [3275 4166] 1270
yd W. of Crime Rigg. Sunk 1958 by N.C.B.

Drift to 27 ft, Five-Quarter 50 in at 246 ft. Drilled to 252 ft.

Cole Hill Farm Bore

Surface 319 ft A.O.D. 6-in NZ 43 S.W. Site [4229 3110] 1400
yd N. of Embleton church. Drilled 1959 by Cementation Co.
Ltd. for N.C.B.

Drift to 126 ft, Magnesian Limestone to 503 ft, Marl Slate to 505 ft, Basal Permian
Sands to 508 ft.

Bottom Hutton 24 in at 563 ft, Top Busty 24 in at 809 ft. Drilled to 866 ft.

Cotefold Close Bore

Surface about 320 ft A.O.D. 6-in NZ 43 S.W. Approximate
site [4319 3276] 1000 yd S.S.W. of Sheraton Grange. Drilled
1918 by A. Kyle Ltd.

Drift to 106 ft, Magnesian Limestone to 618 ft, Marl Slate to 620 ft, Basal Permian
Sand and Breccia to 626 ft.

?Bottom Busty 11 in at 739 ft, ?Marshall Green 9 in at 978 ft. Drilled to 1092 ft.

See Woolacott (1919, p. 164) for further details of Permian sequence.

Coxhoe Brickyard Bore

Surface 305 ft A.O.D.   6-in NZ 33 N.W.   Site [3164 3584]
1550 yd W. of Coxhoe Hall. Sunk 1864. For detailed section
see B. and S. No. 608.

Drift to 90 ft, Maudlin 49 in at 98 ft, Low Main 31 in at 199 ft, Top Hutton (banded)
35 in at 273 ft, Bottom Hutton 35 in at 306 ft, Harvey 30 in at 448 ft. Drilled to
449 ft.

Coxhoe Bridge No. 1 Bore

Surface 324 ft A.O.D.   6-in NZ 33 S.W.   Site [3299 3496] 330 yd
E. of Coxhoe Bridge station. Drilled 1959 by N.C.B. Cores
examined by A. M. Clarke.

	Thickness ft in	Depth ft in		Thickness ft in	Depth ft in
DRIFT			Siltstone, grey; poorly pre-		
Soil	9	9	served plant debris ...	6	206 8
Boulder clay 34	3	35 0	Seatearth-siltstone, grey		
Sand 4	0	39 0	with purple patches;		
Boulder clay 10	0	49 0	ironstone nodules to-		
MAGNESIAN			wards base	8	207 4
LIMESTONE			Shale, black, with coal		
Limestone (not cored) ...148	0	197 0	traces	4	207 8
Limestone, brown, finely			Seatearth-siltstone, purple		
crystalline with numer-			in lower part 4	4	212 0
ous small crystal-lined			Sandstone, purple-grey,		
cavities; becoming grey			fine-grained, small scale		
and porcellanous with			cross-bedding 9	0	221 0
black listric partings;			*Fault*		
thin-bedded towards			Siltstone, purple-grey, thin-		
base 1	0	198 0	bedded, micaceous part-		
MARL SLATE			ings 8	0	229 0
Shale, black, dolomitic,			Mudstone, grey with pur-		
with speckled bedding			ple patches; ironstone		
planes 4	6	202 6	layer 2	6	231 6
BASAL PERMIAN			**Coal,**		
SANDS			inferior 2 in	**2**	**231 8**
Sandstone, blue-grey,			Seatearth-mudstone,		
medium- to coarse-			brownish grey with pur-		
grained with fine-grained			ple patches 4	4	236 0
matrix, thin-bedded; a			*Fault*		
few flat greenish grey			Seatearth-mudstone,		
siltstone fragments in			brownish grey, silty to-		
bottom 2 in 1	3	203 9	wards base 3	0	239 0
MIDDLE COAL			Mudstone, silty at top;		
MEASURES			sporadic plant debris ... 6	0	245 0
Sandstone, pale grey, fine-			Siltstone, grey, thin-		
grained, wispy-bedded;			bedded, with sandy part-		
slightly reddened to-			ings (dip 22°) 3	0	248 0
wards base 2	5	206 2			

97544   T 2

	Thickness ft in	Depth ft in
Mudstone, grey, silty at top; ' mussels ' near base	10 10	258 10
Coal (with 2-in band) 24 in	**2 0**	**260 10**
Fault		
Sandstone, medium grained, cross-bedded; dark micaceous partings with coal scars ...	1 2	262 0
Fault		
Siltstone, grey, thin-bedded, with micaceous partings; comminuted plant debris; scattered rootlets towards base ...	4 0	266 0
Sandstone, wavy-bedded	1 0	267 0
Siltstone, sandy	2 2	269 2
Sandstone, light grey with purple mottling; fine- to medium-grained; wispy-bedded near top, cross-bedded below	12 10	282 0
Shale, black; coal traces ...	6	282 6
Seatearth-mudstone, brownish grey, silty, sandy towards base ...	2 0	284 6
Sandstone, pale grey; black siltstone partings and thin bands (dip 32°)	4 9	289 3
?Fault		
Sandstone, pale grey, medium-grained, massive	2 0	291 3
Sandstone, grey; black silty bands and partings	4 6	295 9
Sandstone, pale grey, medium- to coarse-grained, highly jointed	6	296 3
Fault		
Mudstone, grey, shaly; scattered ' mussels '; pyritized ' fucoid ' markings at 299 ft; pyritized worm trails at 303 ft ...	16 0	312 3
HIGH MAIN MARINE BAND		
Mudstone, micaceous and slightly silty, *Lingula*	2	312 5
Mudstone; coalified plant debris	1	312 6

	Thickness ft in	Depth ft in
Seatearth-mudstone, pale grey; ironstone nodules in silty base	3 6	316 0
Sandstone, pale grey, fine-grained, thin- or wavy-bedded with micaceous partings	20 0	336 0
Siltstone, grey, thin-bedded	8 10	344 10
Mudstone, grey, shaly; silty above	4 5	349 3
Coal, dirty 23 in	**1 11**	**351 2**
Seatearth-mudstone ...	2	351 4
Fault		
Siltstone, grey, thin-bedded, with black carbonaceous partings; plant debris (dip 28°) ...	16 8	368 0
Mudstone, grey, shaly; ' mussels ' at intervals throughout	16 9	384 9
Mudstone; fish scales ...	3	385 0
Seatearth - siltstone, pale grey	2 0	387 0
Siltstone, grey, thin-bedded; rootlets in upper part, sandy below ...	9 9	396 9
Sandstone, pale grey, fine-grained, thin- and wispy-bedded	12 3	409 0
Siltstone, grey, thin-bedded; scattered plant fragments	13 0	422 0
Mudstone, grey, shaly, black and pyritic at base; scattered ironstone nodules with ' mussel ' fragments ...	6 4	428 4
HIGH MAIN		
Coal (with 2-in band) 20 in	**1 8**	**430 0**
Seatearth-mudstone, black	9	430 9
Siltstone, grey, thin-bedded; rootlets at top, plant debris below ...	15 9	446 6
Seatearth-mudstone, dark grey	6 0	452 6
Mudstone, dark grey; plant debris	21 6	474 0
Mudstone, grey, shaly; poorly preserved ' mussels '	4 4	478 4

		Thick-ness ft in	Depth ft in			Thick-ness ft in	Depth ft in

METAL

LOWER COAL MEASURES

METAL				LOWER COAL MEASURES			
Coal ...	4 in			Siltstone, grey, micaceous; plant debris (dip 7 to 11°)		3 0	495 4
Shale ...	4 in						
Coal ...	5 in			Shale, black; abundant plant debris		6	495 10
Band ...	1 in			Coal 4 in[1]		4	496 2
Coal ...	8 in			Seatearth-sandstone, silty		3 3	499 5
Shale, black	10 in			Mudstone, dark grey, shaly		4	499 9
Seggar ...	24 in			Shale, black; coal traces...		3	500 0
Coal ...	24 in	6 8	485 0	Seatearth-mudstone, dark grey		2 0	502 0

Presumed Butterknowle Fault

		Thick-ness ft in	Depth ft in
Breccia, mainly of shale fragments (dip of slickensided planes about 44°)...		7 4	492 4

Coxhoe Colliery, Bells Pit and Bore

Surface 354 ft A.O.D. 6-in NZ 33 N.W. Site [3131 3729] 300 yd E. of Park Hill. For detailed section see B. and S. Nos. 599 and 603.

Drift, thickness not known, Five-Quarter 42 in at 51 ft, Main (banded) 84 in at 108 ft.

Boring below Main—Durham Low Main 26 in at 255 ft, Top Hutton 24 in at 333 ft, Bottom Hutton 28 in at 356 ft, Harvey 24 in at 495 ft.

Coxhoe Colliery Elizabeth (Engine) Pit

Surface 440 ft A.O.D. 6-in NZ 33 N.W. Site [3289 3662] 840 yd N.N.W. of Coxhoe Hall. Sunk 1835 by Wm. Coulson Ltd. For detailed section see B. and S. No. 604.

Drift to 44 ft, Coal 36 in at 81 ft, High Main 10 in at 128 ft, Five-Quarter 46 in at 203 ft, Main (banded) 52 in at 261 ft, Durham Low Main 22 in at 447 ft, Top Hutton 26 in at 531 ft, Bottom Hutton 35 in at 559 ft, Harvey 44 in at 714 ft, Top Tilley (banded) 47 in at 757 ft. Sunk to 785 ft.

Croxdale Colliery, Thornton Pit

Surface 231 ft A.O.D. 6-in NZ 23 N.E. Site [2667 3701] 1200 yd S.W. of Croxdale Hall. For detailed section see B. and S. No. 2627.

Drift to 29 ft, Top Brass Thill 2 in at 45 ft, Bottom Brass Thill (2 leaves) 13 in at 55 ft, Hutton 20 in at 79 ft, Bottom Hutton (banded) 33 in at 118 ft, Harvey 18 in at 271 ft, Tilley (banded) 31 in at 325 ft, ?Top Busty 26 in at 358 ft, Three-Quarter (banded) 24 in at 465 ft, Brockwell 43 in at 516 ft, Victoria 30 in at 584 ft. Sunk to 596 ft.

[1] Identified by the Coal Survey Laboratory as belonging to a horizon between the Tilley and Brockwell.

Croxdale Estate No. 1 Bore

Surface 306 ft A.O.D. 6-in NZ 23 N.E. Site [2806 3832] 870 yd N.E. of Croxdale Hall. Drilled 1904. For detailed section see B. and S. No. 2629.

Drift to 197 ft, Bottom Hutton (2 leaves) 20¾ in at 332 ft, dolerite 14 ft at 421 ft, Harvey (banded) 13 in at 504 ft, Tilley 3 in at 572 ft, Busty 9 in at 605 ft. Drilled to 669 ft.

Dalton Nook Plantation Bore

Surface 88 ft A.O.D. 6-in NZ 43 S.E. Site [4811 3144] 780 yd E.S.E. of Dalton Field House, Dalton Piercy. Drilled 1961 by Craelius Co. Ltd. for N.C.B. For detailed section see Magraw and others (1963, pp. 197–8).

Drift to 95 ft, Magnesian Limestone to 709 ft, Basal Permian Sand at 712 ft. Bottom Harvey 29 in at 941 ft, Brockwell Roof Coal 33 in at 1184 ft, Brockwell 43 in at 1188 ft. Drilled to 1284 ft.

Dalton Piercy Waterworks No. 5 Bore

Surface 218 ft A.O.D. 6-in NZ 43 S.E. Site [4644 3171] 1260 yd E.S.E. of St. Mary's Church, Elwick. Drilled 1958–9 by Le Grand Adsco Ltd. for Hartlepools Water Co.

Drift to 125 ft, Middle Magnesian Limestone to bottom at 351 ft. This is representative of a closely-spaced group of 7 bores, the deepest of which is 450 ft. In No. 6 Bore [4644 3177], 70 yd N. of No. 5, the top of the Lower Magnesian Limestone was met at about 425 ft.

Dalton Pumping Station, Main Well

Surface 339 ft A.O.D. 6-in NZ 44 N.W. Site [4111 4694] at New Hesledon waterworks. Sunk 1905. Outline section given by Anderson (1941, p. 13).

Drift to 27 ft, Magnesian Limestone to 469 ft, Marl Slate (proved in bore from floor of tunnel 30 ft from well bottom) to 471 ft, on Basal Permian Sand.

Deaf Hill Colliery No. 1 Shaft

Surface 481 ft A.O.D. 6-in NZ 33 N.E. Site [3815 3663] 500 yd E.S.E. of Wingate House. Sunk 1891–2. For detailed section see B. and S. No. 2633.

Drift to 69 ft, Magnesian Limestone to 340 ft, Marl Slate to 344 ft, Basal Permian Sands to 351 ft.

Metal and Five-Quarter 84 in at 485 ft, Main 46 in at 585 ft, Bottom Hutton 37 in at 833 ft, Harvey (including 2-in band) 41 in at 969 ft. Sunk to 981 ft.

Deaf Hill Colliery No. 4 Bore

Surface 449 ft A.O.D. 6-in NZ 43 N.W. Site [4016 3534]
230 yd S. of Woodlands Close. Drilled 1962 by John Thom Ltd.
for N.C.B.

Drift to 85 ft, Magnesian Limestone to ?497 ft, Marl Slate to 498 ft, Basal Permian
Sands and Breccias to 504 ft.

Hutton (including 3-in band) 34 in at 613 ft, Harvey 39 in at 760 ft, Top Tilley
(banded) 63 in at 798 ft. Drilled to 1042 ft.

Deaf Hill No. 4 Underground Bore

Up 170 ft and down 100 ft from Top Tilley workings at 341 ft
B.O.D. 6-in NZ 43 S.W. Site [4015 3473] 180 yd N. of White
Hurworth. Drilled 1955 by N.C.B.

Up bore : Bottom Hutton 17 in at 142 ft, Harvey Marine Band 7 in at 61 ft, Harvey
(old workings) 45 in at 30 ft.

Down bore : Bottom Busty 19 in at 49 ft.

Dean and Chapter Colliery No. 3 Shaft

Surface 441 ft A.O.D. 6-in NZ 23 S.E. Site [2824 3301]
1070 yd S.E. of Skibbereen. Sunk 1902 ; deepened 1952 by
N.C.B.

Drift to 37 ft, Three-Quarter 7 in at 43 ft, Five-Quarter 51 in at 76 ft, Main
(2 leaves) 54 in at 156 ft, Maudlin (banded) 21½ in at 239 ft, Durham Low Main
19 in at 299 ft, Top Brass Thill 17½ in at 342 ft, Top Hutton 29 in at 381 ft, Bottom
Hutton 25 in at 406 ft, Harvey 51 in at 624 ft, Tilley (banded) 117½ in at 698 ft,
Top Busty 35½ in at 734 ft, Bottom Busty 32 in at 742 ft, Brockwell (banded) 46 in
at 854 ft, Victoria 8½ in at 927 ft. Sunk to 936 ft.

Dropswell Bore

Surface 435 ft A.O.D. 6-in NZ 33 S.E. Site [3904 3379]
140 yd W.S.W. of Dropswell.

Drift to 37 ft, Magnesian Limestone to 436 ft, ?Marl Slate to 438 ft. Harvey
(banded) 43 in at 595 ft, Top Tilley (banded) 51 in at 629 ft, Top Busty 23 in at
683 ft, Bottom Busty 19 in at 702 ft. Drilled to 834 ft.

Durham Main Colliery

Surface 200 ft A.O.D. 6-in NZ 24 S.E. Site [2742 4350]
1140 yd S.W. of Frankland Farm. For detailed section see
B. and S. No. 2661.

Hutton at 108 ft, Harvey 21 in at 257 ft, Top Busty 22 in at 373 ft, Bottom Busty
22 in at 377 ft. Sunk to 390 ft.

Easington Colliery North Shaft

Surface 197 ft A.O.D. 6-in NZ 44 S.W. Site [4377 4417] 880 yd S.W. of Shot Rock. Sunk 1912 to 1437 ft; deepened 1953–5 to 1586 ft. For detailed section see B. and S. No. 2662.

Drift to 76 ft, Magnesian Limestone to 465 ft, Marl Slate to 468 ft, Basal Permian Sands to 572 ft.

High Main and Metal 126 in at 1032 ft, Main (banded) 78 in at 1129 ft, Durham Low Main 50 in at 1277 ft, Hutton 64 in at 1388 ft.

Easington Colliery Nos. 1B and 2B Underground Bores

Up (2B) 315 ft and down (1B) 406 ft from Durham Low Main workings at 1788 ft B.O.D. 6-in NZ 44 N.E. Site [4821 4669] 4370 yd E.N.E. of Beacon Point. Drilled 1944 by Andrew Kyle Ltd. for N.C.B.

Up bore: High Main (banded) 27 in at 292 ft, Metal (banded) 30 in at 248 ft, Main (banded) 33 in at 165 ft.
Down bore: Hutton 23 in at 106 ft, Harvey 15 in at 278 ft.

Easington Colliery No. 5 Underground Bores

Up 195 ft and down 304 ft from Hutton workings at 1679 ft B.O.D. 6-in NZ 44 N.E. Site [4730 4582] 3200 yd E. of Beacon Point. Drilled 1944 by Messrs. Brydon for Easington Coal Co.

Up bore: Durham Low Main 63 in at 108 ft, Top Brass Thill (banded) 30 in at 80 ft.
Down bore: Harvey (banded) 7 in at 155 ft.

Easington Colliery No. 6 Underground Bores

Up 152 ft and down 119 ft from Main workings at 1233 ft B.O.D. 6-in NZ 44 N.E. Site [4582 4528] 1600 yd E. of Beacon Point. Drilled 1944 by Messrs. Brydon for Easington Coal Co.

Up bore: High Main (banded) 70 in at 129 ft, Metal (banded) 15 in at 87 ft, Five-Quarter 31 in at 47 ft.
Down bore: Bottom Maudlin (including 5-in band) 24 in at 109 ft.

Easington Colliery No. 7 Underground Bores

Up 317 ft and down 556 ft from Durham Low Main workings at 1749 ft B.O.D. 6-in NZ 44 N.E. Site [4969 4649] 5850 yd E.N.E. of Beacon Point. Drilled 1949–50 by A. Kyle Ltd. for N.C.B. Cores examined by G. Armstrong and R. H. Price.

Up bore:

Description	Thickness ft	in	Height above Durham Low Main ft	in
			316	11
Sandstone, argillaceous, with mudstone beds ...	13	9	303	2
Mudstone, sandy, with 'mussels' at base ...	7	1	296	1
HIGH MAIN				
Coal ... 12 in				
Seatearth-mudstone 6 in				
Coal ... 39½ in				
Shale, carbonaceous 0½ in	4	10	291	3
Seatearth-mudstone ...	3	1	288	2
Sandstone, rooty at top ...	6	6	281	8
Mudstone ...	3	9	277	11
Coal and shale ... 4 in		4	277	7
Seatearth-mudstone ...	3	5	274	2
Mudstone; 'mussels' below 267 ft 8 in ...	8	6	265	8
Seatearth-mudstone ...	2	3	263	5
Mudstone; 'mussels' ...	14	3	249	2
Coal ... 4 in		4	248	10
Seatearth-mudstone ...	1	6	247	4
Mudstone, rootlets ...	4	5	242	11
Seatearth-mudstone ...		6	242	5
Mudstone ...	9	0	233	5
Seatearth-mudstone ...		6	232	11
Mudstone; coal bands ...	5	3	227	8
Seatearth-mudstone ...	2	10	224	10
Shale, carbonaceous, and seatearth-mudstone ...	1	8	223	2
Seatearth-mudstone ...	1	2	222	0
Mudstone; some 'mussels'	17	5	204	7
Siltstone; 'mussels' ...	4	5	200	2
Mudstone ...	2	3	197	11
FIVE-QUARTER				
Coal and splint ... 4 in				
Blackstone and splint 10 in				
Seatearth-mudstone 16 in				
Coal and blackstone 15 in	3	9	194	2
Seatearth-mudstone ...	3	11	190	3
Mudstone ...	2	3	188	0
Sandstone, siltstone beds	11	2	176	10
Siltstone; some 'mussels'	21	2	155	8
Mudstone; 'mussels' ...	1	10	153	10
MAIN				
Coal, banded 38 in	3	2	150	8
Seatearth-mudstone ...	5	0	145	8
Shale, sandy ...	4	0	141	8
Sandstone and siltstone ...	3	7	138	1
Mudstone; 'mussels' ...		5	137	8
Siltstone; some ostracods	5	0	132	8
Shale, sandy ...	8	3	124	5
Sandstone, shale beds ...	17	7	106	10
Coal ... 9 in		9	106	1
Seatearth-mudstone ...	2	10	103	3
Siltstone ...	9	4	93	11
Seatearth-mudstone ...	3	9	90	2
Shale, sandy ...	1	2	89	0
Sandstone ...	29	6	59	6
Coal ... 1 in				
Band ... 22 in				
Coal ... 7 in	2	6	57	0
Seatearth-mudstone ...	1	1	55	11
Siltstone ...	19	5	36	6
Mudstone ...		10	35	8
Coal ... 7 in				
Band ... 1 in				
Coal ... 6 in	1	2	34	6
Seatearth-mudstone ...	2	10	31	8
Shale ...	4	1	27	7
Coal ... 1 in		1	27	6
Mudstone and siltstone ...	10	10	16	8
Not cored ...	11	8	5	0
DURHAM LOW MAIN				
Coal ... 60 in	5	0		

Down bore:

Description	Thickness ft	in	Depth below Durham Low Main ft	in
Seatearth-mudstone ...	1	2	1	2
Coal ... 3 in		3	1	5
Seatearth-mudstone ...		4½	1	9½
Sandstone, earthy ...	1	7½	3	5
Shale; 'mussels' ...	6	7	10	0
Ironstone ...		5	10	5
Shale, silty, sandstone beds	26	11	37	4
Sandstone ...	54	9	92	1
Coal ... 8 in		8	92	9

	Thickness ft	Thickness in	Depth below Durham Low Main ft	Depth below Durham Low Main in
Seatearth-mudstone ...	5	7	98	4
Shale	10	8	109	0
Seatearth-mudstone ...	1	0	110	0
Shale	1	11	111	11
Sandstone	1	1	113	0
Shale	3	6	116	6
Sandstone	21	4	137	10
HUTTON				
Coal ... 15 in	**1**	**3**	**139**	**1**
Shale and sandstone ...	1	6	140	7
Seatearth-mudstone, sandy	6	6	147	1
Shale; ' mussels ' between				
159 ft and 174 ft ...	51	6	198	7
Coal ... 1 in		**1**	**198**	**8**
Seatearth-mudstone ...	1	4	200	0
Shale and ironstone ...	4	0	204	0
Shale; ' mussels ' ...	5	6	209	6
Shale, silty, ironstone beds	11	6	221	0
HARVEY MARINE BAND				
Shale; *Lingula* at ba se	4	11	225	11
Seatearth-mudstone ...	1	3	227	2
Coal ... 3 in		**3**	**227**	**5**
Seatearth-mudsto ne ...	2	0	229	5
Sandstone	6	1	235	6
Shale and seatearth-mud-				
stone	2	6	238	0
Sandstone	40	0	278	0
Shale, sandy	12	0	290	0
Coal ... 3 in		**3**	**290**	**3**
Mudstone; ' mussels ' ...	1	7	291	10
HARVEY				
Coal, soft 39 in	**3**	**3**	**295**	**1**
Seatearth-mudstone ...	2	2	297	3
Shale, sandy	3	1	300	4
Sandstone	52	1	352	5
Shale	1	7	354	0
Shale, coal partings ...		6	354	6
Shale and sandstone ...	4	0	358	6
Sandstone	19	6	378	0
Coal ... 2 in		**2**	**378**	**2**
Sandstone	2	0	380	2
Coal,				
inferior 9 in		**9**	**380**	**11**
Seatearth-mudstone ...		2	381	1
Coal ... 7 in		**7**	**381**	**8**
Seatearth-mudstone ...	4	0	385	8
Shale, sandy	2	0	387	8
Sandstone, shale beds ...	3	3	390	11
Shale and mudstone ...	5	0	395	11
TOP BUSTY				
Coal and				
bands... 47 in				
Coal, soft 19 in	**5**	**6**	**401**	**5**
Seatearth-mudstone ...	4	1	405	6
Shale, sandstone beds ...	15	2	420	8
BOTTOM BUSTY				
Coal ... 12 in	**1**	**0**	**421**	**8**
Seatearth-mudstone ...	3	2	424	10
Coal ... 5 in		**5**	**425**	**3**
Seatearth-mudstone ...	4	2	429	5
Shale, silty	8	10	438	3
Sandstone	26	10	465	1
' Sandy fakes ' (not seen)	4	7	469	8
Mudstone, coal bands ...	1	6	471	2
Coal,				
inferior 5 in		**5**	**471**	**7**
Seatearth-mudstone ...	1	7	473	2
Coal and				
black-				
stone ... 9 in		**9**	**473**	**11**
Shale, silty	16	8	490	7
Mudstone, ironstone beds	12	0	502	7
THREE-QUARTER				
Coal ... 14 in				
Coaly				
blaes ... 6 in				
Coal ... 11 in	**2**	**7**	**505**	**2**
Shale, black		3	505	5
Shale, coal partings ...	3	0	508	5
Seatearth-mudstone ...	1	1	509	6
Shale, sandy	11	4	520	10
Sandstone	1	10	522	8
Mudstone		3	522	11
Coal ... 14 in	**1**	**2**	**524**	**1**
Seatearth-mudstone ...		4	524	5
Shale and mudstone ...	5	1	529	6
BROCKWELL				
Blackstone 6 in				
Coal ... 16 in	**1**	**10**	**531**	**4**
Seatearth-mudstone ...		7	531	11
Sandstone (not recovered)	10	9	542	8
Shale, sandy	12	3	554	11
Not recovered		7	555	6

Easington Colliery No. 11 Underground Bore

Up 388 ft from Durham Low Main workings at 1606 ft B.O.D.
6-in NZ 44 N.E. Site [4636 4639] 2370 yd E.N.E. of Beacon
Point. Drilled 1958–9 by N.C.B.

High Main (banded) at least 28 in at 282 ft, Five-Quarter (banded) 42 in at 191 ft,
Main (including 6-in band) 64 in at 140 ft.

Easington Colliery No. 12 Underground Bore

Up 343 ft from Durham Low Main workings at 1532 ft B.O.D.
6-in NZ 44 N.E. Site [4728 4560] 3170 yd E. of Beacon
Point. Drilled 1959 by N.C.B.

High Main 78 in at 303 ft, Metal 20 in at 245 ft, Five-Quarter (banded) 42 in
at 204 ft, Main 60 in at 161 ft, Top Maudlin 30 in at 110 ft.

Easington Colliery No. 20 Underground Bore

Up 295 ft from Durham Low Main at 1607 ft B.O.D. 6-in
NZ 44 N.E. Site [4625 4667] 2420 yd N.E. of Beacon Point.
Drilled 1964 by N.C.B.

High Main (including 2-in band) 72 in at 285 ft. Metal (including 1-in band) 23 in
at 230 ft, Five-Quarter 32 in at 197 ft, Main (banded) 62 in at 151 ft.

East Howle Colliery

Surface about 290 ft A.O.D. 6-in NZ 23 S.E. Site [2918 3401]
290 yd S. of Cooksons Green. Sunk 1872–3. For detailed
section see B. and S. No. 882.

Drift to 21 ft, Top Hutton 24 in at 64 ft, Harvey 31½ in at 255 ft, Top Busty
(banded) 28 in at 368 ft, Bottom Busty 31 in at 386 ft, Brockwell 50 in at 508 ft.
Sunk to 532 ft.

East Hetton Colliery Busty Bore

Down 395 ft from Busty floor at 445 ft B.O.D. 6-in NZ 33
N.W. Site [3423 3671] 350 yd W.N.W. of St. Helen's Church,
Kelloe. Drilled 1940 by A. Kyle Ltd. for East Hetton Collieries
Ltd. Cores examined by G. A. Burnett.

	Thickness ft in	Depth ft in		Thickness ft in	Depth ft in
Fireclay	2 10	2 10	Sandstone, hard	11 0	38 10
Shale, sandy, with thin sandstone ribs in lower part	7 6	10 4	Shale, grey, sandy, hard ...	3 4	42 2
			Bluestone with ironstone ribs	1 8	43 10
Sandstone, white, hard, fine-grained	2 0	12 4	Bluestone with ironstone balls and ' mussels ' ...	8 0	51 10
Sandstone with shaly bands	15 6	27 10	Bluestone, dark	4	52 2

	Thickness ft in	Depth ft in
THREE-QUARTER		
Coal ... 8 in		
Fireclay, coaly ... 1 in		
Coal, dirty, with iron pyrites rib ... 3 in	1 0	53 2
Fireclay-shale, hard, coaly	1 3	54 5
Sandstone, white, fine-grained, jointed ...	3 6	57 11
Bluestone with plants, sandy bands in lower part ...	5 10	63 9
Coal ... 10 in	10	64 7
Fireclay, sandy ...	2 0	66 7
Bluestone, light sandy; 'mussels' ...	4 0	70 7
Sandstone, hard ...	1 1	71 8
Bluestone, sandy ...	1 6	73 2
BROCKWELL OSTRACOD BAND		
Bluestone with ironstone ribs; 'mussels' and ostracods in lowest 2 ft ...	4 8	77 10
Shale-fireclay, sandy, micaceous and more sandy to base ...	3 4	81 2
BROCKWELL ROOF COAL		
Coal, shaly 14 in	1 2	82 4
Fireclay, sandy ...	3 0	85 4
Shale, sandy, with sandstone bands; plants ...	5 9	91 1
BROCKWELL		
Coal with bluestone bands ... 7 in		
Coal with thin dirt bands... 34 in	3 5	94 6
Fireclay, sandy, hard and micaceous ...	9	95 3
Sandstone, shaly, micaceous ...	6 4	101 7
Coal ... 6 in	6	102 1
Fireclay, sandy ...	3 6	105 7
Bluestone, sandy, with sandstone bands ...	2 4	107 11
Sandstone, hard, with micaceous partings ...	10 6	118 5
Fireclay, hard, sandy ...	1 11	120 4
Sandstone, shaly	10	121 2
Bluestone, sandy, with irregular sandstone partings ...	2 3	123 5
Bluestone with hard irony bands ...	3 7	127 0
Sandstone, hard, shaly ...	6 3	133 3
Bluestone with ironstone ribs; 'mussels' in lower part ...	15 10	149 1
Fireclay, dark, with soft bands ...	8	149 9
Sandstone, hard, with thin sandy bluestone bands near top ...	17 9	167 6
Bluestone, sandy, with median 6-in hard sandstone band ...	5 5	172 11
Bluestone, with crushed 'mussels' in upper part, ironstone ribs towards top ...	3 4	176 3
VICTORIA		
Coal ... 4 in	4	176 7
Fireclay-shale, sandy in upper part ...	6 4	182 11
Bluestone with 'mussels'	1 8	184 7
Fireclay ...	1 9	186 4
Sandstone, blue, shaly, with 8-in sandy bluestone band at base ...	2 3	188 7
Sandstone	2 10	191 5
Bluestone, sandy; 'mussels' ...	3 9	195 2
Bluestone, black; crushed 'mussels' and fish scales	2 10	198 0
Mudstone, dark, and bluestone ...	2 2	200 2
Bluestone, sandy; indefinite 'mussels' ...	1 8	201 10
Sandstone, hard, bluey ...	3 1	204 11
Bluestone, black and sandy in lower part ...	1 8	206 7
Sandstone, shaly ...	11 10	218 5
Bluestone, sandy, with median 7-in very hard kingle band ...	6 3	224 8
Sandstone, bluey, with thin coal ribs, 8-in conglomeratic sandstone band at base ...	5 4	230 0
Shale, sandy ...	2 9	232 9

	Thickness		Depth	
	ft	in	ft	in
Sandstone	2	6	235	3
Bluestone, sandy	1	11	237	2
MARSHALL GREEN				
Coal ... 6 in		6	**237**	**8**
Fireclay, sandy, and balls	5	6	243	2
Bluestone, dark sandy, with plant debris ...	5	3	248	5
Bluestone, sandy; ' mussels '	7	4	255	9
Bluestone, black, with ironstone ribs; ' mussels '...		9	256	6
Bluestone with iron balls...		5	256	11
Bluestone, sandy, micaceous; 2-in ironstone band	10	0	266	11
Bluestone	1	6	268	5
Sandstone, shaly at top and bottom	4	4	272	9
Bluestone, sandy in lower part	1	0	273	9
Bluestone, with iron ribs in upper part and iron balls below, bands of sandy bluestone in middle and at base... ...	5	8	279	5
Sandstone, tough, ganister-like, micaceous ...	1	6	280	11
Bluestone, sandy, possible crushed ' mussels ' ...		2	281	1
Coal, hard splint 2 in		2	**281**	**3**
Fireclay-shale, dark, sandy		10	282	1
Sandstone, micaceous ...	4	2	286	3
GANISTER CLAY				
Coal ... 7 in		7	**286**	**10**
Fireclay-shale, sandy fireclay with iron balls in middle part	6	9	293	7
Sandstone, dark bluey, and sandy bluestone ...	1	11	295	6

	Thickness		Depth	
	ft	in	ft	in
Bluestone, sandy, with sandstone ribs	2	3	297	9
Bluestone, sandy	3	2	300	11
Sandstone with micaceous partings	4	6	305	5
Bluestone, sandy, with sandstone ribs	2	6	307	11
Bluestone, sandy, micaceous, with sporadic plant debris	6	1	314	0
Bluestone, dark sandy and planty in lower part ...	2	11	316	11
Sandstone, white, tough micaceous, shaly ...	17	9	334	8
Bluestone, sandy, with sandstone ribs	3	10	338	6
GUBEON MARINE BAND				
Bluestone with *Lingula*	3	9	342	3
GUBEON				
Coal ... 9 in		9	**343**	**0**
Ganister	1	7	344	7
Fireclay, hard, sandy ...		10	345	5
Bluestone, sandy, with iron balls in lower part; 5-in slump at base ...	6	4	351	9
Post, soft, shaly	4	2	355	11
Bluestone, sandy	5	9	361	8
Sandstone with shale partings, 3-in dark sandy bluestone band in upper part	8	10	370	6
Bluestone, dark sandy ...	15	0	385	6
Coal, inferior 7 in		7	**386**	**1**
Fireclay, hard, sandy ...	1	9	387	10
Bluestone, dark, sandy, hard; plants	7	0	394	10

Edderacres Plantation No. 1 Bore

Surface 397 ft A.O.D.   6-in NZ 43 N.W.   Site [4031 3873]
520 yd N.W. of Wellfield Junction Station.   Drilled 1964
by Boyles Bros. Ltd. for N.C.B.

Drift to 114 ft, Magnesian Limestone and Basal Permian Sands to ?500ft.
Metal (banded) 56 in at 681 ft, Five-Quarter (upper leaf, including 11-in inferior coal) 77 in at 692 ft, Main 50 in at 729 ft.   Drilled to 749 ft.

Eden Bore

Surface about 238 ft A.O.D. 6-in NZ 44 S.W. Site [4410 4086]
1300 yd E.S.E. of Little Eden. For detailed section see B. and
S. No. 2666.

Drift to 110 ft, Magnesian Limestone and Marl Slate to 530 ft, Basal Permian Sands
to ?594 ft.

High Main (banded) 66 in at 795 ft, Five-Quarter 40 in at 866 ft, Durham Low
Main 35 in at 1082 ft, Hutton 65 in at 1184 ft. Drilled to 1469 ft.

Elemore Colliery, George Pit

Surface 405 ft A.O.D. 6-in NZ 34 N.E. Site [3565 4564]
860 yd S.W. of St. Michael's Church, Easington Lane. Sunk
1826 to 565 ft. Deepened 1947 to 851 ft. For detailed section
of shaft to 565 ft see B. and S. No. 772.

Outset 6 ft, Drift to 57 ft, Magnesian Limestone and Marl Slate to 133 ft, Yellow
Sands to 196 ft.

High Main (including 2 thin bands) 72½ in at 411 ft, Main 68 in at bottom at
565 ft.

Elemore Colliery, Isabella Pit

Surface 407 ft A.O.D. 6-in NZ 34 N.E. Site [3562 4562]
900 yd S.W. of St. Michael's Church, Easington Lane. Sunk
1825. For detailed section see B. and S. No. 774.

Outset 16 ft, Drift to 64 ft, Magnesian Limestone and Marl Slate to 162 ft, Basal
Permian Sands to 225 ft.

Main 70 in at 470 ft, Hutton 67 in at 766 ft. Sunk to 766 ft.

Elemore Colliery, Prospect Hill Bore

Surface 450 ft A.O.D. 6-in NZ 34 S.E. Site [3616 4243]
650 yd E. of Haswell Moor Farm. Drilled 1964–5 by Boyles
Bros. Ltd. for N.C.B.

Drift to 12 ft, Magnesian Limestone to 132 ft, Marl Slate to 135 ft, Basal Permian
Sands to 231 ft.

Top Tilley 13 in at 965 ft, Bottom Busty (including 2-in band) 30 in at 1079 ft.
Drilled to 1083 ft.

Elemore Colliery Underground Bore

Down 226 ft from Hutton workings at 579 ft B.O.D. 6-in
NZ 34 S.E. Site [3741 4488] 250 yd N.N.W. of Fallowfield.
Drilled 1963 by N.C.B.

Harvey (including 1-in band) 28 in at 116 ft, Tilley (banded) 43 in at 164 ft, Top
Busty 22 in at 212 ft, Bottom Busty 17 in at 219 ft.

Elvet Landsale Colliery, South Engine Pit and Bore

Surface 180 ft A.O.D. 6-in NZ 24 S.E. Site [2762 4160]
720 yd N.N.E. of Oswald House. Sunk 1858. For detailed
section see B. and S. No. 803.

Drift to 23 ft, Hutton 43 in at 202 ft. Sunk to 224 ft.

Boring in Shaft bottom : Harvey 21 in at 353 ft, Busty (banded) 100 in at 505 ft.
Drilled to 535 ft.

Elwick No. 1 Bore

Surface 254 ft A.O.D.   6-in NZ 43 S.E.   Site [4531 3117]
980 yd S. of St. Mary's Church, Elwick.   Drilled 1960–1
by Craelius Co. Ltd. for N.C.B.   For detailed section see Magraw
and others (1963, pp. 196–7).

Drift to 112 ft, Magnesian Limestone to 546 ft, Marl Slate to 547 ft, Basal Permian
Sands and Breccia to 591 ft.

Bottom Busty 24 in at 624 ft, Brockwell Roof Coal 30 in at 702 ft, Brockwell
20 in at 704 ft.   Drilled to 851 ft.

Embleton Old Hall Bore

Surface 303 ft A.O.D.   6-in NZ 43 S.W.   Site [4091 3100]
280 yd W.S.W. of Embleton Old Hall, Embleton.   Drilled 1959
by the Cementation Co. Ltd.   For further details see Magraw
and others (1963, p. 195 and pl. 19).

Drift to 187 ft, Magnesian Limestone to 477 ft, Marl Slate to 482 ft, Basal Permian
Breccia to 486 ft.

Durham Low Main 36 in at 508 ft, Top Harvey 16 in at 785 ft, Bottom Harvey
27 in at 803 ft, Top Busty 24 in at 929 ft, Bottom Busty (including 5-in band) 42 in
at 945 ft, Brockwell (including 2-in band) 28 in at 1082 ft.   Drilled to 1129 ft.

Ferryhill Wood Lane Bore

Surface about 510 ft A.O.D.   6-in NZ 23 S.E.   Site [2937 3293]
400 yd W.S.W. of Cleves Cross.   Drilled 1953 by Boldon Drilling
Co. for N.C.B.

Magnesian Limestone to 53 ft, Marl Slate to 64 ft.

Main 42 in at 144 ft, Durham Low Main 42 in at 315 ft, Top Hutton 28 in at 393 ft,
Bottom Hutton 36 in at 430 ft.   Drilled to 434 ft.

Fifty Rigs (Sheraton) Bore

Surface 395 ft A.O.D.   6-in NZ 43 S.W.   Site [4338 3465]
1060 yd N. of Sheraton Grange.   Drilled 1919 by A. Kyle Ltd.

Drift to 188 ft, Magnesian Limestone to 687 ft, Marl Slate to 688 ft, Basal Permian
Breccia to 692 ft.

Coal Measures proved to 1003 ft.

For further details of the Permian sequence in this bore see Woolacott (1919, pp.
164–5).

Fishburn Colliery South Shaft

Surface 319 ft A.O.D.   6-in NZ 33 S.E.   Site [3612 3181]
1090 yd W.N.W. of Mill House.   Sunk 1912 by Wm. Coulson
Ltd.

Drift to 12 ft, Magnesian Limestone (above 47-ft fault) to 320 ft, Fault from 281 to
320 ft, Marl Slate (below 47-ft fault) to 282 ft, Basal Sandstone to 286 ft.

Bottom Hutton 38 in at 404 ft, Harvey 78 in at 559 ft.   Sunk to 630 ft.

Fishburn ' B ' Bore

Surface 444 ft A.O.D.   6-in NZ 33 S.E.   Site [3779 3400]
890 yd N.W. of Trimdon East House. Drilled 1941 by A. Kyle
Ltd.

Drift to 10 ft, Magnesian Limestone to 345 ft, Marl Slate to 346 ft, Basal Sandstone
to 347 ft.

Harvey 40 in at 454 ft, Top Busty 25 in at 551 ft, Bottom Busty 23 in at 558 ft,
Brockwell 17 in at 705 ft. Drilled to 720 ft.

Fishburn ' D ' Bore

Surface 318 ft A.O.D.   6-in NZ 33 S.E.   Site [3567 3106]
150 yd S. of Lizards. Drilled 1941 by A. Kyle Ltd.

Drift to 36 ft, Magnesian Limestone to 352 ft, Marl Slate to 353 ft, Basal Permian
Sandstone to 354 ft. Permian fossils listed on pp. 174–5.

Top Busty 60 in at 524 ft, Bottom Busty 42 in at 535 ft, Brockwell (2 leaves) 55 in
at 657 ft. Drilled to 687 ft.

Fishburn No. 1 Bore

Surface 390 ft A.O.D. 6-in NZ 33 S.E. Site [3685 3325] 520 yd
W.N.W. of West Carr Side. Drilled 1939 by Wm. Coulson Ltd.

Drift to 27 ft, Magnesian Limestone to 324 ft, Marl Slate to 329 ft, Basal Permian
Sandstone to 346 ft.

Bottom Busty 26 in at 386 ft, Brockwell 25 in at 503 ft.   Drilled to 585 ft.

Fishburn No. 2 Bore

Surface 410 ft A.O.D. 6-in NZ 33 S.E. Site [3745 3379] 790 yd
N. of West Carr Side. Drilled 1941 by Wm. Coulson Ltd. for
Messrs. Henry Stobart & Co. Ltd.

Drift to 15 ft, Magnesian Limestone to 300 ft, Marl Slate to 303 ft, Basal Sandstone
to 318 ft.

Bottom Busty 17 in at 326 ft, Brockwell 16 in at 470 ft, Ganister Clay 15 in at
637 ft. Drilled to 660 ft.

Fishburn No. 3 Bore

Surface 298 ft. A.O.D. 6-in NZ 33 S.E. Site [3718 3084] 660 yd
S. of Mill House. Drilled 1939 by Wm. Coulson Ltd.

Drift to 122 ft, Magnesian Limestone to 388 ft (fossils listed on p. 173).

Top Harvey 33 in at 474 ft, Harvey 30 in at 501 ft, Top Tilley 26 in at 531 ft,
Bottom Tilley (top leaf) 30 in at 586 ft, Bottom Tilley (bottom leaf) 23 in at 593 ft,
Bottom Busty 42 in at 648 ft, Brockwell 49 in at 777 ft. Drilled to 781 ft.

Fishburn No. 4 Bore

Surface 315 ft A.O.D. 6-in NZ 33 S.E. Site [3858 3120] 350 yd
S.E. of Cowburn. Drilled 1940 by Wm. Coulson Ltd. for Messrs.
Henry Stobart & Co. Ltd. Cores of Coal Measures examined
by G. A. Burnett. Permian fossils listed on pp. 173–4.

	Thickness ft in	Depth ft in
GLACIAL DRIFT		
Soil	1 0	1 0
Yellow stony clay ...	4 6	5 6
Clay, dark brown and reddish, stony	48 6	54 0
Clay, stony, with sand partings	10 6	64 6
Clay, brown and red, stony	27 10	92 4
Clay, brown stony, with loamy sand partings ...	9 8	102 0
Clay, strong brown and red, stony	18 0	120 0
MAGNESIAN LIMESTONE		
Marl, yellow, with honeycomb limestone panels	60 0	180 0
Limestone, strong, yellow, honeycombed	99 4	279 4
Limestone, hard, blue and grey; sporadic yellow limestone bands in upper part	97 3	376 7
COAL MEASURES		
Sandstone, strong red; vertical joints	2 11	379 6
Sandstone, soft red and grey; soft shale partings	18 11	398 5
Shale, grey; ironstone bands	2 4	400 9
Fireclay, strong sandy ...	1 9	402 6
Shaly fakes and fireclay-shale	9 3	411 9
Shale, grey and black, with ironstone; a few 'mussels' and *Spirorbis*	7 4	419 1
BOTTOM BRASS THILL		
Coal ... 11 in	**11**	**420 0**
Fireclay	3 9	423 9
Post, grey, sandy shale and fakes	13 3	437 0
Shale, dark grey, with ironstone; sporadic plants and 'mussels', lower 9 ft barren except for rare plants	14 9	451 9
TOP HUTTON		
Coal ... 6 in		
Band ... 1 in		
Coal ... 13 in	**1 8**	**453 5**
Shale, black, carbonaceous, with plants ...	1 6	454 11
Fireclay, and shale with plants	4 2	459 1
Shale, light grey, with post girdles and ironstone bands; plants in upper 4 ft and lower 3 ft ...	26 7	485 8
BOTTOM HUTTON		
Coal ... 24 in		
Band ... 3 in		
Coal, splinty, with black stone ... 21 in	**4 0**	**489 8**
Fireclay-shale, light grey	5 10	495 6
Shale, grey, with 4-ft sandstone near top	15 8	511 2
Shale, soft, dark, with coal partings	9	511 11
Fireclay-shale	3	512 2
Coal with traces of dark shale or fireclay 16 in	**1 4**	**513 6**
Fireclay-shale	9	514 3
Shale	2 10	517 1
Sandstone, white	2 6	519 7
Shale with thin sandstone bands; poor 'mussels' at 530 ft	13 3	532 10
Fireclay	2 0	534 10
Sandstone and shale interbedded	27 8	562 6
Shale with ironstone bands; 'mussels' at 566 ft and from 568 ft	13 6	576 0
Shale, dark grey, sandy, with sporadic sandstone bands and plant debris	45 3	621 3
Shale, sandy, and fireclay with ironstone nodules	3 9	625 0
Shale, sandy, with thin sandy bands	21 2	646 2
Mudstone and shale, 4-in of black shale at base	5 4	651 6
(BOTTOM) HARVEY		
Coal ... 33 in	**2 9**	**654 3**
Seggar-shale with ironstone nodules	7 3	661 6
Shale and mudstone ...	5 2	666 8
Coal with 1-in black band ... 14 in	**1 2**	**667 10**

	Thickness ft	in	Depth ft	in
Fireclay	4	4	672	2
Shale and mudstone with ironstone bands ...	8	3	680	5
Sandstone, shaly	5	1	685	6
Shale, sandy, and fakes ...	30	9	716	3
Coal ... 3 in		3	**716**	**6**
Seggar and coal	1	7	718	1
Sandstone, faky in top 6 ft	17	2	735	3
BOTTOM TILLEY				
Coal ... 4 in				
Black stone ... 1 in				
Coal and black stone ... 16 in	**1**	**9**	**737**	**0**
Fireclay-shale with ironstone nodules	6	0	743	0
Shale and shaly fakes, dark in lower 2 ft with 'mussels' in lower 9 in	25	9	768	9
TOP BUSTY				
Coal with ¼-in pyritic shale at top ... 28 in	**2**	**4**	**771**	**1**
Fireclay-shale, sandy, with pyritic nodules ...	4	0	775	1
Sandstone	3	0	778	1
BOTTOM BUSTY				
Coal ... 24 in	**2**	**0**	**780**	**1**
Fireclay, sandy	1	10	781	11
Faky sandstone with sandy shale partings ...	26	9	808	8
Sandstone, hard, fine-grained, with spar joints	1	11	810	7
Sandstone, medium, lower 6 ft coarser, with shale fragments	19	9	830	4
Shale- and sandstone-conglomerate	2	1	832	5
Sandstone, with shale fragments in bottom 4 in ...	5	3	837	8
FISHBURN 'ESTHERIA' BAND				
Shale with '*Estheria*'	1	10	839	6
THREE-QUARTER				
Coal and splint ... 3 in		**3**	**839**	**9**
Fireclay-shale, sandy ...		11	840	8
Sandstone with shale partings and 3½-ft band at 5 ft above base	19	0	859	8
Shale	2	10	862	6
Coal ... 2 in		**2**	**862**	**8**
Fireclay-shale, sandy at bottom	3	6	866	2
Shale, sandy; plants at 5 ft below top	13	9	879	11
Sandstone with spar joints in upper 3 ft	4	11	884	10
BROCKWELL OSTRACOD BAND				
Shale with ironstone nodules; 'mussels', ostracods and fish remains	5	0	889	10
Fireclay-shale with ironstone nodules	2	4	892	2
BROCKWELL ROOF COALS				
Coal and splint ... 14 in	**1**	**2**	**893**	**4**
Fireclay-shale with ironstone nodules	7	3	900	7
Coal ... 8 in		**8**	**901**	**3**
Shale, sandy with thin sandstone bands and plants; 3-in fireclay at top	18	9	920	0
BROCKWELL				
Coal ... 27 in	**2**	**3**	**922**	**3**
Fireclay-shale, sandy ...	1	9	924	0
Sandstone, shaly sandstone and sandy shale ...	17	0	941	0
Shale, sandy, with 6-in sandy fireclay at top ...	11	9	952	9
Mudstone, dark grey, with ironstone nodules ...	8	2	960	11
Mudstone, dark, with 'mussels'	1	2	962	1
Sandstone, shaly, and sandy shale	24	11	987	0
Shale with 'mussels'; lower 6 in carbonaceous	3	2	990	2
VICTORIA				
Coal ... 7 in		**7**	**990**	**9**
Fireclay-shale, sandy, and shale, with ironstone nodules	9	3	1000	0

	Thickness ft in	Depth ft in
Shale, sandy, with thin sandstone bands ...	9 6	1009 6
Coal ... 6 in	6	**1010 0**
Fireclay-shale, dark sandy, with ironstone nodules	3 2	1013 2
Sandstone, light grey, medium-grained, jointed	14 10	1028 0
Shale, sandy, broken, sporadic plant debris ...	2 0	1030 0
Sandstone, micaceous, with shaly partings in lower part	4 5	1034 5
Shale, micaceous, broken	5 10	1040 3

	Thickness ft in	Depth ft in
Shale, dark carbonaceous, with thin coal films and plant debris; ostracods 4 in above base; parroty shale at 3 in above base	3 0	1043 3
MARSHALL GREEN		
Coal, rather dull ... 3 in	3	**1043 6**
Fireclay, light and dark grey, with coal traces, sandy in lower 1ft ...	14 3	1057 9
Shale, sandy, with thin sandstone bands and shaly sandstone ...	15 3	1073 0

Framwellgate Colliery, Aykley Heads Pit

Surface 238 ft A.O.D. 6-in NZ 24 S.E. Site [2712 4329] 1530 yd S.W. of Frankland Farm. For detailed section see B. and S. No. 2722.

Drift to 16 ft, Durham Low Main 26 in at 62 ft, Hutton 46 in at 139 ft, Harvey 20 in at 282 ft, Busty (banded) 78 in at 405 ft. Sunk to 420 ft.

Framwellgate Moor Colliery

Surface 310 ft A.O.D. 6-in NZ 24 N.E. Site [2698 4511] 1100 yd S.W. of Newton Grange. Sunk 1838. Published graphically in Vertical Sections, Sheet 39, 1871. For detailed section see B. and S. No. 930.

Drift to 143 ft, Durham Low Main 67 in at 155 ft, Top Brass Thill 30 in at 176 ft, Hutton 30 in at 223 ft, ?Harvey 22 in at 341 ft, Busty (banded) 70 in at 471 ft. Sunk to 719 ft.

Frankland Pit

Surface 183 ft A.O.D. 6-in NZ 24 N.E. Site [2937 4515] 830 yd S.S.W. of Union Hall. For detailed section see B. and S. No. 934.

Drift to 46 ft, Durham Low Main 3 in at 81 ft, Top Brass Thill 21 in at 116 ft, Hutton 58 in at 168ft. Sunk to 187 ft.

Garmondsway Pit

Surface 482 ft A.O.D. 6-in NZ 33 S.W. Site [3336 3384] 460 yd N.W. of Mahon House.

Magnesian Limestone to about 180 ft.

Harvey 50 in at 374 ft, Top Busty 43 in at 556 ft, Bottom Busty 55 in at 562 ft.

Grange Colliery New Winning

Surface 245 ft A.O.D. 6-in NZ 34 S.W. Site [3030 4464] 570 yd S.W. of Low Grange. Sunk 1844. For detailed section see B. and S. No. 972.

Drift to 7 ft, Low Main 39 in at 177 ft, Hutton 65 in at 258 ft. Sunk to 282 ft.

Green Lane Cottages Bore

Surface 288 ft A.O.D. 6-in NZ 42 N.W. Site [4095 2992] 1150 yd
W. of St. Mary's Church, Embleton. Drilled 1959 by Cementation Co. Ltd. for N.C.B.

Drift to 186 ft, Magnesian Limestone to 665 ft, Marl Slate to 667 ft, Basal Permian Sands to 668 ft.

Three-Quarter 20 in at 811 ft, Brockwell 31 in at 880 ft. Drilled to 1015 ft.

Harbour House Moor Bore

Surface 45 ft A.O.D. 6-in NZ 24 N.E. Site [2802 4728] 130 yd
S.S.W. of Ford Cottage. Drilled 1854 by Wm. Coulson Ltd. For detailed section see B. and S. No. 1032.

Drift to 90 ft, Harvey 4 in at 201 ft, Busty 22 in at 323 ft. Drilled to 357 ft.

Harbour House No. 66 Bore

Surface 146 ft A.O.D. 6-in NZ 24 N.E. Site [2839 4805] 270 yd
S.E. of Harbour House. Bored 1957 by N.C.B.

Drift to 53 ft, Maudlin 6 in at 66 ft, Durham Low Main 44½ in at 158 ft, Top Brass Thill 30 in at 198 ft, Bottom Brass Thill 4½ in at 201 ft, Hutton (old workings) 34 in at 238 ft. Drilled to 245 ft.

Hart Bore

Surface approximately 70 ft A.O.D. 6-in NZ 43 N.E. Site [4854 3637] 320 yd E.S.E. of Crimdon House. For detailed section see B. and S. No. 2736.

Drift to 37 ft, Magnesian Limestone to 745 ft, Marl Slate to 746 ft, Basal Permian Sands to 750 ft.

Harvey 30 in at 1006 ft, Brockwell 27 in at 1207 ft. Drilled to 1555 ft.

Hart Bushes Bore

Surface about 390 ft A.O.D. 6-in NZ 43 S.W. Approximate site [411 345] 450 yd S. of Hart Bushes Hall, South Wingate. Drilled 1840 by Wm. Coulson Ltd. For detailed section see B. and S. Nos. 2238 and 2928.

Drift to 111 ft, Magnesian Limestone to 585 ft, Marl Slate to 586 ft, ?Basal Permian Sands to 590 ft.

Coal Measures proved to 675 ft.

Hartlepool Lighthouse Bore

Surface 26 ft A.O.D. 6-in NZ 53 S.W. Site [5319 3387] 440 yd E.N.E. of St. Hilda's Church, Hartlepool. Drilled 1961-2 by Craelius Co. Ltd. for N.C.B. For detailed section see Magraw and others (1963, pp. 199-200).

Drift to 22 ft, Upper Magnesian Limestone to 239 ft, Middle Magnesian Limestone (including Hartlepool Anhydrite from 239 to 596 ft) to ?733 ft, Lower Magnesian Limestone to 856 ft, Basal Permian Breccia to 862 ft.

Lower Coal Measures to bottom at 1288 ft.

Permian fossils from this bore are listed on p. 176.

Hartlepool Waterworks Bore

Surface 34 ft A.O.D. 6-in NZ 53 S.W. Site [5075 3339] 870 yd N.N.W. of West Hartlepool railway station. Drilled 1940 by Le Grand Sutcliff and Gell Ltd. for Hartlepools Water Co. For detailed section see Anderson (1941, p. 16) and Trechmann (1942, pp. 321–2). This is representative of a number of bores situated in the floor of a quarry.

Upper Magnesian Limestone to 87 ft, Middle Magnesian Limestone to bottom at 302 ft.

Haswell Colliery, Engine Pit and Bore

Surface about 460 ft A.O.D. 6-in NZ 34 S.E. Site [3738 4224] 1020 yd N. of Harehill Farm. Sunk 1833 to 964 ft. Drilled 1840 to 1109 ft. For detailed section see B. and S. No. 1082.

Superstructure and Drift to 28 ft, Magnesian Limestone to 312 ft, Marl Slate to 313 ft, ?Basal Permian Sands to 325 ft.

Five-Quarter 42 in at 558 ft, Main 25 in at 656 ft, Durham Low Main 43 in at 812 ft, Hutton 65 in at 931 ft, Harvey 18 in at 1056 ft.

Hawthorn Colliery Shaft

Surface 389 ft A.O.D. 6-in NZ 34 N.E. Site [3889 4583] 260 yd W. 23° N. of Hesledon Moor West, South Hetton. Sunk 1954–8. Details of Drift taken from sinker's record. Samples examined by R. H. Allonby, A. M. Clarke, D. Magraw and D. B. Smith.

	Thickness ft in	Depth ft in		Thickness ft in	Depth ft in
DRIFT			Dolomite, cream, buff and		
Soil	2 0	2 0	brown, finely granular,		
Clay with some sand ...	10 0	12 0	generally well-bedded		
Sand, gravel and water ...	5 6	17 6	and with thin laminated		
Clay	1 6	19 0	beds at intervals; harder		
Boulder clay	3 0	22 0	and finer grained below		
Sand and clay, with water	16 0	38 0	295 ft	115 0	343 0
Boulder clay	5 0	43 0			
Sand and gravel	2 0	45 0	Dolomite and dolomitic		
Gravel	10 0	55 0	limestone, grey and		
Boulder clay	10 0	65 0	grey-brown, thin-bed-ded, hard, finely crystal-line, with several thin		
MIDDLE MAGNESIAN LIMESTONE: lagoonal and transitional beds			laminated beds and films of dark grey bituminous		
Dolomite, cream and buff, granular	96 0	161 0	clay	17 11	360 11
			MARL SLATE		
LOWER MAGNESIAN LIMESTONE			Dolomite, grey and dark grey, argillaceous ...	5 1	366 0
Dolomite, cream to pale grey, hard, fine-grained to porcellanous, with many cavities and exten-			**BASAL PERMIAN (YELLOW) SANDS**		
sive autobrecciation ...	67 0	228 0	Sand, grey, incoherent ex-cept at base and top ...	41 0	407 0

MIDDLE COAL MEASURES

	Thickness ft	in	Depth ft	in
Siltstone and sandstone ...	9	0	416	0
Shale and mudstone; plant debris	16	10	432	10

KIRKBY'S MARINE BAND

	Thickness ft	in	Depth ft	in
Mudstone, dark grey, shaly, with foraminifera, pyrite-filled annelid burrows and *Lingula* ...	5	0	437	10
Coal ... 2 in		2	**438**	**0**
Siltstone 17		5½	455	5½
Coal ... 5 in		**5**	**455**	**10½**
Sandstone 39		1½	495	0
Shale, silty; plants ... 12		9	507	9

RYHOPE FIVE-QUARTER

	Thickness ft	in	Depth ft	in
Coal ... 11½ in				
Band ... 1½ in				
Coal ... 26 in	3	3	**511**	**0**
Shale	8	0	519	0
Sandstone; traces of coal at 552 ft; breccia of siltstone fragments at base	138	0	657	0
Siltstone	20	0	677	0
Siltstone and shaly mudstone; 'mussels' *Spirorbis* and '*Estheria*' ...	3	5	680	5
Coal ... 5 in		**5**	**680**	**10**
Sandstone	47	2	728	0
Siltstone; plant debris ...	6	0	734	0
Sandstone, argillaceous from 743 to 752 ft ...	22	0	756	0
Siltstone with thin sandstone laminae	16	7½	772	7½

HIGH MAIN

	Thickness ft	in	Depth ft	in
Coal,				
inferior 7½ in				
Band ... 0½ in				
Coal ... 16½ in				
Band ... 4 in				
Coal ... 9 in	3	1½	**775**	**9**
Shale; plant fragments ...	19	3	795	0
Sandstone, argillaceous ...	5	0	800	0
Shale; roots	11	1	811	1

METAL

	Thickness ft	in	Depth ft	in
Coal ... 15 in	1	3	**812**	**4**
Seatearth, argillaceous ...		8	813	0
Sandstone	20	0	833	0
Siltstone	6	0	839	0
Sandstone	1	8	840	8

FIVE-QUARTER

	Thickness ft	in	Depth ft	in
Coal ... 17 in	1	5	**842**	**1**
Seatearth, argillaceous ...		11	843	0
Siltstone, sandy, grading to silty sandstone from 847 to 852 ft	20	0	863	0
Sandstone	63	0	926	0

MAIN

Old

	Thickness ft	in	Depth ft	in
workings 12 in	**1**	**0**	**927**	**0**
Seatearth-mudstone, 3 ft on mudstone	7	0	934	0
Sandstone, silty in uppermost 4 ft	16	0	950	0
Coal ... 4 in		**4**	**950**	**4**
Mudstone; rooty below 951 ft	5	8	956	0
Sandstone	27	0	983	0
Siltstone	17	0	1000	0
Seatearth, shaly	8	0	1008	0
Siltstone; rooty at top ...	19	0	1027	0
Sandstone, silty	13	0	1040	0
Mudstone, silty		8	1040	8

LOW MAIN

	Thickness ft	in	Depth ft	in
Coal ... 9½ in				
Band ... 11½ in				
Coal ... 26 in	3	11	**1044**	**7**
Seatearth-mudstone ...		5	1045	0
Siltstone; plant fragments	26	0	1071	0
Sandstone, silty from 1077 to 1091 ft ...	40	2	1111	2
Cannel ... 10 in		**10**	**1112**	**0**
Coal, bright 4 in		**4**	**1112**	**4**
Mudstone; many 'mussels'	12	10	1125	2
Coal ... 4 in		**4**	**1125**	**6**
Sandstone, argillaceous; roots	4	6	1130	0
Shale, silty	6	7	1136	7
Coal ... 5 in		**5**	**1137**	**0**
Seatearth-mudstone ...	1	0	1138	0
Shale, partly cannelly from 1150 to 1158 ft ...	37	0	1175	0

HUTTON

Old

	Thickness ft	in	Depth ft	in
workings 12 in	**1**	**0**	**1176**	**0**
Seatearth-mudstone ...	2	0	1178	0
Shale, black and cannelly near base	15	9	1193	9
Coal, with ganister-like sandstone ... 4 in		**4**	**1194**	**1**

	Thickness ft in	Depth ft in
Shale, silty	11 11	1206 0
Sandstone	23 3	1229 3
Coal, shaly 7 in	7	1229 10
Shale, silty below 1236 ft	14 2	1244 0
Sandstone	8 0	1252 0
Siltstone, shaly above 1260 ft	13 0	1265 0
LOWER COAL MEASURES		
Sandstone	47 1	1314 1
HARVEY		
Coal ... 22 in	1 10	1315 11
Mudstone; rooty ...	4 1	1320 0
Sandstone, argillaceous ...	8 0	1328 0
Shale, silty; plant fragments	14 0	1342 0
Sandstone	22 0	1364 0
Shale, silty; plant fragments	2 0	1366 0
TOP TILLEY		
Coal ... 18 in		
Band ... 6 in		
Coal ... 7½ in		
Band ... 6 in		
Coal ... 10 in	3 11½	1369 11½

	Thickness ft in	Depth ft in
Seatearth-mudstone ...	1 9½	1371 9
Siltstone	7 10	1379 7
BOTTOM TILLEY		
Coal ... 5 in	5	1380 0
Shale, silty; plant fragments	8 4	1388 4
Coal ... 2 in	2	1388 6
Siltstone, shaly in uppermost 7½ ft	23 6	1412 0
Sandstone, argillaceous above 1420 ft	17 0	1429 0
TOP BUSTY		
Coal, bright... 19½ in		
Band ... 14 in		
Coal ... 19½ in	4 5	1433 5
Sandstone	4 7	1438 0
BOTTOM BUSTY		
Coal ... 12 in	1 0	1439 0
Seatearth	3 0	1442 0
Sandstone, rooty	4 0	1446 0
Sandstone	43 0	1489 0

Hawthorn Well

Surface 278 ft A.O.D. 6-in NZ 44 S.W. Site [4150 4480] 1500 yd N. of Easington church. Sunk 1906. Outline section given by Anderson (1941, p. 14).

Drift to 117 ft, Magnesian Limestone to 470 ft, Marl Slate to 472 ft.

Middle Coal Measures to bottom at 496 ft.

Hesleden Dene No. 2 (1946) Bore

Surface 92 ft A.O.D. 6-in NZ 43 N.E. Site [4654 3698] 620 yd N.W. of Middlethorpe. Drilled 1945–6 by Thos. Matthews & Son for Horden Collieries Ltd.

Drift to 12 ft, Middle Magnesian Limestone to bottom at 298 ft. For further details of this and adjacent bores see Trechmann (1942, pp. 323–4; 1954, p. 205). Permian fossils from three of these bores are listed on pp. 172–3.

Hett Opencast Site Trial Bore

Surface 337 ft A.O.D. 6-in NZ 23 N.E. Site [2852 3710] 700 yd S. of High Croxdale. Sunk 1956 by N.C.B. Opencast Executive.

Drift to 9 ft, Main 51 in at 50 ft, Durham Low Main 31 in at 212 ft.

Hetton Colliery, Blossom Pit

Surface 329 ft A.O.D.   6-in NZ 34 N.E.   Site [3596 4702]
770 yd E.S.E. of St. Nicholas's Church, Hetton le Hole.   Sunk
1820. For detailed section see B. and S. No. 1154.

Walling and Drift to 21 ft, Lower Magnesian Limestone to 179 ft, Marl Slate to
184 ft, ?Basal Permian Sands to 188 ft.

High Main (banded) 73 in at 505 ft, Main 78 in at 658 ft.   Sunk to 685 ft.

Hetton Colliery, Minor Pit

Surface 329 ft A.O.D.   6-in NZ 34 N.E.   Site [3594 4698]
780 yd S.E. of St. Nicholas's Church, Hetton le Hole.   Sunk
1820–2. Published graphically in Vertical Sections, Sheet 39,
1871. For detailed section see B. and S. No. 1155.

As for Blossom Pit to Main Coal. Durham Low Main 49 in at 781 ft, Hutton
(including 3½-in band) 74 in at 884 ft.   Sunk to 904 ft.

Heugh Hall No. 1 Bore

Surface 392 ft A.O.D.   6-in NZ 33 N.W.   Site [3185 3747]
700 yd N.E. of West Hetton Lodge.   Drilled 1955 by N.C.B.
Cores examined by E. A. Francis.

	Thickness ft in	Depth ft in
DRIFT		
Soil	1 0	1 0
Boulder clay, sandy ...	16 0	17 0
MIDDLE COAL MEASURES		
Shale, black	21 0	38 0
HIGH MAIN MARINE BAND		
Shale, black; *Lingula*	1 0	39 0
Seatearth, sandy above, muddy below	10 0	49 0
Mudstone with ironstone nodules; plant debris ...	2 4	51 4
Sandstone with argillaceous partings; plant debris	1 6	52 10
Coal ... 14 in	1 2	**54 0**
Seatearth, shaly	1 6	55 6
Mudstone with bands of flaggy sandstone ...	26 6	82 0
Coal ... 4 in	4	**82 4**
Seatearth, shaly	2	82 6
Shale, carbonaceous; coalified plant debris in bottom 17 in	4 11	87 5
HIGH MAIN		
Coal ... 13 in	1 1	**88 6**
Seatearth, dark grey ...	1 6	90 0

	Thickness ft in	Depth ft in
Shale, grey, silty and sandy, micaceous; sporadic ferruginous bands; scattered plant debris; leaves and a few rootlets near base	26 0	116 0
Seatearth, grey	1 4	117 4
Mudstone, silty and sandy	2 8	120 0
Ironstone; plant debris ...	3	120 3
Seatearth-mudstone, dark grey and slickensided above, sandy and micaceous below	1 3	121 6
Sandstone, pale grey, fine-grained	3 6	125 0
Shale, dark grey, sandy, with carbonaceous partings and sporadic ferruginous bands; plant debris	3 0	128 0
Mudstone, dark grey, shaly, silty, with a few ferruginous bands; scattered plant debris ...	8 0	136 0
Shale, pale grey	3 4	139 4
Shale, dark grey to black; *Spirorbis, Naiadites* and fish debris	8	140 0

	Thickness ft	Thickness in	Depth ft	Depth in
Shale, black, carbonaceous		8	140	8
Shale, black, cannelly; sporadic fish scales ...		2	140	10
Shale, black, carbonaceout; sporadic plants ...	1	6	142	4
Shale, grey	2	1	144	5
Shale, black, carbonaceous; sporadic fish debris; a few plants	1	7	146	0
Siltstone and sandy shale with micaceous partings; plant debris	4	7	152	11
Sandstone, fine-grained ...		3	153	2
METAL AND FIVE-QUARTER				
Coal ... 8 in				
Mudstone and seatearth 8 in				
Coal ... 75 in	7	7	160	9
Seatearth, grey, shaly ...	2	3	163	0
Siltstone, passing down into sandstone with carbonaceous and pyritic partings; argillaceous bands with plant debris; some ferruginous bands	3	0	166	0
Sandstone, argillaceous, shaly, with muddy partings; sporadic pyritic and micaceous sandy layers	8	6	174	6
Sandstone, pale grey, fine-grained, massive ...	3	6	178	0
Mudstone, shaly and silty; comminuted plant debris	1	6	179	6
Sandstone, light grey; contorted bedding in places	3	11	183	5
Shale, grey, silty, micaceous; some ferruginous bands; plant debris near top and base	19	7	203	0
TOP MAIN				
Coal ... 27 in	2	3	205	3
Seatearth, brown, soft ...	2	5	207	8
MAIN				
Goaf ... 54 in	4	6	212	2
Seatearth, brownish grey, soft	2	3	214	5
Siltstone, pale grey ...	1	3	215	8
Mudstone, pale grey, soft	2	0	217	8
Mudstone, grey, silty, with ironstone nodules; sporadic plant debris ...	24	8	242	4

	Thickness ft	Thickness in	Depth ft	Depth in
Siltstone, grey	1	0	243	4
Shale, sandy; plant debris	6	8	250	0
Sandstone, pale grey, with fine carbonaceous partings; contorted layers in lower part	7	0	257	0
Sandstone with argillaceous and micaceous partings; plant debris in upper part	7	7	264	7
Mudstone and siltstone, contorted	4	4	268	11
Sandstone, argillaceous, sporadic massive or contorted bands	12	6	281	5
Sandstone, pale grey, massive; a few carbonaceous partings	9	1	290	6
MAUDLIN				
Coal ... 8 in		8	291	2
Seatearth, grey		4	291	6
Sandstone, pale grey, fine-grained, with shaly pellets in upper part ...	5	6	297	0
Sandstone, argillaceous ...	5	1	302	1
Mudstone, grey, silty and sandy; carbonaceous and micaceous partings	4	11	307	0
Seatearth	1	6	308	6
Mudstone, grey, silty, with sporadic ironstone bands; sandy beds in lower part; plants and rootlets	7	4	315	10
Sandstone, argillaceous ...	1	7	317	5
Mudstone, silty, with sporadic sandy bands and ironstone layers; plant debris	4	3	321	8
Sandstone, pale grey, fine-grained, micaceous and carbonaceous, with sandy mudstone bands; plant debris towards base	21	4	343	0
Sandstone, fine-grained ...	4	6	347	6
Sandstone, medium-grained, micaceous ...	1	0	348	6
Mudstone, sandy... ...		10	349	4

	Thickness ft in	Depth ft in		Thickness ft in	Depth ft in
Sandstone, pale grey, fine-grained, massive, coarsely micaceous; shale bands, pellets and partings from 355 to 356 ft, carbonaceous from 359½ to 360½ ft	16 8	366 0	Sandstone, argillaceous; plant debris	1 6	370 9
Horizon of Durham Low Main	– –	– –	Seatearth and seatearth-mudstone, ferruginous in part; abundant plants In lowest 3½ ft	9 3	380 0
Seatearth, brownish grey, soft	2 9	368 9	**Coal ... 4 in**	**4**	**380 4**
Shale and seatearth ...	6	369 3	Mudstone, dark grey; rootlets and leaves	1 8	382 0
			Shale, dark grey, sandy, finely micaceous; rootlets	1 0	383 0

Horden Colliery North Pit

Surface about 201 ft A.O.D.   6-in NZ 44 S.W.   Site [4421 4191] 1170 yd E.S.E. of Horden Hall.   Sunk 1900–04. For detailed section see B. and S. No. 2779.   Outline section given by Anderson (1941, p. 15).

Drift to 92 ft, Magnesian Limestone to 463 ft, Marl Slate to 464 ft, Basal Permian Sands to 513 ft.

Ryhope Five-Quarter 33 in at 570 ft, High Main and Metal 53 in at 803 ft, Five-Quarter 39 in at 862 ft, Main 21 at 925 ft, Durham Low Main 43 in at 1084 ft, Hutton (including 1-in band) 65 in at 1198 ft.   Sunk to 1253 ft.

Horden Colliery No. 1 Underground Bores

Up 314 ft and down 149 ft from Hutton workings at 1162 ft B.O.D.   6-in NZ 44 S.E.   Site [4650 4224] 2230 yd E.S.E. of Horden Point.   Drilled 1931 by A. Kyle Ltd. for Horden Collieries Ltd.

Up bore: Main 36 in at 311 ft, Maudlin 48 in at 251 ft, Durham Low Main (banded) 34 in at 150 ft.
Down bore: Harvey (including 6-in band) 36 in at 135 ft.

Horden Colliery No. 3 Underground Bore

Up 293 ft from Hutton workings at 1167 ft B.O.D.   6-in NZ 44 S.E.   Site [4609 4356] 1730 yd E.N.E. of Horden Point. Drilled 1931 by A. Kyle Ltd. for Horden Collieries Ltd.

Five-Quarter 38 in at 288 ft, Main (including 8-in band) 58 in at 245 ft, Durham Low Main 39 in àt 99 ft.

Horden Colliery No. 4 Underground Bore

Down 356 ft from Hutton workings at 979 ft B.O.D.   6-in NZ 44 S.W.   Site [4333 4166] 640 yd N.E. of Little Eden. Drilled 1932 by A. Kyle Ltd. for Horden Collieries Ltd.

Harvey 17 in at 147 ft, Bottom Tilley (banded) 29 in at 221 ft.

Horden Colliery No. 8 Underground Bore

Up 286 ft from Durham Low Main at 863 ft B.O.D. 6-in NZ 44 S.W. Site [4450 4093] 1710 yd E. of Little Eden. Drilled 1936 by A. Kyle Ltd. for Horden Collieries Ltd.

High Main and Metal (banded) 61 in at 280 ft, Five-Quarter 38 in at 201 ft.

Horden Colliery No. 11 Underground Bores

Up 145 ft and down 285 ft from Five-Quarter workings at 1300 ft B.O.D. 6-in NZ 44 S.E. Site [4757 4406] 3430 yd E.N.E. of Horden Point. Drilled 1943–4 by A. Kyle Ltd. for Horden Collieries Ltd.

Up bore: High Main (banded) 65 in at 72 ft, Metal (including 2-in band) 22 in at 32 ft.

Down bore: Hutton (including 5-in band) 65 in at 269 ft.

Horden Colliery No. 15 Underground Bores

Up and down from Durham Low Main workings at 1320 ft B.O.D. 6-in NZ 44 S.E. Site [4974 4374] 5650 yd E. of Horden Point. Drilled 1954 by N.C.B. Cores examined by D. B. Smith and D. E. White.

Up bore:

	detail	Thickness ft	in	Height ft	in
				286	0
Shale, silty, thin sandstone beds		13	3	272	9
Sandstone		3	4	269	5
Shale, silty ...		12	1	257	4
Shale; fish scales ...			11	256	5
HIGH MAIN					
Coal ...	6 in				
Band ...	6 in				
Coal ...	38 in				
Band ...	3 in				
Coal ...	7 in	5	0	251	5
Seatearth-mudstone ...		2	1	249	4
Mudstone and shale, with many plants		3	7	245	9
Siltstone, with many plants		12	2	233	7
Sandstone		21	4	212	3
METAL					
Coal, inferior	8 in				
Seatearth-mudstone	39 in				
Coal, inferior	12 in	4	11	207	4
Seatearth-mudstone, bands of carbonaceous shale...		3	11	203	5
Seatearth-mudstone ...			9	202	8
Shale, silty		9	1	193	7
Sandstone, argillaceous ...		13	7	180	0
Shale		9	7	170	5
Sandstone, argillaceous ...		2	9	167	8
Shale, with fish and 'mussels'		4	10	162	10
FIVE-QUARTER					
Cannel, argillaceous, fish ...	11 in				
Cannel ...	7 in				
Coal ...	36 in				
Shale, carbonaceous	5 in	4	11	157	11
Seatearth-mudstone ...		1	5	156	6
Shale, many plants ...		7	0	149	6
Sandstone		7	6	142	0

Left bore:

	Thickness ft	Thickness in	Height ft	Height in
Siltstone	2	3	139	9
Sandstone	7	9	132	0
Siltstone	9	0	123	0
Shale, with ' mussels ' ...	12	0	111	0

MAIN

		Thickness ft	Thickness in	Height ft	Height in
Coal ...	9 in				
Seatearth-mudstone, coal bands...	23 in				
Coal, inferior	4 in				
Shale, carbonaceous	3 in				
Coal ...	20 in				
Seatearth-mudstone	2 in				
Coal ...	4 in	5	5	105	7
Seatearth-mudstone ...		1	7	104	0
Shale, carbonaceous ...		1	0	103	0
Shale; many plants ...		2	6	100	6
Shale, with fish and ' mussels '		11	0	89	6
Sandstone		4	0	85	6
Shale			3	85	3
Coal, inferior	11 in		11	84	4
Seatearth-mudstone ...		1	4	83	0
Coal ...	9 in		9	82	3
Seatearth-mudstone ...			4	81	11
Shale, carbonaceous ...			5	81	6
Seatearth-mudstone ...			8	80	10
Shale, silty		1	0	79	10
Sandstone		29	4	50	6
Shale			6	50	0
Coal, inferior	9 in		9	49	3
Seatearth-mudstone ...			9	48	6
Shale, many plants ...		6	0	42	6
Sandstone		7	6	35	0
Shale and siltstone ...		8	6	26	6
Shale, carbonaceous ...			7	25	11
Seatearth-mudstone ...		1	3	24	8
Coal ...	18 in	1	6	23	2
Seatearth-mudstone ...		1	0	22	2
Mudstone, silty		17	8	4	6

DURHAM LOW MAIN

		Thickness ft	Thickness in
Coal ...	54 in	4	6

Down bore:

		Thickness ft	Thickness in	Depth ft	Depth in
Seatearth-mudstone ...		1	6	1	6
Shale, plants		1	4	2	10
Sandstone		5	10	8	8
Shale		9	1	17	9
Mudstone, with many ' mussels '			11	18	8
Seatearth-mudstone ...			2	18	10
Shale		31	8	50	6
Sandstone		3	3	53	9
Shale, with ' mussels ' ...			8	54	5
Coal ...	3 in		3	54	8
Seatearth-mudstone ...			2	54	10
Shale, roots		6	0	60	10
Sandstone, argillaceous ...		4	0	64	10
Sandstone		5	10	70	8
Shale, rooty top		11	5	82	1

BOTTOM BRASS THILL

		Thickness ft	Thickness in	Depth ft	Depth in
Coal ...	10 in				
Band ...	0½ in				
Coal ...	15½ in	2	2	84	3
Seatearth-mudstone ...		1	5	85	8
Sandstone		3	0	88	8
Siltstone and shale, with thin sandstone beds ...		45	5	134	1

HUTTON

		Thickness ft	Thickness in	Depth ft	Depth in
Coal ...	14 in	1	2	135	3
Seatearth-mudstone ...		5	8	140	11
Shale		3	7	144	6
Siltstone		3	6	148	0
Mudstone and shale, carbonaceous at base ...		7	3	155	3
Seatearth-mudstone ...		1	0	156	3
Shale, silty		8	10	165	1
Mudstone; ' mussels ' ...			9	165	10
Siltstone, sandy		11	10	177	8
Seatearth-mudstone ...		1	10	179	6
Shale, silty		16	6	196	0
Coal ...	1 in		1	196	1
Shale; roots		1	1	197	2
Sandstone, shale beds ...		9	4	206	6
Mudstone, with ' mussels '			9	207	3
Shale, with many ' mussels ' near base ...		22	9	230	0
Seatearth-mudstone ...		1	3	231	3
Coal ...	2 in		2	231	5
Seatearth-mudstone ...		2	1	233	6
Shale and mudstone ...		10	6	244	0
Seatearth-mudstone ...		1	3	245	3
Mudstone		2	9	248	0
Seatearth-mudstone ...		2	0	250	0
Coal ...	2 in		2	250	2

	Thickness ft in	Depth ft in
Seatearth-mudstone ...	2	250 4
Sandstone ...	17 8	268 0
Shale	6 6	274 6

HARVEY

		Thickness ft in	Depth ft in
Coal ...	5 in		
Band ...	0½ in		
Coal ...	17½ in		
Band ...	1 in		
Coal ...	6 in	2 6	277 0
Seatearth-mudstone ...		2 4	279 4
Shale and siltstone ...		15 4	294 8
Sandstone ...		28 10	323 6
Coal ...	6 in	6	324 0
Seatearth-mudstone ...		1 0	325 0
Shale conglomerate ...		2 0	327 0
Sandstone ...		43 4	370 4
Shale conglomerate ...		4 4	374 8
Sandstone ...		8 2	382 10

TOP BUSTY

		Thickness ft in	Depth ft in
Coal ...	21 in	1 9	384 7
Seatearth-mudstone ...		6	385 1
Sandstone, argillaceous ...		6 11	392 0
Shale; plants		1 1	393 1
Coal ...	3 in	3	393 4
Seatearth-mudstone ...		9	394 0
Shale, sandy		12 0	406 0
Sandstone		1 0	407 0
Shale, sandy		1 0	408 0
Seatearth-mudstone ...		1 0	409 0
Silstone, sandy		1 6	410 6
Seatearth-sandstone ...		6	411 0
Sandstone, argillaceous ...		2 1	413 1

		Thickness ft in	Depth ft in
Siltstone and seatearth-mudstone		3 9	416 10
Sandstone		34 6	451 4
Shale, with ?' mussels ' ...		1 7	452 11
Shale, carbonaceous; fish		1 4	454 3
Coal ...	5 in	5	454 8
Seatearth-mudstone ...		1 4	456 0
Sandstone, siltstone and shale		17 5	473 5
Sandstone		6 1	479 6
Shale, with ' mussels ' and ?ostracods		22 6	502 0
Shale, carbonaceous, with beds of seatearth-mudstone and thin coals ...		1 8	503 8
Coal ...	6 in	6	504 2
Seatearth-mudstone ...		4 2	508 4
Sandstone		5 10	514 2
Shale, carbonaceous ...		2	514 4
Coal ...	2 in	2	514 6
Sandstone		6	515 0
Shale, sandy in top 3 ft ...		3 8	518 8
Coal ...	2 in	2	518 10
Seatearth-mudstone ...		1 2	520 0
Sandstone		10 0	530 0
Shale, sandy; plants ...		4 6	534 6
Sandstone, argillaceous ...		8	535 2
Shale, with fish, ' mussels ', ?ostracods at base ...		6 10	542 0
Coal and band ...	4 in	4	542 4
Seatearth-mudstone ...		8	543 0

Horden Colliery No. 16A Underground Bore

Up 310 ft from Durham Low Main workings at 1022 ft B.O.D. 6-in NZ 44 S.E. Site [4565 4197] 1600 yd S.E. of Horden Point. Drilled 1953 by N.C.B.

High Main 75 in at 279 ft, Main 33 in at 133 ft.

Horden Colliery No. 17 Underground Bore

Up 315 ft from Durham Low Main workings at 1000 ft B.O.D. 6-in NZ 44 S.E. Site [4621 4218] 1950 yd E.S.E. of Horden Point. Drilled 1954 by N.C.B.

Main 27 in at 167 ft, Top Maudlin 33 in at 111 ft.

Horden Colliery No. 19 Underground Bore

Up 480 ft from Durham Low Main workings at 1131 ft B.O.D.
6-in NZ 44 S.E. Site [4738 4258] 3060 yd E. of Horden Point.
Drilled 1956–7 by N.C.B.

High Main Marine Band at 368 ft, High Main 65 in at 276 ft, Five-Quarter 48 in at 206 ft, Main 27 in at 156 ft.

Horden Colliery No. 30 Underground Bore

Up 378 ft from Durham Low Main workings at 1165 ft B.O.D.
6-in NZ 44 S.E. Site [4803 4309] 3730 yd E. of Horden Point. Drilled 1961 by N.C.B.

High Main (banded) 82 in at 285 ft, Metal (including $2\frac{1}{2}$-in band) 26 in at 236 ft, Five-Quarter (banded) 26 in at 211 ft, Main 30 in at 163 ft.

Houghall Estate Shaft

Surface 134 ft A.O.D. 6-in NZ 24 S.E. Site [2816 4059] 400 yd N.E. of Houghall. Sunk 1841. For detailed section see B. and S. No. 1189.

Drift to 32 ft, Durham Low Main 32 in at 115 ft, Hutton 43 in at bottom at 189 ft.

Hutton Henry Colliery, Marley Pit

Surface 404 ft A.O.D. 6-in NZ 43 N.W. Site [4140 3692] 2510 yd S.W. of The Castle, Castle Eden. For detailed section see B. and S. No. 2795.

Drift to 191 ft, Magnesian Limestone to 439 ft, Marl Slate to 441 ft, Basal Permian Sands to 455 ft.

Top Hutton 32 in at 659 ft, Bottom Hutton 39 in at 690 ft, Harvey (including 1-in band) 64 in at 835 ft, Top Tilley (banded) 37 in at 845 ft.

Hutton Henry Colliery, Perseverance Pit

Surface 414 ft A.O.D. 6-in NZ 43 N.W. Site [4152 3665] 2170 yd S.S.W. of The Castle, Castle Eden. For detailed section see B. and S. No. 2794.

Drift to 128 ft, Magnesian Limestone to 440 ft, Marl Slate to 441 ft, Basal Permian Sands to 476 ft.

Bottom Hutton 37 in at 668 ft, Harvey 67 in at 805 ft, Top Tilley 37 in at 815 ft. Sunk to 936 ft, drilled to 993 ft.

Kelloe New Winning Shaft and Bore

Surface 451 ft A.O.D. 6-in NZ 33 N.E. Site [3665 3756] 1450 yd N.W. of Wingate House. Sunk 1856 to 610 ft, drilled 1941 to 1194 ft. For detailed section to Main see B. and S. No. 1218.

Drift to 15 ft, Magnesian Limestone to 334 ft, ?Marl Slate to 337 ft.

Metal and Five-Quarter 83 in at 502 ft, Main 38 in at 589 ft, Bottom Hutton 36 in at 865 ft, Top Tilley (including 5-in band) 39 in at 1002 ft.

Kepier Colliery

Surface 293 ft A.O.D. 6-in NZ 24 S.E. Site [2895 4292] 1630 yd S.S.E. of Frankland Farm. For detailed section see B. and S. No. 1224.

Drift to 101 ft, Durham Low Main 36 in at 184 ft, Brass Thill 14 in at 209 ft, Hutton 47 in at 264 ft. Sunk to 323 ft.

Kepier Colliery, Florence Pit and Bore

Surface 145 ft A.O.D. 6-in NZ 24 S.E. Site [2855 4386] 550 yd S.E. of Frankland Farm. Sunk 1874 to 359 ft, drilled to 556 ft. For detailed section see B. and S. Nos. 1229 and 2804.

Drift to 64 ft, Hutton (banded) 66 in at 86 ft, Harvey 23 in at 211 ft, Bottom Busty 20 in at 353 ft, Three-Quarter 21 in at 413 ft.

Leamside Bore

Surface 180 ft A.O.D. 6-in NZ 34 N.W. Site [3092 4707] 1420 yd N.N.W. of Moor House. Drilled 1957–8 by Cementation Co. Ltd. for N.C.B.

Drift to 47 ft, Main 76 in at 81 ft, Maudlin (banded) 24 in at 144 ft, Durham Low Main 33 in at 225 ft. Drilled to 259 ft.

Littletown Colliery Engine Pit

Surface 321 ft A.O.D. 6-in NZ 34 S.W. Site [3371 4423] 620 yd S.S.W. of Hillside. Sunk 1833. For detailed section see B. and S. No. 1299.

Drift to 72 ft, High Main 10 in at 142 ft, Metal 24 in at 198 ft, Five-Quarter 40 in at 219 ft, Main (banded) 82 in at 287 ft, Durham Low Main 42 in at 446 ft, Hutton 70 in at 541 ft. Sunk to 569 ft.

Littletown Colliery Lady Alice Pit and Bore

Surface 383 ft A.O.D. 6-in NZ 34 S.W. Site [3380 4352] 250 yd N. of Littletown House. Sunk 1831 to 587 ft, drilled to 1555 ft. For detailed section see B. and S. Nos. 1298 and 2825.

Drift to 105 ft, Metal 24 in at 241 ft, Five-Quarter 40 in at 262 ft, Main 52 in at 343 ft, Durham Low Main 16 in at 506 ft, Hutton 68 in at 587 ft, Harvey 14 in at 711 ft, Bottom Busty (banded) 32 in at 844 ft, Marshall Green (banded) 18¼ in at 1101 ft, Ganister Clay (banded) 12 in at 1157 ft.

Low Newton Bore

Surface 180 ft A.O.D. 6-in NZ 24 N.E. Site [2848 4584] 220 yd N.N.E. of Low Newton. Bored 1914 by Wm. Coulson Ltd.

Drift to 172 ft, Harvey 19 in at 275 ft, Busty (banded) 43 in at 417 ft. Drilled to 426 ft.

Low Throston Bore

Surface 98 ft A.O.D.   6-in NZ 43 S.E.   Site [4890 3316]
620 yd S.E. of High Throston.   Drilled 1961 by Craelius
Co. Ltd. for N.C.B.   For further details see Magraw and
others (1963, pp. 198–9).

Made ground to 20 ft, Drift to 85 ft, Magnesian Limestone and Marl Slate to 659 ft,
Basal Permian Sands to 660 ft.

Brockwell Roof Coal 29 in at 799 ft, Brockwell 39 in at 803 ft.   Drilled to
904 ft.

Ludworth Colliery West Shaft

Surface 441 ft A.O.D.   6-in NZ 34 S.E.   Site [3630 4155]
700 yd E.N.E. of Ludworth Tower.   Based on B. and S. No.
1307, amended in part from local information. See also Anderson
(1941, p. 14).

Drift to 120 ft, Magnesian Limestone to 275 ft, Marl Slate to 279 ft, Basal
Permian Sands to 308 ft.

High Main 10 in at 412 ft, Metal and Five-Quarter 137 in at 490 ft, Main 38 in
at 599 ft, Durham Low Main 37 in at 727 ft, Hutton 66 in at 842 ft.   Sunk to
864 ft.

Mainsforth Colliery West (B) Pit

Surface 289 ft A.O.D.   6-in NZ 33 S.W.   Site [3067 3152]
480 yd S.S.E. of Lough House.   For detailed section see B. and
S. No. 1314.

Drift to 11 ft, Permian to 223 ft.

Metal 32 in at 280 ft, Five-Quarter 54 in at 322 ft, Bottom Main 44 in at 396 ft,
Maudlin 48½ in at 473 ft, Durham Low Main 24½ in at 542 ft, Top Hutton 24½ in
at 633 ft, Bottom Hutton 28 in at 668 ft, Harvey (banded) 61½ in at 854 ft, Bottom
Busty (2 leaves) 47 in at 1032 ft, Brockwell (banded) 66 in at 1122 ft.   Sunk to
1136 ft.

Mainsforth Low Main Series No. 2 Bore

Surface 347 ft A.O.D.   6-in NZ 33 S.W.   Site [3112 3200]
690 yd E. of Lough House.   Drilled 1955 by N.C.B.

Drift to 13 ft, Magnesian Limestone to 221 ft, Marl Slate to 226 ft, Basal Permian
Sandstone to 258 ft.

Main (banded) 46 in at 285 ft, Maudlin (banded) 59 in at 346 ft, Durham Low
Main (banded) 36 in at 414 ft, Top Brass Thill 21 in at 449 ft.   Drilled to 461 ft.

Mainsforth Low Main Series No. 9 Bore

Surface 429 ft A.O.D.   6-in NZ 33 S.W.   Site [3202 3305]
1160 yd W. of Highland House.   Drilled 1955 by N.C.B.

Magnesian Limestone to 135 ft, Marl Slate to 139 ft, Basal Permian Sandstone to
223 ft.

Durham Low Main (banded) 38 in at 262 ft, Brass Thill 25 in at 295 ft.   Drilled
to 300 ft.

Middle Stotfold Bore

Surface 137 ft A.O.D. 6-in NZ 42 N.E. Site [4501 2987] 400 yd N.N.W. of Middle Stotfold. Drilled 1961 by Craelius Co. Ltd. for N.C.B. For detailed section see Magraw and others (1963, pp. 196–7).

Drift to 175 ft, Upper Permian Marl to ?382 ft, Magnesian Limestone to 909 ft, Marl Slate to 913 ft.

Carboniferous to bottom at 999 ft.

Mill Hill Bore, Easington

Surface 509 ft A.O.D. 6-in NZ 44 S.W. Site [4122 4248] 1090 yd S. of Easington church. Drilled 1962 by John Thom Ltd.

Drift to 35 ft, Middle Magnesian Limestone to 365 ft. Lower Magnesian Limestone to 511 ft, Marl Slate to 513 ft, Basal Permian Sands to 598 ft.

Coal Measures to bottom at 646 ft.

Detailed sections of the Lower and Middle Magnesian Limestone are given on pp. 113, 125, 137–8. Permian fossils are listed on pp. 170–2.

Miss Chilton's Quarry Bore

Surface 333 ft A.O.D. 6-in NZ 33 S.E. Site [3672 3258] 830 yd S.W. of West Carr Side. Sunk 1837 by Wm. Coulson Ltd. For detailed section see B. and S. No. 886.

Magnesian Limestone to 266 ft, ?Marl Slate to 268 ft, Basal Permian Sandstone to 270 ft.

Harvey 57 in at 425 ft, Bottom Tilley 29 in at 524 ft, Bottom Busty 42 in at 568 ft, Brockwell 34 in at 687 ft. Drilled to 829 ft.

Murton Blue House Lane Bore

Surface 390 ft A.O.D. 6-in NZ 43 S.W. Site [4046 3227] 1270 yd S.S.E. of Hurworth Burn station. Drilled 1959 by Cementation Co. Ltd. for N.C.B.

Drift to 155 ft, Magnesian Limestone to 519 ft, Marl Slate to 520 ft.

Brockwell 17 in at 597 ft, Ganister Clay 12 in at 809 ft. Drilled to 815 ft.

Murton Colliery, Hesledon Curve Underground Bore

Down 258 ft from Hutton workings at about 1110 ft B.O.D. 6-in NZ 44 N.W. Site [4112 4662] 1050 yd N.N.E. of East Batter Law Farm, Cold Hesledon. Drilled 1909.

Harvey (including 5-in band) 24 in at 144 ft, Bottom Tilley 30 in at 188 ft, Busty (banded) 43 in at 258 ft.

Murton Colliery, Hesledon No. 2 Underground Bore

Down 319 ft from Hutton workings at about 1245 ft B.O.D. 6-in NZ 44 N.W. Site [4245 4685] 1060 yd W.N.W. of Kirkby Hill. Drilled ?1909.

Bottom Tilley 24 in at 198 ft, Top Busty 25 in at 270 ft.

Murton Colliery No. 1 Underground Bore

Down from Hutton workings at 1093 ft B.O.D. 6-in NZ 34 N.E. Site [3978 4750] 1200 yd E. of Hesledon Moor East, Murton. Drilled 1950 by A. Kyle Ltd. for N.C.B. Cores examined by G. Armstrong and R. H. Price.

	Thickness ft in	Depth ft in
HUTTON, bottom of		
Coal ... 11 in	11	11
Shale, carbonaceous ...	2	1 1
Seatearth-mudstone, with a 4½-in sandstone bed at 2 ft 7½ in	2 4½	3 5½
Sandstone with thin beds of mudstone	5 10½	9 4
Alternating thin beds of mudstone and sandstone	1 6	10 10
Mudstone with 'mussel' fragments near base, on black shale	18 3	29 1
Cannel and cannelly shale ...	2	29 3
Seatearth-mudstone 11 in on ganisteroid sandstone	2 9	32 0
Sandstone, argillaceous near base	5 0	37 0
Mudstone, dark grey grading down to black; scattered 'mussels' ...	6 9	43 9
Shale, carbonaceous ...	2	43 11
Coal ... 3 in	3	44 2
Seatearth 1 in, on ganister	1 10	46 0
Sandstone, ganisteroid at top	20 8	66 8
Mudstone	11	67 7
Coal ... 6 in	6	68 1
Seatearth-mudstone ...	11	69 0
Sandstone and mudstone interbedded	3 8	72 8
Coal ... 2 in	2	72 10
Mudstone with roots and several sandstone beds near top; 'mussels' below	23 2	96 0
Shale, black; 'mussels' (?horizon of Harvey Marine Band)	8	96 8

	Thickness ft in	Depth ft in
Seatearth-mudstone ...	1 6	98 2
Sandstone, argillaceous at top	48 0	146 2
Mudstone, sandy; plants	5 1	151 3
Mudstone; scattered plants and 'mussels'; many ostracods and Spirorbis sp. at base	8 3	159 6
HARVEY		
Coal ... 33 in	2 9	162 3
Seatearth-mudstone ...	9	163 0
Sandstone; roots at top...	9 3	172 3
Mudstone...	2 6	174 9
Coal ... 3 in	3	175 0
Seatearth-mudstone ...	3 7	178 7
Coal ... 2 in	2	178 9
Seatearth-mudstone ...	2 8	181 5
Sandstone; mudstone beds	4 7	186 0
Mudstone with 1-ft sandstone at 190½ ft ...	6 5	192 5
Coal ... 12 in	1 0	193 5
Seatearth-mudstone ...	5 4	198 9
Coal with 1-in band ... 10 in	10	199 7
Seatearth-mudstone ...	1 11	201 6
Coal ... 11 in	11	202 5
Seatearth-mudstone ...	1 1	203 6
Sandstone, ganister-like ...	3 7	207 1
Seatearth-mudstone ...	1 7	208 8
Coal ... 13 in	1 1	209 9
Seatearth-mudstone ...	6	210 3
Sandstone; roots at top...	32 9	243 0
Coal ... 2 in	2	243 2
Sandstone	5 6	248 8
BUSTY		
Coal ... 7 in		
Band ... 9 in		
Coal, inferior 13 in	2 5	251 1

	Thickness ft in	Depth ft in
Seatearth-mudstone ...	1 11	253 0
Sandstone	3 0	256 0
Seatearth-mudstone ...	1 3	257 3
Coal ... 2 in	**2**	**257 5**
Sandstone, ganister-like ...	3 11	261 4
Sandstone	59 1	320 5
Mudstone, rooty, on shale	6	320 11

THREE-QUARTER

	Thickness ft in	Depth ft in
Coal ... 14 in	1 2	**322 1**
Seatearth-mudstone ...	1 4	323 5
Sandstone; some roots ...	1 10	325 3
Mudstone; roots at top, 'mussels' and ostracods near base ...	8 5	333 8
Coal ... 3 in	**3**	**333 11**
Seatearth-mudstone ...	1 3	335 2
Sandstone, ganister-like ...	1 4	336 6
Mudstone; some roots ...	2 10	339 4
Coal ... 1 in	**1**	**339 5**

	Thickness ft in	Depth ft in
Mudstone, rooty	4 1	343 6
Sandstone, argillaceous; roots	2 2	345 8
Sandstone, argillaceous, with mudstone beds ...	16 10	362 6
Sandstone	21 9	384 3

BROCKWELL

	Thickness ft in	Depth ft in
Coal, inferior 20 in	**1 8**	**385 11**
Sandstone, ganister-like ...	5 8	391 7
Alternations of sandstone and mudstone	17 10	409 5
Siltstone; scattered plants and 'mussels'	4 10	414 3
Seatearth-mudstone ...	9	415 0
Sandstone, argillaceous ...	6 5	421 5
Mudstone; 'mussels' at base	10 2	431 7
Sandstone	5	432 0

Murton Colliery No. 2 Underground Bore

Down from Hutton workings at 1251 ft B.O.D. 6-in NZ 44 N.W. Site [4245 4684] 780 yd W.S.W. of Hesledon East House, Cold Hesledon. Drilled 1951 by A. Kyle Ltd. for N.C.B. Cores examined by G. Armstrong.

	Thickness ft in	Depth ft in
Seatearth-mudstone ...	5 0	5 0
Siltstone; plants	16 10	21 10
Mudstone; 'mussels' ...	9	22 7
Shale, cannelly	4	22 11
Coal, inferior 6 in	**6**	**23 5**
Seatearth-mudstone ...	1 2	24 7
Shale, sandy; plants ...	1 9	26 4
Sandstone, argillaceous ...	1 8	28 0
Siltstone; plants	3 8	31 8
Mudstone; many 'mussels'	5 4	37 0
Coal, inferior 10 in	**10**	**37 10**
Seatearth-mudstone ...	1 6	39 4
Sandstone	22 1	61 5
Siltstone	10	62 3
Shale and mudstone ...	1 1	63 4
Coal ... 8 in	**8**	**64 0**
Seatearth-mudstone ...	3 9	67 9
Shale	2 7	70 4
Sandstone	6 7	76 11

	Thickness ft in	Depth ft in
Mudstone; many 'mussels' from 88 ft to 94 ft 7 in; ironstone beds near base ...	20 7	97 6
HARVEY MARINE BAND		
Mudstone; *Lingula sp.*	1	97 7
Seatearth-mudstone ...	1 0	98 7
Shale; ironstone beds ...	1 11	100 6
Seatearth-mudstone ...	1 10	102 4
Shale; plants	5 8	108 0
Coal and blackstone ... 2 in	**2**	**108 2**
Seatearth-mudstone ...	6	108 8
Shale	6 4	115 0
Sandstone	5 10	120 10
Shale	4 5	125 3
Seatearth-mudstone ...	1 9	127 0
Shale	9	127 9
Seatearth-mudstone ...	1 2	128 11

	Thickness ft	in	Depth ft	in
Shale and mudstone; 'mussels' and ostracods near base	9	4	138	3
Seatearth-mudstone ...		8	138	11
HARVEY				
Coal ... 13 in				
Seatearth-mudstone ... 19 in				
Sandstone 22 in				
Coal ... 10 in	5	4	144	3
Seatearth-mudstone ...	1	9	146	0
Shale; ironstone beds ...	14	0	160	0
Sandstone	32	1	192	1
TOP TILLEY				
Coal ... 6 in				
Seatearth-mudstone ... 35 in				
Coal ... 22 in				
Seatearth-mudstone ... 3 in				
Coal ... 7 in	6	1	198	2
Seatearth-mudstone ...	4	6	202	8
Sandstone, argillaceous ...	5	6	208	2
Shale and mudstone ...	5	9	213	11
BOTTOM TILLEY				
Coal and bands 40 in	3	4	217	3
Seatearth-mudstone ...	5	6	222	9
Shale, sandy ...	7	4	230	1
Sandstone ...	34	3	264	4
TOP BUSTY				
Coal ... 30 in	2	6	266	10
Seatearth-mudstone ...	2	5	269	3
Sandstone, coal scars ...	3	1	272	4
Shale, silty ...	5	2	277	6
BOTTOM BUSTY				
Coal ... 13 in				
Band ... 4 in				
Coal ... 18 in	2	11	280	5

	Thickness ft	in	Depth ft	in
Seatearth-mudstone ...	1	4	281	9
Sandstone ...	3	7	285	4
Shale, splinty at base ...	4	8	290	0
Seatearth-mudstone ...	1	1	291	1
Sandstone, argillaceous ...	1	8	292	9
Shale, 6¾ ft on sandstone	8	1	300	10
Coal ... 6 in		6	301	4
Seatearth-mudstone ...		11	302	3
Shale ...	4	1	306	4
Sandstone ...	1	6	307	10
Shale and mudstone; 'mussels' near base ...	14	8	322	6
Coal, inferior 9 in		9	323	3
Seatearth-mudstone ...	2	8	325	11
Shale ...	2	4	328	3
Sandstone ...	2	5	330	8
Shale; plants ...	2	7	333	3
Sandstone ...	8	5	341	8
Shale, silty ...	3	1	344	9
Mudstone; 'mussels' ...	2	8	347	5
Coal ... 6 in		6	347	11
Seatearth-mudstone ...	3	10	351	9
Shale, silty ...	6	11	358	8
Sandstone, argillaceous ...	1	4	360	0
Shale ...		8	360	8
Coal, inferior 17 in	1	5	362	1
Seatearth-mudstone ...		4	362	5
Sandstone ...	17	7	380	0
Shale ...		4	380	4
BROCKWELL				
Coal, inferior 28 in	2	4	382	8
Seatearth-mudstone ...	4	3	386	11
Sandstone ...	8	5	395	4
Shale and mudstone; scattered 'mussels' and fish debris at base ...	17	0	412	4
Seatearth-mudstone ...	1	10	414	2
Shale ...	7	10	422	0
Mudstone; 'mussels' ...	6	10	428	10
Sandstone ...	7	9	436	7

Murton Colliery No. 3 Underground Bore

Down 372 ft from Harvey workings at 1143 ft B.O.D. 6-in NZ 44 N.W. Site [4041 4592] 230 yd E.N.E. of Little Coop House, Hawthorn. Drilled 1950 by A. Kyle Ltd. for N.C.B.

Top Busty (including 2-in band) 35 in at 127 ft, Three-Quarter 22½ in at 198 ft.

Murton Colliery No. 5 Underground Bore

Down 400 ft from Hutton workings at 893 ft B.O.D. 6-in
NZ 44 S.W. Site [4134 4360] 200 yd N.N.W. of Easington
church. Drilled 1951 by A. Kyle Ltd. for N.C.B.

Harvey (including 13-in band) 49 in at 143 ft, Bottom Tilley 31 in at 202 ft, Bottom
Brockwell 31 in at 398 ft.

Murton Colliery, West Pit

Surface 334 ft A.O.D. 6-in NZ 34 N.E. Site [3986 4718]
830 yd N. of Hesledon Moor East. Sunk 1840–47. For
detailed section see B. and S. No. 1392.

Drift to 39 ft, Magnesian Limestone to 450 ft, Marl Slate to 453 ft, Basal Permian
Sands to 479 ft.

Ryhope Five-Quarter 37 in at 812 ft, High Main 58 in at 1071 ft, Main 74 in at
1228 ft, Durham Low Main 55 in at 1371 ft, Hutton (including 12-in band) 86 in
at 1479 ft.

Naisberry Waterworks No. 1 Bore

Surface approximately 384 ft A.O.D. 6-in NZ 43 S.E. Site
[4662 3363] 610 yd W. of Naisberry, Hart. Drilled 1953 for
Hartlepool Water Co.

Drift to 30 ft, ?Upper Magnesian Limestone to ?60 ft, Middle Magnesian Limestone
to 447 ft, Lower Magnesian Limestone to bottom at 500 ft.

No. 2 Bore, 40 yd N.W. of No. 1, was sunk in 1961 to 550 ft without penetrating
Marl Slate.

Newton Grange Bore

Surface 127 ft A.O.D. 6-in NZ 24 N.E. Site [2848 4610]
500 yd N.N.E. of Low Newton. For detailed section see B. and
S. No. 1464.

Drift to 103 ft, Harvey 21 in at 227 ft. Drilled to 232 ft.

Newton Hall Bore

Surface 230 ft A.O.D. 6-in NZ 24 N.E. Site [2797 4563]
400 yd W. of Low Newton. Bored 1824. For detailed section
see B. and S. No. 1465.

Drift to 233 ft, Hutton (banded) 24 in at 281 ft. Drilled to 300 ft.

North Dalton Waterworks Well

Surface 279 ft A.O.D. 6-in NZ 44 N.W. Site [4084 4778]
260 yd N.N.E. of Glebe House, New Hesledon. Sunk 1905.
Outline section given by Anderson (1941, p. 12).

Boulder clay to 58 ft, Middle Magnesian Limestone to ?177 ft, Lower Magnesian
Limestone to 399 ft, Marl Slate to 401 ft, Basal Permian Sands to 467 ft.

Coal Measures shale to bottom at 469 ft.

North Hetton Colliery, Moorsley Winning

Surface 360 ft A.O.D.   6-in NZ 34 N.W.   Site [3419 4631]
1850 yd E. of Field House.   Sunk 1826–8.   For detailed section
see B. and S. No. 1472.

Drift to 24 ft, Ryhope Little 24 in at 51 ft, High Main 103 in at 203 ft, Main (banded) 82½ in at 376 ft, Low Main (banded) 48¼ in at 503 ft, Hutton 74½ in at bottom at 601 ft.

North Hetton Colliery No. 2 Upcast Pit

Surface c. 360 ft A.O.D.   6-in NZ 34 N.W.   Site [3416 4629]
1810 yd E. of Field House.   Sunk 1857–8.   For detailed section
see B. and S. No. 1473.

Drift to 18 ft, Ryhope Little 42 in at 55 ft, High Main (banded) 85 in at 206 ft, Main 84 in at 378 ft, Low Main 42 in at 510 ft, Hutton 54 in at bottom at 609 ft.

North Pittington Colliery Buddle Pit

Surface 315 ft A.O.D.   6-in NZ 34 S.W.   Site [3317 4409]
1090 yd N.W. of Littletown House.   For detailed section see
B. and S. No. 1540.

Drift to 24 ft, High Main 37 in at 76 ft, Metal (banded) 110 in at 122 ft, Five-Quarter 43 in at 161 ft, Main 41½ in at 202 ft.

North Pittington Colliery Londonderry Pit

Surface 325 ft A.O.D.   6-in NZ 34 S.W.   Site [3350 4432]
1160 yd N.N.W. of Littletown House.   Sunk 1826–8.   For
detailed section see B. and S. No. 1539.

Drift to 30 ft, High Main 27½ in at 168 ft, Metal (banded) 61 in at 215 ft, Five-Quarter 45½ in at 240 ft, Main 53 in at 285 ft, Low Main 48½ in at 446 ft, Hutton 70 in at bottom at 538 ft.

North Sands Bore

Surface 25 ft A.O.D.   6-in NZ 53 N.W.   Site [5027 3534]
900 yd N. of Howbeck Hospital, Hartlepool.   Drilled 1961
by Craelius Co. Ltd. for N.C.B.   For detailed section see
Magraw and others (1963, pp. 200–1).

Made ground to 11 ft, Drift to ?103 ft, Magnesian Limestone at 955 ft, Basal Permian Breccia to 959 ft.

Lower Coal Measures to bottom at 1268 ft.

Nunstainton No. 1 Bore

Surface about 300 ft A.O.D.   6-in NZ 32 N.W.   Site [3182 2919]
220 yd E.N.E. of Nunstainton East.   Sunk 1917.

Drift to 129 ft, Magnesian Limestone to 341 ft, ?Basal Permian Sandstone to 362 ft.

Bottom Busty 34 in at 438 ft, Brockwell (banded) 104 in at 554 ft. Drilled to 890 ft.

Nunstainton No. 2 Bore

Surface about 300 ft A.O.D. 6-in NZ 32 N.W. Site [3183 2988]
870 yd N.N.E. of Nunstainton East. Sunk 1918.

Drift to 141 ft, Magnesian Limestone to 325 ft, Marl Slate to 333 ft, Basal Permian
Sandstone to 335 ft.

Bottom Harvey 30 in at 515 ft, Bottom Tilley 20 in at 641 ft, Top Busty 16 in at
681 ft, Bottom Busty (2 leaves) 32 in at 704 ft. Drilled to 1094 ft.

Offshore No. 1 Bore

Surface of drilling platform 66 ft A.O.D. 6-in NZ 54 S.W.
Site [5334 4043] 4 miles N. of Hartlepool coastguard station.
Drilled 1958 by Foraky Ltd. for N.C.B. For detailed section
see Magraw and others (1963, pp. 201–2 and pl. 19).

Sea-bed at 173 ft, Drift to 245 ft, Upper Magnesian Limestone to 591 ft, Middle
Magnesian Limestone (including Hartlepool Anhydrite 591–1020 ft) to 1114 ft,
Lower Magnesian Limestone to 1154 ft, Basal Permian Sands to 1164 ft.

Metal 42 in at 1201 ft, Top Five-Quarter 30 in at 1240 ft, Durham Low Main 43
in at 1382 ft, Bottom Harvey 32 in at 1673 ft, Brockwell 32 in at 1884 ft, Gubeon
(banded) 18 in at 2150 ft. Drilled to 2158 ft.

Offshore No. 2 Bore

Surface of drilling platform 49 ft A.O.D. 6-in NZ 54 S.W.
Site [5301 4379] 6 miles N. of Hartlepool coastguard station.
Drilled 1958–9 by Foraky Ltd. for N.C.B. For detailed section
see Magraw and others (1963, pp. 202–3 and pl. 19).

Sea-bed at 173 ft, Drift to 253 ft, Upper Magnesian Limestone to 666 ft, Middle
Magnesian Limestone (including Hartlepool Anhydrite 666–1166 ft) to fault at
1186 ft, Lower Magnesian Limestone to fault at 1244 ft.

Five-Quarter (including 3-in band) 43 in at 1393 ft, Durham Low Main (banded)
76 in at 1552 ft, Bottom Harvey 36 in at 1848 ft, Three-Quarter 34 in at 2036 ft,
Brockwell 28 in at 2087 ft. Drilled to 2193 ft.

Offshore No. 13 Bore

Surface of drilling platform 69 ft A.O.D. 6-in NZ 54 N.W.
Site [5151 4566] 4¼ miles E. of Shippersea Bay. Drilled 1964
by Foraky Ltd. for N.C.B.

Sea-bed at 173 ft, Upper Magnesian Limestone to 532 ft, Hartlepool Anhydrite to
937 ft, Middle and Lower Magnesian Limestone to 1059 ft, Basal Permian Sands to
1158 ft.

Ryhope Five-Quarter 35 in at 1342 ft, Ryhope Little 24 in at 1421 ft, High Main
(including 2-in band) 60 in at 1582 ft, Main (banded) 49 in at 1739 ft, Durham Low
Main 60 in at 1873 ft, Harvey 37 in at 2148 ft, Top Busty (banded) 41 in at 2269 ft.
Drilled to 2424 ft.

Pudding Poke Farm Bore

Surface approximately 335 ft A.O.D. 6-in NZ 43 S.W. Site [4313 3280] 1000 yd S.S.W. of Sheraton Grange. Drilled 1960 by George Stow & Co. for Hartlepools Water Co.

Drift to 125 ft, Middle Magnesian Limestone to 424 ft, Lower Magnesian Limestone to bottom at 500 ft.

Rainton Colliery, Alexandrina Pit and Bore

Surface 324 ft A.O.D. 6-in NZ 34 N.W. Site [3327 4637] 900 yd E.N.E. of Field House. Sunk 1823 to 464 ft by Wm. Coulson Ltd., drilled 1896 to 1265 ft. For detailed section see B. and S. Nos. 1589 and 2919.

Drift to 29 ft, High Main (banded) 87 in at 93 ft, Five-Quarter (banded) 38½ in at 188 ft, Main (banded) 77 in at 245 ft, Maudlin 48 in at 323 ft, Low Main (banded) 47½ in at 372 ft, Top Brass Thill 32 in at 405 ft, Hutton 61½ in at 464 ft, Harvey 23 in at 604 ft, Busty 24 in at 738 ft, Brockwell 22 in at 840 ft, Marshall Green 15½ in at 969 ft, Gubeon 10 in at 1085 ft.

Rainton Colliery Lady Seaham Pit

Surface 289 ft A.O.D. 6-in NZ 34 N.W. Site [3244 4532] 800 yd E. of The Rift. Sunk 1844.

Drift to 65 ft, Five-Quarter 48 in at 69 ft, Main 68 in at 110 ft, Low Main 36 in at 265 ft, Brass Thill 39 in at 316 ft, Hutton 69 in at 361 ft. Sunk to 379 ft.

Rainton Meadows Colliery, West Pit

Surface 200 ft A.O.D. 6-in NZ 34 N.W. Site [3246 4778] 1830 yd N. of Field House. Sunk 1821–4. For detailed section see B. and S. No. 1590.

Drift to 25 ft, High Main (banded) 99 in at 142 ft, Five-Quarter 1½ in at 249 ft, Main 68 in at 298 ft 3 in, Maudlin (2 leaves) 45 in at 376 ft, Low Main 48 in at 433 ft, Top Brass Thill 24 in at 465 ft, Hutton 55 in at 512 ft. Sunk to 523 ft.

Ryal Farm Bore

Surface 319 ft A.O.D. 6-in NZ 32 N.E. Supposed site [3625 2969] 400 yd S.W. of Ryal, near Sedgefield. Drilled 1874 by Diamond Rock Boring Co. for Weardale Iron Co.

	Thickness ft in	Depth ft in		Thickness ft in	Depth ft in
DRIFT			MILLSTONE GRIT		
Soil and clay (probably			SERIES		
gravel)	13 4	13 4	Sandstone	13 11	295 10
Clay, stony (probably			Shale	8 8	304 6
boulder clay)	7 5	20 9	Sandstone	41 6	346 0
MAGNESIAN			Shale	6 3	352 3
LIMESTONE			Sandstone	2 0	354 3
Limestone 261	2	281 11	Shale	17 1	371 4

	Thickness ft in	Depth ft in		Thickness ft in	Depth ft in
Grey post ...	8 7	379 11	Shale ...	8 0	607 6
Shale ...	4 1	384 0	Post with shale partings...	9 0	616 6
Limestone ...	1 0	385 0	Shale ...	23 11	640 5
Shale ...	9 10	394 10	Limestone ...	2 2	642 7
Sandstone ...	1 8	396 6	Post, white ...	2 8	645 3
Shale ...	2 0	398 6	Shale ...	6 11	652 2
Sandstone ...	46 8	445 2	Limestone ...	2 4	654 6
Shale ...	3 6	448 8	Shale, blue ...	3 8	658 2
Limestone ...	1 2	449 10	Shale with grey beds ...	2 9	660 11
Shale ...	3 5	453 3	Limestone ...	5	661 4
Post, grey ...	5 0	458 3	Post, grey ...	2 2	663 6
Sandstone ...	47 9	506 0	Post with shaly partings ...	4 0	667 6
Shale ...	1 0	507 0	Shale ...	1 0	668 6
Sandstone ...	4 9	511 9	Post, grey ...	2 1	670 7
Shale ...	10 2	521 11	Shale ...	6	671 1
Fireclay ...	1 10	523 9	Post, grey ...	4 11	676 0
Shale ...	17 1	540 10	Shale with blue beds ...	1 5	677 5
Fireclay ...	3 4	544 2	Limestone ...	1 7	679 0
Shale ...	7 8	551 10	Shale ...	4	679 4
Post, grey ...	6 6	558 4	Limestone (cockleshells)...	8 10	688 2
Shale ...	4 4	562 8	Post, grey, with metal bands ...	2 5	690 7
Limestone ...	11 7	574 3	Post, grey ...	2 0	692 7
Post, grey ...	2 10	577 1	Post, grey, with metal partings ...	4 0	696 7
Shale ...	6	577 7	Dark metal ...	2 6	699 1
Fireclay ...	2 6	580 1	Post ...	1 8	700 9
Post, grey ...	3 2	583 3	Dark metal ...	1 10	702 7
Fireclay ...	5 0	588 3	Post ...	6 5	709 0
Shale ...	8 3	596 6			
Limestone ...	9	597 3			
Post, grey ...	2 3	599 6			

Seaton Carew Bore

Surface approximately 20 ft A.O.D. 6-in NZ 53 N.W. Site [5184 3001] 570 yd N.N.E. of Seaton Carew railway station. Drilled 1887–9 by John Vivian for Messrs. Casebourne and Bird. For detailed section see Bird (1888, pp. 570–1 ; 1890, pp. 263–5).

Drift to 18 ft, Bunter Sandstone to 93 ft, Upper Permian Marl to 522 ft (Billingham Main Anhydrite 25 ft thick at base), Upper Magnesian Limestone to 606 ft, Middle and Lower Magnesian Limestone to 1400 ft.

?Brockwell 14 in at 1580 ft. Drilled to 1815 ft.

Sherburn Hill Colliery East Pit

Surface 390 ft A.O.D. 6-in NZ 34 S.W. Site [3357 4257] 1050 yd N.N.W. of Crime Rigg. Sunk 1835. For detailed section see B. and S. No. 1731.

Drift to 91 ft, Five-Quarter 51 in at 229 ft, Main 6 in at 332 ft, Low Main 37 in at 477 ft. Fault between Brass Thill and Hutton. Hutton 66 in at 536 ft. Sunk to 536 ft.

Sherburn Hill No. 1 Bore

Surface 343 ft A.O.D. 6-in NZ 34 S.W. Site [3318 4105] 1050 yd
S.W. of Crime Rigg. Drilled by N.C.B.

Drift to 183 ft, Main 8 in at 321 ft. Drilled to 325 ft.

Sherburnhouse Colliery North Pit

Surface 303 ft A.O.D. 6-in NZ 34 S.W. Site [3211 4150] 620 yd
N.N.E. of Byers Garth. For detailed section see B. and S.
No. 1741.

Drift to 48 ft, Main 15 in at 123 ft, Low Main 34 in at 261 ft, Hutton 49 in at 352 ft,
Harvey 21 in at 501 ft. Sunk to 671 ft.

Shincliffe Colliery Shaft and Bore

Surface 300 ft A.O.D. 6-in NZ 24 S.E. Site [2977 4007] 2000 yd
E. of Haughall. Sunk 1837 to 447 ft, drilled 1868 to 932 ft.
For detailed section see B. and S. Nos. 1798 and 1799.

Drift to 179 ft, Durham Low Main (2 leaves) 40 in at 315 ft, Hutton (banded) 70½ in
at 411 ft, Harvey 22 in at 548 ft, Top Busty 20 in at 673 ft, Bottom Busty 12 in
at 682 ft.

Shotton Colliery, North Pit

Surface 439 ft A.O.D. 6-in NZ 34 S.E. Site [3973 4128] 660 yd
E. of Shotton Colliery railway station. Sunk 1840. For detailed
section see B. and S. No. 1820. Outline section given by
Anderson (1941, p. 14).

Superstructure 12 ft, Drift to 72 ft, Magnesian Limestone to 390 ft, Marl Slate to
392 ft, Basal Permian Sands to 429 ft.

Ryhope Five-Quarter 31 in at 473 ft, Metal 50 in at 726 ft, Top Five-Quarter 47 in
at 764 ft, Main 27 in at 842 ft, Durham Low Main 36 in at 990 ft, Hutton (including
2 thin bands) 64 in at bottom at 1081 ft.

Shotton Colliery No. 1 Underground Bore

Down 279 ft from Hutton workings at 875 ft B.O.D. 6-in
NZ 44 S.W. Site [4202 4113] 1000 yd W. of Little Eden.
Drilled 1934 by A. Kyle Ltd. for Easington Coal Co.

Harvey 17 in at 130 ft, Top Busty 13 in at 238 ft.

Shotton Colliery No. 6 Underground Bore

Up and down from Durham Low Main workings at 503 ft
B.O.D. 6-in NZ 43 N.W. Site [4201 3929] 730 yd S.E. of
Shotton Hall, Shotton. Drilled 1956 by N.C.B. Cores examined
by D. B. Smith.

Up bore:

	Thickness ft	in	Height ft	in
			364	10
Sandstone, mottled red and grey	1	6	363	4
Mudstone, mottled red and grey		6	362	10
Sandstone, mottled red and grey	2	0	360	10
Siltstone, mottled red ...	2	0	358	10
Sandstone, mottled red ...	9	0	349	10
Shale; plants	5	0	344	10
Mudstone; 'mussels' ...	14	11	329	11
Shale; many '*Estheria*'sp.	1	1	328	10
Shale; fish and 'mussels'	3	1	325	9
Shale; many 'mussels' ...		5	325	4
Ironstone; 'mussels' ...		4	325	9
Shale, black; fish remains		3	324	0
Mudstone, roots	1	11	322	10
Siltstone, roots at top ...	7	7	315	3
Shale and mudstone ...	4	10	310	5
Siltstone, roots	1	1	309	4
Sandstone, argillaceous ...	8	8	300	8
Shale, silty, siltstone and sandstone beds ...	16	1	284	7
Shale; many 'mussels' ...	1	2½	283	4½
Shale, carbonaceous; fish		3	283	1½
HIGH MAIN				
Cannel coal ... 1½ in				
Cannel, pyritic 8 in				
Coal ... 4 in	1	1½	282	0
Shale; 'mussels' and ostracods at base ...	2	5	279	7
Siltstone	3	11	275	8
Shale; many 'mussels' ...	3	0	272	8
Sandstone	2	10	269	10
Siltstone	4	2	265	8
Sandstone	5	5	260	3
Shale and siltstone ...	6	5	253	10
Sandstone, argillaceous ...	8	8	245	2
Siltstone, sandy	2	4	242	10
Shale; fish scales... ...	15	5	227	5
Mudstone, roots	2	7	224	10
Shale and mudstone, carbonaceous at top ...	1	10	223	0
Shale, carbonaceous ...		2	222	10
METAL				
Coal ... 18 in	1	6	221	4
Seatearth-mudstone ...	1	0	220	4
Shale and mudstone ...	7	2	213	2
Siltstone	1	4	211	10
Seatearth-mudstone ...		8	211	2
Shale, carbonaceous at top		8	210	6
Seatearth-mudstone ...		4	210	2
Shale and mudstone ...	14	9	195	5
Sandstone	1	0	194	5
Seatearth-mudstone ...	1	1	193	4
Shale, carbonaceous ...		2	193	2
Seatearth-mudstone ...	4	8	188	6
Mudstone, rooty at top ...	2	1	186	5
Coal ... 1 in	1	1	186	4
Shale, carbonaceous ...		10	185	6
Shale, silty; plants ...	11	2	174	4
Shale, carbonaceous ...		4	174	0
Shale and mudstone ...		7	173	5
Shale, carbonaceous ...		3	173	2
Mudstone, silty; plants ...	14	3	158	11
Shale, carbonaceous ...		3	158	8
Siltstone	5	0	153	8
Shale, carbonaceous ...		6	153	2
Coal ... 2 in		2	153	0
Shale, carbonaceous ...	1	6½	151	5½
Coal and shale ... 3 in		3	151	2½
Shale, carbonaceous ...		7½	150	7
FIVE-QUARTER				
Coal, several bands 65 in	5	5	145	2
Seatearth-mudstone ...	1	2	144	0
Shale, carbonaceous ...	1	6	142	6
MAIN				
Coal, several bands 68 in	5	8	136	10
Seatearth-mudstone ...	4	2	132	8
Shale, carbonaceous ...		1	132	7
Seatearth-mudstone ...	2	11	129	8
Siltstone		4	129	4
Sandstone, argillaceous ...	12	0	117	4
Siltstone	1	0	116	4
Shale; fish at base ...	3	9	112	7
Shale, carbonaceous with bands of coal	1	1	111	6
Seatearth-mudstone ...	1	8	109	10
Mudstone, silty	1	6	108	4
Seatearth-mudstone ...	1	8	106	8
Shale, carbonaceous, with bands of coal	2	3	104	5
Seatearth-mudstone ...	1	1	103	4
Siltstone	2	6	100	10
Sandstone		6	100	4
Siltstone		10	99	6
Sandstone	4	11	94	7

	Thickness ft in	Height ft in
Siltstone	8	93 11
Shale	5 5	88 6
Siltstone	8	87 10
Seatearth-mudstone ...	2 10	85 0
Shale; plants	3	84 9
Seatearth-mudstone ...	10	83 11
Mudstone, roots	2 1	81 10
Siltstone	2 8	79 2
Sandstone	1 10	77 4
Shale, silty	4 11	72 5
Sandstone and excavation	72 5	

Down bore:

	Thickness ft in	Depth ft in
Mudstone, rooty	8	8
Shale; plants; many 'mussels' below 16 ft...	19 5	20 1
Sandstone and siltstone ...	5 7	25 8
Shale; 'mussels' ...	2 0	27 8
Seatearth-mudstone ...	4	28 0
Siltstone, roots at top ...	1 6	29 6
Seatearth-mudstone ...	4	29 10
Sandstone ...	1 8	31 6
Shale, silty; plants ...	4 0	35 6
Siltstone and sandstone ...	3 10	39 4
Shale; plants and 'mussels'	6 10	46 2
Shale, black; many 'mussels'	6	46 8
Shale; 'mussels' ...	4	47 0
Coal, banded 12 in	**1 0**	**48 0**
Seatearth-mudstone ...	2	48 2
Shale and mudstone ...	1 2	49 4
Sandstone	4 8	54 0
Sandstone, argillaceous ...	4 0	58 0
Shale; plants and 'mussels'	1 5	59 5
Shale, carbonaceous ...	3	59 8
Siltstone; plants	1 2	60 10
Sandstone	7	61 5
Shale	7	62 0
Sandstone	19 6	81 6
Shale and sandstone ...	9 11	91 5
Shale, cannelly; fish scales	3	91 8
Coal, banded 17 in		
Cannel, shaly 2 in	**1 7**	**93 3**
Shale, carbonaceous ...	2	93 5

	Thickness ft in	Depth ft in
Shale and mudstone; plants and 'mussels' ...	4 5	97 10
Siltstone	1 5	99 3
Shale, silty	6 3	105 6
Siltstone, with shale and sandstone beds ...	9 8	115 2
Sandstone	1 6	116 8
Siltstone	3	116 11
Shale; 'mussels' ...	1 9	118 8
Shale, carbonaceous ...	4	119 0

BOTTOM HUTTON

	Thickness ft in	Depth ft in
Coal 51 in		
Shale, cannelly 3 in		
Cannel coal 6 in		
Cannel, shaly 3 in	**5 3**	**124 3**
Seatearth-mudstone ...	2 9	127 0
Mudstone and shale; 'mussels' at base ...	7 10	134 10
Coal 4 in	**4**	**135 2**
Mudstone and shale ...	4 10	140 0
Siltstone and shale ...	8 11	148 11
Shale; 'mussels' and ostracods	2 1	151 0
Coal 4 in	**4**	**151 4**
Mudstone; roots... ...	7	151 11
Siltstone	5	152 4
Sandstone	8 8	161 0
Shale, silty	10	161 10
Sandstone	1 2	163 0
Shale; 'mussels' ...	3 6	166 6
Sandstone	10	167 4
Shale	4 9	172 1
Coal 4 in	**4**	**172 5**
Seatearth-siltstone ...	1 2	173 7
Siltstone; plants	1 7	175 2
Sandstone	1 1	176 3
Siltstone and shale ...	1 6	177 9
Sandstone	1 3	179 0
Shale; 'mussels'; roots at 188¼ ft	17 6	196 6
Shale; many 'mussels' ...	13 11	210 5

HARVEY MARINE BAND

	Thickness ft in	Depth ft in
Shale, black; fish debris and *Lingula sp.*	1 4	211 9
Seatearth-mudstone ...	1	211 10
Coal 14 in	**1 2**	**213 0**
Seatearth-siltstone ...	1 0	214 0
Siltstone; plants	8 2	222 2

	Thickness ft in	Depth ft in			Thickness ft in	Depth ft in
Sandstone	1 10	224 0	Shale		2 4	249 4
Coal 4 in	4	**224 4**	**Coal** 6 in		**6**	**249 10**
Seatearth-mudstone ...	11	225 3	Mudstone, silty; roots ...		2 2	252 0
Siltstone and mudstone ...	12 7	237 10	Siltstone		1 6	253 6
			Seatearth-mudstone ...		3 6	257 0
HARVEY			Siltstone		1 3	258 3
Coal 12½ in			Seatearth-mudstone ...		2	258 5
Band 1½ in			Shale, silty; plants ...		2 11	261 4
Coal 4 in	1 6	**239 4**	Sandstone ...		2 2	263 6
Seatearth-mudstone ...	2 7	241 11	Siltstone, shaly ...		2 9	266 3
Sandstone, argillaceous ...	2 11	244 10	Shale, silty ...		3 ·4	269 7
Siltstone	2 2	247 0	Sandstone, massive ...		62 8	332 3

Shotton Colliery No. 7 Bore

Surface 295 ft A.O.D. 6-in NZ 43 N.W. Site [4080 3888] 2110 yd S.W. of The Castle, Castle Eden. Drilled 1958 by Cementation Co. Ltd. for N.C.B.

Drift to 37 ft, Magnesian Limestone to about 390 ft, Basal Permian Sands to 464 ft.

Metal (banded) 63 in at 702 ft, Five-Quarter (banded) 107 in at 715 ft, Main (banded) 55 in at 746 ft, Bottom Hutton 47 in at 1022 ft. Drilled to 1027 ft.

South Hetton Colliery, Downcast Shaft

Surface 410 ft A.O.D. 6-in NZ 34 N.E. Site [3811 4526] 350 yd E. of Holy Trinity Church, South Hetton. Sunk 1831–3 to 1075 ft, bored 1865 to 1933 ft. For detailed section see B. and S. No. 1828. Outline section given by Anderson (1941, p. 13). Depths of levels re-measured 1956.

Drift to 50 ft, Magnesian Limestone and Marl Slate to 340 ft, Basal Permian Sands to 370 ft.

Ryhope Five-Quarter 34 in at 397 ft, High Main (including 4-in band) 36 in at 649 ft, Main 80 in at 827 ft, Durham Low Main (including 12-in band) 64 in at 970 ft, Top Brass Thill 44 in at 1004 ft, Hutton (including 5-in band) 77 in at 1066 ft, Harvey 26 in at 1213 ft, Tilley (banded) 41 in at 1265 ft.

South Hetton Colliery No. 6 Underground Bore

Down from Main workings at 226 ft B.O.D. 6-in NZ 34 S.E. Site [3726 4425] 410 yd S. 33° W. of High Fallowfield, South Hetton. Drilled 1951 by A. Kyle Ltd. for N.C.B. Cores examined by R. H. Price and G. Armstrong.

	Thickness ft in	Depth ft in		Thickness ft in	Depth ft in
Seatearth-mudstone, sandy	1 0	1 0	Sandstone	2 2	10 0
Coal ... 6 in	6	**1 6**	Mudstone; ' mussels ' ...	12 6	22 6
Seatearth-mudstone, sandy	2 3	3 9	Shale, black; fish remains	1 4	23 10
Sandstone; rootlets ...	1 1	4 10	**Coal,**		
Mudstone, sandy ...	3 0	7 10	inferior 11 in	**11**	**24 9**

	Thickness ft	in	Depth ft	in
Seatearth-mudstone ...	1	8	26	5
Sandstone; roots at top...	53	5	79	10
Mudstone, sandy ...	5	8	85	6
Sandstone ...	2	6	88	0
Mudstone, sandy ...	5	5	93	5
Sandstone ...	62	3	155	8
Seatearth-mudstone (floor of old DURHAM LOW MAIN workings) ...	2	4	158	0
Siltstone ...	1	10	159	10
Sandstone ...	7	3	167	1
Shale; 'mussels' ...	2	2	169	3
Sandstone ...	16	8	185	11
Mudstone; 'mussels' ...	13	5	199	4
Seatearth-mudstone ...		2	199	6
Shale and siltstone ...	5	1	204	7
Mudstone; 'mussels' ...	4	0	208	7
Coal ... 15 in	1	3	209	10
Seatearth-mudstone ...	1	4	211	2
Shale, sandy ...	3	6	214	8
Shale, sandstone beds ...	27	5	242	1
HUTTON Old workings 73 in	6	1	248	2
Shale, carbonaceous ...		6	248	8
Seatearth-mudstone ...	1	4	250	0
Siltstone ...	12	0	262	0
Shale and mudstone; 'mussels' ...	2	1	264	1
Coal and blackstone ... 3 in		3	264	4
Seatearth-mudstone ...	1	2	265	6
Siltstone and sandstone ...	16	11	282	5
Mudstone; 'mussels' ...	5	6	287	11
Coal ... 5 in		5	288	4
Seatearth-mudstone ...	1	3	289	7
Sandstone ...	30	9	320	4
Seatearth-mudstone, sandy	3	5	323	9
Sandstone and shale ...	2	5	326	2
Mudstone; 'mussels' ...	24	6	350	8
Seatearth-mudstone ...	1	10	352	6
Shale ...	4	6	357	0
Coal ... 2 in		2	357	2
Seatearth-mudstone ...	1	3	358	5
Sandstone ...	5	1	363	6
Sandstone and mudstone interbedded; some 'mussels' ...	19	0	382	6
Siltstone ...	9	11	392	5
Mudstone; ostracods ...		6	392	11
HARVEY Coal ... 32 in	2	8	395	7

	Thickness ft	in	Depth ft	in
Seatearth-mudstone ...	2	2	397	9
Shale ...	9	5	407	2
Coal ... 4 in		4	407	6
Seatearth-mudstone ...	2	2	409	8
Shale, silty ...	18	1	427	9
Sandstone, argillaceous ...	4	8	432	5
Shale, slickensided (?small fault) or seatearth ...	11	6	443	11
Siltstone ...	16	7	460	6
Seatearth-mudstone ...	1	0	461	6
Shale, silty ...	2	2	463	8
Sandstone ...	11	3	474	11
Coal ... 7 in		7	475	6
Seatearth-mudstone ...	1	9	477	3
Mudstone ...	1	5	478	8
Coal ... 6 in		6	479	2
Seatearth-mudstone ...	1	0	480	2
Shale ...	9	10	490	0
Sandstone ...	14	11	504	11
TOP BUSTY Coal ... 21 in				
Seatearth-mudstone 8 in				
Coal ... 3 in	2	8	507	7
Seatearth-mudstone ...	4	1	511	8
BOTTOM BUSTY Coal ... 17 in				
Seatearth-mudstone 6 in				
Coal ... 7 in	2	6	514	2
Seatearth-mudstone ...	2	9	516	11
Sandstone ...	53	5	570	4
Mudstone ...	6	7	576	11
THREE-QUARTERS Coal ... 11 in				
Seatearth-mudstone 11 in				
Coal ... 5 in	2	3	579	2
Seatearth-mudstone ...		3	579	5
Sandstone and siltstone ...	9	1	588	6
Mudstone; 'mussels' ...	5	3	593	9
Shale; 'mussels' ...	7	2	600	11
Seatearth-mudstone ...	1	5	602	4
Sandstone and shale ...	1	1	603	5
Mudstone ...	1	5	604	10
Coal ... 10 in		10	605	8
Seatearth-mudstone ...	1	4	607	0
Coal ... 6 in		6	607	6
Seatearth-mudstone ...		6	608	0
Shale ...	1	7	609	7
Sandstone ...	3	4	612	11
Siltstone ...	1	7	614	6

	Thickness ft in	Depth ft in		Thickness ft in	Depth ft in
			Sandstone	3 1	624 4
BROCKWELL			Siltstone	6 2	630 6
Coal ... 18 in	1 6	616 0	Mudstone; 'mussels' ...	3 10	634 4
Seatearth-mudstone ...	5	616 5	Sandstone, argillaceous ...	2 8	637 0
Shale	4 10	621 3	Shale, sandy	6 6	643 6

South Hetton Colliery No. 7 Underground Bore

Down 623 ft from Main workings at 540 ft B.O.D. 6-in NZ 34 S.E. Site [3961 4484] 300 yd W. of Great Coop House. Drilled 1950 by A. Kyle Ltd. for N.C.B.

Low Main (old workings) at 157 ft, Hutton (old workings) at 266 ft, Harvey about 31 in at 405 ft, Top Busty 16 in at 518 ft, Bottom Busty 17 in at 530 ft.

South Hetton Colliery No. 10 Underground Bore

Down 484 ft from Durham Low Main at 452 ft B.O.D. 6-in NZ 34 S.E. Site [3919 4215] 120 yd W.S.W. of Sandy Carrs.

Hutton 67 in at 101 ft, Harvey 32 in at 216 ft, Top Busty 18 in at 348 ft, Bottom Busty 10 in at 358 ft, Three-Quarter (including 20-in band) 28 in at 433 ft, Brockwell 18 in at 474 ft.

South Kelloe Colliery Shaft

Surface 494 ft A.O.D. 6-in NZ 33 N.W. Site [3373 3705] 420 yd N.N.E. of Coxhoe Hall. Sunk 1844.

Drift to 29 ft, Magnesian Limestone to 115 ft.

High Main 14 in at 244 ft, Metal (banded) 42 in at 327 ft, Five-Quarter 43 in at 332 ft, Main (banded) 58 in at bottom at 389 ft.

South Mainsforth 'D' Bore

Surface 348 ft A.O.D. 6-in NZ 33 S.W. Site [3142 3004] 1200 yd S.E. of Chilton East House. Drilled 1959 by Cementation Co. Ltd. for N.C.B.

Drift to 149 ft, Magnesian Limestone to 367 ft, Marl Slate to 374 ft, Basal Permian Sandstone to 404 ft.

Top Brass Thill 25 in at 430 ft, Bottom Brass Thill 6 in at 438 ft, Top Hutton 13 in at 465 ft, Bottom Hutton 35 in at 511 ft. Drilled to 515 ft.

South Mainsforth 'J' Bore

Surface 301 ft A.O.D. 6-in NZ 32 N.W. Site [3104 2933] 650 yd W.N.W. of Nunstainton East. Drilled 1959 by Cementation Co. Ltd. for N.C.B. Cores examined by A. M. Clarke.

	Thickness ft in	Depth ft in
DRIFT		
Soil	1 0	1 0
Clay and boulders ...	53 0	54 0
MAGNESIAN LIMESTONE		
Limestone (not cored) ...	185 0	239 0
Dolomitic limestone, grey, alternating semi-porcellanous and finely crystalline layers, flaggy with black partings; irregular and angular cavities, some lined with calcite	19 0	258 0
Dolomitic limestone, grey and buff-grey, auto-brecciated; irregular fragments of randomly orientated bedded limestone set in white calcite, net-veined; some pyrite in veins	11 0	269 0
Limestone, auto-brecciated as above but set in slickensided dark grey argillaceous matrix ...	3 0	272 0
Dolomitic limestone, grey, semi-porcellanous, flaggy; stylolitic partings ½ to 2 in apart; thin-bedded below 286 ft	21 0	293 0
MARL SLATE		
Shale, hard, dark grey, poorly fissile but finely laminated, with mottled brown bedding planes; limestone 1 in at base ...	2 4	295 4
BASAL PERMIAN BRECCIA		
Breccio-conglomerate; angular and flattened fragments of cream limestone, soft white marl, blue-grey siltstone, quartzite and sporadic crinoid ossicles in bimodal coarse-grained (' millet-seed ') and fine-grained sandstone matrix	3 2	298 6

	Thickness ft in	Depth ft in
LOWER COAL MEASURES		
Sandstone, pale grey in top 7 in, red below; fine- to medium-grained, cross-bedded, coarse and massive in bottom 4 ft ...	21 7	320 1
Mudstone, red, purple, yellow and grey, shaly; few plants at top ...	3 0	323 1
Sandstone, red and buff-yellow, medium- to coarse - grained, conglomeratic in top 2 ft and from 329 to 331 ft, cross-bedded to 329 ft; silty mudstone layers 1 to 2 in thick in bottom 2 ft	9 11	333 0
Seatearth-mudstone, purple and grey, sandy at base	3 0	336 0
Sandstone, purple, medium-grained, cross-bedded to 343 ft continuing massive to 360½ ft; coarse-grained and massive 360½ ft to base; coal scars in bottom 5½ ft	48 11	384 11
<small>BOTTOM TILLEY</small>		
Coal ... 20 in	**1 8**	**386 7**
Seatearth-siltstone, grey, argillaceous at top ...	1 5	388 0
Siltstone, grey, thin-bedded with sandy bands and partings; scattered rootlets throughout	10 0	398 0
Mudstone, grey, shaly, ironstone nodules below 414 ft	35 6	433 6
Shale, black, carbonaceous; fish scales at top, coalified plant debris below	1 3	434 9
<small>TOP BUSTY</small>		
Coal ... 10 in	**10**	**435 7**
Seatearth-siltstone, brownish grey, sandy; irregular sandstone intercalations towards base	5 11	441 6

	Thickness ft in	Depth ft in

BOTTOM BUSTY

	Thickness ft in	Depth ft in
Coal ... 29 in	2 5	443 11
Seatearth-sandstone, grey, silty	3 8	447 7
Sandstone, grey, fine- to medium-grained; thin-bedded in top 2 ft, with rootlets, small-scale cross-bedding 449½ to 461 ft; black micaceous partings 461 to 472½ ft; coarse-grained feldspathic and massive with irregular coal scars 472½ to 486½ ft; fine- to medium-grained massive with red mottling 486½ to 497 ft; small-scale cross-bedding 497 to 501 ft; fine- to medium-grained massive to base, with clay-ironstone and mudstone pellets in bottom 8 in...	59 2	506 9
Mudstone, grey, shaly; rare plant fragments in upper part, pyritic below; carbonaceous with coaly partings in bottom 1 in	5 2	511 11

THREE-QUARTER

	Thickness ft in	Depth ft in
Coal 10 in	10	512 9
Shale, carbonaceous ...	3	513 0
Seatearth-siltstone, black, argillaceous, micaceous	7	513 7
Sandstone, pale grey, fine-grained, thin-bedding, dark micaceous partings	5 8	519 3
Siltstone, grey, fissile; scattered plant fragments	3 9	523 0
Mudstone, grey, shaly; *Gyrochorte sp.* at top, 'mussels' below; black shaly with fish scales at base	3 0	526 0
Coal 2 in		
Carbonaceous shale 1 in		
Coal 1 in	4	526 4
Seatearth-mudstone, grey	1 8	528 0

97544

	Thickness ft in	Depth ft in
Siltstone, grey, argillaceous, grading down into white fine-grained sandstone; rootlets and plant debris throughout	4 6	532 6
Mudstone, grey, shaly, with thin sandy bands at top; plants in lower part, coalified in bottom 3 in	4 6	537 0
Seatearth-mudstone, dark grey	4	537 4
Siltstone, thick-bedded, grey, sandy bands at top; rootlets at top, plant debris below	7 8	545 0

BROCKWELL OSTRACOD BAND

	Thickness ft in	Depth ft in
Mudstone, grey, shaly; 'mussels'; black shaly with fish scales and spines towards base; carbonaceous and pyritic in bottom 1 in ...	3 2	548 2
Seatearth-mudstone, brownish grey	9	548 11

BROCKWELL

	Thickness ft in	Depth ft in
Coal 34 in		
Shale, carbonaceous 5 in		
Seatearth-mudstone 3 in		
Coal 25 in	5 7	554 6
Seatearth-mudstone, grey	4 0	558 6
Mudstone, grey, shaly; abundant plant remains	6	559 0
Coal, inferior 6 in	6	559 6
Seatearth-mudstone; ironstone nodules towards base	4 1	563 7
Mudstone, grey, shaly; coalified plant debris ...	7	564 2

BOTTOM BROCKWELL

	Thickness ft in	Depth ft in
Coal 4½ in		
Shale, coaly 3 in		
Coal 1 in		
Mudstone, carbonaceous 1½ in		

	Thickness ft in	Depth ft in		Thickness ft in	Depth ft in
Coal	1 in		Mudstone, carbonaceous	0½ in	
Mudstone, carbonaceous	1 in		Coal	10 in 2 7	566 9
Coal	2 in		Seatearth-sandstone, grey, fine-grained, curly-bedded in top 4 ft ...	6 5	573 2
Mudstone, carbonaceous	3½ in		Sandstone, pale grey, fine-grained, thin-bedded; thin siltstone bands and partings	4 10	578 0
Coal	3 in				

South Wingate (Rodridge) Colliery Shaft and Bore

Surface 448 ft A.O.D. 6-in NZ 43 S.W. Site [4164 3478]
230 yd N. of Catlaw Hall, South Wingate. Sunk 1841–3 to
Bottom Busty by Wm. Coulson Ltd. Drilled 1843 to 1223 ft.
For detailed section see B. and S. No. 1609.

Drift to 150 ft, Magnesian Limestone to 589 ft, Marl Slate to 593 ft, Basal Permian Sands to 624 ft.

Top Harvey 24 in at 680 ft, Bottom Harvey 34 in at 690 ft, Top Tilley (banded) 58 in at 726 ft.

Stony Hall Bore

Surface about 300 ft A.O.D. 6-in NZ 32 N.W. Site [3260 2953]
80 yd S.E. of Stony Hall. For detailed section see B. and S. No. 1483.

Drift to 144 ft, Magnesian Limestone to 288 ft, ?Basal Permian Sandstone to 299 ft.

Top Busty 22 in at 409 ft, Bottom Busty (banded) 63 in at 422 ft, Brockwell (banded) 107 in at 530 ft. Drilled to 532 ft.

Ten-o'-Clock Barn No. 1 Bore

Surface 297 ft A.O.D. 6-in NZ 32 N.E. Site [3992 2980]
1300 yd E. of Butterwick East Farm. Drilled 1943 by A. Kyle
for Messrs. Henry Stobart & Co. Ltd.
Permian fossils are listed on p. 175.

	Thickness ft in	Depth ft in		Thickness ft in	Depth ft in
DRIFT			Clay and stones	43 0	256 0
Soil	1 6	1 6	Sand	2 0	258 0
Boulder clay with stones ...	126 6	128 0	Clay, reddish, and stones	14 0	272 0
Sand	39 6	167 6	Sand and gravel	2 0	274 0
Boulder clay ...	2 6	170 0	Clay, red sandy, and gravel	7 0	281 0
Sand	5 0	175 0	Sand and gravel	6 0	287 0
Boulder clay ...	24 6	199 6	Clay, yellow	4 0	291 0
Sand and gravel ...	13 6	213 0	Sand	6 0	297 0

	Thickness ft in	Depth ft in
MAGNESIAN LIMESTONE		
Limestone, decomposed ...	61 0	358 0
Limestone	150 6	508 6
Limestone, hard, broken in lower 20 ft	36 6	545 0
Limestone, grey	63 0	608 0
LOWER COAL MEASURES		
Sandstone, grey	11 0	619 0
Marl, red; ' Estheria '? at base (dip 20°)	8 0	627 0
Fireclay, red, with ironstone	3 0	630 0
Siltstone, red, micaceous; high dip...	5 0	635 0
Fireclay, sandy	6 2	641 2
Sandstone, red	4 0	645 2
Sandstone, grey, with red stained top and some red bands	54 0	699 2
Horizon of Brockwell ...	– –	– –
Fireclay, light grey ...	4 0	703 2
Sandstone, reddish grey ...	21 9	724 11
Fireclay, reddish, sandy ...	1 4	726 3
Sandstone, red	4 9	731 0
Fakes, strong dark ...	4 0	735 0
Blaes, dark, with ironstone band with ' mussels ' near base	13 9	748 9
Fakes and faky sandstone	3 8	752 5
Sandstone, reddish grey, with 6-in dark fakes 2½ ft above base ...	12 1	764 6
Sandstone, grey, hard ...	1 0	765 6
Sandstone, faky	3 2	768 8
Fakes, dark	5 5	774 1
Blaes, dark, with ironstone bands	9 7	783 8
Blaes, dark	2 1	785 9
VICTORIA		
Coal ... 9½ in		
Dirty rib 0½ in		
Coal ... 12 in	1 10	787 7
Fireclay, lower part sandy	4 6	792 1
Coal with dirty ribs 9 in	9	792 10
Sandstone, faky	2 0	794 10
Sandstone, grey	2 11	797 9
Sandstone, reddish grey, with thin bands of dark fakes in upper part ...	28 1	825 10

	Thickness ft in	Depth ft in
Fakes, dark, reddish ...	6	826 4
Sandstone, faky	8	827 0
Fakes, dark	3 2	830 2
Blaes, dark; traces of ' mussels '	10	831 0
Blaes, dark, with ironstone bands	4 0	835 0
Blaes, dark	8	835 8
MARSHALL GREEN		
Coal ... 16 in	1 4	837 0
Fireclay, sandy	10	837 10
Sandstone, grey	8 6	846 4
Sandstone, reddish grey ...	3 8	850 0
Sandstone, grey	16 4	866 4
Sandstone, reddish grey ...	9 5	875 9
Blaes	1 4	877 1
Fakes	6	877 7
GANISTER CLAY		
Coal ... 1 in	1	877 8
Blaes, dark	1 5	879 1
Sandstone, faky	1 5	880 6
Fakes, dark	7	881 1
Sandstone, grey, very hard	1 11	883 0
Sandstone, faky	3 0	886 0
Fakes, dark	7 9	893 9
Blaes, dark, with ironstone balls from 902½ to 906¼ ft	13 11	907 8
Fakes, dark	2 1	909 9
Blaes, dark	1 0	910 9
Fakes, dark	1 0	911 9
Sandstone, faky	8	912 5
Fakes, dark	1 9	914 2
Blaes, dark, with ironstone balls in upper part ...	7 5	921 7
GUBEON		
Coal ... 13 in		
Dirty rib 0½ in		
Coal ... 3½ in	1 5	923 0
Fireclay, sandy	1 1	924 1
Sandstone	1 4	925 5
Sandstone, faky, thin sandy blaes at top and base ...	8 10	934 3
Sandstone, very hard ...	4 6	938 9
Fakes, dark	1 10	940 7
Blaes, stony, dark, with sandy partings	11 10	952 5

Ten-o'-Clock Barn No. 2 Bore

> Surface 300 ft A.O.D. 6-in NZ 33 S.E. Site [3916 3035]
> 330 yd W.N.W. of Ten-o'-Clock Barn. Drilled 1945 by
> A. Kyle Ltd.

Drift to 267 ft, Magnesian Limestone to 518 ft, Marl Slate to 519 ft.

Harvey 38 in at 637 ft, Bottom Busty (banded) 37 in at 765 ft, Brockwell (banded) 33 in at 827 ft. Drilled to 968 ft.

Thornley Colliery, No. 1 Pit

> Surface 456 ft A.O.D. 6-in NZ 33 N.E. Site [3664 3946]
> 900 yd N.N.W. of White House. Sunk 1833–5 by Jas. Johnson
> and Son to 1151 ft, drilled 1856 to 1331 ft. For detailed
> section see B. and S. No. 1971.

Drift to 18 ft, Magnesian Limestone to 153 ft, Marl Slate to 155 ft, Basal Permian Sands to 191 ft.

Metal and Five-Quarter 93 in at 507 ft, Main 31 in at 614 ft, Durham Low Main 32 in at 739 ft, Bottom Hutton 40 in at 866 ft, Harvey 44 in at 994 ft, Top Tilley (banded) 35 in at 1032 ft, Bottom Busty 24 in at 1099 ft.

Thornley Colliery No. 3 Underground Bore

> Down 275 ft from Harvey workings at 498 ft B.O.D. 6-in
> NZ 34 S.E. Site [3608 4060] 250 yd N.W. of Fatclose House.
> Drilled 1953 by N.C.B.

Top Busty 14 in at 97 ft, Bottom Busty 27 in at 133 ft, Brockwell 22 in at 212 ft.

Thorpe Bulmer Bore

> Surface 280 ft A.O.D. 6-in NZ 43 N.E. Site [4570 3608]
> 300 yd N.W. of Thorpe Bulmer. Drilled 1939–40 by Jas.
> Johnson and Sons Ltd. for Horden Collieries Ltd. For further
> details see Trechmann (1942, pp. 313–5).

Drift to 146 ft, Magnesian Limestone to 705 ft, Marl Slate to 708 ft, Basal Permian Breccia to 717 ft.

Durham Low Main 6 in at 786 ft, Harvey Marine Band at about 995 ft. Drilled to 1083 ft.

Thorpe Well and Bore

> Surface 366 ft A.O.D. 6-in NZ 44 S.W. Site [4277 4387]
> 400 yd E.S.E. of Holm Hill. Sunk to 483 ft, drilled 1901
> to 684 ft. Outline section given by Anderson (1941, p. 14).

Drift to 12 ft, Magnesian Limestone to 523 ft, Basal Permian Sand to 684 ft.

Middle Coal Measures said to be proved.

Thrislington Colliery Jane Pit and Bore

Surface 410 ft A.O.D. 6-in NZ 33 S.W. Site [3096 3388] 610 yd N.E. of Thrislington Hall. Sunk 1867 to Harvey, continued 1867 by bore in Florence Pit.

Drift to 18 ft. Magnesian Limestone to 132 ft.

Main (old workings) 66 in at 188 ft, Durham Low Main 30 in at 341 ft, Top Brass Thill 22 in at 383 ft, Bottom Hutton 22 in at 436 ft, Harvey 52 in at 623 ft, Top Busty 35 in at 761 ft, Bottom Busty (2 leaves) 27 in at 772 ft, Brockwell 14 in at 901 ft. Drilled to 948 ft.

Tinkler's Gill Bore

Surface 261 ft A.O.D. 6-in NZ 43 S.W. Site [4157 3054] 750 yd S.E. of Embleton Old Hall, Embleton. Drilled 1959 by N.C.B. Log to 431 ft is driller's record. Permian strata below 431 ft examined by D. B. Smith. Carboniferous strata examined by A. M. Clarke.

	Thickness ft in	Depth ft in
DRIFT		
Soil	1 6	1 6
Boulder clay	15 6	17 0
Boulder clay with panels of sand	39 0	56 0
Boulder clay	52 0	108 0
Gravel	47 0	155 0
Boulder clay with heavy sand panels	20 0	175 0
?UPPER PERMIAN MARL		
' Boulder clay ', red [heavy red wash from 195 to 225 ft]	75 0	250 0
Marl	8 0	258 0
?UPPER MAGNESIAN LIMESTONE		
Limestone, soft, with hard panels	55 0	313 0
Limestone, harder ...	17 0	330 0
?MIDDLE MAGNESIAN LIMESTONE		
Limestone	32 0	362 0
Limestone, sand panels ...	23 0	385 0
Limestone, heavy marl panels	40 0	425 0
Limestone	6 0	431 0
Dolomite, soft, finely granular	14 0	445 0
Dolomite, oolitic ...	5 0	450 0
Dolomite, finely granular	5 0	455 0

	Thickness ft in	Depth ft in
Dolomite, oolitic, some compound ooliths ...	19 0	474 0
Dolomite, soft, finely granular	15 0	489 0
Dolomite, soft, finely granular, with thin oolitic beds; scattered lamellibranchs including *Astartella vallisneriana*	51 0	540 0
Limestone, dolomitic, fine-grained	15 0	555 0
LOWER MAGNESIAN LIMESTONE		
Limestone, dolomitic, and calcareous dolomite, very fine-grained to semi-porcellanous, partly auto-brecciated...	83 0	638 0
Limestone, dolomitic, and calcareous dolomite, very fine-grained, with alternating thin nodular dark grey and lighter grey beds	47 0	685 0
Limestone, dolomitic, finely interbedded with dolomitic shale; contorted	3 0	688 0
Limestone, dolomitic, as 638 to 685 ft; bedding planes are less nodular below	28 5	716 5

	Thickness ft	in	Depth ft	in
MARL SLATE				
Shale, grey, dolomitic, with blende and galena on bedding planes ...	1	5	717	10
BASAL PERMIAN BRECCIA				
Breccia of small pebbles in matrix of grey sandstone containing large frosted grains; see Plate VII	3	11	721	9
MIDDLE COAL MEASURES				
Sandstone, light grey in top 2 ft, pink below ...	15	3	737	0
Seatearth-mudstone, purple-grey	4	0	741	0
Sandstone, purple-grey to pink	29	0	770	0
Mudstone; ' mussels ' ...	2	6	772	6
Sandstone	15	0	787	6
Mudstone; ' mussels ' ...	4	7	792	1
Coal ... 4 in		4	**792**	**5**
Seatearth-mudstone ...	1	7	794	0
Sandstone	10	3	804	3
Mudstone; ' mussels ' ...	5	9	810	0
Sandstone	11	0	821	0
Mudstone; ' mussels ' ...	3	5	824	5
TOP HUTTON (UPPER LEAF)				
Coal ... 18 in	1	6	**825**	**11**
Seatearth-mudstone ...	1	5	827	4
Mudstone; plants ...	4	10	832	2
TOP HUTTON (LOWER LEAF)				
Coal ... 14 in	1	2	**833**	**4**
Seatearth-sandstone ...	2	10	836	2
Sandstone	24	0	860	2
BOTTOM HUTTON				
Coal ... 25 in	2	1	**862**	**3**
Shale, carbonaceous ...		6	862	9
Seatearth-mudstone ...	2	3	865	0
Mudstone; ' mussels ' near base	10	5	875	5
Coal ... 6 in		6	**875**	**11**
Seatearth-mudstone ...	1	8	877	7
Coal ... 19 in	1	7	**879**	**0**
Seatearth-mudstone ...	2	10	882	0
Sandstone and siltstone ...	5	0	887	0
Sandstone	8	0	895	0
Siltstone	6	1	901	1
Mudstone; ' mussels ' ...	2	11	904	0
Coal ... 6 in		6	**904**	**6**
Seatearth-siltstone ...	3	0	907	6
Sandstone	20	6	928	0
Siltstone; plants	5	0	933	0
Mudstone; ' mussels ' and ostracods	14	6	947	6
HARVEY MARINE BAND				
Mudstone, dark grey, silty, with *Lingula sp.*		6	948	0
Seatearth-mudstone, silty	1	9	949	9
Coal ... 6 in		6	**950**	**3**
Seatearth-mudstone ...	4	3	954	6
Sandstone	9	6	964	0
Siltstone	8	0	972	0
Sandstone	60	2	1,032	2
BOTTOM HARVEY				
Coal ... 26 in	2	2	**1,034**	**4**
Seatearth-mudstone ...	2	0	1,036	4
Siltstone	2	2	1,038	6
Mudstone, carbonaceous		2	1,038	8
Seatearth-mudstone ...	10	4	1,049	0
Siltstone	2	0	1,051	0
Seatearth-mudstone ...	3	7	1,054	7
Coal ... 7 in		7	**1,055**	**2**
Seatearth-mudstone ...	6	10	1,062	0
Mudstone; plants ...	6	8	1,068	8
Siltstone	9	0	1,077	8
TOP TILLEY (UPPER LEAF)				
Coal ... 12 in	1	0	**1,078**	**8**
Seatearth-siltstone, fragmented, heavily slickensided along many minor faults	3	4	1,082	0
Siltstone, many polished joint and minor fault planes (probably near major fault)	4	0	1,086	0

Trimdon Grange Colliery New Pit

Surface 495 ft A.O.D. 6-in NZ 33 N.E. Site [3679 3572] 1470 yd S.W. of Wingate House. Sunk 1872. For detailed section see B. and S. No. 2035.

Drift to 66 ft, Magnesian Limestone and Marl Slate to 278 ft.

Metal and Five-Quarter 129 in at 360 ft, Main (including 9-in band) 68 in at 451 ft, Bottom Hutton 35 in at 716 ft, Harvey 45 in at 858 ft, Busty (including 3-in band) 35 in at bottom at 901 ft.

Tudhoe Colliery, West Pit

Surface 313 ft A.O.D.   6-in NZ 23 N.E.   Site [2685 3569]
1330 yd W.N.W. of Mount Huley.   Sunk 1866 to 525 ft,
drilled to 1134 ft.   For detailed section see B. and S.
No. 2040.

Drift to 31 ft, Top Brass Thill 10 in at 66 ft, Bottom Brass Thill (2 leaves) 14 in at 80 ft, Top Hutton 24 in at 105 ft, Bottom Hutton 27 in at 133 ft, Harvey 26 in at 301 ft, Busty (banded) 63 in at 409 ft, Brockwell 42 in at 525 ft.

Tudhoe Park Site No. 2 Bore

Surface 447 ft A.O.D.   6-in NZ 23 S.E.   Site [2791 3485]
30 yd W. of High Butcher Race.   Drilled 1958 by N.C.B.

Drift to 15 ft, Maudlin 15 in at 31 ft, Durham Low Main 26 in at 110 ft, Top Hutton (banded) 15 in at 186 ft.   Drilled to 192 ft.

Tursdale Colliery South Shaft

Surface 281 ft A.O.D.   6-in NZ 33 N.W.   Site [3017 3593]
260 yd S.W. of Standalone.   Sunk 1859.   For detailed section
see B. and S. No. 2043.

Drift to 86 ft, Durham Low Main 24 in at 128 ft, Top Hutton 22 in at 204 ft, Bottom Hutton 17 in at 226 ft, Harvey 27 in at 408 ft, Top Busty 33 in at 546 ft, Bottom Busty 24 in at 549 ft, Brockwell (banded) 49 in at bottom at 672 ft.

Warren Cement Works Bore

Surface 18 ft A.O.D.   6-in NZ 53 S.W.   Site [5175 3449]
930 yd W.N.W. of Hartlepool railway station.   Drilled 1923
by Wm. Coulson Ltd. for Warren Cement Works Co.   For
detailed section see Trechmann (1925, p. 140).

Made ground 4 ft, Drift to 83 ft, ?Upper Magnesian Limestone to 101 ft, Middle Magnesian Limestone (including Hartlepool Anhydrite 101 ft to 354 ft) to bottom at 371 ft.

West Cornforth (Thrislington) Colliery, Coke Oven Pit

Surface 284 ft A.O.D.   6-in NZ 33 S.W.   Site [3044 3405]
690 yd N. of Thrislington Hall.   Sunk 1840.   For detailed
section see B. and S. No. 570.

Drift to 72 ft, High Main 4 in at 270 ft, Five-Quarter (banded) 66 in at 362 ft, Main (2 leaves) 65 in at bottom at 449 ft.

West Hetton Colliery Engine Pit

Surface 350 ft A.O.D.   6-in NZ 33 N.W.   Site [3247 3702]
1420 yd N.N.W. of Coxhoe Hall.   Sunk 1837–8 by Wm.
Coulson Ltd.   For detailed section see B. and S. No. 2180.

Drift to 57 ft, Five-Quarter (banded) 66 in at 123 ft, Main (banded) 56 in at 179 ft. Sunk to 191 ft.

West Murton Blue House Bore

Surface 320 ft A.O.D.   6-in NZ 33˙ S.E.   Site [3954 3198]
670 yd N.E. of Butterwick Moor.   Drilled 1961 by N.C.B.

Drift to 164 ft, Magnesian Limestone to 441 ft, Marl Slate to 443 ft, Basal Permian Sandstone to 445 ft.

Harvey 39 in at 504 ft, probable Top Busty 21 in at 596 ft, probable Bottom Busty 21 in at 607 ft.   Drilled to 680 ft.

West Murton Blue House Farm Bore

Surface 330 ft A.O.D.   6-in NZ 43 S.W.   Site [4000 3166]
1400 yd W.N.W. of Embleton Old Hall, Embleton.   Drilled 1959 by Cementation Co. Ltd. for N.C.B.

Drift to 145 ft, Magnesian Limestone to 460 ft, Marl Slate to 463 ft.

Bottom Harvey 27 in at 627 ft, Bottom Busty (including 2-in band) 39 in at 771 ft, Brockwell (including 3-in band) 23 in at 918 ft.   Drilled to 925 ft.

Wheatley Hill Colliery Nos. 1 and 2 Shafts

Surface 435 ft A.O.D.   6-in NZ 33 N.E.   Site [3858 3929]
1970 yd E.N.E. of White House, Wheatley Hill.   Sunk 1869.
For detailed section see B. and S. No. 2200.

Drift to 72 ft, Magnesian Limestone and Marl Slate to 379 ft.
Metal and Five-Quarter (banded) 132 in at 659 ft, Main 49 in at 750 ft, Bottom Hutton 51 in at 992 ft, Harvey 48 in at 1140 ft.   Sunk to 1192 ft.

Whin Houses Bore

Surface 307 ft A.O.D.   6-in NZ 33 S.E.   Site [3997 3058]
690 yd N.E. of Ten-o'-Clock Barn.   Drilled 1941 by A. Kyle Ltd.

Drift to 204 ft, Magnesian Limestone to 462 ft (fossils listed on pp. 175–6).

Harvey 32 in at 663 ft, Top Busty 28 in at 795 ft, Bottom Busty 53 in at 804 ft, Brockwell 23 in at 948 ft.   Drilled to 990 ft.

Whin Houses Farm Bore

Surface 356 ft A.O.D.   6-in NZ 33 S.E.   Site [3988 3111]
2¼ miles E.S.E. of Fishburn.   Drilled 1959 by N.C.B.

Drift to 180 ft, Magnesian Limestone to 479 ft, Marl Slate to 482 ft, Basal Permian Sandstone to 483 ft.

Bottom Hutton 27 in at 596 ft, Harvey 33 in at 737 ft, Top Busty 22 in at 842 ft, Bottom Busty (banded) 40 in at 868 ft.   Drilled to 901 ft.

Whitwell 'A' Pit Underground Bore

Down 426 ft from Hutton workings at approximately 93 ft B.O.D.   6-in NZ 34 S.W.   Site [3104 4060] 520 yd N.N.E. of Whitwell House.   Drilled 1851.   For detailed section see B. and S. No. 2233.

Harvey 14 in at 121 ft, Bottom Busty 19¼ in at 247 ft, Three-Quarter 10 in at 305 ft, Brockwell not recorded.

Whitwell Colliery ' A ' Pit

Surface 260 ft A.O.D.   6-in NZ 34 S.W.   Site [3095 4061]
500 yd N. of Whitwell House.   Sunk 1836–7.   For detailed
section see B. and S. No. 2232.

Drift to 126 ft, Low Main 36 in at 249 ft, Hutton (banded) 74 in at 353 ft.   Sunk
to 365 ft.

Whitwell Colliery ' B ' Pit

Surface 349 ft A.O.D.   6-in NZ 33 N.W.   Site [3126 3972]
1050 yd N. of Cassop Grange.   Sunk 1838 by Wm. Coulson
Ltd.   For detailed section see B. and S. No. 2236.

Drift to 106 ft, Main 47 in at 145 ft, Durham Low Main 34 in at 287 ft, Hutton
(banded) 77½ in at 387 ft.   Sunk to 393 ft.

Whitwell Colliery ' C ' Pit

Surface 306 ft A.O.D.   6-in NZ 33 N.W.   Site [3104 3993]
1050 yd N.N.E. of Whitwell South Farm.   Sunk 1855–6.
For detailed section see B. and S. No. 2239.

Drift to 73 ft, Main 46½ in at 116 ft, Durham Low Main 26 in at 255 ft, Hutton
about 50 in at bottom at 358 ft.

Whitwell House Bore

Surface 264 ft A.O.D.   6-in NZ 34 S.W.   Site [3058 4038]
410 yd N.W. of Whitwell House.   Drilled 1910 by Wm.
Coulson Ltd.

Drift to 100 ft, Low Main 21 in at 221 ft, Hutton 13 in at 309 ft, Harvey (banded)
26 in at 454 ft, Busty (banded) 31 in at 605 ft, Three-Quarter 16 in at 679 ft,
Brockwell 27½ in at 719 ft.   Drilled to 741 ft.

Wingate Grange, Lord Pit

Surface 407 ft A.O.D.   6-in NZ 33 N.E.   Site [3969 3721]
460 yd N.E. of Wingate Grange.   Sunk 1839–40 to Hutton,
subsequently deepened to Harvey.   For detailed section see
B. and S. No. 2280.

Drift to 92 ft, Magnesian Limestone and Marl Slate to 360 ft, Basal Permian
Sands to 361 ft.

Metal (banded) 55 in at 442 ft, Five-Quarter 49 in at 448 ft, Main (including
5-in band) 60 in at 538 ft, Bottom Hutton 32 in at 778 ft, Harvey (including 2-in band)
44 in at 923 ft.

REFERENCES

ANDERSON, W. 1941. Water supply from underground sources of north-east England. *Wartime Pamphlet Geol. Surv.,* **19,** pt. 3.

BIRD, W. J. 1888. The south Durham salt bed and associated strata. *Trans. Manch. Geol. Soc.,* **19,** 564–84.

—— 1890. Note on the Seaton Carew Boring. *Trans. Manch. Geol. Soc.,* **20,** 263–5.

MAGRAW, D., CLARKE, A. M. and SMITH, D. B. 1963. The stratigraphy and structure of part of the south-east Durham coalfield. *Proc. Yorks. Geol. Soc.,* **34,** 153–208.

TRECHMANN, C. T. 1913. On a mass of anhydrite in the Magnesian Limestone at Hartlepool, and on the Permian of south-eastern Durham. *Quart, J. Geol. Soc.,* **69,** 184–218.

—— 1925. The Permian Formation in Durham. *Proc. Geol. Assoc.,* **36,** 135–45.

—— 1942. Borings in the Permian and Coal Measures around Hartlepool. *Proc. Yorks. Geol. Soc.,* **24,** 313–27.

—— 1954. Thrusting and other movements in the Durham Permian. *Geol. Mag.,* **91,** 193–208.

WOOLACOTT, D. 1919. Borings at Cotefield Close and Sheraton, Durham. *Geol. Mag.,* **6,** 163–70.

Appendix II

LIST OF GEOLOGICAL SURVEY PHOTOGRAPHS

Copies of these photographs are deposited in the Library of the Geological Survey and Museum, South Kensington, London, S.W. 7, and of the Geological Survey Northern England Office, Ring Road Halton, Leeds, 15. Prints and lantern-slides may be supplied at a standard tariff.

PLEISTOCENE AND RECENT

L 54–5	Brierton Lane Gravel Pit at Seaton Carew
L 56	General view of drift topography at Whelly Hill
L 57	Whangdon Hill moraine at Whelly Hill
L 58	Calcreted gravels in stream bank at Dalton Piercy
L 59–60	Peat bed with tree stumps at Hartlepool beach
L 61	Peat bed with log at Hartlepool beach
L 62	Dry valley at Trimdon
L 63	Dry valley at Fishburn
L 80–1	High terrace at 190 ft A.O.D., 3 miles N. of Durham
L 82	Finchale Brick Pit
L 83–8	Laminated clays at Finchale Brick Pit
L 89–101	General and close up views of Durham Gravel Pit
L 102	Terraces of River Wear
L 103–5	The Sliddings: laminated clay in high bank above the River Wear
L 106	Glacial topography at Bucks Hill, Durham
L 107	Tursdale Beck
L 109	Glacial drainage channel at Ferryhill
L 110–7	Ferryhill Gap
L 123	Drift topography at The Broom, Ferryhill
L 124–5	Lacustrine flats at Bradbury Carr
L 126–8	Marine erosion and landslip features at Black Halls Rocks
L 141	View N.N.E. from Yoden Village, Horden
L 149–50	Dry valley near Kelloe Law
L 151	Shotton Brick Pit
L 152–4	Gorge eroded along fault plane at Gunner's Pool, Castle Eden Dene
L 155	Raised beach at Shippersea Bay
L 157–8	Overflow channel at Townfield Hill, Easington Colliery
L 584–7	General and close up views of Durham Gravel Pit. L 584 reproduced as Pl. XIII

TRIAS

L 52–3	Ripple-bedded Triassic sandstone at Seaton Carew

PERMIAN

A 5247–8	Magnesian Limestone at Bishop Middleham Quarries
A 5249–50	Magnesian Limestone at Raisby Hill Quarries
A 5251–2	Lower Magnesian Limestone at Coxhoe Colliery Quarry
A 5253–4	Lower Magnesian Limestone at Coxhoe Bank Quarry
A 5255–7	Lower Magnesian Limestone at Wingate Quarries
A 5258–63	Bedded dolomites, reef limestones and sea stacks at Black Halls Rocks. A 5261 reproduced as Pl. XIIB
L 47–8, 74	Magnesian Limestone with algal structures at Whelly Hill Quarry
L 49–50	Magnesian Limestone wave-cut platform at Hartlepool
L 51	Magnesian Limestone on beach at Hartlepool Scar
L 72	Quarry in Magnesian Limestone at Pittington
L 73	Quarry at Raisby Hill
L 75	Quarry in Lower Magnesian Limestone at Sherburn, Durham
L 76–9	Quarry in Lower Magnesian Limestone on Basal Permian (Yellow) Sands at Crime Rigg, Durham
L 108	The Permian Escarpment
L 120–2	Lower Magnesian Limestone with baryte at Chilton Quarry
L 129–31	Algal domes and wave-cut platform at Black Halls Rocks
L 132	Cliffs of brecciated Concretionary Limestone, 1100 yd S.S.E. of Black Halls Rocks. Reproduced as Pl. XIIIC
L 133–7	Bedded fore-reef and algal structures at Black Halls Rocks. L 135 reproduced as Pl. XIIA
L 138–9	Cross-sections of algal domes at Cross Gill
L 141	Dipping reef-structures in old quarry at Yoden Village, Horden
L 142–8	Algal domes in Middle Magnesian Limestone at Red Slide, Hesleden Dene. L 145 reproduced as Pl. XA
L 156	Old quarry in Middle Magnesian Limestone, Townfield Hill, Easington Colliery
L 159	Old quarry in Middle Magnesian Limestone, Easington East Lea
L 160–1	Magnesian Limestone with subsidence cracks caused by mining in Coal Measures, north of Easington Colliery
L 583	Reef dolomite of Magnesian Limestone at Hawthorn Quarry, Hawthorn. Reproduced as Pl. IX

COAL MEASURES

L 64–5	Coal Measures cropping out along River Wear at Finchale Priory
L 66–8	Low Main Post, River Wear at Finchale Priory
L 69	River Wear at Durham
L 70–1	View from Battery at Durham
L 118–9	Coal Measures unconformably overlain by Permian at Ferryhill Gap

INDEX

ABBOTT, T. G., 155, 182
Acanthocladia, 132, 140, 147; *A. anceps*, 108, 111, 117, 132–6, 139, 143–5, 169–70, 174, 176, 178
Acre Rigg, 61
Acrolepis sedgwickii, 180
Adam Lythgoe Ltd., 261
'*Adhaerentina*', 180–1; '*A*'. *permiana*, 111, 126, 129, 132–5, 139, 146, 170–7
Adit, Drift, 170, 178
—— , sites of, 30–1
AGAR, R., 246, 255
Agathammina, 180
Albrighton Coal, 23
Alethopteris, 80
Algae, 95–6
——, calcareous, 96, 119, 121
—— filaments, 133, 146, 169–71
—— laminae, 133–9, 141–4, 147–50
—— nodules, pellets, 133, 138, 171
—— sheets, 139
Algites virgatus, 157, 177, 180
Algoolites, 136–7
ALLONBY, R. H., 183
Alluvium, lacustrine and flood plain, 7, 250–1
——, river, 222, 229, 234, 237–8, 251–4
Alston Block, 193
Amerston Hall (Trial) Bore, 156, 164, 264, **266**
Ammodiscus, 22, 146, 173–5, 180–1; '*A*'. *roessleri*, 112, 129, 171–7
Ammonema, 86
Amphibia, 108
Analyses, chemical, 105–6, 113, 115–6, 191
——, modal, 191–2
ANDERSON, F. W., 135
ANDERSON, W., 18, 87, 91, 142, 182, 246, 263–5
Anhydrite, 122–3, 132, 145, 151–3, 155, 161, 165, 167–8, **263**
Ankerite, 12, 168
Annelids (worms), 56; annelid burrows, 12, 36, 86
Anthracoceras cambriense Marine Band, 11
—— *vanderbeckei* Marine Band, 11, 27
Anthracomya, 55; *A. adamsi*, 25
Anthraconaia, 20, 25, 27, 52, 60, 64–5, 68, 82–3; *A. curtata*, 27, 59–60; *A. fugax*, 25, 43; *A. lenisulcata*, 13; *A. librata*, 28, 83; *A. modiolaris*, 14, 26–7, 59; *A.*

Anthraconaia (*cont.*)
  pulchella, 27–8; *A. pumila*, 64, 68; *A. salteri*, 59, 67; *A. sp. nov.*, 67; *A. sp. nov. cf. wardi*, 28, 83
—— *lenisulcata* Zone, 14, 23
—— *modiolaris* Zone, 25–7, 50
Anthraconauta, 20
Anthracosia, 20, 27, 57, 60, 80, 82; *A. acutella*, 28, 82; *A. aquilina*, 16, 27, 59, 82; *A.* cf. *aquilinoides*, 28, 80; *A. atra*, 28–9, 82–3, 86; *A. beaniana*, 16, 27, 67; *A. caledonica*, 28, 65, 80; *A. concinna*, 28; *A. disjuncta*, 27; *A. elegans*, 82; *A. lateralis*, 59; *A. ovum*, 16, 27, 59–61, 64–5, 67; *A. phrygiana*, 16, 27, 59–60, 64–5; *A. regularis*, 16, 26, 53, 57, 59; *A. retrotracta*, 67; *A. simulans*, 28, 83; *A. subrecta*, 27, 60, 64
—— *phrygiana* group, 27
Anthracosphaerium, 20; *A.* cf. *affine*, 27, 59; *A. dawsoni*, 23, 26, 37, 50; *A. exiguum*, 27, 59, 67; *A. propinquum*, 28, 80; *A. radiatum*, 28, 83; *A. turgidum*, 27, 59, 64, 67
Anthrapalaemon, 80
Apatite, 12
Argutor, 212
Arley–Kilburn coals, 23
Armorican Orogeny, 8
ARMSTRONG, G., 10, 56, 87, 260, 265
Arvicola, 216
Astartella ('*Astarte*'), 126, 171–3, 175, 177; *A. vallisneriana*, 117, 125–6, 129, 131–2, 172, 175, 177
Augite, 190, 192
Autobrecciation, 107, 109–10, 113–4, 148–50
Aycliffe Gap, 254
Aykley Heads, 227
Aykleyheads Pit, 30, 50–1, 59, 66, 70, **293**

Back-reef, 120, 133
BADEN-POWELL, D. F. W., i, 208
BAGNOLD, R. A., 97–8, 182
Bakevellia, 117, 131–2, 140–1, 170–1, 175–6, 182; *B. antiqua*, 117, 119, 126, 128–9, 132–5, 139–41, 143–5, 164, 170–6, 178, 182: *B. ceratophaga*, 117, 119, 129–30, 132–6, 139, 141, 143–5, 169, 171–5, 178, 182
Band, meaning of, 32
BANNISTER, F. A., 183

335

01188   42420

ISBN 0 11 884242 0